Springer Series in Materials Science

Volume 274

The Springer Series in Materials Science covers the complete spectrum of materials research and technology, including fundamental principles, physical properties, materials theory and design. Recognizing the increasing importance of materials science in future device technologies, the book titles in this series reflect the state-of-the-art in understanding and controlling the structure and properties of all important classes of materials.

More information about this series at http://www.springer.com/series/856

Paolo M. Ossi
Editor

Advances in the Application of Lasers in Materials Science

Springer

Editor
Paolo M. Ossi
Dipartimento di Energia
Politecnico di Milano
Milan, Italy

ISSN 0933-033X ISSN 2196-2812 (electronic)
Springer Series in Materials Science
ISBN 978-3-030-07261-2 ISBN 978-3-319-96845-2 (eBook)
https://doi.org/10.1007/978-3-319-96845-2

Softcover re-print of the Hardcover 1st edition 2018

This Springer imprint is published by the registered company Springer Nature Switzerland AG
The registered company address is: Gewerbestrasse 11, 6330 Cham, Switzerland

Preface

The influence of lasers in materials science continues to rapidly advance along with the progress in the development of laser sources. For example, ultrafast lasers once employed mainly in science for their ability to reveal the dynamics of excited atoms and molecules now have the average power and reliability needed for manufacturing, both additive and subtractive, from metals, to polymers, to biomaterials. From the exquisite control needed for laser surgery to the extreme power necessary for the laser propulsion of satellites to the outer reaches of the galaxy, understanding the interactions between laser radiation and matter is fundamentally central to emerging applications.

Laser interactions play a special role in materials science. Pulsed lasers effortlessly produce highly nonequilibrium conditions to synthesize new materials, permitting access to new states of matter that challenge our understanding based upon bulk phase diagrams. These states are transient, dynamically evolve, and are especially challenging to theoretically describe and computationally model. However, thanks to advances in the development of laser spectroscopic tools in materials science, we are gaining new insight at the necessary timescales in the form of time-resolved, in situ diagnostics provided by laser interactions with these excited states of matter. As new laser-based characterization techniques evolve, new understanding allows a more complete understanding of laser processes important to materials science.

With this perspective, this volume therefore presents a readable set of critical reviews meant to introduce both the fundamentals of laser–materials interactions and emerging applications of laser synthesis, processing, characterization, and modeling in materials science. Each chapter in the book was written by a lecturer at the 5th School on Lasers in Materials Science (SLIMS 2016) held July 10–17, 2016 at Venice International University. One of the goals of this biennial school is to provide Ph.D. students and young research scientists working in the field of laser–materials interactions with robust fundamental knowledge that is often lacking in their training, so that they may profitably interact with colleagues working in areas neighboring their own research fields. Although the laser community organizes several international conferences allowing focused exchange of information on similar topics, such a structured training opportunity that was specifically geared toward young researchers was lacking before the SLIMS series was established. The chapters reflect the content and didactic nature of the lectures as well as the spirited questions and ensuing discussions at the school. Active participation of the students and lecturers, sequestered together in the quiet and beautiful confines of San Servolo Island during the weeklong school, shaped the tone and scope of each chapter. Each chapter, written by an active researcher in their respective field, serves to provide both a didactic review of the relevant fundamental science and an overview of the emerging trends in applications and technology with current references. For similar reviews of emerging topic areas in the past few areas, the reader is strongly encouraged to refer also to two other volumes (Volumes 130 and 191) published in 2010 and 2014 by Springer in the same series that contain chapters originating from complementary sets of lectures from earlier schools.

The school directors are grateful to the lecturers for the attention they put in the preparation of truly didactic presentations at the school, for their valuable time spent at the school, and for the additional work spent writing comprehensive reviews of these topic areas into self-contained chapters designed for the general reader. The book is dedicated both to the pioneers in laser interactions in materials science, such as the late Roger Kelly (in honor of whom special student awards are presented at each school), but mainly to the new generation of students who are following in their footsteps.

Magurele, Romania Maria Dinescu
Lasers, INFLPR
Oak Ridge, TN, USA David Geohegan
Center for Nanophase Materials Sciences
Oak Ridge National Laboratory
Trento, Italy Antonio Miotello
Physics Department, University of Trento
Milan, Italy Paolo M. Ossi
Dipartimento di Energia, Politecnico di Milano

Contents

Contributors

Oskar Armbruster Department of Physical Chemistry, University of Vienna, Vienna, Austria

Emanuel Axente Lasers Department, National Institute for Lasers, Plasma and Radiation Physics (INFLPR), Magurele, Ilfov, Romania

Manoj Kumar Bhuyan RIKEN Center for Advanced Photonics, RIKEN, Wako, Saitama, Japan

Adriana Bigi Department of Chemistry "G. Ciamician", University of Bologna, Bologna, Italy

Aziz Boulesbaa Functional Hybrid Nanomaterials Group Center for Nanophase Materials Sciences, Oak Ridge National Laboratory, Oak Ridge, TN, USA

Nadezhda M. Bulgakova HiLASE Centre, Institute of Physics of the Czech Academy of Sciences, Dolní Břežany, Czech Republic

Anna Paola Caricato Department of Mathematics and Physics "E. De Giorgi", University of Salento, Lecce, Italy

Marina Casazza Fondazione I.R.C.C.S. Istituto Neurologico Carlo Besta, Milan, Italy

Marta Castillejo Instituto de Química Física Rocasolano, CSIC, Madrid, Spain

Anna Gabriella Ciriolo Dipartimento di Fisica, Politecnico di Milano, Milano, Italy

Emilio Ciusani Fondazione I.R.C.C.S. Istituto Neurologico Carlo Besta, Milan, Italy

Ugo de Grazia Fondazione I.R.C.C.S. Istituto Neurologico Carlo Besta, Milan, Italy

Rebeca de Nalda Instituto de Química Física Rocasolano, CSIC, Madrid, Spain

Thibault J.-Y. Derrien HiLASE Centre, Institute of Physics of the Czech Academy of Sciences, Dolní Břežany, Czech Republic

Michele Devetta Istituto di Fotonica e Nanotecnologie - CNR, Milano, Italy

Maria Dinescu Lasers Department, National Institute for Lasers, Plasma, and Radiation Physics, Magurele, Romania

Gerd Duscher Functional Hybrid Nanomaterials Group Center for Nanophase Materials Sciences, Oak Ridge National Laboratory, Oak Ridge, TN, USA

Gyula Eres Functional Hybrid Nanomaterials Group Center for Nanophase Materials Sciences, Oak Ridge National Laboratory, Oak Ridge, TN, USA

Davide Faccialà Istituto di Fotonica e Nanotecnologie - CNR, Milano, Italy

Enza Fazio Dipartimento di Scienze Matematiche e Informatiche, Scienze Fisiche e Scienze della Terra, Messina, Italy

Wangyao Ge Department of Electrical and Computer Engineering, Duke University, Durham, NC, USA

Prabhash Prasannan Geetha Dipartimento di Fisica, Politecnico di Milano, Milano, Italy; Istituto di Fotonica e Nanotecnologie - CNR, Milano, Italy

David B. Geohegan Functional Hybrid Nanomaterials Group Center for Nanophase Materials Sciences, Oak Ridge National Laboratory, Oak Ridge, TN, USA

Miao He Department of Materials Science and Engineering, University of Virginia, Charlottesville, VA, USA

Wolfgang Kautek Department of Physical Chemistry, University of Vienna, Vienna, Austria

Yoann Levy HiLASE Centre, Institute of Physics of the Czech Academy of Sciences, Dolní Břežany, Czech Republic

Xufan Li Functional Hybrid Nanomaterials Group Center for Nanophase Materials Sciences, Oak Ridge National Laboratory, Oak Ridge, TN, USA

Liangbo Liang Functional Hybrid Nanomaterials Group Center for Nanophase Materials Sciences, Oak Ridge National Laboratory, Oak Ridge, TN, USA

Thomas Lippert Research with Neutrons and Muons Division, Paul Scherrer Institut, Villigen PSI, Switzerland; Laboratory of Inorganic Chemistry, ETH Zürich, Zürich, Switzerland

Sergey A. Lizunov HiLASE Centre, Institute of Physics of the Czech Academy of Sciences, Dolní Břežany, Czech Republic; Institute of Thermophysics SB RAS, Novosibirsk, Russia

Andrea Lucotti Dipartimento di Chimica, Materiali e Ing. Chimica "G. Natta", Politecnico di Milano, Milan, Italy

Masoud Mahjouri-Samani Functional Hybrid Nanomaterials Group Center for Nanophase Materials Sciences, Oak Ridge National Laboratory, Oak Ridge, TN, USA

Alberto Mazzi Center for Materials and Microsystems, Povo (Trento), Italy

Ion N. Mihailescu Lasers Department, National Institute for Lasers, Plasma and Radiation Physics (INFLPR), Magurele, Ilfov, Romania

Antonio Miotello Dipartimento di Fisica, Università degli Studi di Trento, Povo (Trento), Italy

Aida Naghilou Department of Physical Chemistry, University of Vienna, Vienna, Austria

Fortunato Neri Dipartimento di Scienze Matematiche e Informatiche, Scienze Fisiche e Scienze della Terra, Messina, Italy

Michele Orlandi Dipartimento di Fisica, Università degli Studi di Trento, Povo (Trento), Italy

Paolo M. Ossi Dipartimento di Energia, Politecnico di Milano, Milan, Italy

Mohamed Oujja Instituto de Química Física Rocasolano, CSIC, Madrid, Spain

Alexandra Palla-Papavlu Lasers Department, National Institute for Lasers, Plasma, and Radiation Physics, Magurele, Romania

Nainesh Patel Department of Physics and National Centre for Nanosciences and Nanotechnology, University of Mumbai, Mumbai, India

Irina Alexandra Paun CETAL Department, National Institute for Lasers, Plasma, and Radiation Physics, Magurele, Romania; Faculty of Applied Sciences, University Politechnica of Bucharest, RO, Bucharest, Romania

Claude R. Phipps Stanford University, Stanford, CA, USA; Photonic Associates, LLC, Santa Fe, NM, USA

Alex A. Puretzky Functional Hybrid Nanomaterials Group Center for Nanophase Materials Sciences, Oak Ridge National Laboratory, Oak Ridge, TN, USA

Aditya Pusala Dipartimento di Fisica, Politecnico di Milano, Milano, Italy

Juergen Reif Brandenburgische Technische Universität – BTU Cottbus-Senftenberg, Cottbus, Germany

Carmen Ristoscu Lasers Department, National Institute for Lasers, Plasma and Radiation Physics (INFLPR), Magurele, Ilfov, Romania

Chris Rouleau Functional Hybrid Nanomaterials Group Center for Nanophase Materials Sciences, Oak Ridge National Laboratory, Oak Ridge, TN, USA

Marco Santoro Dipartimento di Scienze Matematiche e Informatiche, Scienze Fisiche e Scienze della Terra, Messina, Italy

Mikel Sanz Instituto de Química Física Rocasolano, CSIC, Madrid, Spain

Maxim V. Shugaev Department of Materials Science and Engineering, University of Virginia, Charlottesville, VA, USA

Felix Sima Center for Advanced Laser Technologies (CETAL), National Institute for Lasers, Plasma and Radiation Physics (INFLPR), Magurele, Ilfov, Romania

Salvatore Spadaro Dipartimento di Scienze Matematiche e Informatiche, Scienze Fisiche e Scienze della Terra, Messina, Italy

Salvatore Stagira Dipartimento di Fisica, Politecnico di Milano, Milano, Italy; Istituto di Fotonica e Nanotecnologie - CNR, Milano, Italy

Adrienne D. Stiff-Roberts Department of Electrical and Computer Engineering, Duke University, Durham, NC, USA

Flavian Stokker-Cheregi Lasers Department, National Institute for Lasers, Plasma, and Radiation Physics, Magurele, Romania

Koji Sugioka RIKEN Center for Advanced Photonics, RIKEN, Wako, Saitama, Japan

Wesley Tennyson Functional Hybrid Nanomaterials Group Center for Nanophase Materials Sciences, Oak Ridge National Laboratory, Oak Ridge, TN, USA

Mengkun Tian Functional Hybrid Nanomaterials Group Center for Nanophase Materials Sciences, Oak Ridge National Laboratory, Oak Ridge, TN, USA

Matteo Tommasini Dipartimento di Chimica, Materiali e Ing. Chimica "G. Natta", Politecnico di Milano, Milan, Italy

Sebastiano Trusso CNR-Istituto per i Processi Chimico-Fisici, Messina, Italy

Caterina Vozzi Istituto di Fotonica e Nanotecnologie - CNR, Milano, Italy

Kai Wang Functional Hybrid Nanomaterials Group Center for Nanophase Materials Sciences, Oak Ridge National Laboratory, Oak Ridge, TN, USA

Kai Xiao Functional Hybrid Nanomaterials Group Center for Nanophase Materials Sciences, Oak Ridge National Laboratory, Oak Ridge, TN, USA

Mina Yoon Functional Hybrid Nanomaterials Group Center for Nanophase Materials Sciences, Oak Ridge National Laboratory, Oak Ridge, TN, USA

Chiara Zanchi Dipartimento di Chimica, Materiali e Ing. Chimica "G. Natta", Politecnico di Milano, Milan, Italy; Dipartimento di Energia, Politecnico di Milano, Milan, Italy

Leonid V. Zhigilei Department of Materials Science and Engineering, University of Virginia, Charlottesville, VA, USA

Vladimir P. Zhukov HiLASE Centre, Institute of Physics of the Czech Academy of Sciences, Dolní Břežany, Czech Republic; Institute of Computational Technologies SB RAS, Novosibirsk, Russia

Chapter 1
Laser Synthesis, Processing, and Spectroscopy of Atomically-Thin Two Dimensional Materials

David B. Geohegan, Alex A. Puretzky, Aziz Boulesbaa, Gerd Duscher, Gyula Eres, Xufan Li, Liangbo Liang, Masoud Mahjouri-Samani, Chris Rouleau, Wesley Tennyson, Mengkun Tian, Kai Wang, Kai Xiao and Mina Yoon

D. B. Geohegan (✉) · A. A. Puretzky · A. Boulesbaa · G. Duscher · G. Eres · X. Li · L. Liang
M. Mahjouri-Samani · C. Rouleau · W. Tennyson · M. Tian · K. Wang · K. Xiao · M. Yoon
Functional Hybrid Nanomaterials Group Center for Nanophase Materials Sciences, Oak Ridge National Laboratory, 1 Bethel Valley Road, Oak Ridge, TN 37831-6488, USA
e-mail: geohegandb@ornl.gov
URL: http://cnms.ornl.gov

A. A. Puretzky
e-mail: puretzkya@ornl.gov

A. Boulesbaa
e-mail: aboules@csun.edu

G. Duscher
e-mail: gduscher@utk.edu

G. Eres
e-mail: eresg@ornl.gov

X. Li
e-mail: xli@honda-ri.com

L. Liang
e-mail: liangl1@ornl.gov

M. Mahjouri-Samani
e-mail: mzm0185@auburn.edu

C. Rouleau
e-mail: rouleaucm@ornl.gov

W. Tennyson
e-mail: rouleaucm@ornl.gov

M. Tian
e-mail: mtian1@vols.utk.edu

K. Wang
e-mail: wangk@ornl.gov

P. M. Ossi (ed.), *Advances in the Application of Lasers in Materials Science*, Springer Series in Materials Science 274,
https://doi.org/10.1007/978-3-319-96845-2_1

Abstract Atomically-thin two-dimensional (2D) materials display widely varying electronic and vibronic properties compared to their bulk counterparts. Laser interactions with 2D materials are central to their development. Here we attempt to overview recent progress and define the current challenges in the broad range of laser interactions involved in the synthesis, processing, and optical characterization of 2D materials as the field has emerged from graphene and h-BN to encompass a multitude of other atomically-thin semiconducting, superconducting, thermoelectric, etc. 2D materials as "building blocks" for future energy applications and devices. Here, we first focus on challenges in the synthesis and processing of mainly semiconducting 2D layers for optoelectronics, and the advantages offered by non-equilibrium laser processing. Then, we review the optical characterization techniques that are being developed to serve as remote probes of their electronic and vibronic properties, as well as their structure, stacking, and atomistic alignment. Together, examples will be shown how these developments are already being merged to fulfill the promise for tailored synthesis and assembly of these exquisite materials with real-time in situ control of structure and optoelectronic properties.

1.1 Introduction

The discovery that the properties of layered materials such as graphite, boron nitride, and molybdenum disulfide can become radically altered when they are isolated in thicknesses of just one-, two-, or a few-layers [1] has generated an explosion of research over the past decade into atomically-thin two-dimensional (2D) materials. The revelation that certain transition metal chalcogenides (TMCs), (i.e., MX_2 where M is a metal, and X is a chalcogen) become direct band gap semiconductors when isolated as monolayers (e.g., ~1.9 eV for MoS_2) added the necessary "building blocks" of semiconductors to monolayered graphene (a pseudometal) and h-BN (an insulator) toward the fulfillment of a vision for stackable, flexible optoelectronics and other functional van der Waals (vdW) heterostructures [2, 3]. Figure 1.1 shows an example of such a vdW heterostructure combining two semiconducting TMC monolayers that self-align in reliable registry when GaSe (p-type) grows from the vapor phase on $MoSe_2$ (n-type) to form a photovoltaic p-n junction. Many other varieties of single-element (e.g., black phosphorus and silicene) and multi-element ultrathin layers (e.g., MXene) have been added to the growing list of atomically-thin building blocks that can be combined to form similar heterostructures [4–6].

Numerous excellent reviews of graphene [7] and materials 'beyond graphene' [8–10] including TMDs [11] have summarized their device [4] and broad energy applications [12] as well as challenges in their synthesis and preparation [13, 14],

K. Xiao
e-mail: xiaok@ornl.gov

M. Yoon
e-mail: myoon@ornl.gov

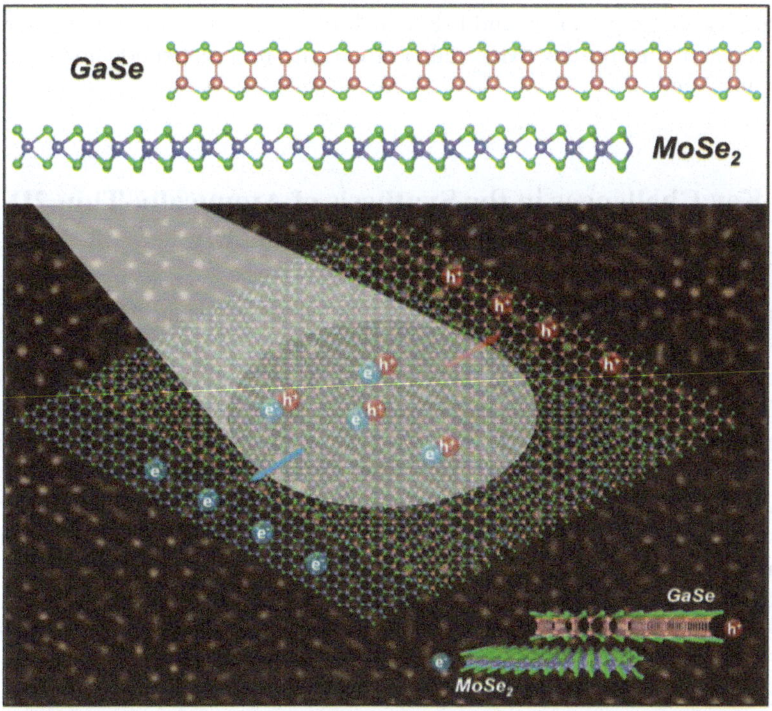

Fig. 1.1 Schematic showing two atomically-thin crystalline layers of GaSe and $MoSe_2$ that form a photovoltaic p-n junction, and exhibit photovoltaic behavior. The layers have different lattice constants but reliably align by van der Waals forces to form Moiré patterns when viewed by atomic-resolution Z-contrast scanning transmission electron microscopy (background) [38]

including their heterostructures and heterojunctions [15–18] and hybrids with other nanomaterials [19]. Bulk applications of TMC's for catalysis and composites are enabled by exfoliation techniques [20–22].

As for graphene (and single-wall carbon nanotubes), optical spectroscopy [23–25], including ultrafast dynamics [26] and Raman spectroscopy [27–32] has played a crucial role in revealing the electronic and vibronic properties of monolayers and stacked vdW heterostructures for refinement of theory and comparison with electronic transport measurements and photoresponse [33–38]. The TMCs are especially interesting because their optical properties are tunable with composition [39]. Optically-probing TMCs also reveals the strong role played by the substrate in determining their properties [40], and how strain can engineer their properties [41, 42].

Although 'top down' exfoliation of bulk crystals yields layers that are relatively free of defects, their lateral dimensions are typically smaller than those necessary for electronics. Reliable wafer-scale integration of these electronic materials requires 'bottom up' growth of atomically-thin layers of comparable quality to the 'top down' method, and as for graphene and h-BN such processes are currently an extremely active research area for the TMCs, other atomically-thin semiconductors, and topological insulators.

1.2 Key Challenges in the Synthesis of Atomically-Thin 2D Materials with Controllable Functionality

Understanding and controlling the growth of atomically-thin 2D materials with desired functional properties is hindered by the same fundamental gaps in synthesis science that crosscut many materials systems [43]. As outlined in a recent report by the U.S. Dept. of Energy on *Basic Research Needs for Synthesis Science*, tailoring the design and synthesis of materials for a desired functionality will require a new effort to "*Integrate emerging theoretical, computational, and* in situ *characterization tools to achieve directed synthesis with real-time adaptive control*" [43]. Nanomaterials with sensitive optoelectronic properties offer a golden opportunity for in situ optical characterization of both their structure and properties, both during and after synthesis. Such has been the case with single wall carbon nanotubes (SWCNT), where optical diagnostics coupled with atomic-resolution electron microscopy and predictive theory have stimulated world experts converge every 2 years to compare advances in characterization techniques, in situ diagnostics experiments, and modeling in order to redefine and publish a "top ten" list of remaining SWCNT synthesis science questions [44]. Atomically-thin 2D materials provide an even richer platform for directed synthesis of materials with designed functionality, and similar fundamental questions currently confront this challenge. Here we define a draft "top ten" list of crucial questions central to 2D materials synthesis.

A 'top ten' list of outstanding synthesis questions for 2D materials

1. What are the *chemical species and kinetic pathways* responsible for nucleation and for growth?
2. Is *nucleation* homogeneous or heterogeneous, and can it be suppressed to enable the growth of large single crystals?
3. What controls the *growth kinetics and shape* of the crystals during growth?
4. Why do *defects* form and how do they evolve during growth?
5. What is the role of *substrate interaction* (including van der Waals forces, covalent bonding, and hybrid orbital interactions determined by its crystallinity) in determining the shape, kinetics, defects, and termination of 2D crystals as they grow?
6. How does *strain* induced by the substrate or other crystalline layers influence growth?
7. Can *second layer* nucleation, growth, and orientation be understood to control size, stacking angle, and registry?
8. What are the roles of *dopants, 'catalysts', and adsorbates* in nucleation, growth, and crystalline perfection of 2D crystals? How do they affect defect formation?

9. What are the *collective effects* determining the growth of crystal ensembles and continuous layers (considering Ostwald versus Smoluchowski ripening)?
10. Which *growth method* (e.g., CVD, PLD, MOCVD, PECVD, MBE) is the most suitable to achieve directed synthesis of functional properties (e.g., electronic mobility, photoresponsivity, catalytic activity, etc.) consistent with practical issues of speed and scalability?

As implied by question 10, the most important practical challenge necessary to drive the field is an important question (#11) “*What is the “killer application” that will drive the development of 2D materials synthetic techniques*”? Currently, there are two main drivers and needs in 2D materials synthesis: (1) **Scalable synthesis** of wafer-scale 2D crystals with large-domain sizes for high electron mobility materials useful for electronics, or of loose 2D materials for bulk applications (e.g., for catalysis, batteries, composites), and (2) Control over heterogeneity in the crystals, meaning point defects and unwanted substitutional dopants, grain boundaries, and control over the number of layers and their orientation/stacking. The “top 10” questions above clearly focus on the fundamental problems in these two areas.

Laser interactions during and after synthesis of 2D materials provide the keys to answer these questions. Laser interactions can alter the synthesis of the crystals during their growth by providing well-defined bursts of energy to identify and control kinetic processes and control chemical species. Similarly, controllable laser processing treatments can modify the properties of existing crystals to alter the phase, composition, crystallinity, defect levels, and layer number. In addition to their use in synthesis and processing, laser interactions play an equal or larger role in the remote probing of 2D crystals through laser spectroscopy. Laser spectroscopy probes the most fundamental properties of atomically-thin 2D layers to remotely check theoretical predictions of their electronic band gap and band structure, vibrational modes, defects, and excited state dynamics. Practically, it can identify 2D materials, their number of layers, how they are stacked, and their quality as inferred from their Raman spectra, photoluminescence, and absorbance. When laser interactions are combined both in synthesis and processing, and as spectroscopic probes, they offer the capability to serve as in situ controls of the synthesis process.

This chapter will attempt to briefly review some of the laser synthesis and processing techniques that are rapidly evolving as non-equilibrium methods to explore these most exquisitely-thin films. It will also endeavor to summarize some of the most useful laser spectroscopy techniques. Along the way, examples of the applications of laser spectroscopy as in situ diagnostics of the growth environment and growing 2D crystals will be highlighted to illustrate the importance of laser interactions as real time probes to understand the essential links between synthesis, structure, and properties. This chapter cannot serve as a comprehensive review of atomically-thin 2D materials and their bright future for applications, for that the reader is referred to the excellent recent reviews referenced above. Instead it will attempt to review a few examples that illustrate the wide range of laser interactions in the synthesis, processing, and diagnostics of atomically-thin 2D materials that are key to developing and understanding these unique materials.

1.3 Laser-Based Synthesis and Processing of 2D Materials

1.3.1 Pulsed Laser Deposition of 2D Materials

Pulsed laser deposition (PLD) has been used for many years to explore the synthesis of thin films, offering the advantages of the transfer of stoichiometry from targets with complex compositions, epitaxial growth of films and superlattices with unit cell thickness control, suprathermal kinetic energy of plume species that allows the formation of metastable phases, and digital delivery of reactants. PLD would appear well suited to the synthesis of atomically-thin layers of two-dimensional crystals, however the high instantaneous fluxes that are typical for the process tend to result in high nucleation densities, resulting in the formation of networks of nanoscale domains. Instead, current research in the exploration of 2D materials has focused on techniques capable of growing single crystals of 2D materials with lateral dimensions >10 microns that are sufficient for the measurement of electronic transport (with lithographically-patterned contacts), optical measurements using micron-sized spot sizes in an optical microscope, and removal of the crystals to TEM grids or polymer stamps for alignment with other crystals. The growth of these single crystals is typically approached with continuous vapor transport techniques in order to slowly grow single crystals layer-by-layer from their edges in the manner that graphene crystals were grown to millimeter dimensions, using boats of precursor powders within a tube furnace, with molecular beam epitaxy, or by metal organic vapor deposition. Nevertheless, the commercialization of PLD systems in the past two decades has addressed some of the key objections to the technique for the synthesis of electronics-grade materials, including the ejection of particulates from the target, non-uniformity of the deposit, and lack of online process control. In addition, PLD does not introduce other reactants, or involve hazardous gases that are often associated with techniques such as MOCVD. In all of these techniques, the film stoichiometry is typically adjusted to compensate for the loss of more volatile elements during deposition, for example the loss of the chalcogen atoms in the family of 2D metal dichalcogenides.

Recently, a number of groups have explored the digital control of PLD to grow stoichiometric, atomically-thin 2D films of TMDs for optoelectronic applications, especially MoS_2. This builds from early work that demonstrated the viability of PLD for the stoichiometric deposition of MoS_2, which was first reported for tribological applications as early as 1988 using ns-pulse 532 nm lasers (~1 J/cm^2) in vacuum [45–47]. Today, very similar recipes have emerged for the PLD of stoichiometric MoS_2 with controllable number (1–10) of layers using ns-pulse 248-nm ablation (1–3 J/cm^2) of typically sulfur-rich [e.g., Mo:S (1:4)] mixed powder targets onto a variety of substrates (including typically (0001) Al_2O_3) at 700–850 °C in vacuum [48–53]. In these experiments, the base background pressure ($<1 \times 10^{-6}$ Torr) may rise to 0.1 mTorr during ablation, and films can be either cooled directly, or post-annealed in a sulfur atmosphere in order to repair sulfur vacancies and increase crystallinity [48]. Similar approaches have been developed for other 2D TMDs, such as WTe_2 [54]. The mobilities and small grain sizes of within the films (e.g., ~20 nm for MoS_2 [49], 20–50 nm for GaSe [55, 56]) need to be improved for electronic applications, yet are comparable to those grown by CVD. Therefore, PLD appears to be a promising method to digitally control the thickness of 2D TMDs over large areas with surprising reliability in stoichiometry. PLD has the additional advantage of reactive species in the ablation plasma that, in addition, have high kinetic energy capable of penetrating several centimeters through background gases at pressures up to 0.2 Torr. By varying the background gas pressure and target-substrate distance, the kinetic energy of the species arriving at the substrate can be adjusted, as well as the type of species, ranging from fast ions, to slower atoms and molecules, to clusters and nanoparticles. As shown in Fig. 1.2a, gated intensified CCD-array photographs or ion-probe current waveforms can be used to adjust the progression of the plasma plume in order to produce pure nanoparticle "building blocks" if desired, rather than atomic or molecular species, that can crystallize into different morphologies by the process of crystallization by particle attachment [57]. Such an approach was utilized to deposit amorphous GaSe nanoparticles at room temperature (Fig. 1.2b) that crystallized into highly photoresponsive 2D nanosheets in either lateral (at 600 °C, Fig. 1.2c) or vertical (at 550 °C, Fig. 1.2d) geometries, depending on deposition rate and temperature. These amorphous nanoparticles are formed by condensation during the thermalization of the gas-phase plume of reactants, and with proper tuning of the conditions have sufficient kinetic energy to impact the substrate to deposit at room temperature into nanoparticle aggregates, or crystallize at higher temperatures into different nanostructures [57]. These stoichiometric nanoparticles, deposited at room temperature, can be digitally delivered as precursors for the growth of very large monolayers, bilayers, and few-layers of 2D TMDs by post annealing within confined geometry between two substrates in "digital transfer growth", as shown in Figs. 1.2e–h [58]. Therefore, PLD has the potential to grow large 2D TMD crystals through the appropriate choice of deposition conditions.

1.3.2 Laser Techniques for "Top-Down" and "Bottom Up" Defect Engineering of 2D Crystals

The kinetic energy of plume atoms and ions in PLD is an advantage that can be used to controllably induce structural defects and compositional changes in the top-down processing of 2D crystals. Recently, pulses of KrF-laser irradiation vaporized a sulfur target in vacuum to deliver precise quantities of suprathermal sulfur atoms and molecules to $MoSe_2$ single crystals in vacuum at 700 °C in order to replace Se atoms to form $MoSe_{1-x}S_x$ alloys as shown in Fig. 1.3a. Photoluminescence and Raman spectra of the alloy crystals revealed a smooth transition in the photoluminescence peak wavelength and a corresponding change in the Raman spectra, confirming a smooth conversion from $MoSe_2$ to MoS_2 after 400 pulses. The technique was used in conjunction with lithographic patterning of SiO_2 protective ad-layers as shown in Fig. 1.3b to form the first patterned arrays of lateral semiconductor heterojunctions

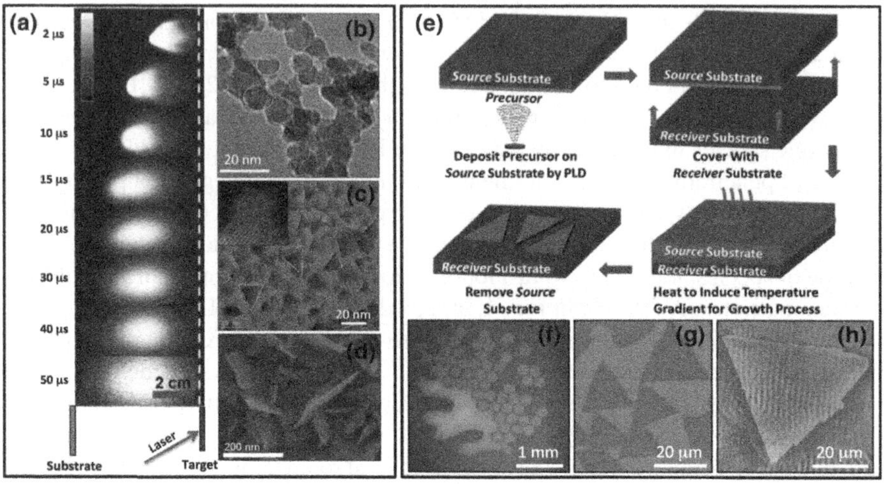

Fig. 1.2 Pulsed laser deposition of two-dimensional materials and digital transfer growth. **a** Gated ICCD-photography and other plasma diagnostics were adjusted [57] to deposit **b** primarily stoichiometric nanoparticles of GaSe and $MoSe_2$ at room temperature. Deposition at high temperatures and different repetition rates results in direct formation of either **c** small ~20 nm-grain, triangular, multilayered nanosheet networks at 600 °C or **d** vertically-standing nanoflakes at 550 °C. [56] These same amorphous nanoparticles can be **e** deposited onto *source* substrates at room temperature, then annealed between two substrates to grow large nanocrystals in confined spaces on a *receiver* substrate. The crystals grow directly opposite the regions where they are deposited, allowing **f** complex patterns of **g** large (20–100 μm) triangular crystals to grow as monolayer domains or **h** multilayers, depending on the amount of nanoparticle feedstock that was laser-deposited [58]

within monolayer 2D semiconductors—an essential step for the lateral integration of 2D semiconductor building blocks with different electronic and optoelectronic properties for high-density, ultrathin devices [59]. Laser plumes therefore can serve as sources of fast ions and neutrals to introduce defects and substitutional dopants in a top-down approach.

Alternatively, laser-vaporization can be used to controllably introduce defects from the 'bottom-up' during synthesis. A novel CO_2 laser-heater source was used to intentionally evaporate $MoSe_2$ powder non-stoichiometrically in order to grow single crystals of monolayer $MoSe_{2-x}$ on a substrate with tunable levels of Se vacancies up to very high levels (up to 20%) [60]. These highly-non-stoichiometric single crystals exhibited a remarkable transition in electrical transport properties from n-type to p-type character with increasing vacancy concentration. The vibrational modes of the crystal were also altered, and Raman spectroscopy of the crystals revealed the emergence of a new phonon mode, which was confirmed by theory and could then be used as a fingerprint of vacancy concentration in the crystals. Interestingly, laser evaporation of a selenium target could be used to repair the Se-vacancy sites introduced during synthesis to nearly recover the properties of pristine $MoSe_2$ crystals. Thus, laser vaporized fluxes can be used in both top-down and bottom-up synthesis

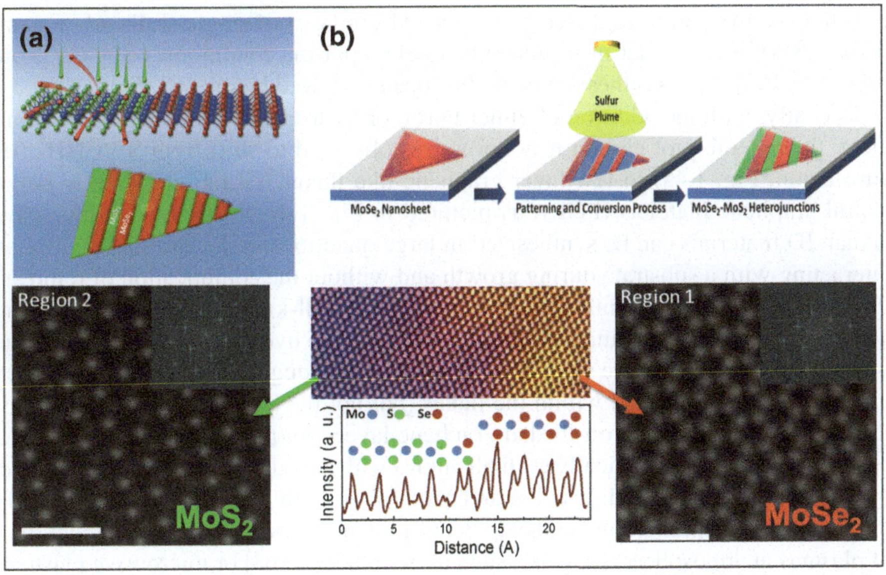

Fig. 1.3 Pulsed laser conversion of two-dimensional materials and the formation of multiple lateral semiconductor heterojunctions. **a** Energetic atoms of sulfur (green) replace selenium atoms (red) in a triangular single crystal of $MoSe_2$. After conversion, a map of the Raman spectrum reveals the regions of the crystal corresponding to MoS_2 and $MoSe_2$. **b** Schematic of the conversion process, starting with a triangular $MoSe_2$ single crystal monolayer domain. Laser ablation of a sulfur target produces energetic S atoms. Electron-beam lithography patterning/deposition of a protective SiO_2 layer (blue) shields regions of the $MoSe_2$ crystal. After 400 pulses at 700 °C, the stepwise conversion of $MoSe_2$ to MoS_2 is complete. The lower panels show atomic-resolution HAADF-STEM contrast images that confirm the transformation of the crystalline monolayer and the formation of lateral semiconductor heterojunctions within a single step induced by the energetic laser-generated plume of S atoms [59]

and processing strategies to digitally control the heterogeneity and properties of 2D crystals.

1.3.3 Substrateless Growth of 2D Materials by Laser Vaporization

The nonequilibrium conditions of laser ablation plasmas provide a versatile synthetic method to explore the condensation of novel clusters and nanoparticles, such as carbon fullerenes (e.g., C_{60}, C_{84}, etc.). Given sufficient time, temperature, and spatial confinement, larger nanostructures such as single wall carbon nanohorns (SWCNHs) can self-assemble in the laser plasma. Time-resolved growth experiments using pulsed laser plasmas have shown that these pure carbon SWCNH nanostructures grow at nm/ms rates (equivalent to microns per second) [61], which are as fast as the high growth rates measured for the catalyst-assisted synthesis of single-wall carbon nanotubes (SWCNTs), either in similar substrateless plasma conditions, or by CVD on substrates [62]. This synthesis process has been reviewed before [63, 64].

Recently, with the addition of either boron or hydrogen to the reaction environment, the assembly of carbon nanohorns was observed to shift from SWCNHs (in pure argon, Fig. 1.4a) to few-layer graphene-like flakes (GLFs) and thicker polyhedral graphene platelets (PGPs) as pictured in Fig. 1.4b–d [65, 66]. In this way, planar 2D materials can be synthesized in large quantities in the gas phase, without interacting with a substrate during growth and without the complication of removal from the substrate after synthesis. While boron is a well-known graphitization agent during carbonization treatments [67], its similar role to hydrogen in the stabilization of the planar GLFs in these studies is unclear, yet is thought to be related to the fact that boron remains loaded within the flakes (but not the SWCNHs) at 2–4% levels [65]. Interestingly, the boron-loaded graphene layers are not stacked as in graphite, but are arranged turbostratically at small angles with the adjoining layers, as revealed by selected area electron diffraction and Moiré fringes in STEM images (Fig. 1.4d) [65]. Note that similar graphene sheets can be grown without substrates in other types of plasmas at atmospheric pressure, such as arc plasmas [68] or microwave plasmas [69]. More generally, other 2D materials such as the transition metal dichalcogenides have been grown cleanly in microwave plasmas from just the base metals and chalcogens powders, without solvents, proving the generality of the technique [70]. Such so-called 'high temperature' synthesis routes that employ plasma torches or high-power lasers have been successfully developed to scale the production of SWCNTs and boron nitride nanotubes (BNNTs) [71–73] to tens of grams per hour in continuous flow reactors at high rates. So, as in the exploration of novel thin films by PLD, laser vaporization plasmas are well suited to explore the synthesis of pure and doped layered 2D materials that self assemble into flakes in the vapor phase.

1.3.4 Laser Thinning of Layered Two-Dimensional Materials

Control over layer number is often difficult to achieve during synthesis, so laser processing is currently being explored to selectively remove layers of 2D materials by laser thinning. Theoretically, Tomanek predicted that pristine monolayer graphene could be photoexfoliated intact from bulk graphite through the use of ultrafast (45 fs) laser pulses to rapidly produce an expanded electron gas capable of causing ejection of polarized layers by Coulomb ejection before the electron thermalization time [74].

Experimentally, however, success in laser thinning of few-layer crystals has relied upon selective ablation or etching of the topmost layers. Dhar, et al., showed that the ablation threshold of the topmost layers of few-layer graphene rises sharply as the number of layers is decreased, allowing processing windows to selectively leave a desired number of layers intact on a substrate. They proposed that the steep dependence of the threshold ablation energy arises from a dimensional crossover of the specific heat of a layered material for layers $N < 7$, where flexural modes

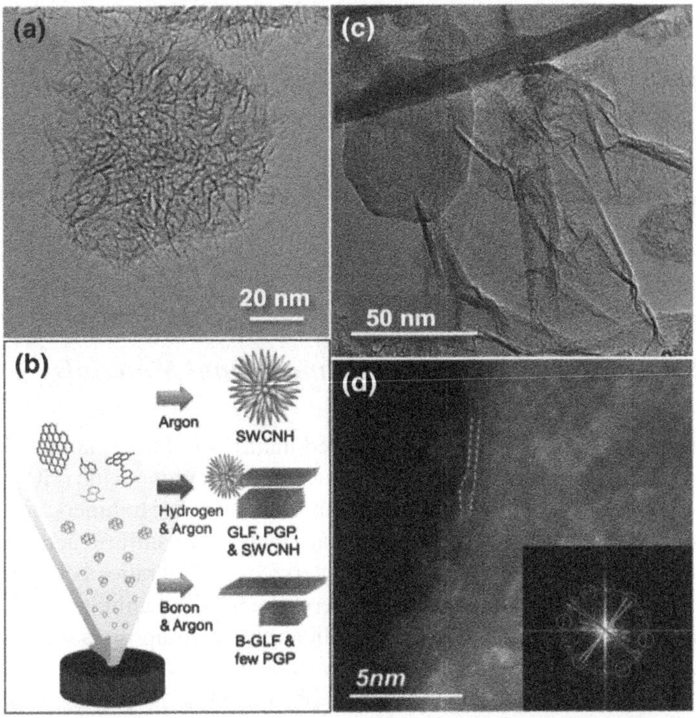

Fig. 1.4 Large volume laser vaporization synthesis of 2D materials: B-doped graphene. **a** Single-wall carbon nanohorn (SWCNH) aggregate produced by high-power laser vaporization of pure carbon in argon. **b** Schematic showing the shift in product distribution as hydrogen or boron is added to the growth environment, either in the gas phase (5% H_2 in Ar mix) or in the target (5% B in C). Ablating carbon into 5%H_2/Ar at 760 Torr shifts the product distribution to predominantly polyhedral graphite platelets (PGP) and graphene like flakes (GLF) as pictured in the TEM image of **c**. Boron is even more effective at shifting the products to few-layer GLF which incorporate boron at levels of 2–4 at.% and are turbostratically rotated as indicated by the Moiré patterns in atomic-resolution HAADF-STEM images shown in **d**

in thin-layered materials overwhelm bulk acoustic modes, leading to flaking of the topmost layers—a process that should occur in other layered materials with weak interlayer coupling [75]. Recently, similar layer-by-layer exfoliation of graphene has been achieved in air using picosecond laser thinning in a scanning fashion, as indicated in Fig. 1.5, where each 30 μm laser spot receives ~600 picosecond pulses at 2 MHz repetition rate [76]. However, even continuous wave sources such as those found within commercial micro-Raman spectroscopy apparatus can achieve selective layer-by-layer thinning as shown by Castellanos-Gomez, et al., for MoS_2 in air [77]. The ability to utilize the unique Raman spectroscopy signatures of the different layers in situ is an enabling diagnostic in these studies to assess the layer number, quality, and defects as outlined below.

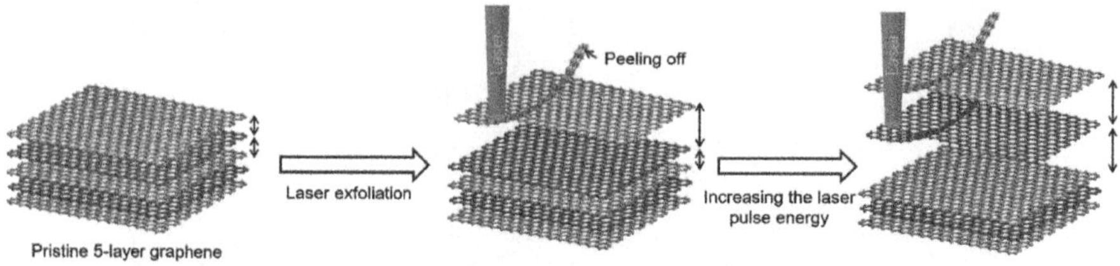

Fig. 1.5 Laser thinning of 2D materials. Schematic of layer-by-layer removal of 2D crystals, indicating different fluences for different layers. From [76]

1.3.5 Laser Conversion of Two-Dimensional Materials

Laser conversion of 2D materials has focused mainly on the formation of graphene or reduced graphene oxide (rGO) by the reduction of graphite oxide (GO). Graphitic oxide materials were originally produced in 1958 by the Hummers and Offeman method which oxidizes graphite by chemical reactions in sulfuric acid solutions to yield a graphitic oxide (GO) material characterized by C:O ratios between ~2.1 and 2.9 [78]. The Hummers method is advantageous because it facilitates the exfoliation of graphite in an industrial process for applications in composites, however unlike graphite or graphene, the material is highly defective and not electrically conductive. Following the experimental realization of the unique properties of single layer graphene in 2004 [1, 4] attempts to produce graphene from GO were explored, and conversion of epitaxial graphene to GO was explored to open a band gap for semiconducting properties [79]. Chemical, thermal, and photolytic methods were developed to remove the oxygen and hydrogen that were introduced in the Hummers (or modified) method, producing a spectrum of materials generally referred to as reduced graphene oxide (rGO). GO begins to lose oxygen at temperatures of 100 °C, and can be reduced at 180–250 °C to tune its electrical properties [79, 80]. Raman spectroscopy is a convenient and rapid method to assess the level of defects in GO and rGO materials compared to that in graphene (see below) through the comparison of the D-, G-, and 2D (or G')—bands of these materials compared to the characteristic Raman spectrum of pristine graphene [81].

Laser reduction of GO to graphene is remarkably successful, and can be accomplished with both pulsed and CW irradiation, although the mechanisms are unclear. Repeated pulses of nanosecond UV-radiation in air or vacuum are most effective, with C:O ratios up to 40 produced by Sokolov et al., through repeated 0.14 J/cm^2 KrF-laser (248 nm) in 10^{-6} Torr vacuum, which was much more effective compared to the same measurements in pure N_2 [82]. Arul, et al., measured a 0.01 J/cm^2 ablation threshold for GO under similar conditions in air [83]. In the process, layers are removed by ablation while layers below are becoming reduced. This procedure allowed the direct-laser writing of 100–500 Ω/□ conductive graphene patterns on GO films [82]. Arul, et al., recently compared ns (248 nm), fs (800 nm), and CW (788 nm) irradiation in ambient air, finding that repeated ns-laser irradiation is the most effective at graphenization [83]. The greater thermal component of ns-laser ablation compared to that for fs-lasers is thought to provide a localized heat-affected zone that is key to the process. Remarkably, El Kady, et al., showed that high-surface area (1500 m^2/g) and high conductivity (1738 S/m) rGO suitable for electrochemical capacitors could be laser scribed on thin-film GO using commercial low-power DVD-optical drives [84]. Using this technique, they patterned GO on flexible PET polymer substrates to make miniaturized micro-supercapacitors with very high power density (200 W/cm^3) by encapsulation with an ionic electrolyte [85]. Gao, et al., utilized a CO_2 laser printer to reduce as-prepared hydrated GO films to produce rGO that operated as supercapacitors with or without external electrolyte, using the trapped water as an ionic conductor [86]. Interestingly, focused solar radiation has been shown to simultaneously exfoliate GO to produce rGO with a Raman spectra nearly identical to graphite, with the mechanisms of thermal decomposition of GO and the expulsion of CO_2 and H_2O vapors thought to lead to the deflagration of the layers to overcome the van der Waals forces between the layers in order to simultaneously exfoliate them [87].

Similar methods of laser graphitization for the formation of porous graphene or rGO are now being applied using polymers as precursors. Back in 1991, Schumann et al., showed that non-conducting polymer foils of Kapton and PBI could similarly be processed with repeated KrF-laser (>0.02 J/cm^2) nanosecond pulses to form highly conductive (10 S/cm) carbon layers [88]. Now, with new applications generated for graphene and rGO, different polymers are being explored as precursors for their formation, and the laser conversion of polymers by graphitization has been demonstrated in particular for polymers such as polyimide, using CO_2-laser direct-write [89].

In addition to polymers, there are a great variety of graphitizable carbonaceous precursors that have been studied over the past many years that can be envisioned to form graphene [90]. Lasers offer both the possibility for both thermal and photolytic conversion of these precursors.

1.3.6 Laser Crystallization and Annealing of TMDs

Similarly, explorations of the synthesis of atomically-thin 2D TMD crystalline materials using laser processing of deposited amorphous precursor films or nanoparticles (e.g., as shown in Fig. 1.2) are now being pursued. The approaches follow those that have been used for years for thin film chalcogenide photovoltaics [91], however the motivations are limited to 2D, and not for the growth of large 3D grains. In both cases, the advantages of laser processing over broadband lamps include the wide variety of wavelengths available to provide selective absorption in particular layers, and pulse widths ranging from femtoseconds to CW. This versatility allows the exploration of both highly non-equilibrium processes driven by targeted photolytic pathways, all the way to thermal processing, within a single platform.

Practically, pulsed laser crystallization of amorphous precursors has the potential to grow 2D crystals directly on polymer substrates for the rapidly developing field of flexible electronics. Recently, McConney et al., demonstrated that few-layer, hexagonal MoS_2 and WS_2 crystalline films could be grown on PDMS polymer substrates using CW 514-nm irradiation of sputtered precursor films within a commercial Raman spectrometer [92]. The same Raman spectrometer and laser system was also used after processing to non-perturbatively assess the crystallinity and layer number of the films, using a lower laser power. Heterostructures could be formed by successive evaporation and laser treatments of other amorphous films. To understand the mechanisms of the crystallization process, a novel approach utilizing in situ monitoring of the Raman spectra of these films as well as their silicon post substrates during CW laser processing at different laser powers was recently employed to simultaneously estimate the crystallization rate versus temperature and to infer an activation energy barrier for the process [93]. Crystallinity was found to improve with higher CW annealing temperatures and dwell times, including those where temperatures exceeded temperature limitations of plastic substrates, so new approaches utilizing pulsed laser processing or photonic curing may need to be developed.

Pulsed lasers offer high instantaneous powers and temperatures of thin films during brief periods, followed by thermal diffusion over longer periods, thereby mitigating the temperature rise of the underlying substrates. Such approaches can be used to preferentially anneal one layer on top of another, or improve the contact between metals and 2D TMD crystals. For example, using picosecond pulse trains (12 ps, 80 MHz, 355 nm) to anneal Au/Ti films contacts deposited on MoS_2 field-effect transistors, significant increases in the device mobility were achieved while maintaining low (<200 °C) substrate temperatures compatible with the flexible PEN substrate [94]. The improvement was attributed to lowering the Schottky contact barrier through the laser-driven intercalation of Ti between the layers of MoS_2 [94].

1.3.7 Laser-Induced Phase Conversion of Two-Dimensional Crystals

Single monolayers of various 2D crystals can be polymorphic, having a variety of different crystalline structures for the same elemental composition. For example, transition metal dichalcogenides all have a metal atom layer (e.g., Mo or W) sandwiched between two chalcogenide layers (e.g., S, Se, or Te), but have different structure and are referred to using the well-known phases (e.g., 2H, 1T, or 1T') of the bulk crystals from which they can often be exfoliated. In the 2D TMDs the hexagonal 2H phase is the semiconducting phase, but the 1T and 1T' phases are metallic with octahedral or distorted octahedral geometry, respectively [95].

Laser irradiation can alter the phase of monolayer crystals, as shown in recent experiments by Cho et al. involving laser processing of $MoTe_2$. They showed the *phase patterning* of $MoTe_2$ layers by laser irradiation at 532 nm, which converted a spatial region of the stable 2H. phase irreversibly to the metallic 1T' phase, forming an in-plane homojunction [96]. The mechanism driving the phase transformation was concluded to be the formation of Te vacancies that were observed in atomic-resolution electron microscopy. Interestingly, in multilayers of $MoTe_2$ the topmost layer could be selectively converted. The formation of these metallic 1T'-$MoTe_2$ regions within a single layer of semiconducting 2H-$MoTe_2$ allowed patterning of source and drain regions for the fabrication of an all-$MoTe_2$ transistor. The metallic regions also allowed ohmic contacts to be made, increasing the carrier mobility by a factor of ~50.

1.3.8 Future Directions of Laser Synthesis and Processing of Atomically-Thin 2D Materials

Atomically-thin 2D materials present new conceptual challenges both experimentally and theoretically. Computational simulation of electronic and thermal processes in two dimensions is complicated by a lack of understanding of materials properties, including the wide variations in electronic structure of the material itself in one-, two-, and few-layered forms, which is governed by quantum confinement effects and the effects of interlayer coupling and interactions with the substrate and environment. Understanding and successful developing computational models that correctly describe the competition between van der Waals interactions, charge transfer interactions, and the role of hybrid orbital interactions due to stacking are essential to develop an understanding of laser interactions with these interesting model systems in order to develop predictive methodologies for their synthesis and processing.

Much of what we know about these interactions comes from optical measurements of their electronic and vibrational properties. These measurement techniques will be briefly reviewed in the following sections. It is important to note that these techniques serve to guide theoretical developments, which in turn become predictive tools to guide the experiments.

1.4 Optical Techniques for 2D Material Characterization

1.4.1 Overview

The goal of this section is to outline the most common experimental techniques used to characterize the optical properties of 2D materials. The experimental approaches were performed using a custom-built setup that will be described in detail as diagrammed in Fig. 1.6a. The setup consists of coupled inverted and upright microscopes that provides high-throughput optical coupling to a spectrometer that is equipped with a CCD detector for spectroscopic measurements and a photomultiplier for time-

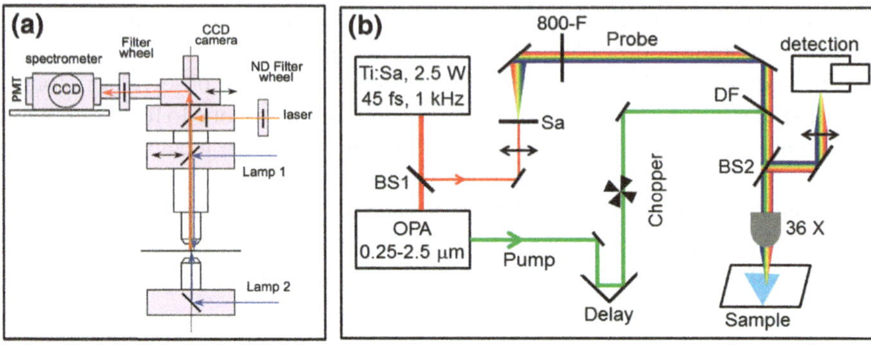

Fig. 1.6 Schematics of experimental setups used for comprehensive optical characterization of 2D materials at the Center for Nanophase Materials Sciences (CNMS) Oak Ridge National Laboratory (ORNL). **a** Custom built setup based on two coupled microscopes, inverted and upright with high optical throughput coupling to a spectrometer, which is equipped with a CCD detector and a photomultiplier for time-correlated single photon counting. A variety of different lasers and lamps are employed to provide efficient micro-measurements of Raman scattering, photoluminescence (PL), PL-lifetimes, absorption, second harmonic generation (SHG) in wide range of temperatures (low temperatures from 4 to 300 K and high-temperatures from 300 to 1300 K) and under bias. An inverted microscope is Olympus IX 81; lamp 1 is an upright microscope imaging lamp; lamp 2 is a laser driven white light source (EQ-99XFC, Energetiq); a spectrometer (Spectra Pro 2300i, Acton, $f = 0.3$ m) is equipped with 150, 600, and 1800 groves/mm gratings; a CCD camera is Pixis 256BR, Princeton Instruments; a photomultiplier (PMT) (PPD-850, Horiba Scientific). **b** Femtosecond micro-pump-probe setup for measurements of ultrafast dynamics in 2D materials

correlated single photon counting. A variety of different lasers and lamps (see below and the Fig. 1.6 caption) are employed to provide efficient ~ micron-resolution measurements of absorption, Raman scattering, and photoluminescence (PL) spectra, PL-lifetimes, and polarization-dependent mapping of second harmonic generation (SHG) in wide range of temperatures including low temperatures from 4 to 300 K and high-temperatures from 300 to 1300 K and under bias. A computer-controlled X-Y stage allows spectra to be acquired at multiple points for spatial mapping all of these spectral profiles for comparison with atomic force microscopy spatial profiles.

Micro-absorption Measurements The absorption spectra of 2D materials are measured using a micro-absorption setup operating in the transmission mode. The white light source (lamp 2 in Fig. 1.6a) is coupled out to a 25 μm diameter optical fiber. The white light is focused on a sample surface to a 1.5 μm spot using two microscope objectives: a 5× collimating objective, NA (numeric aperture) = 0.1 and a 100× long working distance objective, NA = 0.8 in an inverted microscope. The transmitted light is collected by a 50× objective (NA = 0.5) in an upright microscope coupled to an inverted microscope, and is analyzed by a spectrometer equipped with a CCD camera.

Microscale Second Harmonic Generation Measurements Second harmonic generation (SHG) measurements are conducted using a 40 fs Ti:Sapphire laser

(Micra, Coherent) operating at 800 nm and 80 MHz repetition rate. The laser beam was passed through a half-wave plate mounted in a rotation stage and is directed into an upright microscope and focused onto a sample surface to a ~1 μm spot using a 100x microscope objective (Fig. 1.6a). The typical laser energy at the sample surface is ~0.2 mW. The SHG light is collected in backscattering configuration using the same objective and is directed to a monochromator equipped with a 150 grooves/mm grating and a CCD camera. Before entering the monochromator the SHG light is passed through a shortpass cut-off filter (650 nm) and a polarizer to filter out the fundamental excitation light at 800 nm and to select the SHG polarization parallel to that of the excitation light. The SHG mapping is performed using a motorized microscope X-Y stage.

Micro-photoluminescence Measurements PL spectra are measured using an upright microscope (Fig. 1.6a). The PL is excited with a continuous wave (cw) diode-pumped solid-state laser (Excelsior, Spectra Physics, 532 nm, 100 mW) or 405 nm diode laser (Ondax) through an upright microscope. The typical incident laser power on a sample varies depending on a 2D sample with a typical minimum power at ~1 μW to reduce possible laser heating of the samples during PL spectra acquisition. The PL light is analyzed by a spectrometer (Fig. 1.6a).

Micro-PL Lifetime Measurements Time-resolved PL is measured by time-correlated single-photon counting (TCSPC) (PPD-850, Horiba Scientific with Picosecond Photon Detection Module, and Fluorohub model JY IBH: Horiba). The PPD-850 is mounted to a second port of the spectrometer (Fig. 1.6a). The 2D samples are excited using a second harmonic (400 nm) of a Ti:sapphire laser (Coherent, Mira 900) (800 nm, 5 ps pulses, 76 MHz repetition rate). To match the TCSPC repetition rate requirements the laser repetition rate is reduced to 5 MHz using a pulse picker (Coherent). The output of the pulse picker is frequency doubled using an ultrafast harmonic generator (Coherent 5–050) and is directed into a microscope to illuminate the samples through a microscope objective.

Micro-Raman measurements An important feature of the experimental setup shown in Fig. 1.6a is high optical throughput that allows measurements of Raman and PL spectra with a very low excitation laser power and fast switching between the Raman and PL measurements by computer controlled exchange of the three spectrometer gratings. However, for low-frequency Raman measurements (<50 cm^{-1}) a triple spectrometer based micro-Raman setup (T64000, Jobin-Yvon) is used that is equipped with three 1800 grooves per millimeter gratings (spectral resolution ~0.7 cm^{-1}, FWHM) and a liquid nitrogen cooled CCD (Symphony, Horiba JY). In this case a linearly polarized laser beam is focused onto a sample to a ~1 μm spot using a microscope objective with the typical laser power on the samples ~0.3 mW. The polarized Raman measurements are conducted by inserting a polarizer in the scattered beam and a half-wave plate to rotate the scattered light polarization to optimize the grating's response.

Femtosecond pump-probe spectroscopy Figure 1.6b shows a schematic of our fs-pump-probe setup, which is based on a titanium sapphire oscillator (Micra, Coherent) with its output seeded by a Ti:Sa Coherent Legend (USP-HE) amplifier operating at 1 kHz repetition rate. The Legend amplifier provides pulses centered at 800 nm, with ~45 fs duration and 2.2 mJ energy per pulse. Pump pulses are the second harmonic (3.1 eV) of the fundamental output of the Legend amplifier or the signal output of an optical parametric amplifier (TOPAS, Coherent). A small portion (~2 μJ) of the laser beam from the Legend amplifier is focused onto a 2 mm-thick sapphire window to generate a white-light continuum probe, which covers the spectral region from 1.6 to 2.8 eV. At the sample, the pump and probe spot diameters are ~7 μm. A spectrograph (Shamrock 303i, Andor) coupled with a EMCCD (Newton, Andor) is used for detecting the reflected white light probe.

1.4.2 Raman Spectroscopy of 2D Materials

Stacked monolayers of two-dimensional (2D) materials present a new class of hybrid materials with tunable optoelectronic properties determined by their stacking orientation [2], order, and atomic registry [2, 97]. Atomic-resolution Z-contrast scanning transmission electron microscopy (AR-Z-STEM) and electron energy loss spectroscopy (EELS) can be used to determine the exact atomic registration between different layers in few-layer 2D stacks, however characterization of these traits using fast optical techniques is essential for rapid development of the field.

An integrated experimental/theoretical Raman scattering approach can help to determine many crucial properties of 2D materials beyond graphene, such as the number of layers [98–100], the layer-layer stacking [101–106], and the defect concentration [60]. For transition metal dichalcogenides (TMDs) and black phosphorus, recent experimental and theoretical studies have demonstrated that low-frequency interlayer Raman modes that correspond to layer-layer vibrations with each layer moving as a whole unit are much more sensitive to the interlayer coupling, and can be more effective at determining layer thickness and stacking compared to the typically-studied high-frequency intralayer Raman modes (see reviews [29, 107]). Establishing a correlation between low-frequency modes and layer thickness/stacking is very important for enabling rapid characterization of 2D layered materials. For example, low-frequency Raman spectroscopy and modeling of twisted $MoSe_2$[102] and MoS_2[103] monolayers indicate the existence of unique bilayers with variable coupling between the monolayers that occurs in the narrow range of twist angles. High-frequency Raman modes also can provide important information on 2D materials, e.g., recently we showed that the high-frequency modes in MoS_2 and $MoSe_2$ monolayers exhibit unique dependence on S or Se vacancy concentration, which was corroborated by first-principles modeling [60].

Low-frequency Raman spectroscopy of 2D materials is a rapidly developing field of research, which is accelerated due to recent development of volume Bragg gratings for ultra-narrow optical filters with bandwidth ~1 cm^{-1}, which allows efficient cut off of the excitation laser light without employing expensive triple monochromators for Raman measurements [108].

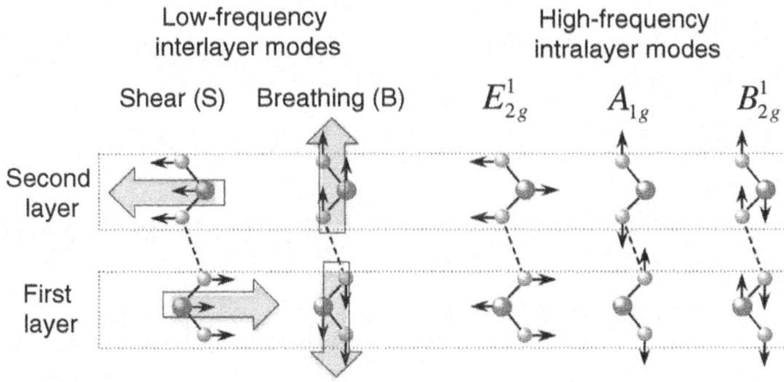

Fig. 1.7 Schematic of vibrations corresponding to low-frequency interlayer (left) and high-frequency intralayer (right) Raman modes

Currently, many different 2D materials have been investigated using low-frequency Raman spectroscopy including graphene [109–112], TMDs [98–103, 113–116], Black Phosphorus [117, 118], ReS_2 [119], and others. Here, we briefly illustrate the main aspects of low- and high-frequency Raman spectroscopies for determining number of layers, stacking configurations, and defects in layered 2D materials.

Low-frequency versus high-frequency Raman modes For the commonly studied high-frequency (HF) intralayer Raman modes, which involve vibrations from the intralayer chemical bonds, the restoring forces are dominated by the strength of these chemical bonds with only minor influence of the interlayer interactions (Fig. 1.7). Consequently, the HF modes are not very susceptible to the interlayer coupling. In contrast, for low-frequency (LF) interlayer Raman modes, each layer vibrates as a rigid unit and thus their frequencies are only determined by the weak interlayer restoring forces (Fig. 1.7). The weak nature of interlayer interactions typically renders their requencies below 100 cm^{-1}. Depending on the direction of vibrations, they are categorized into two types: the in-plane shear (S) modes and out-of-plane breathing (B) modes (Fig. 1.7).

For LF vibrations, each layer can be simplified as a ball and the interlayer coupling can be treated as a spring (Fig. 1.8) [29, 100]. Thus a general *linear chain model (LCM)* can be formulated to describe the frequencies of LF phonon modes at different number of layers, N, for any 2D layered materials with the mass, μ, per unit area. The frequency of the j-th branch S (or B) mode is directly related to layer number, N, and interlayer shear (or breathing) force constants (K^S or K^B) (Fig. 1.8).

Low-frequency Raman: fingerprints of thickness The LF Raman peak frequencies change much more dramatically with respect to the number of layers (>30 cm^{-1}), while HF modes show only small shifts about 3 cm^{-1} from monolayer to bulk [117]. The thickness dependence of LF peak frequencies follows the equations described by linear chain model: either monotonous increase or decrease with the layer number,

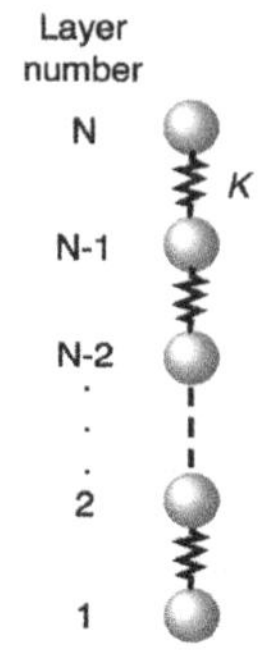

$$\omega(\mathrm{S}_{N,N-j}) = \omega(\mathrm{S}_{\mathrm{bulk}})\sin(j\pi/2N)$$

$$\omega(\mathrm{B}_{N,N-j}) = \omega(\mathrm{B}_{\mathrm{bulk}})\sin(j\pi/2N)$$

$$\omega(\mathrm{S}_{\mathrm{bulk}}) = (1/\pi c)\sqrt{K^{\mathrm{S}}/\mu}$$

$$\omega(\mathrm{B}_{\mathrm{bulk}}) = (1/\pi c)\sqrt{K^{\mathrm{B}}/\mu}$$

Fig. 1.8 Linear chain model (LCM) of a 2D material provides a simple prediction of Raman frequencies for an arbitrary number of layers, N

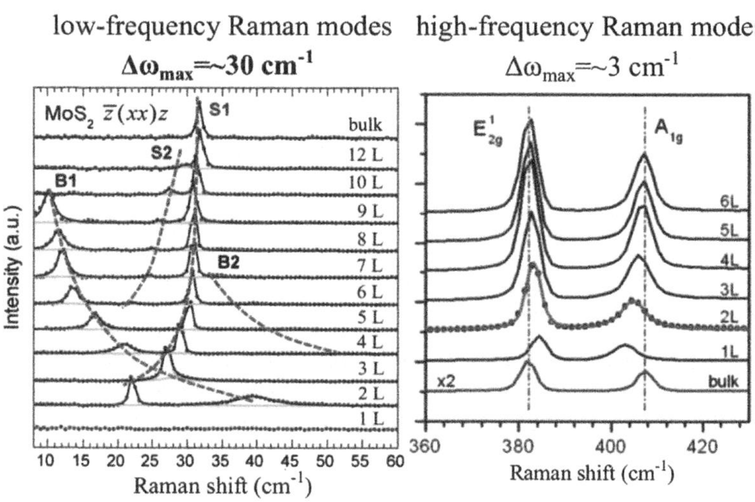

Fig. 1.9 Low- and high-frequency Raman modes for MoS_2 for different numbers of layers showing dramatic changes in the LF Raman spectra on the number of layers and only minor changes in the corresponding HF Raman spectra [100, 113]

rendering them potentially thickness fingerprints for 2D materials. For example, in the case of MoS_2 the HF E^1_{2g} and A_{1g} Raman modes show only small shifts relative to bulk MoS_2 (Fig. 1.9 right) [113], but the LH breathing (B) and shear (S) modes (Fig. 1.9 left) exhibit large characteristic shifts and spectral patterns [100], which make them perfect indicators of the number of layers in 2D materials.

Low-frequency Raman: fingerprints of stacking The most stable stacking configuration for 2L TMDs is 2H (also named as AA'). In the case of $MoSe_2$ this correspond to the following alignment of atoms: Mo over 2Se and 2Se over Mo as shown in Fig. 1.10 (left). This stacking configuration is the most common in natural $MoSe_2$ crystals and 2D layers of $MoSe_2$ obtained by their exfoliation. The second stacking configuration the formation energy of which is only 1 meV/atom higher relative to 2H is 3R (or AB) stacking with the following atoms alignment: Mo over Se and

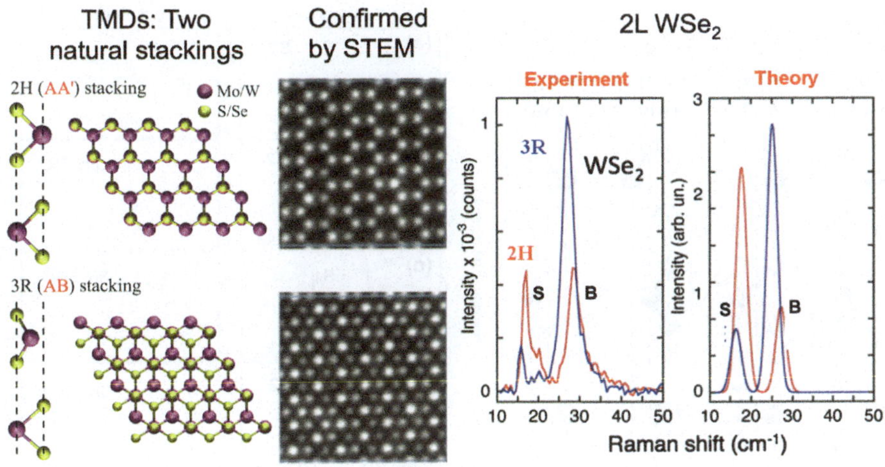

Fig. 1.10 Low-frequency Raman fingerprints of 2H (AA′) and 3R(AB) stacking configurations in bilayers (2L) of TMDs (e.g., MoS_2, $MoSe_2$, WS_2, WSe_2). Side and top views of TMD atomic structures for 2H and 3R stackings (left). Atomic resolution STEM images of 2L $MoSe_2$ (middle). Experimental and theoretical LF Raman spectra of 2L WSe_2 for 2H (red) and 3R (blue) stacking configurations (right)

the other Mo and Se over the center of the hexagons (Fig. 1.10 left). This stacking configuration commonly appears in chemical vapor deposition synthesized TMD crystals. Both 2H and 3R stacking configurations were verified using AR-Z-STEM and EELS (Fig. 1.10 middle). The other three high symmetry stacking configurations, AB', A'B, and AA, have formation energies per atom 5.7, 13, and 13.7 meV higher than that for 2H and do not appear during synthesis, although can be observed in twisted bilayers at specific twist angles [102].

Due to the small difference between the formation energies of the 2H and 3R stacking configurations of TMDs it is hard to expect a substantial shift in their Raman spectra, but the intensities of the LF Raman lines should show large changes due to different symmetries of the TMD crystals in these two stackings. As shown in Fig. 1.10 (right) both experimental and simulated Raman spectra show dramatic intensity changes of LF shear (S) and breathing (B) modes for WSe_2 when the stacking changes from 2H to 3R [101].

The number of different stacking patterns increases rapidly with the number of layers. For example, 6 different stacking configurations are possible for 3-layer $MoSe_2$. The four of these that can be distinguished by LF Raman spectroscopy are shown in Fig. 1.11. Three-layer $MoSe_2$ has two S modes: S1 and S2, which show unique intensity dependence on the stacking, while HF modes do not show any substantial difference (Fig. 1.11) [101]. Similar dependence of LF S modes versus stacking also exists for 3L graphene [105]. The observed intensities of the LF Raman modes versus stacking can be explained by density functional theory (DFT) Raman modeling, group theory, and a generalized bond polarizability model [101, 104].

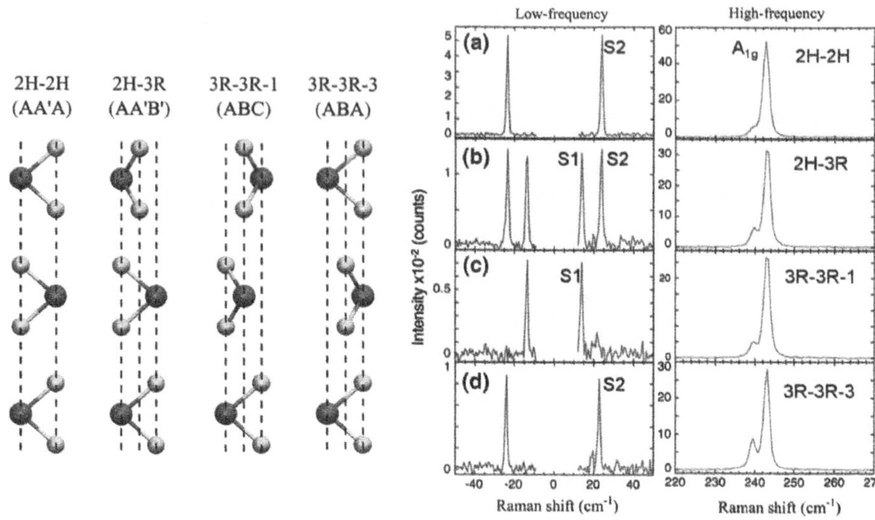

Fig. 1.11 Low-frequency Raman fingerprints of different stacking configurations (left) in trilayers (3L) of $MoSe_2$ showing characteristic LF Raman spectra-fingerprints for each stacking pattern

High-frequency Raman: Defect Concentration indicator The key problem in finding spectroscopic signatures of defects in 2D materials is synthesis of 2D crystals with specific type and number of defects. Recently, a novel laser evaporation approach was used to directly evaporate $MoSe_2$ powder to controllably vary the flux of vapor to synthesize $MoSe_{2-x}$ monolayer crystals with variable number of single Se vacancies [60]. The concentration and type of vacancies were determined using Z-contrast atomic resolution STEM imaging (Fig. 1.12). This allowed us to link Raman spectra to the specific concentration of Se vacancies in $MoSe_2$ monolayers. The observed characteristic red shift of the A_{1g} peak in Raman spectra of $MoSe_2$ monolayer with increase of Se vacancy concentration can be predicted by ab initio calculations (Fig. 1.12) [60]. As a result, the Raman shift of the A_{1g} peak and other characteristic changes in the Raman spectra can be used as an indicator of this particular type of defects and as well as a measure of their concentration.

In summary (1) Frequencies of LF Raman modes can be used to remotely determine the number of layers of 2D materials, which can be modeled by a simple linear chain model and first-principles DFT calculations. (2) Intensities of LF Raman modes can reveal the atomistic stacking configuration of 2D materials in accordance with ab initio Raman modeling, group theory and the generalized bond polarizability model. (3) The commonly studied HF Raman modes reveal the effects of defects, and can potentially be calibrated to vacancy concentrations in TMDs, and possibly other 2D materials in accordance with DFT calculations and the empirical phonon confined model.

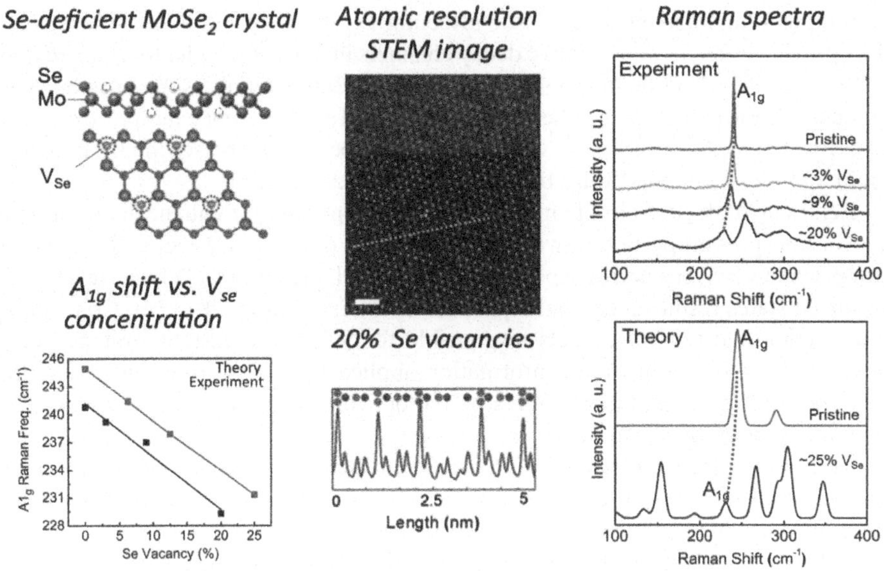

Fig. 1.12 Raman spectroscopy of Se-deficient $MoSe_2$ monolayers. The defect type and concentration were determined using atomic resolution STEM (middle). Experimental and calculated Raman spectra at different concentrations of Se vacancies (right column). Raman shift of the A_{1g} peak versus concentration of Se vacancies (left column, bottom)

1.4.3 Photoluminescence Spectroscopy of 2D Materials

Layered 2D materials have unique optoelectronic properties, which are quite different from their bulk counterparts. For example, excitons in 2D materials are confined to a monolayer and the dielectric screening is reduced due to change in the dielectric environment. As a consequence, the electron-hole (e-h) interaction is enhanced that results in strong e-h binding, an increased band gap, and even the possibility to observe charged excitons (trions) at room temperature. In addition, due to the absence of the weak interlayer interaction in monolayers some of them exhibit a transition from an indirect band gap in the bulk to a direct band gap for monolayers that produces a strong enhancement of their photoluminescence (PL) [120, 121]. However, despite this enhancement observations of this PL quantum yield in TMD monolayers remain fairly rare, e.g., only 0.1–1% in MoS_2 monolayers [120] implying the existence of efficient nonradiative decay channels.

Since 2D materials are nearly all 'surface'and usually supported on a substrate, their PL is very sensitive to the type of the supports [122, 123] and is quite different compared to suspended material [124]. The main effect of a substrate is charge transfer, resulting in electron- or hole-doping of 2D layers that is responsible for the observations of PL from charged excitons (trions) [125].

Similarly, another important issue affecting PL in 2D materials is defects since they provide efficient nonradiative decay channels in 2D semiconductors (see review [126]) An interesting feature, which is specific to 2D materials is the possibility to compensate the defects using chemical treatments [127, 128]. For example, treatment of exfoliated MoS_2 monolayers with an organic superacid increases the PL quantum yield up to 95% and the PL lifetime from 0.3 to 10.8 ns [127].

Removing defects from 2D materials is very important for enhancement of their carrier transport properties, however in some cases defects are beneficial since they can provide new very narrow spectral features in PL spectra of 2D materials due to localized states resulting in single quantum emitters (SQE) [129–132]. Controlling the location and type of defects responsible for SQEs is a current strategy being developed for future quantum information applications. Therefore, understanding the types of defects and their effects on the optical properties of 2D materials is a very important topic of research.

Here, we will use an example of Se-deficient $MoSe_2$ to discuss the clear defect-related bands that appear in low- temperature PL spectra. Although, there are many different ways to introduce defects in 2D materials, e.g., ion beam irradiation [133], plasma treatments [134], etc., these approaches commonly introduce many different types of defects, which makes very difficult to assign the spectral features to the specific defects in PL spectra. However, as discussed above recently we showed that it is possible to synthesize 2D materials with specific defects, i.e., $MoSe_2$ monolayers with variable concentration of Se vacancies (up to 20 at.%) [60]. Although the room temperature PL band of Se-deficient $MoSe_2$ exhibits only a small blue shift without the appearance of any pronounced additional bands related to defects, the low-temperature PL spectra at 4 K show an intense defect-related band at 905 nm (1.37 eV) (Fig. 1.13). Moreover, with increasing concentrations of Se vacancies this band becomes the main feature of the low-temperature PL spectrum as shown in Fig. 1.13c. In addition, the low-temperature PL spectrum shows 3 bands related to free excitons (756 nm, 1.64 eV), trions (780 nm, 1.59 eV), and bound excitons (811 nm, 1.53 eV). PL mapping at 4 K shows that the defects are distributed uniformly along the $MoSe_2$ monolayer surface (Fig. 1.13f) except at the edges, where the defect concentration drops significantly, and as a result the free exciton emission dominates (Fig. 1.13g). Since the edges of the crystal is where growth occurs, this result indicates that Se-vacancy defects in $MoSe_2$ crystals do not develop during the earlier stages of growth.

Another way to create specific defects in TMDs is to substitute metal atoms. For example, Mo atoms in $MoSe_2$ can be isoelectronically substituted with W atoms that provide traps for free excitons, which results in the appearance of new bands in the low- temperature PL spectra of these monolayers [135].

In summary (1) Advanced synthesis methods permit monolayer growth with variable concentration of specific defects. (2) The combination of atomic resolution STEM characterization with measured low-temperature PL spectra and theory allows the assignment of PL bands to specific defects. (3) Low-temperature PL measurements provide characteristic fingerprints of specific defects in 2D materials. (4)

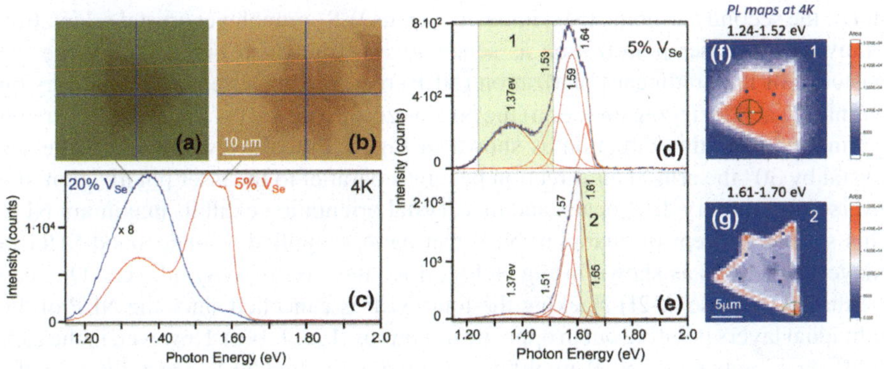

Fig. 1.13 Low temperature PL spectroscopy of $MoSe_2$ monolayers with the variable number of specific defects (Se vacancies). **a**, **b** Optical microscope images of $MoSe_2$ monolayers with 20% and 5% of Se vacancies, respectively and **c** their corresponding PL spectra measured at 4 K. **d** Deconvoluted low-temperature PL spectrum shows 4 bands related to free excitons (1.64 eV), trions (1.59 eV), bound excitons (1.53 eV) and defects (1.37 eV), which positions and intensities depend on the location within the $MoSe_2$ crystal as shown in (**e**). **f**, **g** Low-temperature PL maps plotted by integrating the PL intensities within the highlighted regions 1 and 2 in (**d**, **e**) corresponding to the defect (1) and the free-exciton (2) bands and showing much smaller concentration of Se-vacancies at the edges of the $MoSe_2$ crystal

Low-temperature PL mapping reveals the spatial distribution of defects in 2D crystals.

1.4.4 Second Harmonic Generation Microscopy of 2D Materials

Second Harmonic Generation (SHG) provides another powerful method to characterize 2D materials. The SHG intensity is very sensitive to crystal symmetry. In the dipole approximation SHG can be generated only in non-centrosymmetric crystals so it provides a useful method to assess crystal symmetries and structural phase transitions. The SHG can provide information on many important parameters of 2D materials including crystal orientations [136–140], layer stacking configurations [141–144], size of single-crystal domains [144], grain boundaries [144, 145], and even electronic structure [137]. The attractiveness of SHG for the study 2D crystals is also based on the strong SHG signal obtained despite the atomic layer thicknesses of the crystals. For example, the measured nonlinear susceptibility in WS_2 monolayers is three orders of magnitude larger than that for common nonlinear optical crystals [138, 140].

SHG is being developed to understand the growth orientations of TMD cyrstals on different substrates. Here we first examine WS_2 monolayer crystal growth on patterned substrates using SHG polarization microscopy. Figure 1.14 shows the polar plot of the second harmonic (SH) intensity versus WS_2 monolayer crystal orientation relative to the laser polarization as schematically shown in Fig. 1.14b. Since the existence of the nonlinear polarization (NLP) requires the absence of the inversion symmetry, at 0° (zig-zag direction) the NLP is zero since the crystal shows inversion symmetry along this direction as shown in Fig. 1.14b. However, if we rotate the crystal by 30° the arm-chair direction becomes parallel to the laser polarization, the inversion symmetry disappears, and this crystal orientation exhibits maximum NLP. This simple concept of changing NLP can be also applied to understand SHG in bilayers of TMDs as shown in Fig. 1.14c, d for the case of WS_2 bilayers. One can see that in the case of 2H stacking the total NLP is cancelled since the NLP of the individual layers pointing out in opposite directions (Fig. 1.14c). However, in the case of 3R stacking the total NLP amplifies since the individual NLPs are pointing in the same direction. The SH signal for the number of layers >2 also can be qualitatively predicted based on the NLP behavior of the individual layers.

Figure 1.15 illustrates an example of using SHG to reveal the growth mode of WS_2 monolayer crystals on patterned substrates where strain is introduced by growth over posts of different height. For small strains, the uniform SH map and the equilateral triangular shape of the crystal indicates that the crystal can maintain its single crystallinity. However, for large strains the many small crystal grains formed with different orientation are evident in the map of SHG intensities (whereas SEM or AFM images of the crystal would not reveal this).

In summary (1) The nearly three orders of magnitude higher nonlinear susceptibility of 2D TMDs compared to common nonlinear crystals makes SHG a very attractive technique to characterize 2D materials. (2) SHG can provide information on 2D crystal orientations, layer stacking configurations, size of single-crystal domains, and grain boundaries.

1.4.5 Ultrafast Spectroscopy of 2D Materials

Ultrafast spectroscopy of 2D materials addresses the very important aspect of the dynamics of elementary excitations and carriers in these materials. Recently, this topic has been reviewed by Ceballos et al. [26] Ultrafast processes in many different 2D materials have been studied including the most investigated case of MoS_2 [146–151] as well as many others: $MoSe_2$ [152] WS_2 [153–156], WSe_2 [157, 158], ReS_2[159], and Black Phosphorus [160]. In the simplest case, laser excitation of 2D materials can create narrow distributions of excited electrons and holes, which then thermalize, cool, and then can combine to form quasiparticles such as excitons and trions, which can recombine to yield the characteristic PL signals that were previously discussed. Of course, quasiparticles can be directly generated by the laser excitation, and the ratio of direct to indirect formation depends upon interactions with the substrate, adsorbates, and other layers, etc. as previously discussed. The characteristic time scales for these processes in atomically-thin 2D materials are expected to be different from those measured in their bulk counterparts due to the

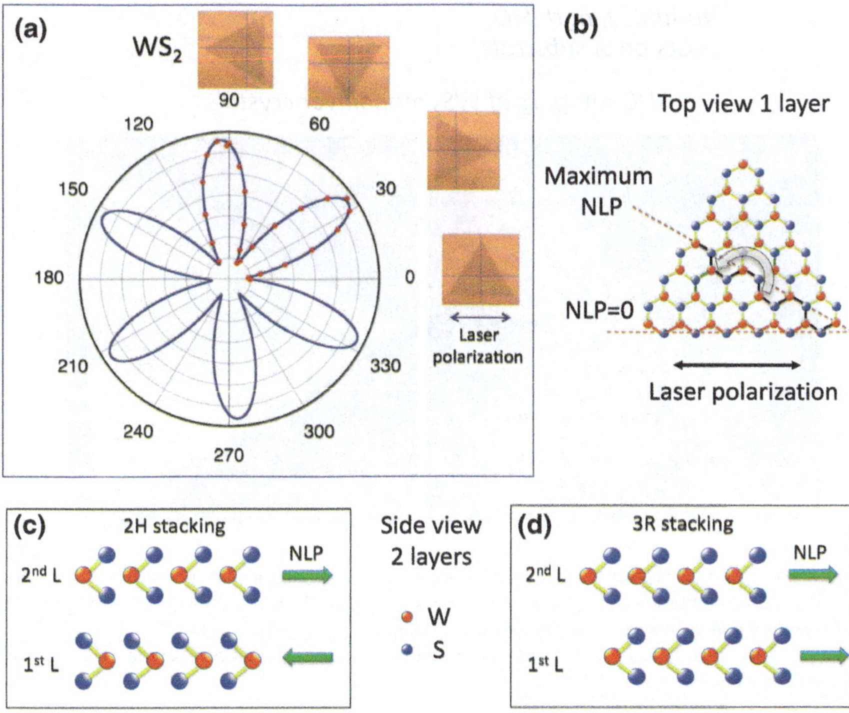

Fig. 1.14 SHG polarization microscopy of a WS_2 monolayer crystal. SHG measurements were conducted using a 40 fs Ti:Sapphire laser (Micra, Coherent) at 800 nm and 80 MHz repetition rate. The laser beam was passed through a half-wave plate mounted in a rotation stage and was directed into an upright microscope (Olympus) and focused onto a sample surface using a 100x microscope objective (Numerical Aperture: NA = 0.9) to a ~ 1 μm spot. The laser energy at the sample surface was ~0.1 mW. The SHG light was collected in backscattering configuration using the same objective and was directed to a monochromator (Spectra Pro 2300i, Acton, $f = 0.3$ m) that was coupled to the microscope and equipped with a 150 grooves/mm grating and a CCD camera (Pixis 256BR, Princeton Instruments). Before entering the monochromator the SHG light was passed through a short-pass cut-off filter (650 nm) and a polarizer to filter out the fundamental excitation light at 800 nm and to select the SHG polarization parallel to that of the excitation light. **a** Polar plot of the second harmonic (SH) intensity versus crystal orientation relative to the laser polarization as schematically shown in (**b**), dots and the blue solid line show the experimental points and fit with $\sin^2(3(\phi + \phi_0))$, respectively. Inserts show optical microscope images of a WS_2 monolayer crystal at 0°, 30°, 60°, and 90° relative to the laser polarization. At 0° (zig-zag direction) the nonlinear polarization (NLP) is zero since the crystal shows inversion symmetry along this direction (see b). However, at 30° (arm-chair direction) the inversion symmetry disappears and this crystal orientation exhibits maximum NLP. **c** Cancelation of NLPs in bilayer WS_2 in the case of 2H stacking configuration resulting in zero SH intensity. **d** Doubling of NLP in the case of 3R WS_2 stacking leading to strong increase of SH signal. The SH signal for the number of layers >2 can be qualitatively understood based on the NLP behavior (see also [145])

aforementioned strong enhancement of interactions between electrons and holes due to reduced dielectric screening. Therefore one should expect considerable shorten-

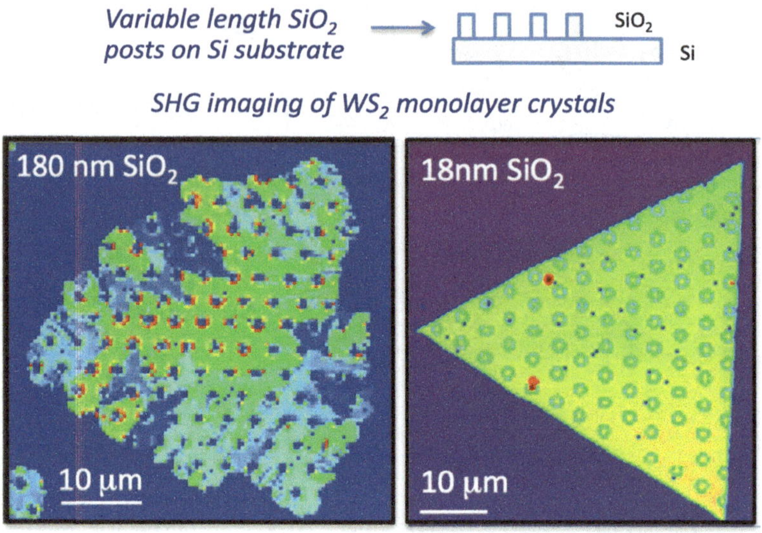

Fig. 1.15 An example illustrating how SHG mapping of two triangular WS_2 monolayers can infer crystalline grain orientations. When crystals grow over short 18 nm-tall posts patterned on SiO2 substrates, they maintain a constant crystalline orientation (right), which is not the case (left) when crystals grow around tall posts, and many smaller grains of varied orientation result in SHG intensity variations consistent with Fig. 1.14

ing of the lifetimes for such major interactions as electron-electron, electron-phonon, electron-defect, and electron-quasiparticle scattering.

The timescales for these processes are revealed using ultrafast laser excitation and probing. For example, using 10 fs visible laser pulses in their pump-probe setup, Nie et al., revealed that carrier thermalization in few-layer MoS_2 occurs within only 20 fs after excitation, followed by cooling with the characteristic time of ~0.6 ps [149]. Recently, Ceballos et al., showed that the time-scale for exciton formation in TMD monolayers is ~0.3–0.5 ps [161]. When excitons are formed (directly or indirectly) they diffuse and become subject to a variety of loss processes including radiative recombination that results in observable PL, as well as nonradiative recombination, exciton-exciton annihilation, etc. It should be emphasized that the major problem in ultrafast pump-probe spectroscopy experiments is the interpretation of the observed transient signals, which is commonly based on exclusion of a possible set of competing processes. For example, fs-transient absorption measurements showed that decay of the A exciton in MoS_2 monolayers exhibited three characteristic decay times, i.e., 2–4 ps, 30–80 ps, and 300–1000 ps. These three lifetimes were tentatively assigned to trapping of excitons by defects, carrier-phonon scattering, and direct electron-hole recombination, respectively [147]. However, other measurements showed different lifetimes (see discussion in Review [26]). This situation is not surprising since the exciton decay dynamics in monolayers are heavily dominated by defects, and the heterogeneity inherent in samples used in different studies can exhibit different types and concentrations of these defects.

2D materials also provide a unique opportunity for controlling the spin of photoexcited electrons using circularly polarized light, where spin can be used to process information for spintronics applications (see Review [162]). This is based on the spin-valley selection rules where right- and left-polarized light can excite only specific valleys labeled K and –K in hexagonal lattices of TMDs [163]. To measure valley dynamics in TMDs polarized ultrafast measurements were employed that revealed very short spin lifetimes (in the order of a few ps) for variety of different TMDs including MoS_2 [148, 164] and WS_2 [153] (see also Review [26]). However, the valley-polarized exciton dynamics changes considerably in stacked heterobilayers of TMDs. Recently Rivera et al., demonstrated that valley-specific *interlayer* excitons created with polarized optical excitation have very long valley lifetime of 40 ns in WSe_2-$MoSe_2$ bilayers [165].

Another interesting phenomenon that has been revealed by ultrafast laser spectroscopy is a very efficient electron-hole separation between different layers in stacked hetero-bilayers of TMDs, which occurs through extremely fast (in the order of few tens of fs) charge transfer between the monolayers [166–168]. This is surprising since the layers are well separated and the van der Waals interaction between the layers is relatively weak. Recently, Zheng et al., using ab initio non-adiabatic molecular dynamic calculations revealed that ultrafast interlayer hole transfer could occur through phonon-assisted mechanism within only 20 fs for the case of type II band alignment in MoS_2/WS_2 bilayers [169].

Here, we consider an interesting example of a hybrid 2D/0D system that also exhibits ultrafast charge transfer [170]. Figure 1.16 shows schematic of the creation of the A exciton in a WS_2 monolayer followed by ultrafast electron transfer to a monolayer of CdSe-core/ZnS-shell quantum dots (QD) deposited on the top of the WS_2 monolayer, which results in formation of hybrid excitons where holes reside in the WS_2 layer and electrons are transferred to reside in the QDs (Fig. 1.16 a, b). In this case, the QDs had a CdSe core (~4 nm), ZnS shell (~2 nm), and ~2 nm octadecylamine capping ligand for a total average size of ~8 nm. By using a tunable 40-fs pump laser from an OPA, and the broadband 40-fs probe beam at variable time delays (as shown in Fig. 1.6b) a new absorption band could be observed within the region occupied by the quantum dots on the 2D crystals (and not from either component alone) after pump laser creation of the hybrid excitons [170]. By observing the depletion dynamics of 2D-X_A exciton ground state absorption and the 1S transition in quantum dots, the timescale for this process was shown to be comparable to that of the pump laser pulse (blue dashed line in Fig. 1.16c), i.e., ~45 fs. Such efficient transfer of energy, leading to weaker-bound hybrid excitons, may be useful for photovoltaic and sensing applications.

In summary: (1) Ultrafast laser spectroscopy and especially, fs-laser pump-probe approach reveal the time scales for formation and decay processes of basic elementary excitations and charge carriers in 2D materials such as formation of excitons, biexitons, and trions (charged exciton); electron-electron scattering, electron-phonon scattering, electron-defects scattering, electron-quasiparticles scattering, etc., and

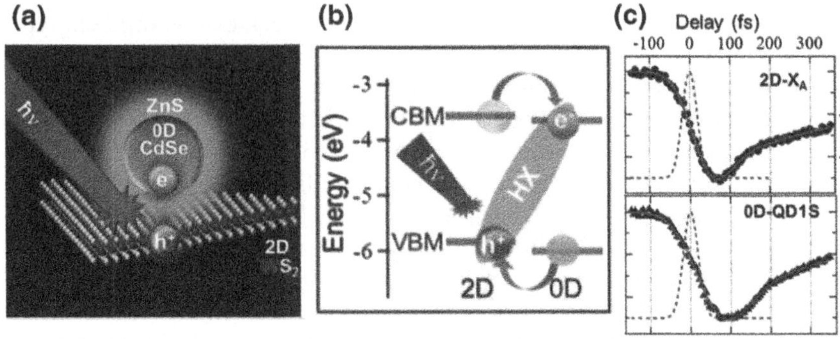

Fig. 1.16 **a** Schematic of a hybrid exciton formation in a 2D/0D heterostructure composed of a WS_2 monolayer and a single layer of CdSe-core/ZnS-shell quantum dots. **b** Energy level diagram showing formation of hybrid excitons. **c** Depletion dynamics of 2D-X_A exciton and 1S transition in quantum dots upon X_A excitation [170]

can establish the preferential decay channels of these quasiparticles. (2) The majority of these processes in 2D materials occurs faster compared to those in their bulk counterparts. (3) Dynamics of spin selective excitation were investigated in TMD monolayers through addressing different valleys in hexagonal TMD lattices using circular polarized fs-laser pulses (very fast spin relaxation, ~a few ps) and in TMD heterobilayers, where remarkably long lifetimes of valley-specific *interlayer* excitons were measured (~40 ns). (4) It was found that a very efficient electron-hole separation between different layers in stacked hetero-bilayers of TMDs occurs through extremely fast (in the order of few tens of fs) charge transfer between the monolayers. (5) However, many competing ultrafast processes in 2D materials and difficulties with comprehensive theoretical interpretation of the excited state dynamics are hindering wider application of this important approach to characterization of 2D materials.

1.5 Summary

In summary, we have discussed the advantages and advances of laser interactions to understand and control the synthesis, processing, and characterization of atomically-thin 2D materials. The key questions for the synthesis and processing of 2D materials were presented, and mainly involve the *heterogeneity* inherent in these materials resulting from different synthesis techniques—that is, point defects, grain boundaries, substitutional or intercalated dopants or impurities, uncontrolled layer nucleation, adsorbates and substrate interactions. Laser interactions provide non-equilibrium conditions and kinetic pathways to address these heterogeneities, both during synthesis and in post-processing. At the same time, laser interactions are being developed to provide real-time and remote spectroscopic characterization of these heterogeneities. In close conjunction with predictive theory and atomistic characterization techniques such as STM and AR-Z-STEM that can measure atomic positions and composition, laser spectroscopy is providing the link between predicted electronic and vibrational spectra and the optoelectronic properties resulting from heterogeneity. The abilities to: (1) atomistically "see" all the atoms in a materials system, (2) tractably computationally model them with first principles theory, and (3) remotely characterize their optoelectronic properties with laser spectroscopy presents a unique practical platform to help achieve a priority research direction in synthesis science to "*integrate emerging theoretical, computational, and* in situ *characterization tools to achieve directed synthesis with real-time adaptive control* [43]." Through the use of laser spectroscopy with real-time measurement of growth kinetics and materials properties, mechanistic control over 2D materials synthesis can be obtained through the development of growth models [171] and automated exploration of parameter space [172]. Ultimately, through machine learning and artificial intelligence, this platform may provide a means to autonomously explore materials synthesis [173] in the "building block" approach envisioned for 2D van der Waals heterostructures [2].

Acknowledgements The authors gratefully acknowledge support for the explorations of synthesis science of 2D materials, which is sponsored by the U.S. Department of Energy, Office of Science, Basic Energy Sciences, Materials Sciences and Engineering Division, and for support for the explorations of their heterogeneity and associated functionality through the development of laser spectroscopic and other characterization techniques at the Center for Nanophase Materials Sciences, a nanoscale science research center which is sponsored by the U.S. DOE-BES Scientific User Facilities Division.

References

1. K.S. Novoselov, A.K. Geim, S.V. Morozov, D. Jiang, Y. Zhang, S.V. Dubonos, I.V. Grigorieva, A.A. Firsov, Science **306**(5696), 666–669 (2004)
2. A.K. Geim, I.V. Grigorieva, Nature **499**(7459), 419–425 (2013)
3. J.D. Cain, E.D. Hanson, F.Y. Shi, V.P. Dravid, Curr. Opin. Solid St. M **20**(6), 374–387 (2016)
4. C. Tan, X. Cao, X.-J. Wu, Q. He, J. Yang, X. Zhang, J. Chen, W. Zhao, S. Han, G.-H. Nam, M. Sindoro, H. Zhang, Chem. Rev. **117**(9), 6225–6331 (2017)
5. C.N.R. Rao, U. Maitra, Annu. Rev. Mater. Res. **45**, 29–62 (2015)
6. G.R. Bhimanapati, Z. Lin, V. Meunier, Y. Jung, J. Cha, S. Das, D. Xiao, Y. Son, M.S. Strano, V.R. Cooper, L.B. Liang, S.G. Louie, E. Ringe, W. Zhou, S.S. Kim, R.R. Naik, B.G. Sumpter, H. Terrones, F.N. Xia, Y.L. Wang, J. Zhu, D. Akinwande, N. Alem, J.A. Schuller, R.E. Schaak, M. Terrones, J.A. Robinson, ACS Nano **9**(12), 11509–11539 (2015)
7. M. Garcia-Hernandez, J. Coleman, 2D Mater **3**(1) (2016)
8. S.Z. Butler, S.M. Hollen, L.Y. Cao, Y. Cui, J.A. Gupta, H.R. Gutierrez, T.F. Heinz, S.S. Hong, J.X. Huang, A.F. Ismach, E. Johnston-Halperin, M. Kuno, V.V. Plashnitsa, R.D. Robinson, R.S. Ruoff, S. Salahuddin, J. Shan, L. Shi, M.G. Spencer, M. Terrones, W. Windl, J.E. Goldberger, ACS Nano **7**(4), 2898–2926 (2013)
9. R. Lv, H. Terrones, A.L. Elías, N. Perea-López, H.R. Gutiérrez, E. Cruz-Silva, L.P. Rajukumar, M.S. Dresselhaus, M. Terrones, Nano Today **10**(5), 559–592 (2015)
10. S. Das, J.A. Robinson, M. Dubey, H. Terrones, M. Terrones, Annu. Rev. Mater. Res. **45**, 1–27 (2015)

11. A.V. Kolobov, J. Tominaga, *Two-Dimensional Transition-Metal Dichalcogenides*. (Springer International Publishing, 2016)
12. F. Bonaccorso, L. Colombo, G.H. Yu, M. Stoller, V. Tozzini, A.C. Ferrari, R.S. Ruoff, V. Pellegrini, Science **347**(6217) (2015)
13. H. Li, J. Wu, Z. Yin, H. Zhang, Acc. Chem. Res. **47**(4), 1067–1075 (2014)
14. Y. Cui, B. Li, J. Li, Z. Wei, Sci. China Phys. Mech. Astron. **61**(1), 016801 (2018)
15. A. Pant, Z. Mutlu, D. Wickramaratne, H. Cai, R.K. Lake, C. Ozkan, S. Tongay, Nanoscale **8**(7), 3870–3887 (2016)
16. D.M. Hamann, E.C. Hadland, D.C. Johnson, Semicond. Sci. Technol. **32**(9), 093004 (2017)
17. R. Dong, I. Kuljanishvili, J. Vac. Sci. Technol. B Nanotechnology Microelectron.: Mater. Proc. Meas. Phenom. **35**(3), 030803 (2017)
18. W. Xia, L. Dai, P. Yu, X. Tong, W. Song, G. Zhang, Z. Wang, Nanoscale **9**(13), 4324–4365 (2017)
19. X. Huang, C. Tan, Z. Yin, H. Zhang, Adv. Mater. **26**(14), 2185–2204 (2014)
20. M. Chhowalla, H.S. Shin, G. Eda, L.-J. Li, K.P. Loh, H. Zhang, Nat. Chem. **5**(4), 263–275 (2013)
21. C.R. Ryder, J.D. Wood, S.A. Wells, M.C. Hersam, **10**(4) 3900–3917 (2016)
22. C. Tan, H. Zhang, Chem. Soc. Rev. **44**(9), 2713–2731 (2015)
23. M. Wasala, H.I. Sirikumara, Y.R. Sapkota, S. Hofer, D. Mazumdar, T. Jayasekera, S. Talapatra, J. Mater. Chem. C. **5**, 11214–11225 (2017)
24. X. Congxin, L. Jingbo, J. Semiconductors **37**(5), 051001 (2016)
25. W. Zhang, Q. Wang, Y. Chen, Z. Wang, A.T. Wee, 2D Mater **3**(2), 022001 (2016)
26. F. Ceballos, H. Zhao, Adv Funct Mater **27**(19), 1604509 (2017)
27. H.B. Ribeiro, M.A. Pimenta, C.J. de Matos, J. Raman Spectrosc. 49,76–90 (2018)
28. R. Saito, A.R. Nugraha, E.H. Hasdeo, S. Siregar, H. Guo, T. Yang, Phys. Status Solidi (b) **252**(11), 2363–2374 (2015)
29. X. Zhang, Q.-H. Tan, J.-B. Wu, W. Shi, P.-H. Tan, Nanoscale **8**(12), 6435–6450 (2016)
30. X.L. Li, W.P. Han, J.B. Wu, X.F. Qiao, J. Zhang, P. H. Tan, Adv Funct Mater **27**(19), 1604468 (2017)
31. J. Ji, S. Dong, A. Zhang, Q. Zhang, Phys. E **80**, 130–141 (2016)
32. X. Zhang, X.-F. Qiao, W. Shi, J.-B. Wu, D.-S. Jiang, P.-H. Tan, Chem. Soc. Rev. **44**(9), 2757–2785 (2015)
33. H.Y. Yu, X.D. Cui, X.D. Xu, W. Yao, Natl. Sci. Rev. **2**(1), 57–70 (2015)
34. C. Tan, Z. Liu, W. Huang, H. Zhang, Chem. Soc. Rev. **44**(9), 2615–2628 (2015)
35. X. Wei, F.-G. Yan, C. Shen, Q.-S. Lv, K.-Y. Wang, Chin. Phys. B **26**(3), 038504 (2018)
36. J. Wang, H. Fang, X. Wang, X. Chen, W. Lu, W. Hu, Small **13**(35), 1700894 (2017)
37. A. Eftekhari, Appl. Mater. Today **8**, 1–17 (2017)
38. X. Li, M.W. Lin, J.H. Lin, B. Huang, A.A. Puretzky, C. Ma, K. Wang, W. Zhou, S.T. Pantelides, M.F. Chi, I. Kravchenko, J. Fowlkes, C.M. Rouleau, D.B. Geohegan, K. Xiao, Sci Adv **2** (4), e1501882 (2016)
39. T. Heine, Acc. Chem. Res. **48**(1), 65–72 (2014)
40. Y. Sun, R. Wang, K. Liu, Appl. Phys. Rev. **4**(1), 011301 (2017)
41. R. Roldan, A. Castellanos-Gomez, E. Cappelluti, F. Guinea, J. Phys-Condens Mat. **27**(31), 313201 (2015)
42. S.W. Wang, H. Medina, K.B. Hong, C.C. Wu, Y.D. Qu, A. Manikandan, T.Y. Su, P.T. Lee, Z.Q. Huang, Z.M. Wang, F.C. Chuang, H.C. Kuo, Y.L. Chueh, ACS Nano **11**(9), 8768–8776 (2017)
43. Basic Research Needs for Synthesis Science, Report of the Basic Energy Sciences Workshop on Synthesis Science for Energy Relevant Technology, May 2–4 (2016), https://science.energy.gov/bes/community-resources/reports/
44. Presented at the Workshop on Nucleation and Growth Mechanisms of Single Wall Carbon Nanotubes, Bandera, Texas, 2015 (unpublished)
45. M.S. Donley, P.T. Murray, S.A. Barber, T.W. Haas, Surf. Coat. Tech. **36**(1–2), 329–340 (1988)

46. M.S. Donley, N.T. Mcdevitt, T.W. Haas, P.T. Murray, J.T. Grant, Thin Solid Films **168**(2), 335–344 (1989)
47. N.T. McDevitt, J.E. Bultman, J.S. Zabinski, Appl. Spectrosc. **52**(9), 1160–1164 (1998)
48. F. Ullah, V. Senthilkumar, S.H. Kim, C.T. Le, H. Rock, D.Y. Lee, S. Park, A.I. Ali, Y.S. Kim, J. Nanosci. Nanotechno. **16**(10), 10284–10289 (2016)
49. M.I. Serna, S.H. Yoo, S. Moreno, Y. Xi, J.P. Oviedo, H. Choi, H.N. Alshareef, M.J. Kim, M. Minary-Jolandan, M.A. Quevedo-Lopez, ACS Nano **10**(6), 6054–6061 (2016)
50. C.R. Serrao, A.M. Diamond, S.L. Hsu, L. You, S. Gadgil, J. Clarkson, C. Carraro, R. Maboudian, C.M. Hu, S. Salahuddin, Appl. Phys. Lett. **106**(5), 052101 (2015)
51. G. Siegel, Y.P.V. Subbaiah, M.C. Prestgard, A. Tiwari, APL Mater. **3**(5), 056103 (2015)
52. A. Barvat, N. Prakash, B. Satpati, S.S. Singha, G. Kumar, D.K. Singh, A. Dogra, S.P. Khanna, A. Singha, P. Pal, J. Appl. Phys. **122**(1), 015304 (2017)
53. Y.T. Ho, C.H. Ma, T.T. Luong, L.L. Wei, T.C. Yen, W.T. Hsu, W.H. Chang, Y.C. Chu, Y.Y. Tu, K.P. Pande, E.Y. Chang, Phys. Status Solidi-R **9**(3), 187–191 (2015)
54. M. Gao, M.H. Zhang, W. Niu, Y.Q. Chen, M. Gu, H.Y. Wang, F.Q. Song, P. Wang, S.C. Yan, F.Q. Wang, X.R. Wang, X.F. Wang, Y.B. Xu, R. Zhang, Appl. Phys. Lett. **111**(3), 031906 (2017)
55. S.R. Jian, J.Y. Juang, C.W. Luo, S.A. Ku, K.H. Wu, J. Alloy. Compd. **542**, 124–127 (2012)
56. M. Mahjouri-Samani, R. Gresback, M.K. Tian, K. Wang, A.A. Puretzky, C.M. Rouleau, G. Eres, I.N. Ivanov, K. Xiao, M.A. McGuire, G. Duscher, D.B. Geohegan, Adv. Funct. Mater. **24**(40), 6365–6371 (2014)
57. M. Mahjouri-Samani, M.K. Tian, A.A. Puretzky, M.F. Chi, K. Wang, G. Duscher, C.M. Rouleau, G. Eres, M. Yoon, J. Lasseter, K. Xiao, D.B. Geohegan, Nano Lett. **17**(8), 4624–4633 (2017)
58. M. Mahjouri-Samani, M. Tian, K. Wang, A. Boulesbaa, C.M. Rouleau, A.A. Puretzky, M.A. McGuire, B.R. Srijanto, K. Xiao, G. Eres, G. Duscher, D.B. Geohegan, ACS Nano **8**(11), 11567–11575 (2014)
59. M. Mahjouri-Samani, M.W. Lin, K. Wang, A.R. Lupini, J. Lee, L. Basile, A. Boulesbaa, C.M. Rouleau, A.A. Puretzky, I.N. Ivanov, K. Xiao, M. Yoon, D.B. Geohegan, Nat. Commun. **6**, 7749 (2015)
60. M. Mahjouri-Samani, L.B. Liang, A. Oyedele, Y.S. Kim, M.K. Tian, N. Cross, K. Wang, M.W. Lin, A. Boulesbaa, C.M. Rouleau, A.A. Puretzky, K. Xiao, M. Yoon, G. Eres, G. Duscher, B.G. Sumpter, D.B. Geohegan, Nano Lett. **16**(8), 5213–5220 (2016)
61. A.A. Puretzky, D.J. Styers-Barnett, C.M. Rouleau, H. Hu, B. Zhao, I.N. Ivanov, D.B. Geohegan, Appl. Phys A-Mater **93**(4), 849–855 (2008)
62. K. Hata, D.N. Futaba, K. Mizuno, T. Namai, M. Yumura, S. Iijima, Science **306**(5700), 1362–1364 (2004)
63. D.B. Geohegan, A.A. Puretzky, C. Rouleau, J. Jackson, G. Eres, Z. Liu, D. Styers-Barnett, H. Hu, B. Zhao, I. Ivanov, in *Laser-Surface Interactions for New Materials Production* (Springer Berlin Heidelberg, 2010), pp. 1–17
64. D.B. Geohegan, A.A. Puretzky, M. Yoon, G. Eres, C. Rouleau, K. Xiao, J. Jackson, J. Readle, M. Regmi, N. Thonnard, in *Lasers in Materials Science* (Springer International Publishing, 2014), pp. 143–173
65. W.D. Tennyson, M.K. Tian, A.B. Papandrew, C.M. Rouleau, A.A. Puretzky, B.T. Sneed, K.L. More, G.M. Veith, G. Duscher, T.A. Zawodzinski, D.B. Geohegan, Carbon **123**, 605–615 (2017)
66. R. Yuge, S. Bandow, M. Yudasaka, K. Toyama, S. Iijima, T. Manako, Carbon **111**, 675–680 (2017)
67. L.E. Jones, P.A. Thrower, Carbon **29**(2), 251–269 (1991)
68. B. Qin, T.F. Zhang, H.H. Chen, Y.F. Ma, Carbon **102**, 494–498 (2016)
69. A. Dato, V. Radmilovic, Z. Lee, J. Phillips, M. Frenklach, Nano Lett. **8**(7), 2012–2016 (2008)
70. A. Chaturvedi, A. Slabon, P. Hu, S.L. Feng, K.K. Zhang, R.R. Prabhakar, C. Kloc, J. Cryst. Growth **450**, 140–147 (2016)

71. M.W. Smith, K.C. Jordan, C. Park, J.W. Kim, P. T. Lillehei, R. Crooks, J. S. Harrison, Nanotechnology **20**(50), 505604 (2009)
72. A. Fathalizadeh, T. Pham, W. Mickelson, A. Zettl, Nano Lett. **14**(8), 4881–4886 (2014)
73. K.S. Kim, C.T. Kingston, A. Hrdina, M.B. Jakubinek, J.W. Guan, M. Plunkett, B. Simard, ACS Nano **8**(6), 6211–6220 (2014)
74. Y. Miyamoto, H. Zhang, D. Tomanek, Phys. Rev. Lett. **104**(20), 208302 (2010)
75. S. Dhar, A.R. Barman, G.X. Ni, X. Wang, X.F. Xu, Y. Zheng, S. Tripathy, Ariando, A. Rusydi, K.P. Loh, M. Rubhausen, A.H. Castro Neto, B. Ozyilmaz, T. Venkatesan, AIP. Adv. **1**(2), 022109 (2011)
76. Z. Lin, X.H. Ye, J.P. Han, Q. Chen, P.X. Fan, H. J. Zhang, D. Xie, H.W. Zhu, M.L. Zhong, Sci. Rep-UK **5**, 11662 (2015)
77. A. Castellanos-Gomez, M. Barkelid, A.M. Goossens, V.E. Calado, H.S.J. van der Zant, G.A. Steele, Nano Lett. **12**(6), 3187–3192 (2012)
78. W.S. Hummers, R.E. Offeman, J. Am. Chem. Soc. **80**(6), 1339–1339 (1958)
79. X.S. Wu, M. Sprinkle, X.B. Li, F. Ming, C. Berger, W.A. de Heer, Phys. Rev. Lett. **101**(2), 026801 (2008)
80. Z.Q. Wei, D.B. Wang, S. Kim, S.Y. Kim, Y.K. Hu, M.K. Yakes, A.R. Laracuente, Z.T. Dai, S.R. Marder, C. Berger, W.P. King, W.A. de Heer, P.E. Sheehan, E. Riedo, Science **328**(5984), 1373–1376 (2010)
81. A.C. Ferrari, J. Meyer, V. Scardaci, C. Casiraghi, M. Lazzeri, F. Mauri, S. Piscanec, D. Jiang, K. Novoselov, S. Roth, Phys. Rev. Lett. **97**(18), 187401 (2006)
82. D.A. Sokolov, C.M. Rouleau, D.B. Geohegan, T.M. Orlando, Carbon **53**, 81–89 (2013)
83. R. Arul, R.N. Oosterbeek, J. Robertson, G.Y. Xu, J.Y. Jin, M.C. Simpson, Carbon **99**, 423–431 (2016)
84. M.F. El-Kady, V. Strong, S. Dubin, R.B. Kaner, Science **335**(6074), 1326–1330 (2012)
85. M.F. El-Kady, R.B. Kaner, Nat. Commun. **4**, 1475 (2013)
86. W. Gao, N. Singh, L. Song, Z. Liu, A.L.M. Reddy, L.J. Ci, R. Vajtai, Q. Zhang, B.Q. Wei, P.M. Ajayan, Nat. Nanotechnol. **6**(8), 496–500 (2011)
87. V. Eswaraiah, S.S.J. Aravind, S. Ramaprabhu, J. Mater. Chem. **21**(19), 6800–6803 (2011)
88. M. Schumann, R. Sauerbrey, M.C. Smayling, Appl. Phys. Lett. **58**(4), 428–430 (1991)
89. J. Lin, Z.W. Peng, Y.Y. Liu, F. Ruiz-Zepeda, R. Q. Ye, E.L.G. Samuel, M.J. Yacaman, B.I. Yakobson, J.M. Tour, Nat. Commun. **5**, 5714 (2014)
90. P.J.F. Harris, Crit. Rev. Solid State **30**(4), 235–253 (2005)
91. B.J. Simonds, H.J. Meadows, S. Misra, C. Ferekides, P.J. Dale, M.A. Scarpulla, J. Photon Energy **5**(1), 050999 (2015)
92. M.E. McConney, N.R. Glavin, A.T. Juhl, M.H. Check, M.F. Durstock, A.A. Voevodin, T.E. Shelton, J.E. Bultman, J.J. Hu, M.L. Jespersen, M.K. Gupta, R.D. Naguy, J.G. Colborn, A. Haque, P.T. Hagerty, R.E. Stevenson, C. Muratore, J. Mater. Res. **31**(7), 967–974 (2016)
93. A.V. Rafael, R. Rahul, M. Christopher, B. Elisabeth, A.R. Joshua, M. Benji, R.G. Nicholas, 2D Mater **5**(1), 011009 (2018)
94. H. Kwon, W. Choi, D. Lee, Y. Lee, J. Kwon, B. Yoo, C.P. Grigoropoulos, S. Kim, Nano Res. **7**(8), 1137–1145 (2014)
95. K.A.N. Duerloo, Y. Li, E.J. Reed, Nat. Commun. **5**, 4214 (2014)
96. S. Cho, S. Kim, J.H. Kim, J. Zhao, J. Seok, D.H. Keum, J. Baik, D.H. Choe, K.J. Chang, K. Suenaga, S.W. Kim, Y.H. Lee, H. Yang, Science **349**(6248), 625–628 (2015)
97. Y. Ma, P. M. Ajayan, S. Yang, Y. Gong, Small 1801606 (2018)
98. H.L. Zeng, B.R. Zhu, K. Liu, J.H. Fan, X.D. Cui, Q.M. Zhang, Phys. Rev. B **86**(24), 241301(R) (2012)
99. X. Zhang, W.P. Han, J.B. Wu, S. Milana, Y. Lu, Q.Q. Li, A.C. Ferrari, P.H. Tan, Phys. Rev. B **87**(11), 115413 (2013)
100. Y.Y. Zhao, X. Luo, H. Li, J. Zhang, P.T. Araujo, C.K. Gan, J. Wu, H. Zhang, S.Y. Quek, M.S. Dresselhaus, Q.H. Xiong, Nano Lett. **13**(3), 1007–1015 (2013)
101. A.A. Puretzky, L.B. Liang, X.F. Li, K. Xiao, K. Wang, M. Mahjouri-Samani, L. Basile, J.C. Idrobo, B.G. Sumpter, V. Meunier, D.B. Geohegan, ACS Nano **9**(6), 6333–6342 (2015)

102. A.A. Puretzky, L.B. Liang, X.F. Li, K. Xiao, B.G. Sumpter, V. Meunier, D.B. Geohegan, ACS Nano **10**(2), 2736–2744 (2016)
103. S.X. Huang, L.B. Liang, X. Ling, A.A. Puretzky, D.B. Geohegan, B.G. Sumpter, J. Kong, V. Meunier, M.S. Dresselhaus, Nano Lett. **16**(2), 1435–1444 (2016)
104. X. Luo, X. Lu, C.X. Cong, T. Yu, Q.H. Xiong, S. Y. Quek, Sci. Rep-UK **5**, 14565 (2015)
105. C.H. Lui, Z.P. Ye, C. Keiser, E.B. Barros, R. He, Appl. Phys. Lett. **106**(4), 041904 (2015)
106. L.B. Liang, A.A. Puretzky, B.G. Sumpter, V. Meunier, Nanoscale **9**(40), 15340–15355 (2017)
107. L. Liang, J. Zhang, B.G. Sumpter, Q. Tan, P.H. Tan, V. Meunier, ACS Nano **11**(12), 11777–11802 (2017)
108. A.L. Glebov, O. Mokhun, A. Rapaport, S. Vergnole, V. Smirnov, L.B. Glebov, Proc. SPIE **8428**, 1–11 (2012)
109. P.H. Tan, W.P. Han, W.J. Zhao, Z.H. Wu, K. Chang, H. Wang, Y.F. Wang, N. Bonini, N. Marzari, N. Pugno, G. Savini, A. Lombardo, A.C. Ferrari, Nat. Mater. **11**(4), 294–300 (2012)
110. C.H. Lui, T.F. Heinz, Phys. Rev. B **87**(12), 121404 (2013)
111. J.B. Wu, X. Zhang, M. Ijas, W.P. Han, X.F. Qiao, X.L. Li, D.S. Jiang, A.C. Ferrari, P.H. Tan, Nat. Commun. **5**, 5309 (2014)
112. J.B. Wu, Z.X. Hu, X. Zhang, W.P. Han, Y. Lu, W. Shi, X.F. Qiao, M. Ijias, S. Milana, W. Ji, A.C. Ferrari, P.H. Tan, ACS Nano **9**(7), 7440–7449 (2015)
113. C. Lee, H. Yan, L.E. Brus, T.F. Heinz, J. Hone, S. Ryu, ACS Nano **4**(5), 2695–2700 (2010)
114. K.H. Liu, L.M. Zhang, T. Cao, C.H. Jin, D.A. Qiu, Q. Zhou, A. Zettl, P.D. Yang, S.G. Louie, F. Wang, Nat. Commun. **5**, 4966 (2014)
115. X. Lu, M.I.B. Utama, J.H. Lin, X. Luo, Y.Y. Zhao, J. Zhang, S.T. Pantelides, W. Zhou, S.Y. Quek, Q.H. Xiong, Adv. Mater. **27**(30), 4502–4508 (2015)
116. C.H. Lui, Z.P. Ye, C. Ji, K.C. Chiu, C.T. Chou, T.I. Andersen, C. Means-Shively, H. Anderson, J.M. Wu, T. Kidd, Y.H. Lee, R. He, Phys. Rev. B **91**(16), 165403 (2015)
117. X. Ling, L.B. Liang, S.X. Huang, A.A. Puretzky, D.B. Geohegan, B.G. Sumpter, J. Kong, V. Meunier, M.S. Dresselhaus, Nano Lett. **15**(6), 4080–4088 (2015)
118. X. Luo, X. Lu, G.K.W. Koon, A.H.C. Neto, B. Ozyilmaz, Q.H. Xiong, S.Y. Quek, Nano Lett. **15**(6), 3931–3938 (2015)
119. X.F. Qiao, J.B. Wu, L.W. Zhou, J.S. Qiao, W. Shi, T. Chen, X. Zhang, J. Zhang, W. Ji, P.H. Tan, Nanoscale **8**(15), 8324–8332 (2016)
120. K.F. Mak, C. Lee, J. Hone, J. Shan, T.F. Heinz, Phys. Rev. Lett. **105**(13), 136805 (2010)
121. A. Splendiani, L. Sun, Y.B. Zhang, T.S. Li, J. Kim, C.Y. Chim, G. Galli, F. Wang, Nano Lett. **10**(4), 1271–1275 (2010)
122. Y.Y. Li, Z.M. Qi, M. Liu, Y.Y. Wang, X.R. Cheng, G.B. Zhang, L.S. Sheng, Nanoscale **6**(24), 15248–15254 (2014)
123. M. Buscema, G.A. Steele, H.S.J. van der Zant, A. Castellanos-Gomez, Nano Res. **7**(4), 561–571 (2014)
124. N. Scheuschner, O. Ochedowski, A.M. Kaulitz, R. Gillen, M. Schleberger, J. Maultzsch, Phys. Rev. B **89**(12), 125406 (2014)
125. K.F. Mak, K.L. He, C. Lee, G.H. Lee, J. Hone, T.F. Heinz, J. Shan, Nat. Mater. **12**(3), 207–211 (2013)
126. Z. Lin, A. McCreary, N. Briggs, S. Subramanian, K.H. Zhang, Y.F. Sun, X.F. Li, N.J. Borys, H.T. Yuan, S.K. Fullerton-Shirey, A. Chernikov, H. Zhao, S. McDonnell, A.M. Lindenberg, K. Xiao, B.J. LeRoy, M. Drndic, J.C.M. Hwang, J. Park, M. Chhowalla, R.E. Schaak, A. Javey, M.C. Hersam, J. Robinson, M. Terrones, 2D Mater **3**(4), 022002 (2016)
127. M. Amani, D.H. Lien, D. Kiriya, J. Xiao, A. Azcatl, J. Noh, S.R. Madhvapathy, R. Addou, K.C. Santosh, M. Dubey, K. Cho, R.M. Wallace, S.C. Lee, J.H. He, J.W. Ager, X. Zhang, E. Yablonovitch, A. Javey, Science **350**(6264), 1065–1068 (2015)
128. H.Y. Nan, Z.L. Wang, W.H. Wang, Z. Liang, Y. Lu, Q. Chen, D.W. He, P.H. Tan, F. Miao, X.R. Wang, J.L. Wang, Z.H. Ni, ACS Nano **8**(6), 5738–5745 (2014)
129. T. Kato, T. Kaneko, ACS Nano **8**(12), 12777–12785 (2014)
130. M. Koperski, K. Nogajewski, A. Arora, V. Cherkez, P. Mallet, J.Y. Veuillen, J. Marcus, P. Kossacki, M. Potemski, Nat. Nanotechnol. **10**(6), 503–506 (2015)

131. Y.M. He, G. Clark, J.R. Schaibley, Y. He, M.C. Chen, Y.J. Wei, X. Ding, Q. Zhang, W. Yao, X.D. Xu, C.Y. Lu, J.W. Pan, Nat. Nanotechnol. **10**(6), 497–502 (2015)
132. A. Srivastava, M. Sidler, A.V. Allain, D.S. Lembke, A. Kis, A. Imamoglu, Nat. Nanotechnol. **10**(6), 491–496 (2015)
133. W.M. Parkin, A. Balan, L.B. Liang, P.M. Das, M. Lamparski, C.H. Naylor, J.A. Rodriguez-Manzo, A.T.C. Johnson, V. Meunier, M. Drndic, ACS Nano **10**(4), 4134–4142 (2016)
134. P.K. Chow, R.B. Jacobs-Gedrim, J. Gao, T.M. Lu, B. Yu, H. Terrones, N. Koratkar, ACS Nano **9**(2), 1520–1527 (2015)
135. X. Li, A.A. Puretzky, X.H. Sang, K.C. Santosh, M.K. Tian, F. Ceballos, M. Mahjouri-Samani, K. Wang, R.R. Unocic, H. Zhao, G. Duscher, V.R. Cooper, C.M. Rouleau, D.B. Geohegan, K. Xiao, Adv. Funct. Mater **27**(19), 1603850 (2017)
136. N. Kumar, S. Najmaei, Q.N. Cui, F. Ceballos, P.M. Ajayan, J. Lou, H. Zhao, Phys. Rev. B **87**(16), 161403(R) (2013)
137. L.M. Malard, T.V. Alencar, A.P.M. Barboza, K.F. Mak, A.M. de Paula, Phys. Rev. B **87**(20), 201401(R) (2013)
138. C. Janisch, Y.X. Wang, D. Ma, N. Mehta, A.L. Elias, N. Perea-Lopez, M. Terrones, V. Crespi, Z.W. Liu, Sci. Rep-UK **4**, 5530 (2014)
139. W.J. Jie, X. Chen, D. Li, L. Xie, Y.Y. Hui, S.P. Lau, X.D. Cui, J.H. Hao, Angew. Chem. Int. Edit. **54**(4), 1185–1189 (2015)
140. J. Ribeiro-Soares, C. Janisch, Z. Liu, A.L. Elias, M.S. Dresselhaus, M. Terrones, L.G. Cancado, A. Jorio, 2D Mater **2**(4), 045015 (2015)
141. Y.L. Li, Y. Rao, K.F. Mak, Y.M. You, S.Y. Wang, C.R. Dean, T.F. Heinz, Nano Lett. **13**(7), 3329–3333 (2013)
142. C.J. Kim, L. Brown, M.W. Graham, R. Hovden, R.W. Havener, P.L. McEuen, D.A. Muller, J. Park, Nano Lett. **13**(11), 5660–5665 (2013)
143. T. Jiang, H.R. Liu, D. Huang, S. Zhang, Y.G. Li, X.G. Gong, Y.R. Shen, W.T. Liu, S.W. Wu, Nat. Nanotechnol. **9**(10), 825–829 (2014)
144. W.T. Hsu, Z.A. Zhao, L.J. Li, C.H. Chen, M.H. Chiu, P.S. Chang, Y.C. Chou, W.H. Chang, ACS Nano **8**(3), 2951–2958 (2014)
145. J.X. Cheng, T. Jiang, Q.Q. Ji, Y. Zhang, Z.M. Li, Y.W. Shan, Y.F. Zhang, X.G. Gong, W.T. Liu, S.W. Wu, Adv. Mater. **27**(27), 4069–4074 (2015)
146. R. Wang, B.A. Ruzicka, N. Kumar, M.Z. Bellus, H.Y. Chiu, H. Zhao, Phys. Rev. B **86**(4), 045406 (2012)
147. H.Y. Shi, R.S. Yan, S. Bertolazzi, J. Brivio, B. Gao, A. Kis, D. Jena, H.G. Xing, L.B. Huang, ACS Nano **7**(2), 1072–1080 (2013)
148. Q.S. Wang, S.F. Ge, X. Li, J. Qiu, Y.X. Ji, J. Feng, D. Sun, ACS Nano **7**(12), 11087–11093 (2013)
149. Z.G. Nie, R. Long, L.F. Sun, C.C. Huang, J. Zhang, Q.H. Xiong, D.W. Hewak, Z.X. Shen, O.V. Prezhdo, Z.H. Loh, ACS Nano **8**(10), 10931–10940 (2014)
150. H.N. Wang, C.J. Zhang, F. Rana, Nano Lett. **15**(1), 339–345 (2015)
151. H.N. Wang, C.J. Zhang, W.M. Chan, C. Manolatou, S. Tiwari, F. Rana, Phys. Rev. B **93**(4), 045407 (2016)
152. N. Kumar, Q.N. Cui, F. Ceballos, D.W. He, Y.S. Wang, H. Zhao, Phys. Rev. B **89**(12), 125427 (2014)
153. C. Mai, Y.G. Semenov, A. Barrette, Y.F. Yu, Z.H. Jin, L.Y. Cao, K.W. Kim, K. Gundogdu, Phys. Rev. B **90**(4), 041414(R) (2014)
154. J.Q. He, D.W. He, Y.S. Wang, Q.N. Cui, F. Ceballos, H. Zhao, Nanoscale **7**(21), 9526–9531 (2015)
155. A. Boulesbaa, B. Huang, K. Wang, M.W. Lin, M. Mahjouri-Samani, C. Rouleau, K. Xiao, M. Yoon, B. Sumpter, A. Puretzky, D. Geohegan, Phys. Rev. B **92**(11), 115443 (2015)
156. A. Boulesbaa, V.E. Babicheva, K. Wang, I.I. Kravchenko, M.W. Lin, M. Mahjouri-Samani, C.B. Jacobs, A.A. Puretzky, K. Xiao, I. Ivanov, C.M. Rouleau, D.B. Geohegan, A Photonics **3**(12), 2389–2395 (2016)
157. Q.N. Cui, F. Ceballos, N. Kumar, H. Zhao, ACS Nano **8**(3), 2970–2976 (2014)

158. C. Poellmann, P. Steinleitner, U. Leierseder, P. Nagler, G. Plechinger, M. Porer, R. Bratschitsch, C. Schuller, T. Korn, R. Huber, Nat Mater **14**(9), 889–894(2015)
159. Q.N. Cui, J.Q. He, M.Z. Bellus, M. Mirzokarimov, T. Hofmann, H.Y. Chiu, M. Antonik, D.W. He, Y.S. Wang, H. Zhao, Small **11**(41), 5565–5571 (2015)
160. K.P. Wang, B.M. Szydlowska, G.Z. Wang, X.Y. Zhang, J.J. Wang, J.J. Magan, L. Zhang, J.N. Coleman, J. Wang, W.J. Blau, ACS Nano **10**(7), 6923–6932 (2016)
161. F. Ceballos, Q.N. Cui, M.Z. Bellus, H. Zhao, Nanoscale **8**(22), 11681–11688 (2016)
162. S.A. Wolf, D.D. Awschalom, R.A. Buhrman, J.M. Daughton, S. von Molnar, M.L. Roukes, A.Y. Chtchelkanova, D.M. Treger, Science **294**(5546), 1488–1495 (2001)
163. X.D. Xu, W. Yao, D. Xiao, T.F. Heinz, Nat. Phys. **10**(5), 343–350 (2014)
164. C. Mai, A. Barrette, Y.F. Yu, Y.G. Semenov, K.W. Kim, L.Y. Cao, K. Gundogdu, Nano Lett. **14**(1), 202–206 (2014)
165. P. Rivera, K.L. Seyler, H.Y. Yu, J.R. Schaibley, J.Q. Yan, D.G. Mandrus, W. Yao, X.D. Xu, Science **351**(6274), 688–691 (2016)
166. X.P. Hong, J. Kim, S.F. Shi, Y. Zhang, C.H. Jin, Y.H. Sun, S. Tongay, J.Q. Wu, Y.F. Zhang, F. Wang, Nat. Nanotechnol. **9**(9), 682–686 (2014)
167. F. Ceballos, M.Z. Bellus, H.Y. Chiu, H. Zhao, ACS Nano **8**(12), 12717–12724 (2014)
168. K. Wang, B. Huang, M. Tian, F. Ceballos, M.W. Lin, M. Mahjouri-Samani, A. Boulesbaa, A.A. Puretzky, C.M. Rouleau, M. Yoon, H. Zhao, K. Xiao, G. Duscher, D.B. Geohegan, ACS Nano **10**(7), 6612–6622 (2016)
169. Q. Zheng, W.A. Saidi, Y. Xie, Z. Lan, O.V. Prezhdo, H. Petek, J. Zhao, Nano Lett. **17**(10),6435–6442 (2017)
170. A. Boulesbaa, K. Wang, M. Mahjouri-Samani, M. Tian, A.A. Puretzky, I. Ivanov, C.M. Rouleau, K. Xiao, B.G. Sumpter, D.B. Geohegan, J. Am. Chem. Soc. **138**(44), 14713–14719 (2016)
171. A.A. Puretzky, I.A. Merkulov, C.M. Rouleau, G. Eres, D.B. Geohegan, Carbon **79**, 256–264 (2014)
172. P. Nikolaev, D. Hooper, N. Perea-Lopez, M. Terrones, B. Maruyama, ACS Nano **8**(10), 10214–10222 (2014)
173. P. Nikolaev, D. Hooper, F. Webber, R. Rao, K. Decker, M. Krein, J. Poleski, R. Barto, B. Maruyama, NPJ Comput. Mater. **2**, 16031 (2016)

Chapter 2
The Role of Defects in Pulsed Laser Matter Interaction

Oskar Armbruster, Aida Naghilou and Wolfgang Kautek

Abstract Laser nano- and micromachining exhibits multiple technological applications particularly in cyber physical production systems, where the integrity of production processes has to be maintained (e.g. in the implementation of Industry 4.0). Key characteristics of laser processing, such as the ablation threshold was commonly related to the fundamental parameters (fluence, pulse number, and irradiated area) by empirical adaptation and optimization. The description of the pulse number dependence (incubation, laser-induced defects) and the irradiation area dependence (intrinsic defects) of the threshold fluence still resorts to phenomenological models. This deficit can be resolved by a combined description of pulse number and beam radius based on a model involving high density and low density defects in the solid material. This extended defect model can describe single and multiple nanosecond and femtosecond pulse ablation experiments on various technological materials such as polystyrene, monocrystalline silicon, and stainless steel. While this model allows a quantification of the laser-induced threshold fluence in dependence of pulse number (incubation) and irradiated area (beam radius), the physical mechanisms involved in the interaction between light and defects, be it intrinsic or laser-generated, are still marginally understood. Further experimental and theoretical effort in this direction is aiming at the provision of the deterministic understanding required in any field applying intense laser light with multiple pulses and various spot sizes, which is essential in e.g. laser cutting, drilling, marking, engraving, hardening, and welding.

2.1 Introduction

Laser induced damage (LID) is a topic as old as the laser itself. The gain medium of the first operational laser was a ruby rod with silver coatings on both ends [1]. One year later, the laser pioneers indicated that the laser light burns holes into the

O. Armbruster · A. Naghilou · W. Kautek (✉)
Department of Physical Chemistry, University of Vienna, Währinger Strasse 42, 1090 Vienna, Austria
e-mail: wolfgang.kautek@univie.ac.at

P. M. Ossi (ed.), *Advances in the Application of Lasers in Materials Science*, Springer Series in Materials Science 274,
https://doi.org/10.1007/978-3-319-96845-2_2

silver coating [2]. Years after, LID has become one of the key characteristics of optical components. With the widespread use of lasers and especially high-power lasers, determining a LID-threshold (LIDT) for optical materials has developed into an important field of research. Since 1969, an annual conference exclusively devoted to LID is held in Boulder, CO. The conference series was initiated by Glass and Guenther, and is now organized by SPIE.

Quite contrary, LID can also be an intended process, e.g. in laser machining. The ablation of a metallic coating which is considered as damage on laser mirrors can be intentional in laser processing. Thus, laser-induced damage and laser processing rely on the same physical mechanisms.

The interaction between an optical beam and a material depends on the laser parameters (wavelength λ, pulse duration τ, pulse repetition frequency f), the beam radius w, the ambient conditions, the mounting conditions (e.g. bracing), and the optical, mechanical, and thermal properties of the material irradiated.

Even for transparent materials, the damage threshold is most likely linked to thermal phenomena for nanosecond pulses up to continuous wave (CW) irradiation [3]. In this scenario, the onset of modification and/or ablation is related to melting and/or vaporization of the material. For highly transparent materials (e.g., fused silica, diamond, quartz, sapphire), other damage mechanisms come into play before primarily thermal effects take place. These are dielectric breakdown (ps to ns), avalanche ionization (ps), and multiphoton absorption (sub-ps).

Generally, absorption of optical energy will ultimately cause a rise in temperature, leading to thermal expansion, strain, birefringence, movement of internal defects, cracking, melting, and catastrophic shattering. Short pulse durations may also cause nonlinear absorption, electro-optic effects, second harmonic generation, optical parametric oscillation, and self-focusing. These effects may add to the amount of energy absorbed and reduce the LIDT. A combination of these mechanisms may add up to a change of the beam shape, induce birefringence, and shatter or melt the component [3].

2.2 Intrinsic Defects

Defects are known to severely lower the local LIDT [4–7] and can be distinguished in two main categories. Structural defects (cracks, scratches, voids, etc.) and impurities. On the one hand, defects may be present in the raw material itself. On the other hand, surface processing, such as polishing, also introduces defects. A near-surface layer undergoes modifications due to chemical reactions, incorporation of polishing materials, and surface marks from the polishing aid. This cumulates in reduced mechanical strength, enhanced absorption, and light trapping. From a crystallographic point of view, every surface is a defect. Thus, the electronic structure at a surface is significantly different from the bulk. The calculated electronic structure of a finite SiO_2 crystal [8] shows the presence of surface quantum states within the band gap that permit single-photon ionization by 351 nm (3.5 eV) photons.

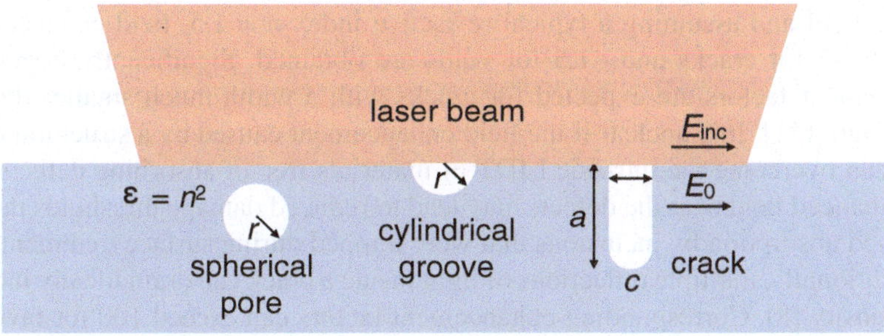

Fig. 2.1 Representative geometries for electric field enhancement near pores, scratches, and incipient cracks. Typical dimensions are $r = 0.1$ μm, $c = 0.1$ μm, and $a = 1$ μm [4]

Optical damage can arise on the surface and within the bulk of a material. As surfaces are often exhibiting a higher density of microscopic defects than bulk material, their LIDT is usually lower and may heavily depend on the exact surface treatment. Scratches or traces of contamination can lower the LIDT significantly [4].

Light absorption is made up of two parts: material intrinsic absorption and defect absorption [9]. The progress in optical materials engineering continuously reduces the number and size of absorbers in optical materials used for laser applications. Still, with ever-increasing optical intensities, even nanoscale absorbing defects continue to be a major source of damage [3].

A typical type of damage, which can be observed on e.g. dielectric coatings of laser mirrors and in bulk media is the occurrence of small isolated defects, which cause increased scattering losses. They can be linked to nanoscale defects introduced already in the manufacturing process. Such point defects may be too small to be seen even under a microscope, and nevertheless act as seeds for laser-induced damage, creating preferential damage sites [10].

The interaction of laser light with structural defects and absorption has two main aspects in relation to laser damage: Energy absorption can generate thermal and mechanical effects (e.g. heating, melting, evaporation, fracturing). The field strength of the laser pulse can be enhanced in the vicinity of the defect, which consequently affects modification of the host material [3].

2.2.1 *Field Enhancement by Structural Defects*

Structural defects (cracks, voids, inclusions, local stress concentration) can significantly increase the local electric field strength and thus lower the macroscopically measured LIDT. The prototypical work of Bloembergen [4] presents the field enhancement properties of cracks, grooves, and voids (Fig. 2.1).

Enhancement factors are calculated for voids, grooves, and cracks (Fig. 2.1) with dimensions of a few λ. Considering that the intensity scales quadratically with the electric field and assuming a typical refractive index $n = 1.5$, field enhancements between ~5 for cracks and ~1.5 for voids are obtained. Significantly larger field enhancement factors are expected for cracks with a width much smaller than the wavelength [11]. It is unclear if the field enhancement caused by a structural defect alone can overcome the intrinsic LIDT of materials free of absorbing defects [12]. The enhanced field near the defects may lead to reduced damage thresholds through increased absorption by inclusions that were trapped during surface treatment.

Additionally, multiple reflections of light inside a crack can dramatically increase the intensity [8]. Corresponding enhancement factors can exceed 100 for favorable conditions (polarization, feature dimensions, angle of incidence).

In experiments with a nanosecond dye laser, optical grade polished dielectrics were studied. The damage threshold was measured by photoacoustic probe beam deflection and the effect of surface roughness on the onset of damage was investigated [7]. The authors propose that the correlation of rough materials with low damage thresholds is due to multiphoton surface ionization as the principal energy coupling process. Polished materials exhibit a more random behavior, as would be expected for avalanche breakdown. The defect states can also, via single photon absorption, lead to local heating with the consequence of thermal strain or thermal desorption. This increases the defect density and, in turn, facilitates absorption.

2.2.2 Field Enhancement by Impurities

Besides structural defects, embedded particles can also serve as centers of field enhancement. Field enhancement factors near a particle increase with diminishing size of metallic particles and with increasing size of dielectric particles [13]. The latter is due to focusing of light by large dielectric spheres. Non-absorbing defects can also serve as positive or negative lenses for the incident laser light [14] due to their refractive index difference from the surrounding medium. The light intensity enhancement is dependent on the refractive index and geometric arrangement of the defects.

Nonlinear self-focusing arises from a heterogeneous refractive index of a material. Temperature rise, stress, and distortion due to laser radiation can change the refractive index of a material and make it act as a focusing lens, which also leads to an electric field enhancement [15].

2.2.3 Thermal Damage by Absorber Impurities

The study of localized absorber-driven damage considering energy transfer from a laser heated defect to the host material by heat conduction was first described for metallic absorbers [16], and then for absorbers with various thermal and optical properties and geometries [17–20].

Assuming an ideal boundary (no thermal resistance) between absorber (*a*) and matrix (*h*), the heat equation for the spherically symmetric problem is given by

$$\frac{1}{\delta_a}\frac{\partial T_a}{\partial t} = \frac{1}{r^2}\frac{\partial}{\partial r}\left(r^2\frac{\partial T_a}{\partial r}\right) + \frac{S}{k_a}\ \text{for}\ r < a \tag{2.1}$$

and

$$\frac{1}{\delta_h}\frac{\partial T_h}{\partial t} = \frac{1}{r^2}\frac{\partial}{\partial r}\left(r^2\frac{\partial T_h}{\partial r}\right)\ \text{for}\ r > a \tag{2.2}$$

where a is the radius of the particle, δ is the thermal diffusivity, κ is the thermal conductivity, r is the radial coordinate, S is the source term, t is time, and T is the temperature. The interface between the absorber and the matrix is characterized by the boundary conditions

$$T_a(a) = T_h(a) \tag{2.3}$$

and

$$k_a\left(\frac{\partial T_a}{\partial r}\right)_{r=a} = k_h\left(\frac{\partial T_h}{\partial r}\right)_{r=a.} \tag{2.4}$$

Solving the heat diffusion problem can be simplified by assuming homogeneous energy deposition (and thus temperature) inside the absorber [20, 21]. This is valid for small, strong absorbers (metals) and larger, weak absorbers (dielectrics). This approximation allows to simplify the boundary conditions. The energy deposited inside the absorber is equal to the increase in the absorber's heat content corrected by the heat flux through its surface:

$$\gamma\, I(t) = \frac{4}{3}\rho_a\, C_a\, a\left(\frac{\partial T}{\partial t}\right)_{r=a} - 4k_h\left(\frac{\partial T}{\partial r}\right)_{r=a} \tag{2.5}$$

Here, C_a is the heat capacity, γ is the absorptivity, I is the optical intensity, and ρ_a is the density. A simple expression for the temperature of the absorber heated by a temporally rectangular pulse is given by Chan [21]

$$T = \frac{\gamma\, F\, a}{4\, k_h \tau}\left[1 - \exp\left(-\frac{4\,\delta\,\tau}{a^2}\right)\right] \tag{2.6}$$

with

$$\delta = \frac{3}{4}\frac{k_h}{\rho_a C_a} \tag{2.7}$$

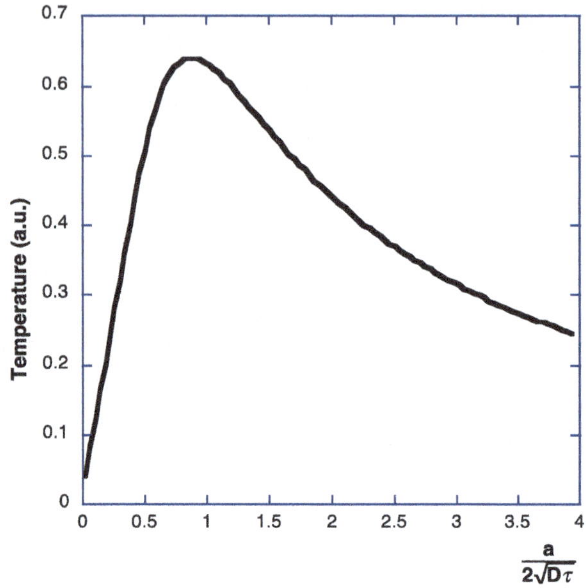

Fig. 2.2 Inclusion temperature induced by a rectangular laser pulse for absorbers of different sizes. Particle size a is normalized by the effective thermal diffusion length. D is the thermal diffusivity and τ is the laser pulse duration. An absorber with a size comparable to the effective diffusion length is heating the host material most efficiently [22]

where F is the fluence of the pulse. Figure 2.2 shows the temperature as a function of absorber size normalized by the effective thermal diffusion length.

A finite absorber sized similar to the effective diffusion length is most effective in respect to heating due to an optimum surface to volume ratio. LIDT and host thermal conductivity scale inversely due to the slower heat removal [17, 19].

Absorption is initiated at defects embedded in the host material. This can lead to a higher absorptivity in the adjacent volume, thus increasing the deposited energy and leaving the material more susceptible to catastrophic and macroscopic failure by consecutive pulses [23]. Experimental evidence was obtained for gold nanoparticles embedded in a silica layer [24]. Small craters appeared at the locations of the nanoparticles upon ultraviolet (UV) pulsed laser irradiation with fluences just above the LIDT. The crater is significantly wider than the particle size as revealed by atomic force microscopy (AFM, Fig. 2.3). This phenomenon was interpreted as absorption taking place outside the absorbing particle, and is termed absorption delocalization.

Several possible mechanisms and variants of absorption delocalization exist. It was suggested that heating of the host material in proximity to the absorber causes the absorption coefficient to increase [25]. Furthermore, the band gap changes with temperature (e.g. from 8.5 to 7.0 eV for temperatures from 4 to 1873 K [26]) promoting photoexcitation. Electron-phonon relaxation causes the surrounding material to be heated as well, initiating a chain reaction and ultimately delocalized material removal.

Another mechanism is termed photoionization thermal explosion [27–29]. The absorbers are thermally heated to high temperatures of the order of 10^4 K so that energetic black body radiation is emitted. This radiation can be absorbed by the host,

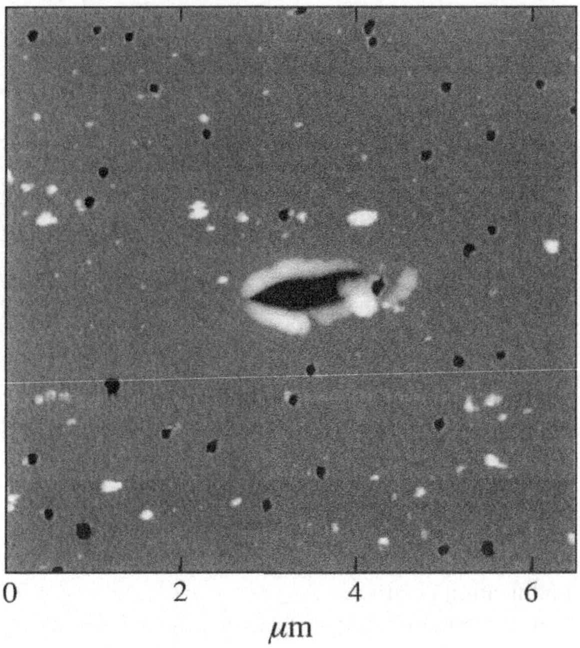

Fig. 2.3 AFM image of a silica film with embedded gold absorbers (8.4 nm) after irradiation by a single 4.6 J cm^{-2} laser pulse [24]

creating free carriers. These are further heated by the laser pulse, eventually leading to damage. Besides the emission of energetic photons, hot particles can also emit electrons which, in turn, can absorb laser radiation.

2.2.4 Irradiation Area Dependence of Laser-Induced Threshold Fluences

Although the fluence is specified as laser pulse energy per beam area, it may not be an independent quantity. Actually, the size of the illuminated area influences the threshold fluence. This effect was studied to a limited extent, and mostly in the context of damage to (transparent) optical materials, such as glasses [30, 31] and optical coatings [32]. An early work discussing the impact of the illuminated area on the modification threshold was performed with a nanosecond pulsed ruby laser on dielectric coatings [32]. The spot size was varied from 52 to 250 μm and the impact on the modification threshold was examined. A decrease in threshold fluence by a factor of at least two was observed (Fig. 2.4).

The experimental findings are supported by a model accounting for the stochastic nature of defect distributions in the absorption volume, lowering the threshold from F_i of the intrinsic (defect-free) material to a defect-mediated threshold F_d. Possible mechanisms for a lowered threshold in the vicinity of defects are discussed above.

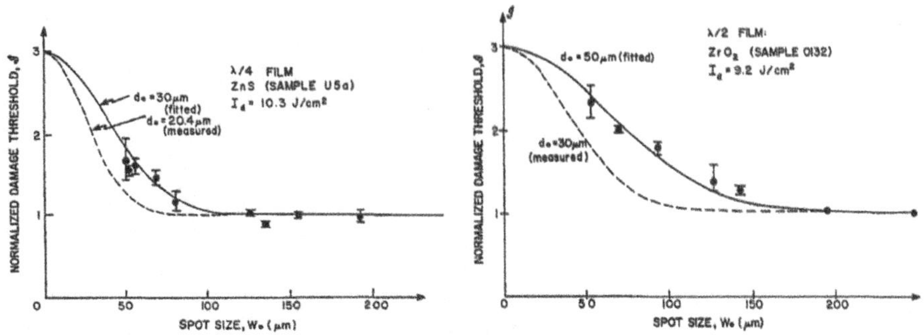

Fig. 2.4 Spot-size dependence for single half-wave-thick (left) and quarter-wave-thick (right) films of ZnS on glass substrate. The experimentally observed threshold fluence decreases by more than a factor of two [32]

The model relies on the assumption that the surface of a material is covered with randomly (Poisson) distributed point-sized defects. If a defect is irradiated with a fluence greater than the threshold for defect-mediated modification F_d, surface modification occurs.

According to Poisson's statistics, the probability P_n to find n randomly distributed defects on a surface with an area A is given by

$$P_n = \frac{(\sigma A)^n}{n!} \exp(-\sigma A) \tag{2.8}$$

where σ is the areal density of defects on the surface. From this, it can be concluded that the probability to find zero defects in the area A is given by

$$P_0 = \exp(-\sigma A). \tag{2.9}$$

On the other hand, it is trivial that the probability to find $0 \leq n \leq \infty$ defects in the area A is unity. Thus, the probability to have at least 1 defect in A is

$$P_{1\ldots\infty} = \sum_{n=1}^{\infty} P_n = \sum_{n=0}^{\infty} P_n - P_0 = 1 - \exp(-\sigma A) \tag{2.10}$$

The fluence F of a Gaussian beam is given by

$$F(r) = F_0 \exp\left(-2\frac{r^2}{w^2}\right) \tag{2.11}$$

where w is the exp(−2) Gaussian beam radius, r is the radial distance from the center of the beam, and F_0 is the fluence in the center of the beam. Solving this equation to r and calculating the area $A = r^2\,\pi$ yields the area in which the fluence is greater than F for a beam with a maximum fluence of F_0:

$$A(F, F_0) = \frac{w^2 \pi}{2} \ln\left(\frac{F_0}{F}\right) \tag{2.12}$$

Thus, $A_d(F_0) = A(F_d, F_0)$ is the area in which the fluence is sufficient to modify the material in case of the presence of a defect.

Combining 2.10 and 2.12 yields the probability to irradiate at least one defect with a fluence greater than F_d:

$$P_d = 1 - \left(\frac{F_0}{F_d}\right)^{\left(-\frac{1}{2} w^2 \pi \sigma\right)} \tag{2.13}$$

The lines in Fig. 2.4 were calculated on the basis of this model and provide a good description of the experimental data. Furthermore, the mean distance between defects determined via electron microscopy (dashed lines) is in good agreement with the defect density obtained by fitting 2.13 (solid lines).

2.3 Laser-Generated Defects

It has been acknowledged since the early days of laser-matter interaction studies [33] that the threshold fluence at pulse numbers $N > 1$, $F_{th}(N)$, is lower than the single pulse threshold fluence $F_{th}(1)$ for most experimental conditions and materials [10], including metals [34–36], semiconductors [37, 38], and dielectrics [39–43]. In other words, repeated irradiation below the single pulse threshold fluence can still cause damage. One of the early works in this field was performed on single crystal quartz and two types of fused silica under single and multiple pulse laser irradiation at 1064 and 532 nm [44]. A memory of preceding laser pulses has been indicated. The observed macroscopic failure after many pulses was attributed to arise from the accumulation of microscopic changes in the material brought about by the preceding pulses. It was also shown that such an accumulation mechanism requires a minimum incident intensity to initiate or support the growth of microscopic changes or damages. The authors propose a binomial distribution but indicate that this model does not fit their data and a peaked distribution would be more suitable.

Later, this memory effect was denoted *incubation* during experiments with poly(methyl methacrylate) (PMMA) and polyimide. Calculations considering a dynamic model and the sum of the concentration of absorbed photons was suggested [45].

Perhaps the most well-known model for incubation is the phenomenological description proposed by Jee et al. [34]. In this work, the incubation behavior of single crystal metal surfaces was investigated with nanosecond pulses at 1064 nm wavelength. The authors justified their incubation model with the following: "If we redraw the damage fluence curve in another way, we can get a damage accumulation curve. This accumulation curve is used to examine the effect of cumulative laser action on metal surfaces. The y axis is the log of the product of the N-pulse dam-

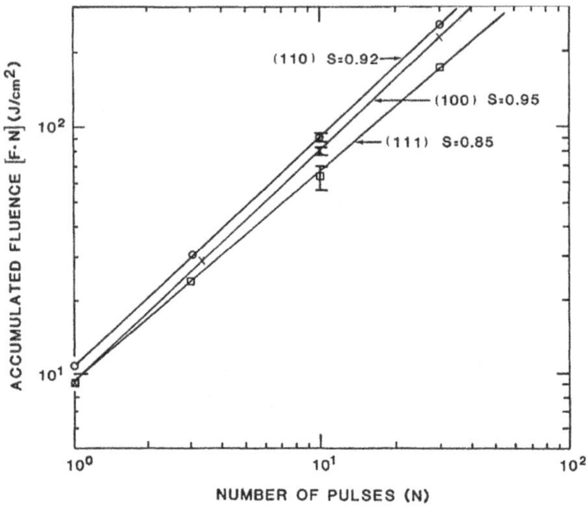

Fig. 2.5 The original "accumulation curve" presented by Jee et al. [34]. Damage fluence versus pulse number curves for various crystallographic orientations of chemically polished Cu surfaces

age threshold, F_N, and the incident pulse number, N [$\log(F_N\, N)$]. The x axis is the log of the pulse number ($\log N$). The plot [...] shows straight lines for the various chemically polished Cu surfaces. From this linear segment of the damage accumulation curve, we can derive the cumulative equation $F_N = F_I\, N^{S-1}$ where F_I is the single-shot damage threshold and S is the slope of the accumulation curve." This is depicted in Fig. 2.5. Actually, this incubation model is purely phenomenological and is not based on a physical process. It however follows the trend observed in numerous experiments.

The majority of laser processing studies are merely restricted to fitting the experimentally obtained threshold fluences $F_{th}(N)$ according to this damage accumulation model [34]:

$$F_{th}(N) = F_{th}(1)\, N^{-\xi} \tag{2.14}$$

This equation allows a quantification of the incubation phenomenon by the empirical incubation parameter ξ. However, it does not provide an insight into the underlying physical and chemical mechanisms. $\xi = 0$ describes the case where $F_{th}(N)$ is identical to $F_{th}(1)$ (the material does not incubate). On the other hand $\xi > 0$ describes the case where $F_{th}(N) < F_{th}(1)$.

An alternative incubation model was proposed for fused silica (amorphous silicon dioxide, a-SiO_2) and yttrium lithium fluoride ($LiYF_4$, YLF), where the threshold fluence starts to level off at $F_{th}(\infty)$ for high pulse numbers (Fig. 2.6) [46].

An irradiation at a fluence of $F_{th}(\infty)$ would require an infinite number of pulses to initiate defect accumulation: [46]

$$F_{th}(N) = F_{th}(\infty) + \lfloor F_{th}(1) - F_{th}(\infty) \rfloor \, \exp \lfloor -\xi(N-1) \rfloor \tag{2.15}$$

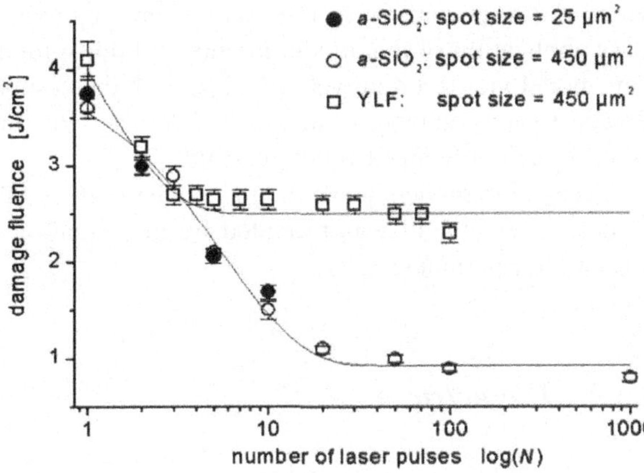

Fig. 2.6 Semi-logarithmic representation of threshold fluence versus pulse numbers for a-SiO$_2$ (hollow symbols) and YLF (full symbols). Solid lines represent the result from the fit of the incubation model [46]

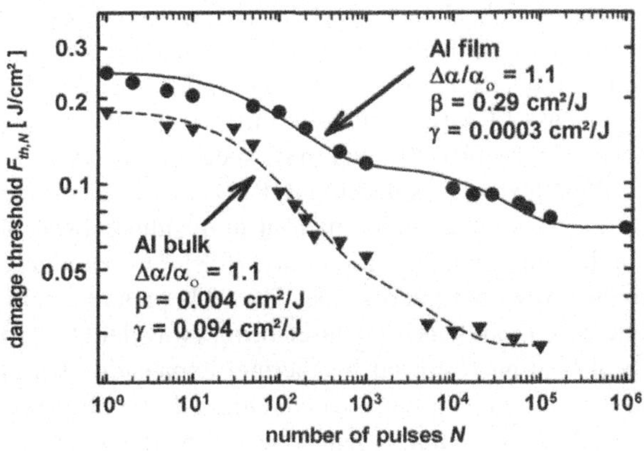

Fig. 2.7 Laser damage (ablation) threshold fluence F_{th} as a function of the number N of pulses for aluminum film (dots) and bulk sample (triangles). The lines are fits with the parameters given in the figure [47]

In the fluence region $F < F_{\text{th}}(\infty)$, no macroscopic damage would occur at any N. In this model, the $F_{\text{th}}(\infty)$ plateau is reached at lower N for materials with higher ξ.

Both above mentioned models are empirical in nature. They allow a quantification of the incubation by a parameter ξ.

A recent incubation model allows the description of more complex $F_{\text{th}}(N)$ dependences (Fig. 2.7), however, requiring many free parameters [47]. It was assumed that incubation is controlled by an increase of the energy deposition per pulse (absorption) and/or a decrease in the critical energy needed to damage/ablate the material during the pulse train. In the absence of relaxation, the N-pulse threshold fluence is described by Sun et al. [47]

$$F_{\text{th}}(N, F) = \frac{F_{\text{th}}(1) - \left\lfloor F_{\text{th}}(1) - F_{\text{th}}(\infty)\left(1 + \frac{\Delta\alpha}{\alpha_0}\right)\right\rfloor \{1 - \exp[-\gamma\, F(r)(N-1)]\}}{1 + \frac{\Delta\alpha}{\alpha_0}\{1 - \exp[-\beta\, F(r)(N-1)]\}} \tag{2.16}$$

where $\Delta\alpha/\alpha$, β, and γ are free parameters depending on the material. An exemplary application of this model to bulk and thin-film aluminum is given in Fig. 2.7. This model yields a relatively good fit with the complex shaped experimental data. However, physical insights into the difference of the incubation mechanism between thin-film and bulk are still not possible.

These three models leave the mechanisms of incubation still widely speculative. Some deeper insight was provided by investigations at a number of materials at certain laser parameters.

2.3.1 Dielectrics

It is generally accepted that incubation of dielectrics involves the formation of color centers which alter the optical absorptivity [48]. This interpretation is based on experimental evidence obtained from nanosecond UV irradiation of PMMA [40, 49, 50] and materials for UV optics [51, 52] (e.g. fused silica, CaF_2, LiF, MgF_2, BaF_2) by means of Fourier transform infrared spectroscopy (FT-IR) [40], UV transmission spectroscopy [40, 49], and mass spectrometry [50].

Photothermal displacement (PTD) was employed to study the incubation behavior of dielectric coatings by measuring residual absorption after the illumination with single nanosecond UV laser pulses [48]. The incubated volume was not visible, either by eye or with a microscope. The PTD patterns showed enhanced absorption of an intensity modulated Ar^+ laser beam (wavelength 514 nm) which leads to periodic local heating followed by thermal expansion. The top hat profile of the excimer laser beam is well reproduced, although the observed structure is about 20% wider (Fig. 2.8). This enlargement is caused by light scattering and/or color center diffusion. The absolute increase in absorption is proportional to the energy deposited in the irradiated volume. The microscopic nature of the light absorbing defects generated by the UV radiation is presently not understood in detail [48].

Similarly, UV nanosecond pulse laser irradiated PMMA was investigated with UV and infrared spectroscopy (Fig. 2.9) [40]. The UV spectrum of irradiated PMMA shows a steep increase in absorption over a wide range of wavelengths, up to the visible. After 1000 pulses, the transmission at 248 nm has dropped to less than 6% of its initial value. This indicates significant chemical changes revealed by IR spectroscopy. The most important effect is the formation of unsaturated species also revealed by a strong fluorescence. Unsaturated compounds are known to absorb at longer UV wavelengths than their saturated derivatives. Sub-threshold irradiation of PMMA also leads to a reduction in molecular weight by backbone cleavage [50]. It was assumed that C=C bonds at chain ends represent defect centers increasing the absorption coefficient thus lowering the ablation threshold. The mass spectrum of the products of subthreshold ablation following incubation is identical to that of the single pulse ablation above threshold and standard pyrolysis spectra. These results are strong indicators that subthreshold ablation that follows incubation is a thermally driven phenomenon. Measurements of the elastic and the adhesive properties by

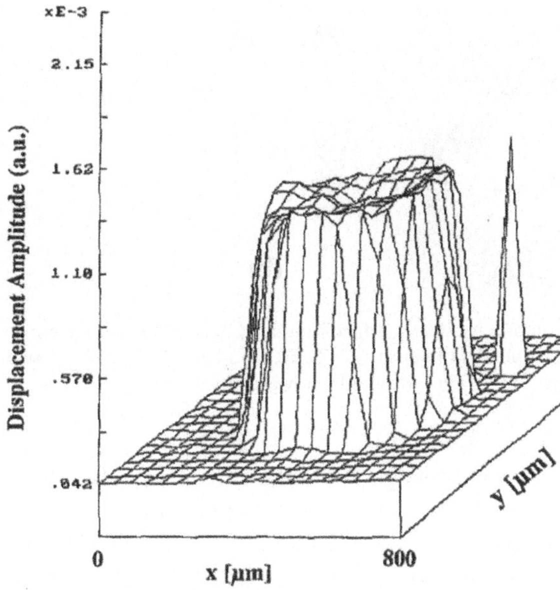

Fig. 2.8 Two-dimensional photothermal displacement pattern imaging an incubated area on a TiO_2 coating (ion-plated on a fused silica substrate) that was irradiated by a single laser shot of 248 nm and 35 mJ cm^{-2} (pulse length 15 ns). The damage threshold of this coating was 65 mJ cm^{-2}. The modulation frequency of the photothermal pump beam was 100 kHz [48]

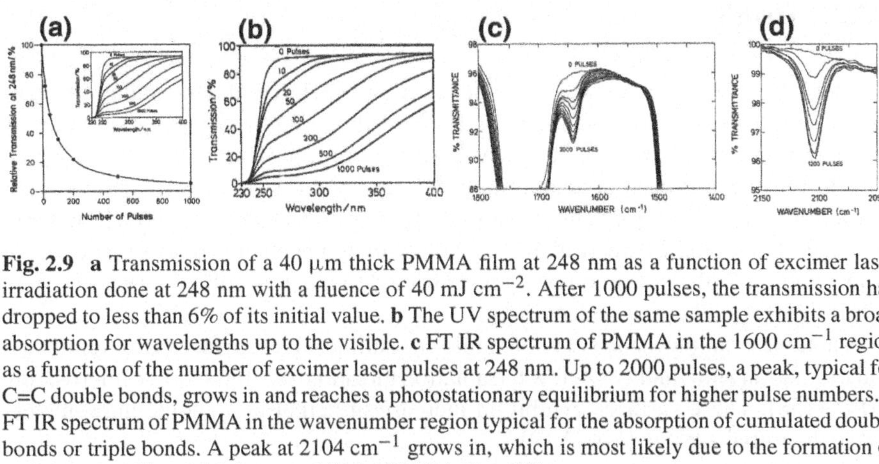

Fig. 2.9 **a** Transmission of a 40 μm thick PMMA film at 248 nm as a function of excimer laser irradiation done at 248 nm with a fluence of 40 mJ cm^{-2}. After 1000 pulses, the transmission has dropped to less than 6% of its initial value. **b** The UV spectrum of the same sample exhibits a broad absorption for wavelengths up to the visible. **c** FT IR spectrum of PMMA in the 1600 cm^{-1} region as a function of the number of excimer laser pulses at 248 nm. Up to 2000 pulses, a peak, typical for C=C double bonds, grows in and reaches a photostationary equilibrium for higher pulse numbers. **d** FT IR spectrum of PMMA in the wavenumber region typical for the absorption of cumulated double bonds or triple bonds. A peak at 2104 cm^{-1} grows in, which is most likely due to the formation of carbon monoxide, which is trapped in the PMMA film [40]

means of a Si AFM tip of incubated PMMA revealed decreased pull-off forces attributed to a chemical modification and/or bond breaking [53].

Further, poly(dimethylsiloxane) (PDMS) was irradiated with nanosecond pulsed UV laser light [54]. The evolution of μ-Raman spectra from the cross-section of the

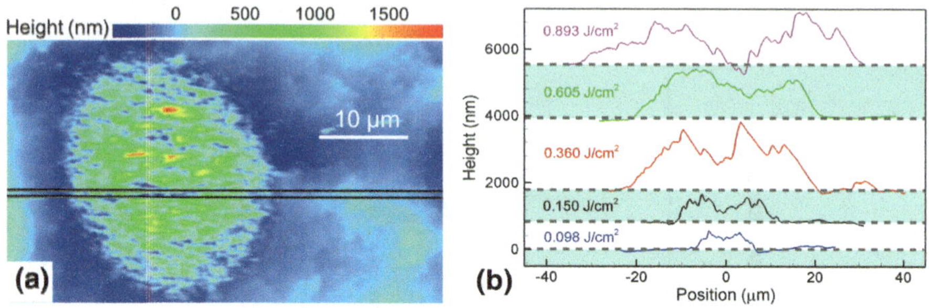

Fig. 2.10 **a** AFM scan of a Ag (001) surface irradiated by a 100-fs laser pulse at an absorbed laser fluence of 0.15 J cm^{-2} (incident fluence: (4.87 ± 0.08) J cm^{-2}). **b** AFM line scans of several spots generated by irradiation at various absorbed fluences [58]

sample showed noticeable changes from the surface to bulk. These measurements were supplemented by X-ray computer micro-tomography volume reconstruction. In tomography images, low-density regions, most probably containing voids with a total volume of 3–4% were observed. The void formation is also visible as a swelling of the material. The μ-Raman spectra were dominated by features characteristic of crystalline silicon, amorphous carbon, and graphite.

2.3.2 *Metals*

Although first described almost three decades ago [34], the understanding of incubation on metals is at best rudimentary. It is mostly based on computer simulations [55] and non-optical phenomena such as stress-strain measurements [34, 56, 57]. Very little work has been done qualitatively, e.g. by cross-sectional scanning electron microscopy (SEM) [58] and transmission electron microscopy (TEM) [59]. Only few quantitative studies are available, and those do not reveal the microscopic nature of incubation, as the measurements are restricted to macroscopic observations such as transient photothermal deflection [60].

A notable example is the study of single crystalline silver surfaces after femtosecond irradiation with fluences below the ablation threshold by atomic force microscopy (AFM) [58]. These surfaces exhibit an increase in height up to some micrometers above the initial surface (Fig. 2.10).

Classical atomistic molecular dynamics (MD) and continuum-level two-temperature model (TTM) approaches were combined and applied to simulate the swelling. They revealed a complex interplay of fast laser melting, rapid resolidification, and dynamic relaxation of laser-induced stresses that lead to the formation of a subsurface porous region (Fig. 2.11) covered by a nanocrystalline surface layer. The generation of the porous region is responsible for the experimentally observed swelling of the surface.

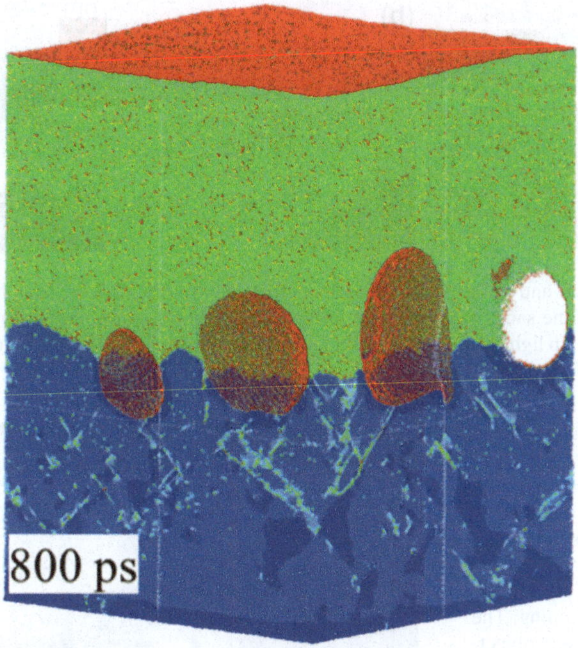

Fig. 2.11 Snapshot of the atomic configuration after 800 ps generated in a TTM-MD simulation of an Ag (001) target irradiated by a 100 fs laser pulse at an absorbed fluence of 85 mJ cm^{-2}. The atoms are colored by their potential energies, with the scale from–2.84 eV (blue) to –2.65 eV (red). This scale ensures that most atoms in the crystalline part of the target are blue, the atoms in the molten part are green, and the atoms on free surfaces are red [58]

Surface swelling and subsurface voids were experimentally observed on monocrystalline aluminum. Laser-treated samples (Fig. 2.12a, b) were cross-cut by focused ion beam (FIB) milling and then inspected by SEM (Fig. 2.12c, d) [61]. The observed features are similar to those predicted by the TTM-MD simulation for silver discussed above.

Also single pulse picosecond experiments below the threshold performed on copper reveal a complicated sub-surface morphology [59]. After laser irradiation, the sample was embedded and TEM lamellae were cut with a FIB mill (Fig. 2.13). These non-uniform, altered structures may influence the optical, thermal, and mechanical properties and thus modify and incubate the material [58]. A quantitative analysis confirming the contribution of the subsurface morphology to the macroscopically observed incubation is, however, still pending.

Titanium nitride (TiN) is an extremely hard ceramic material with metallic properties (good electrical conductivity, high thermal conductivity, etc.). X-ray photoelectron spectroscopy (XPS) and Auger electron spectroscopy (AES) [62, 63] was performed on TiN films irradiated with femtosecond laser pulses below the ablation threshold. Figure 2.14 shows an optical micrograph next to an oxygen-map obtained

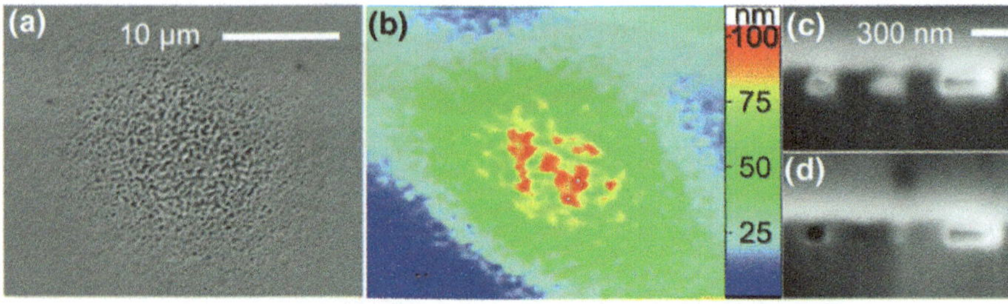

Fig. 2.12 SEM (**a**) and AFM (**b**) images of swelling due to ultrafast irradiation at 0.79 J cm^{-2}. **c** SEM images of the same spot after FIB milling. **d** The bottom dark-gray area is the aluminum sample, while the top lighter gray stems from a protective tungsten layer. Same as (**c**) after additional 50 nm milling [61]

Fig. 2.13 TEM image of the platinum-coated copper interface. There are two sub-surface defects at a depth of approximately 50 nm. The high resolution TEM inset shows the details of the material twinning. The bright right-hand side object is a sub-surface void filled with coating deposit [59]

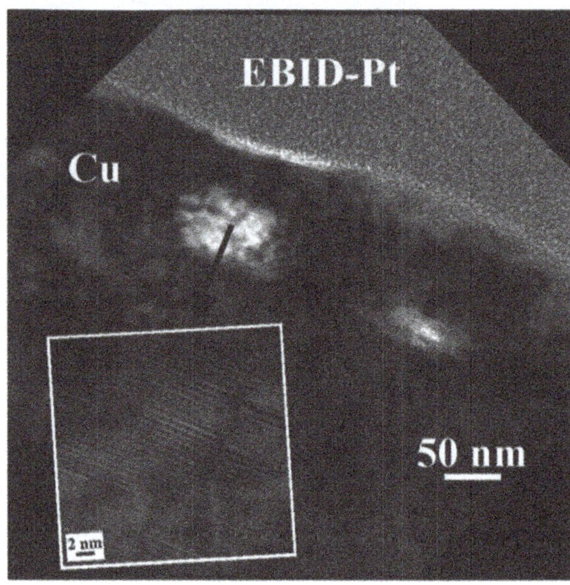

by scanning Auger electron microscopy of a sub-threshold irradiated spot on TiN. The rim of the irradiated zone exhibits a significantly increased content of oxygen and appears dark in optical microscopy.

A strong decrease of the N 1 s XPS signal was observed in the processed area ($\approx$5%) compared to the unaffected surface ($\approx$27%). The O 1 s signal on the other hand increased from $\approx$30% in the unaffected surface to 42% in the irradiated region. Depth profiling in the laser-processed area evidenced that the modified region of substoichiometric TiO_{2-x} exhibited a depth of more than 270 nm. This is in agreement with AES depth profiles. At a depth of 24 nm, the laser modified region exhibits a reduced titanium content in regions where the laser-induced oxidation was observed. The Ti concentration rises again for depths larger than 72 nm. Due to the short time

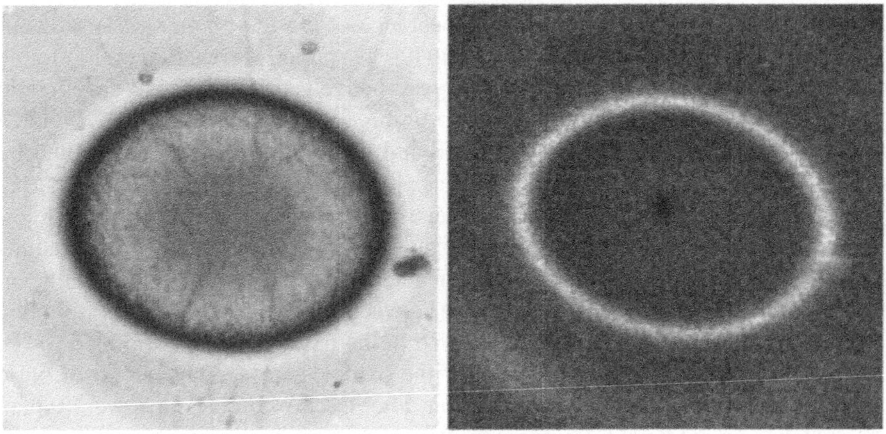

Fig. 2.14 Oxygen map recorded by scanning Auger electron microscopy (left) and AFM topography (right) for femtosecond laser irradiated titanium nitride. The oxygen-rich rim appears black in optical microscopy [62]

for heating of the irradiated surface, the formation of localized point-defects such as oxygen vacancies is more probable compared to the formation of more extended crystallographic shear planes.

2.3.3 Semiconductors

Semiconductors have electrical and optical properties in between of dielectrics and metals. Their laser interaction behavior hence shows features known from metals and dielectrics. Semiconductors may exhibit color center formation as demonstrated by time resolved photoelectron spectroscopy on silicon, zinc telluride, and cadmium telluride [64]. Also amorphization and photothermal oxidation play an important role as revealed by cross-sectional transmission electron microscopy [65].

Significantly altered Raman [66] and Auger electron [37, 67] spectra were obtained for indium phosphide (InP) within the laser-irradiated zone as compared to the pristine material. The Raman spectrum taken at the rim appeared smooth with two broad features around 300 and 440 cm^{-1} typical for amorphous InP. The broad features were induced by the loss of long-range order reducing the coherence length of the phonons. The band around 440 cm^{-1} is attributed to the phosphorous-phosphorous bond vibration. The Raman spectra from the center of the laser spot showed LO- and TO-phonon modes indicative of crystalline material. The initially molten InP resolidified in a polycrystalline state. This increased disorder was supported by the fact that the width of both peaks increased by a factor of 3–5 compared to that of the single-crystalline LO peak width. The peak at 174 cm^{-1} was attributed to defects introduced during recrystallization. The high density of defects after recrystallization was consistent with the disappearance of the LO-phonon-plasmon coupled modes (LOPCM). One plausible origin lies in a partial segregation of the atomic species during the molten transition state. Alternatively, the peak at 174 cm^{-1} could arise from the oxidation process observed in this surface area. The oxygen-related signal in Auger electron spectra increased and the indium-related lines shifted by some 3 eV toward lower energies. This was also supported by the observed characteristic changes in the peak ratios and shapes of the indium and phosphorus lines, indicating an increased indium oxide and phosphorus oxide content. Investigations of the thermal oxidation of InP showed that at temperatures above 340 °C significant oxidation rates were achieved. In the range between 350 and 550 °C, red phosphorus, P_2O_5, $InPO_4$ and In_2O_3 have been observed in the oxide films. Exceeding 550 °C, $InPO_4$ was found as major constituent of the oxidized surface layers. All of these compounds may be formed in the conditions after laser excitation. However, Raman spectra taken on oxidized InP wafers are known to exhibit a variety of Raman peaks which, depending on oxidation parameters, are associated with different oxide compounds. A single peak at 174 cm^{-1} is not consistent with any of these spectra.

2.4 Conclusion

Laser nano- and micromachining has attracted vivid attention since decades [68] and enabled multiple technological applications [69–71]. However, a key characteristic of laser processing, the ablation threshold, was commonly related to the fundamental parameters such as fluence, pulse number, and irradiated area by empirical adaptation and optimization. The description of the pulse number dependence (incubation, laser-induced defects) [34] and the irradiation area dependence (intrinsic defects) [30, 32, 72, 73] of the threshold fluence still resorts to phenomenological models.

Based on the assumption of sharp threshold fluences and statistical considerations (Sect. 2.4, 2.13), the experimentally observed threshold fluence in dependence of the beam radius $F_{\mathrm{th}}(w)$ is given by DeShazer e al. [32]

$$F_{\mathrm{th}}(w) = F_d + (F_i - F_d)\left(\frac{F_i}{F_d}\right)^{-\frac{1}{2}w^2 \pi \sigma} \tag{2.17}$$

where F_{i} is the threshold fluence of an intrinsic (defect-free) region of material and F_{d} is the threshold fluence in the presence of a defect. While this model describes the reduction of threshold fluence with beam radius for a given number of pulses, an understanding of the interplay between threshold fluence and pulse number cannot be extracted.

On the other hand, incubation models, such as 2.14 or 2.15 allow a quantification of the incubation phenomenon by the empirical incubation parameter ξ, however, do not consider beam size effects.

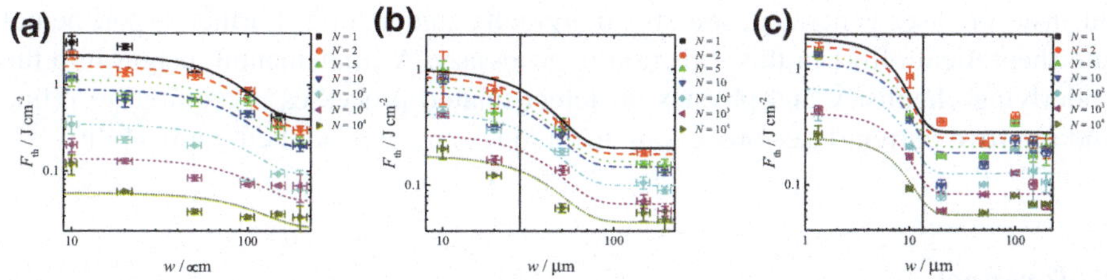

Fig. 2.15 Ablation threshold fluence F_{th} in dependence of beam radius w for various pulse numbers N of **a** high-impact polystyrene (HIPS), **b** silicon <111>, and **c** SAE 304 stainless steel under sub-100 fs laser irradiation [74, 75]

The combined description of pulse number and beam radius is based on the defect model, whereas the incubation model given in 2.14 is applied to F_{i} and F_{d} in 2.17, yielding

$$F_{\text{th}}(w, N) = F_d(N) + [F_i(N) - F_d(N)]\left[\frac{F_i(N)}{F_d(N)}\right]^{-\frac{1}{2}w^2\pi\sigma} \tag{2.18}$$

with

$$F_i(N) = F_i(1)\, N^{-\xi_i} \tag{2.19}$$

and

$$F_d(N) = F_d(1)\, N^{-\xi_d}. \tag{2.20}$$

This allows the description of the interplay between beam radius and pulse number on threshold fluences. Additionally, the spot size dependence of the incubation parameter could be modeled as

$$\xi(w) = \xi_d + (\xi_i - \xi_d)\frac{\frac{F_i(1)}{F_d(1)} - \frac{1}{2}w^2\pi\sigma\left(\frac{F_i(1)}{F_d(1)} - 1\right)}{\frac{F_i(1)}{F_d(1)} + \left(\frac{F_i(1)}{F_d(1)}\right)^{\frac{1}{2}w^2\pi\sigma} - 1}. \tag{2.21}$$

Thus, the incubation parameter is, much like the threshold fluence at a given pulse number, dependent on the beam radius.

This extended defect model [74] could describe single and multiple pulse ablation experiments on high-impact polystyrene (HIPS) [74], monocrystalline silicon <111> [75, 76], and SAE 304 stainless steel [75, 76] with nano- [76] and femtosecond [74–76] laser pulses (Fig. 2.15).

While this model allows a quantification of the laser-induced threshold fluence in dependence of pulse number (incubation) and irradiated area (beam radius), the physical mechanisms involved in the interaction between light and defects, be it intrinsic or laser-generated, are still marginally understood. Further experimental and theoretical effort in this direction is necessary. A fundamental insight into the underlying chemistry and physics of defect-related processes is required to refine and improve current laser-processing techniques and expand their applicability.

References

1. T.H. Maiman, Stimulated optical radiation in ruby. Nature **187**, 493–494 (1960)
2. F.J. McClung, R.W. Hellwarth, Giant optical pulsation from ruby. J. Appl. Phys. **33**, 828–829 (1962)
3. D. Ristau, Laser-Induced Damage in Optical Materials (CRC Press, 2014)
4. N. Bloembergen, Role of cracks, pores, and absorbing inclusions on laser induced damage threshold at surfaces of transparent dielectrics. Appl. Opt. **12**, 661–664 (1973)
5. R.H. Picard, D. Milam, R.A. Bradbury, Statistical analysis of defect-caused laser damage in thin films. Appl. Opt. **16**, 1563–1571 (1977)
6. D. Schafer, V. Gopner, R. Wolf, G. Zscherpe, Defect-induced laser damage of optical thin-films. Thin Solid Films, **116**, 176–176 (1984)
7. J. Reif, S. Petzoldt, A.P. Elg, E. Matthias, The role of defects in laser surface damage thresholds of fluoride-crystals. Appl. Phys. A **49**, 199–204 (1989)
8. H. Bercegol, P. Grua, D. Hébert, J.-P. Morreeuw, Progress in the understanding of fracture related laser damage of fused silica, in Laser-Induced Damage in Optical Materials: 2007 (SPIE, Bellingham, WA, 2007)
9. J. Yu, X. Xiang, S. He, X. Yuan, W. Zheng, H. Lü, X. Zu, Laser-induced damage initiation and growth of optical materials. Adv. Condens. Matter Phys. **2014**, 10 (2014)
10. R. Paschotta, Encyclopedia of Laser Physics and Technology (Wiley, 2008)
11. F.Y. Génin, A. Salleo, T.V. Pistor, L.L. Chase, Role of light intensification by cracks in optical breakdown on surfaces. J. Opt. Soc. Am. A **18**, 2607–2616 (2001)
12. L.B. Glebov, Intrinsic Laser-Induced Breakdown of Silicate Glasses, in Laser-induced damage in optical materials, (SPIE, Bellingham, WA, 2002)
13. F. Bonneau, P. Combis, J. Vierne, G. Daval, Simulations of Laser Damage of SiO2 Induced by a Spherical inclusion, in Laser-induced damage in optical materials: 2000 (SPIE, Bellingham, WA, 2001)
14. M.D. Feit, A.M. Rubenchik, Laser intensity modulation by nonabsorbing defects, in Proceedings of the 2nd International Conference on Solid State Lasers for Application to ICF, pp. 475–480 (1997)
15. J. Yu, X. Xiang, S. He, X. Yuan, W. Zheng, H. Lü, X. Zu, Laser-induced damage initiation and growth of optical materials. Adv. Condens. Matter Phy. **2014**, 364627 (2014)
16. R.W. Hopper, D.R. Uhlmann, Mechanism of inclusion damage in laser glass. J. Appl. Phys. **41**, 4023–4037 (1970)
17. T.W. Walker, A.H. Guenther, P. Nielsen, Pulsed laser-induced damage to thin-film optical coatings—part II: theory. IEEE J. Quantum Electron. QE **17**, 2053–2065 (1981)
18. M.R. Lange, J.K. McIver, A.H. Guenther, T.W. Walker, Pulsed laser induced damage of an optical material with a spherical inclusion: Influence of the thermal properties of the materials, in Laser Induced Damage in Optical Materials (Washington, DC: U.S. Government Printing Office, 1982)
19. M.R. Lange, J.K. McIver, A.H. Guenther, T.W. Walker, The influence of the thermal and mechanical properties of optical materials in thin film form on their damage resistance to pulsed lasers. Thin Solid Films **118**, 49–60 (1984)
20. M.Z. Fuka, J.K. McIver, A.H. Guenther, Laser induced damage in optical materials: 1989, in *Effects of Thermal Conductivity and Index of Refraction Variation on the Inclusion Dominated Model of Laser-Induced Damage*, ed. by A.J. Bennett, L.L. Chase, A.H. Guenther, B.E. Newnam, M.J. Soileau (SPIE, Bellingham, WA, 1990), pp. 576–583

21. C.H. Chan, Effective absorption for thermal blooming due to aerosols. Appl. Phys. Lett. **26**, 628–630 (1975)
22. M.D. Feit, A.M. Rubenchik, Implications of nanoabsorber initiators for damage probability curves, pulselength scaling, and laser conditioning. Proc. SPIE **5273**, 74–82 (2004)
23. S.G. Demos, M.R. Kozlowski, M. Staggs, L.L. Chase, A. Burnham, H.B. Radousky, Mechanisms to explain damage growth in optical materials, in Proceedings of the 32nd Annual Boulder Damage Symposium, pp. 277–284 (2001)
24. S. Papernov, A.W. Schmid, Correlations between embedded single gold nanoparticles in SiO2 thin film and nanoscale crater formation induced by pulsed-laser radiation. J. Appl. Phys. **92**, 5720–5728 (2002)
25. I.V. Aleshin, S.I. Anisimov, A.M. Bonch-Bruevich, Y.A. Imas, V.L. Komolov, Optical breakdown of transparent media containing microinhomogeneities. Soviet Physics-JETP **43**, 631–636 (1976)
26. K. Saito, A.J. Ikushima, Absorption edge in silica glass. Phys. Rev. B **62**, 8584–8587 (2000)
27. Y.K. Danileiko, A.A. Manenkov, V.S. Nechitailo, The mechanism of laser-induced damage in transparent materials, caused by thermal explosion of absorbing inhomogeneities. Soviet J. Quantum Electron. **8**, 116−118 (1978)
28. Y.K. Danileiko, A.A. Manenkov, V.S. Nechitailo, A.M. Prokhorov, V.Y. Khaimov-Mal'kov, The role of absorbing impurities in laser-induced damage of transparent dielectrics. Soviet Physics-JETP **36**, 541−543 (1973)
29. M.F. Koldunov, A.A. Manenkov, I.L. Pokotilo, Theoretical analysis of the conditions for a thermal explosion and of a photoionization instability of transparent insulators containing absorbing inclusions. Sov. J. Quantum Electron. **18**, 345–349 (1988)
30. S. Martin, A. Hertwig, M. Lenzner, J. Krüger, W. Kautek, Spot-size dependence of the ablation threshold in dielectrics for femtosecond laser pulses. Appl. Phys. A **77**, 883–884 (2003)
31. G. Mann, S. Pentzien, J. Krüger, Beam diameter dependence of surface damage threshold of fused silica fibers and preforms for nanosecond laser treatment at 1064 nm wavelength. Appl. Surf. Sci. **276**, 312–316 (2013)
32. L.G. DeShazer, B.E. Newnam, K.M. Leung, Role of coating defects in laser-induced damage to dielectric thin films. Appl. Phys. Lett. **23**, 607–609 (1973)
33. G.L.M. Gibson, Retinal damage from repeated subthreshold exposures using a ruby laser photocoagulator, U.S. Government Research and Development Reports, **71**(1971) 50
34. Y. Jee, M.F. Becker, R.M. Walser, Laser-induced damage on single-crystal metal surfaces. J. Opt. Soc. Am. B **5**, 648–659 (1988)
35. J. Güdde, J. Hohlfeld, J.G. Müller, E. Matthias, Damage threshold dependence on electron–phonon coupling in Au and Ni films. Appl. Surf. Sci. **127–129**, 40–45 (1998)
36. F. Di Niso, C. Gaudiuso, T. Sibillano, F.P. Mezzapesa, A. Ancona, P.M. Lugarà, Influence of the repetition rate and pulse duration on the incubation effect in multiple-shots ultrafast laser ablation of steel. Phys. Procedia **41**, 698–707 (2013)
37. J. Bonse, J.M. Wrobel, J. Krüger, W. Kautek, Ultrashort-pulse laser ablation of indium phosphide in air. Appl. Phys. A **72**, 89–94 (2001)
38. J. Bonse, S. Baudach, J. Krüger, W. Kautek, M. Lenzner, Femtosecond laser ablation of silicon–modification thresholds and morphology. Appl. Phys. A **74**, 19–25 (2002)
39. S. Küper, M. Stuke, Femtosecond uv excimer laser ablation. Appl. Phys. B **44**, 199–204 (1987)
40. S. Küper, M. Stuke, UV-Excimer-laser ablation of polymethylmethacrylate at 248 nm: characterization of incubation sites with fourier transform IR- and UV-spectroscopy. Appl. Phys. A **49**, 211–215 (1989)
41. M. Lenzner, J. Krüger, W. Kautek, F. Krausz, Incubation of laser ablation in fused silica with 5-fs pulses. Appl. Phys. A **69**, 465–466 (1999)
42. A. Rosenfeld, M. Lorenz, R. Stoian, D. Ashkenasi, Ultrashort-laser-pulse damage threshold of transparent materials and the role of incubation. Appl. Phys. A **69**, S373–S376 (1999)
43. A. Hertwig, S. Martin, J. Krüger, W. Kautek, Surface damage and color centers generated by femtosecond pulses in borosilicate glass and silica. Appl. Phys. A **79**, 1075–1077 (2004)

44. L.D. Merkle, M. Bass, R.T. Swimm, Multiple pulse laser-induced bulk damage, in Crystalline and Fused Quartz at 1.064 and 0.532 Âμm, Optical Engineering, **22**, 224405–224405 (1983)
45. E. Sutcliffe, R. Srinivasan, Dynamics of UV laser ablation of organic polymer surfaces. J. Appl. Phys. **60**, 3315 (1986)
46. D. Ashkenasi, M. Lorenz, R. Stoian, A. Rosenfeld, Surface damage threshold and structuring of dielectrics using femtosecond laser pulses: the role of incubation. Appl. Surf. Sci. **150**, 101–106 (1999)
47. Z. Sun, M. Lenzner, W. Rudolph, Generic incubation law for laser damage and ablation thresholds. J. Appl. Phys. **117**, 073102 (2015)
48. E. Matthias, Z.L. Wu, Non-destructive reading of laser-induced single-shot incubation in dielectric coatings. Appl. Phys. A **56**, 95–98 (1993)
49. J. Kleinschmidt, J.U. Walther, Analysis of polymer ablation. Phys. Status Solidi (a), **131**, 167–178 (1992)
50. G.B. Blanchet, P. Cotts, C.R. Fincher, Incubation: Subthreshold ablation of poly-(methyl methacrylate) and the nature of the decomposition pathways. J. Appl. Phys. **88**, 2975–2978 (2000)
51. K. Hata, M. Watanabe, S. Watanabe, Nonlinear processes in UV optical materials at 248 nm. Appl. Phys. B **50**, 55–59 (1990)
52. E. Eva, K. Mann, Calorimetric measurement of two-photon absorption and color-center formation in ultraviolet-window materials. Appl. Phys. A **62**, 143–149 (1996)
53. B. Hopp, T. Smausz, J. Kokavecz, N. Kresz, Z. Bor, S. Hild, O. Marti, Investigation of incubation in ArF excimer laser irradiated poly(methyl-methacrylate) using pulsed force mode atomic force microscopy. J. Appl. Phys. **96**, 5548–5551 (2004)
54. K. Kolev, D. Tatchev, S. Armyanov, E. Valova, K. Baert, A. Hubin, Incubation phenomena during UV-pulse excimer laser processing of silicone elastomers. J. Appl. Polym. Sci. **134**, 44541 (2017)
55. Z. Lin, R.A. Johnson, L.V. Zhigilei, Computational study of the generation of crystal defects in a bcc metal target irradiated by short laser pulses. Phys. Rev. B **77**, 214108 (2008)
56. C.S. Lee, N. Koumvakalis, M. Bass, Spot-size dependence of laser-induced damage to diamond-turned Cu mirrors. Appl. Phys. Lett. **41**, 625–627 (1982)
57. Y. Jee, M.F. Becker, R.M. Walser, Multiple-pulse laser-induced damage to metal mirror surfaces. Proc. SPIE **895**, 236–244 (1988)
58. C. Wu, M.S. Christensen, J.-M. Savolainen, P. Balling, L.V. Zhigilei, Generation of subsurface voids and a nanocrystalline surface layer in femtosecond laser irradiation of a single-crystal Ag target. Phys. Rev. B **91**, 035413 (2015)
59. J.V. Oboňa, V. Ocelik, J.C. Rao, J.Z.P. Skolski, G.R.B.E. Römer, A.J. Huis in 't Veld, J.T.M. De hosson, modification of Cu surface with picosecond laser pulses. Appl. Surf. Sci. **303,** 118–124 (2014)
60. M.F. Becker, C. Ma, R.M. Walser, Predicting multipulse laser-induced failure for molybdenum metal mirrors. Appl. Opt. **30**, 5239–5246 (1991)
61. J.-M. Savolainen, M.S. Christensen, P. Balling, Material swelling as the first step in the ablation of metals by ultrashort laser pulses. Phys. Rev. B **84**, 193410 (2011)
62. J. Bonse, H. Sturm, D. Schmidt, W. Kautek, Chemical, morphological and accumulation phenomena in ultrashort-pulse laser ablation of TiN in air. Appl. Phys. A **71**, 657–665 (2000)
63. J. Bonse, J. Krüger, Probing the heat affected zone by chemical modifications in femtosecond pulse laser ablation of titanium nitride films in air. J. Appl. Phys. **107**, 054902 (2010)
64. R.T. Williams, J.P. Long, M.N. Kabler, Photoelectron spectroscopy of laser-excited states in semiconductors. Opt. Eng. **28**, 1085–1095 (1989)
65. A. Borowiec, H.K. Haugen, Subwavelength ripple formation on the surfaces of compound semiconductors irradiated with femtosecond laser pulses. Appl. Phys. Lett. **82**, 4462–4464 (2003)
66. J. Bonse, J.M. Wrobel, K.-W. Brzezinka, N. Esser, W. Kautek, Femtosecond laser irradiation of indium phosphide in air: Raman spectroscopic and atomic force microscopic investigations. Appl. Surf. Sci. **202**, 272–282 (2002)

67. J.M. Wrobel, J. Bonse, J. Krüger, W. Kautek, Femtosecond pulse laser machining of InP wafers. Proc. SPIE **3933**, 280–287 (2000)
68. D. Bäuerle, *Laser Processing and Chemistry* (Springer, Heidelberg, 2011)
69. L. Cerami, E. Mazur, S. Nolte, C.B. *Schaffer, Femtosecond Laser Micromachining*, in ed. by R. Thomson, C. Leburn, D. Reid. Ultrafast Nonlinear Optics (Springer International Publishing, 2013, pp. 287–321
70. R. Osellame, G. Cerullo, R. Ramponi, *Femtosecond Laser Micromachining* (Springer, Berlin, 2012)
71. R. Poprawe, *Lasertechnik für die Fertigung* (Springer, Berlin Heidelberg, 2005)
72. B.E. Newham, L.G. DeShazer, Spot-size dependence of laser-induced damage thresholds of dielectric thin films. IEEE J. Quantum Electron. QE-8 611–611 (1972)
73. B.-M. Kim, M.D. Feit, A.M. Rubenchik, E.J. Joslin, J. Eichler, P.C. Stoller, L.B. Da Silva, Effects of high repetition rate and beam size on hard tissue damage due to subpicosecond laser pulses. Appl. Phys. Lett. **76**, 4001–4003 (2000)
74. A. Naghilou, O. Armbruster, M. Kitzler, W. Kautek, Merging spot size and pulse number dependence of femtosecond laser ablation thresholds: modeling and demonstration with high impact polystyrene. J. Phys. Chem. C **119**, 22992–22998 (2015)
75. O. Armbruster, A. Naghilou, M. Kitzler, W. Kautek, Spot size and pulse number dependence of femtosecond laser ablation thresholds of silicon and stainless steel. Appl. Surf. Sci. **396**, 1736–1740 (2017)
76. A. Naghilou, O. Armbruster, W. Kautek, Femto- and nanosecond pulse laser ablation dependence on irradiation area: the role of defects in metals and semiconductors. Appl. Surf. Sci. **418**, 487–490 (2017)

Chapter 3
Surface Functionalization by Laser-Induced Structuring

Juergen Reif

Abstract This chapter reviews the possibilities of functionalizing solid surfaces by laser irradiation—i.e. modifying the surfaces with respect to, e.g., wettability, optical properties, friction/wear-control, corrosion resistance, electrical properties. The functionalization occurs via the modification of surface morphology by the laser-induced formation of (regular) nano- to micro-textures. We present the different approaches to generate these patterns as well as their implication for specific functions obtained by this structuring and, finally, show some examples of applications.

3.1 Introduction

Materials surfaces play a paramount role in many aspects such as the optical appearance, mechanical properties like friction and wear, the interaction with the environment like wettability, corrosion, or adhesion. Also, chemical activity, e.g. in catalysis, is strongly dependent on the surface. The modification of solid surfaces to obtain new, functional properties is of considerable impact on the quality of modern engineering products and innovation [1, 2]. Such modification may be achieved, on the one hand, by coating the surface with a thin film providing the desired performance [3–5]. On the other hand, there is the possibility of modifying the actual surface morphology by the formation of new surface textures with feature size ranging from several hundred nanometers to the several ten microns, sometimes combining to form complex multi-scale hierarchical structures [6]. The patterns and their function often have their paradigm in nature and biology, with the most popular examples of the lotus leaf and its dirt-repelling hydrophobicity or of the colorful appearance of butterfly wings [7]. There are several techniques to achieve such surface patterning, including a combination of successive grinding, etching and chemical coating [8], or lithography [9], sometimes based on a template directly taken from a real

---

J. Reif (✉)
Brandenburgische Technische Universität – BTU Cottbus-Senftenberg,
Platz der Deutschen Einheit 1, 03046 Cottbus, Germany
e-mail: reif@b-tu.de

P. M. Ossi (ed.), *Advances in the Application of Lasers in Materials Science*, Springer Series in Materials Science 274,
https://doi.org/10.1007/978-3-319-96845-2_3

biological antetype [10]. A particular method is the replication from a mold formed by, e.g., diamond-machining [11]. Whereas most of these procedures employ multiple processing steps, surface nano-/micro-texturing can also be achieved by either self-organized pattern formation during thin film epitaxial growth [12] or by direct exposure to energetic irradiation, namely by ion, electron, or laser beams.

For charged particles beams, there are two typical ways of structuring: (a) direct-write lithography [13–16], and (b) relaxation from thermodynamic instability after ion-beam sputtering, where a large variety of regular to complex nano-patterns are attributed to a self-organized structure formation [17–19]. Surface morphology modification by intense laser pulses can be, similarly, differentiated into two main lines: (a) ablative lithography of multi-beam induced interference patterns [20], and (b) single beam formation of Laser-Induced Periodic Surface Structures (LIPSS) [21–28]. The latter two will be discussed in detail in the present chapter. We will start by showing some typical relations between surface textures—including laser-generated ones—and functionality. Then, we explicitly consider laser patterning and present several experimental approaches for their realization, together with current models explaining the interaction.

3.2 Functionality of Textured Surfaces

More or less regular surface patterns on the sub- to several micrometer scale result in numerous surface functionalities ranging from specific coloring over modified wettability [29], such as hydro- and oleo-phobicity or -philicity, special tribologic properties, wear-resistance, catalysis, modified adhesion. They can serve as templates for biological cell growth [30] and can stimulate surface enhanced Raman scattering [31, 32]. In the following, we present some typical surface functionalities and the patterns at their basis.

3.2.1 Wettability

One of the most important surface-morphology induced functionalities is, certainly, the modification of wettability, i.e. the controlled adjustment of liquid-repellent resp. liquid-attracting properties [1, 7]. The correlation is bio-inspired by the surface of the lotus leaf [7, 24, 33] (Fig. 3.1) or the rose petal [10].

Wettability, i.e. the capability of a solid to attract and spread or to repel a liquid on its surface, is of great importance both in daily life and, particularly, in technology. High wettability—for water called "hydrophilicity"—is required if a surface is to be covered by a thin liquid film, e.g. a lubricant or wet paint, to obtain an all-over coverage. Also the uniform spread of oil or water on a warm Teflon pan is due to this quality [34], despite its "cleaning effect" (the resulting wetting film provides a smooth, separating interface between the pan and the contents). The opposite

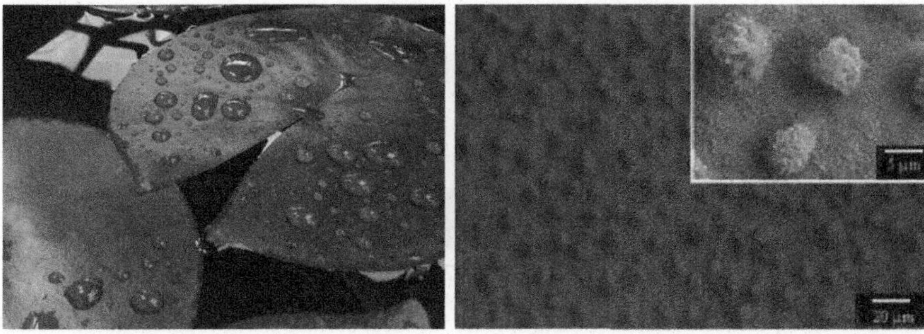

Fig. 3.1 Super-hydrophobic lotus leaf (from [33]). Left: photograph; right: SEM image of the lotus leaf surface

behavior of liquid repelling—for water called "hydrophobicity"—is important for effects like self-cleaning, e.g. of windshields, anti-icing of airplane wings, reduced corrosion, bacterial resistance.

The phenomenon of wettability is determined by the mutual interaction of the three phases: solid, liquid, and gaseous. It is, mainly, determined by attractive forces inside the liquid (cohesion) compared to interface forces between solid-liquid (adhesion) and liquid-gas. Typically, the latter are significantly weaker than the liquid internal binding, resulting in the confinement of a free liquid droplet because work against the internal binding is needed to increase the droplet surface. This effect is called surface tension; it controls the shape and size of a free liquid droplet, as described already in 1805 by Young [35]. At the droplet-surface interface, the competition of adhesive forces between the surface and the liquid, one the one hand, with cohesive forces within the liquid, on the other hand, is responsible for a minimization of the contact area. The shape of the droplet at the triple boundary (solid, liquid, gas) is then determined by a minimization of the total interface energy, γ [36], combining surface tension (liquid–gas, LG), droplet-surface adhesion (SL) and gas-surface adhesion (SG). It is characterized by the contact angle, θ, which is >90° for liquid repulsion (hydrophobicity) and <90° for attraction (hydrophilicity). At an *ideal* surface the contact angle is determined by Young's Equation:

$$\cos\theta = \frac{\gamma_{\mathrm{SG}} - \gamma_{\mathrm{SL}}}{\gamma_{\mathrm{LG}}} \tag{1}$$

This is, schematically, shown on the left panel in Fig. 3.2. For a *real* surface, however, we have to account for surface roughness, modifying the surface-liquid contact by, e.g., either increasing or reducing the efficient contact area, thus modifying the contact angle. These two possibilities are depicted on the center and right panels in Fig. 3.2.

It becomes immediately evident that the specific wettability is strongly influenced by the morphological structure of the surface roughness. Another important influence is due to the surface active area together with the surface charge distribution.

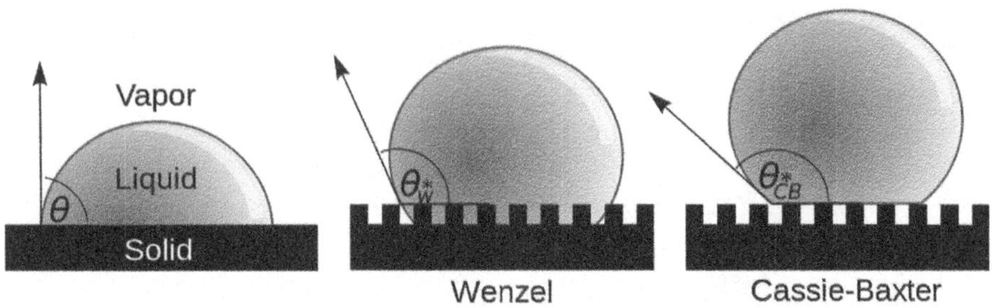

Fig. 3.2 Wetting states: Classical (Young) model for ideal surface (left); rough surface: Wenzel model, Cassie-Baxter model (super-hydrophobic). The wetting state is characterized by the contact angle θ

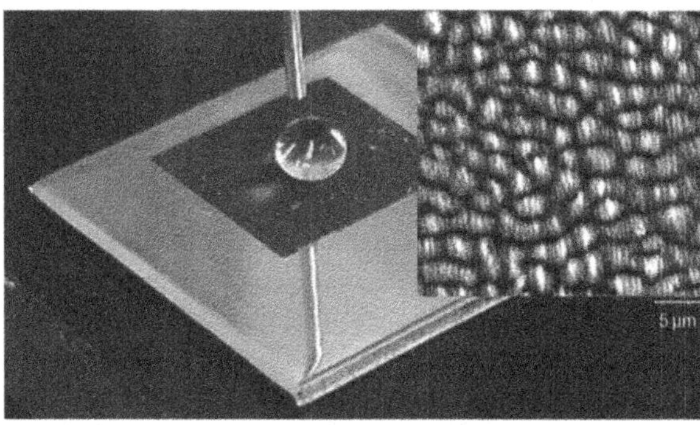

Fig. 3.3 Super-hydrophobic laser-processed steel surface

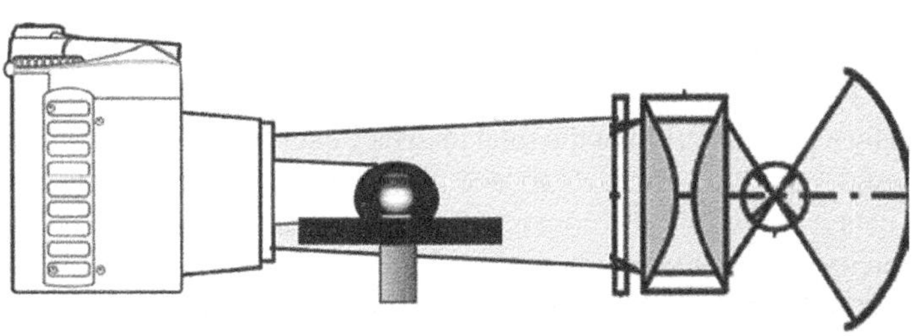

Fig. 3.4 Principle of sessile drop method for measuring contact angles: the liquid droplet sitting on the solid surface is homogenously illuminated from the back; the resulting shadowgraph is recorded by a camera

Experimentally, super-hydrophobicity, similar to the lotus leaf, can be achieved by hierarchical (multi-scale) laser-modification of stainless steel [37] (cf. Fig. 3.3).

The contact angle can be, usually, measured optically by the sessile drop method [38] as shown in Fig. 3.4.

A particularly sensitive variation of the method consists in measuring the dynamic contact angle by changing the droplet volume. Then, a hysteresis of two different contact angles is obtained: advancing by increasing the droplet volume and receding by reducing the droplet volume [39]. Generally, the receding contact angle is smaller and the advancing contact angle is larger than or equal to the static one.

3.2.2 Color

Similar to the feature of wettability, also the connection between surface micro-/nanostructure and color is bio-inspired. The most familiar examples are the multi-colored wings of butterflies [7, 40] (Fig. 3.5).

The phenomenon is based on the dispersion of white light transmitted (cf. Fig. 3.6) or reflected by optical gratings: due to interference, light incident on a grating of periodicity (line spacing) d is wavelength-dispersed and the fraction of wavelength λ continues to propagate under diffraction angle α for the first diffraction order (2):

$$\sin a = \lambda/d \tag{2}$$

For given angles of illumination and observation, thus, different values of d result in different colors observed. The effective values of d do not only depend on the physical spacing but also on the orientation of the grating with respect to the plane of incidence (Fig. 3.7a) which is clearly obvious by the multi-color directional reflection from a compact disk with its circular grating pattern (Fig. 3.7b).

Fig. 3.5 Butterfly [Robert Clark, *National Geographic*] and SEM pictures of microstructures on a butterfly wing [40]

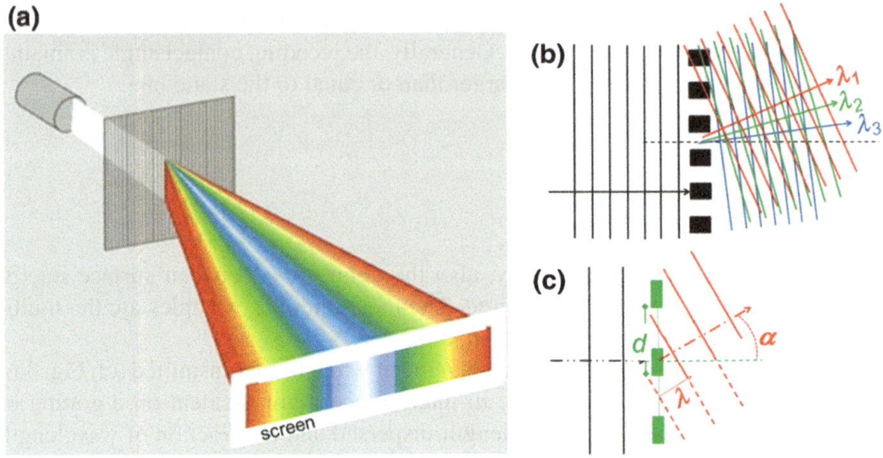

Fig. 3.6 Dispersion of white light by an optical grating. **a** Schematic (transmission); **b** Principle: incident white light (wave fronts and direction of propagation are indicated) is split and different wavelengths are diffracted into different directions; **c** Detail: for a line spacing d, light of wavelength λ is diffracted under angle α

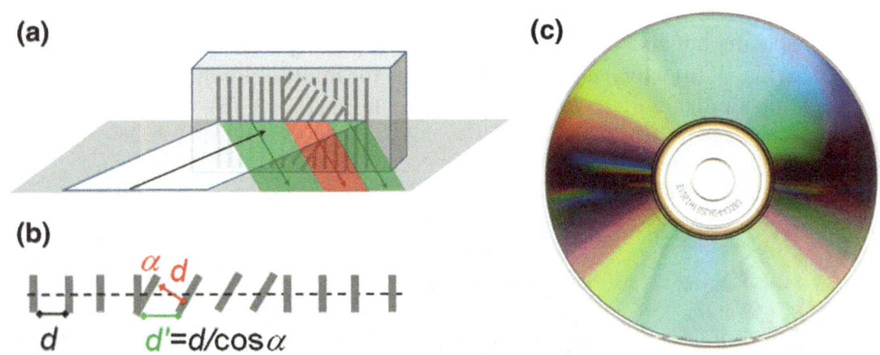

Fig. 3.7 Diffraction from identical but differently oriented—with respect to the plane of incidence—gratings. **a** Schematic; **b** Dependence of the effective grating spacing d' on grating orientation; **c** Diffraction from a compact disk (CD) with annular gratings

Indeed, such multi-color diffraction effect has been produced also by Laser-induced periodic structures (LIPSS) by Dusser et al. [41] (Fig. 3.8), where they could show, in addition, that the effect can be used for hidden marking.

A particular type of structure-induced "coloring" is the formation of deep black surfaces, such as carbon black [42, 43], or highly absorbing surfaces for applications in solar cells [44, 45]. The micro-/nano-structures responsible for that high absorption act two-fold: one the one hand, they simply increase the active surface considerably. But more important, on the other hand, is the multiple reflection and final trapping

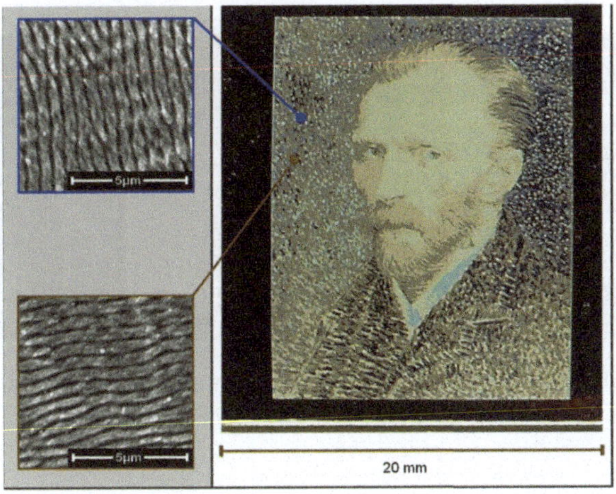

Fig. 3.8 Multi-color diffraction from a laser-structured surface with different LIPSS directions (from [41])

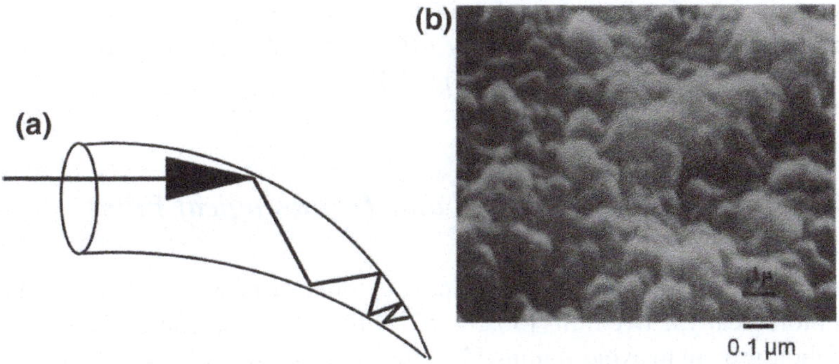

Fig. 3.9 Light-trapping by complex structures. **a** Wood's Horn; **b** "Black Chrome" (from [44])

inside the micro-/nano-structures like in a Wood's Horn (Fig. 3.9a). Typical structures consist of multiple pores [45] or multi-scale arrays of deep grooves or notches.

The creation of structural colors, notably black surfaces, by LIPSS has also been successfully reported [32]. As a particular application-related variant, "Black Silicon" has been produced, amongst others, by Eric Mazur's group [46].

3.2.3 Field Enhancement

Whereas, so far, surface functionality mostly because of depth modulation (depression) was considered, also sharp protrusions can have an important effect. In the presence of electric fields (static or electromagnetic) such tips or tip-arrays cause

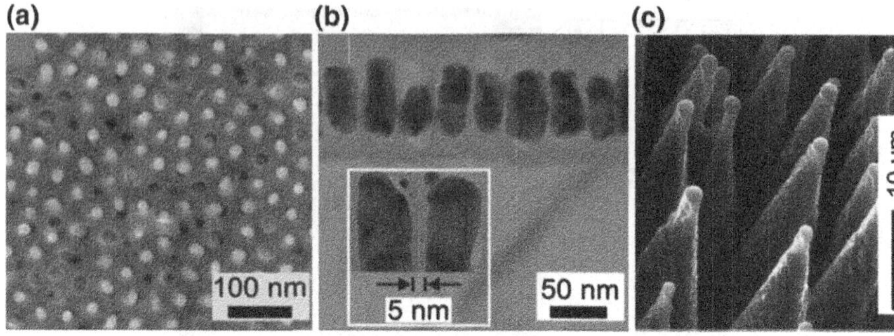

Fig. 3.10 Nanopillar arrays. **a** Ag-pillars on top of porous anodic alumina; **b** detail of the pillars in (**a**) (from [51]); **c** spikes on Si (100), produced by femtosecond LIPSS (from [53])

a considerable local field enhancement [47, 48]. This can be exploited as a very efficient, brilliant source of fast electron bunches [48, 49], e.g. for applications in ultrafast electron diffraction, crystallography, and microscopy [50]. The strong field enhancement in a tip-array can also increase the efficiency of surface enhanced Raman scattering (SERS) considerably [31, 51, 52]. Again, such tip-arrays can be easily fabricated also by LIPSS [53] (Fig. 3.10).

3.2.4 Templates for Biological and Technological Films

The main biological functionality of nanostructured surfaces is related to the formation of biologically active films [30, 54, 55] (Fig. 3.11)—or, contrarily, the suppression of their growth to avoid fouling [56]; the nanostructures can even serve to make titanium implants antibiotic [57].

An example for the formation of active films is the exploitation of ZnO nanostructures on silicon substrate as templates for the development of topography-mediated neuronal cultures [58]. It was demonstrated that the ZnO-templates can support neuronal cell growth and proliferation. Another effect is the stabilization and immobilization of enzymes on nano-structured surfaces [59, 60]. Natural nanostructures, such as diatoms with their outer shell of micro- and nano-porous silica, have been proposed as templates for bio/chemical sensors and biomimetic membranes [61]. Again, surface structures as templates for biofilms have also been produced by laser induced periodic surface structuring [62].

Not only for bio-films, also in Materials Science, can surface textures be used as templates to increase the functionality of films grown on these surfaces. As an example, epitaxially grown self-organized semiconductor nanostructures can serve as templates for very regular nanomagnet arrays [12].

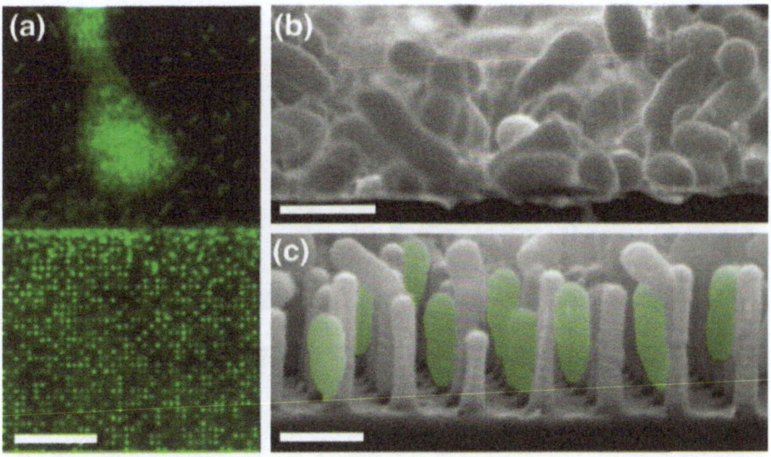

Fig. 3.11 Comparison of P. aeruginosa adhesion on structured and unstructured regions. **a** Fluorescence microscopy shows the different localized adhesion on flat region (upper) and structured region (lower), with an abrupt transition. **b** and **c** are corresponding cross-sectional SEM images, showing the strong difference in attachment morphology. The aligned cells in (**c**) are false-colored to highlight their orientation. Scale bars are 10 μm in (**a**) and 1 μm in (**b**) and (**c**). (From [30])

3.3 Laser Patterning

In the following, we present the main methods to use laser techniques for regular surface texturing. In one way or the other, they are all closely related to the field of laser ablation: the light of, usually, strong laser pulses is absorbed by the target material and the resulting material response ranges from (surface) melting to massive material removal. In general, one can conceive two scenarios of responsible processes for pattern formation:

(1) Local ablation after a patterned irradiation, very similar to lithography, using electron or ion beams [13–16], or local direct writing; one typical example is mask projection of excimer laser irradiation [63] (a; cf. Fig. 3.12), another approach is generating the pattern by multi-beam interference (b; Sect. 3.1).
(2) Surface structuring by irradiation with a single beam (laser-induced periodic surface structures; LIPSS) [21–27].

3.3.1 Multi-beam Interference and Ablation

The most obvious way to generate mask-less patterned irradiation, as required for scenario (1b), is the illumination by interference of two or more beams, hitting the sample from different directions [20, 64–66] (Fig. 3.13).

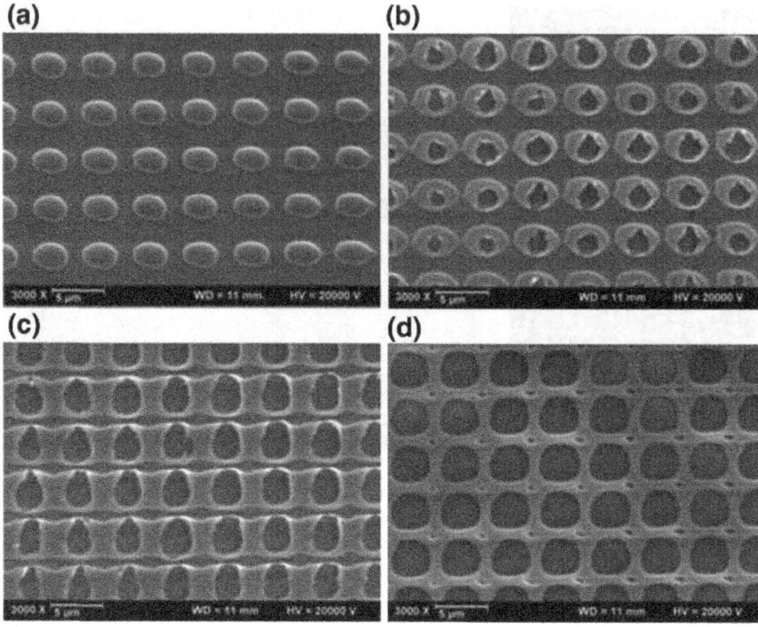

Fig. 3.12 Patterned 90 nm thick SiOx film; laser parameters: 248 nm, 1 pulse; structure period: 4.7 μm (from [63]). Fluence: **a** 180 mJ/cm^2, **b** 190 mJ/cm^2, **c** 200 mJ/cm^2, **d** 220 mJ/cm^2

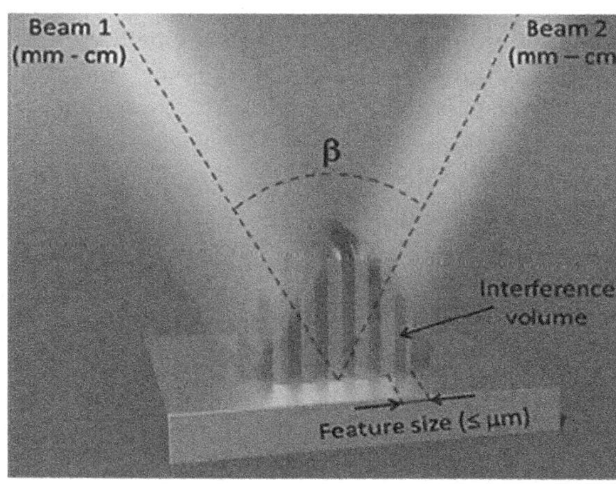

Fig. 3.13 Incident fluence modulation by interference pattern (from [64])

A typical example of such multi-beam-interference patterning is shown in Fig. 3.14 [66, 67]: an array of several parallel coherent beams is derived from a single laser in a diffractive optical element (DOE) and, then, focused to interfering overlap at the sample.

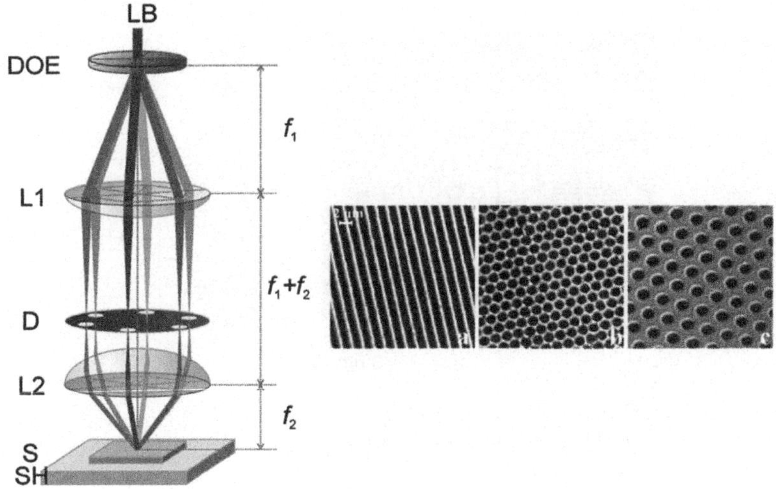

Fig. 3.14 Multi-beam-interference patterning. The left panel presents the principal set-up with the incident laser beam split into, here, six parallel beams by a combination of DOE and imaging lens. After spatial filtering, the beams are refocused to overlap on the sample. The right panels show micrographs of three typical patterns, obtained by interference of two, three, and six beams, respectively (from [67])

3.3.2 Single-Beam Laser Induced Periodic Surface Structures (LIPSS)

The phenomenon of surface morphology modification by incidence of a single laser beam was first observed already in 1965, shortly after the realization of the first laser, by Birnbaum [68]. In the beginning, this "surface damage" was mostly considered as an unwanted, though inevitable, trouble in materials processing as well as for optical components used with high power lasers [69, 70]. It was not before two decades later, that the phenomenon was studied in its own right [21, 71–77]. The next progress was the formation of sub-wavelength structures induced by ultrafast lasers [27, 78], again almost two decades later, initiating still ongoing research. This time, the emphasis is no longer on the damage aspect but, instead, growing interest is focused on possible, positive applications of the generated surface patterns, as indicated above in Sect. 3.2.

Nevertheless, the first experiments of femtosecond-LIPSS were concentrated on the effect as such. The first experiments investigated the surface morphology produced on a single interaction spot with diameters ranging from a few to more than 100 μm (cf. Fig. 3.15).

The effect is observed on a wide range of solid state materials [22], such as dielectric insulators [27], semiconductors [79], metals [80], and polymers [77, 81]. The irradiation is always close to the macroscopic ablation threshold, and the surface-morphology modification appeared as well after very few pulses [82, 83] (or even a single pulse [84–86]) as only after many thousand pulses [27].

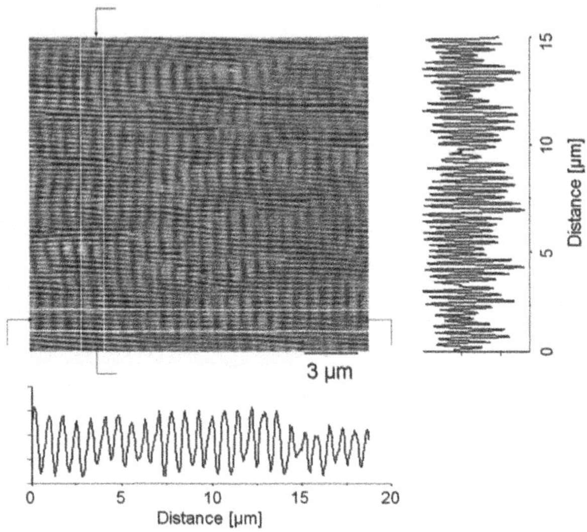

Fig. 3.15 Typical LIPSS pattern at the bottom of a 100-μm spot on a BaF_2 crystal surface, irradiated by 43,000 pulses from an amplified Ti:Sapphire Laser (800 nm wavelength; 100 fs pulse duration; 1 J/cm^2 fluence {90% of ablation threshold}; polarization vertical). Note that two periodic structures of different periodicity Λ (one of the order of the laser wavelength λ, the other one of about $\lambda/10$) are superimposed perpendicularly. The horizontal and vertical AFM traces, recorded in the indicated stripes, show the periodicities $\Lambda_1 \approx 690$ nm, $\Lambda_2 \approx 280$ nm. (From [27])

All experiments show that the LIPSS orientation is strongly determined by the polarization of the laser beam (Fig. 3.16a–e), with a few exceptions where surface scratches or similar irregularities can override the polarization dependence (Fig. 3.16f, g).

The laser-generated morphology can exhibit a variety of different appearances:

(1) Very narrow, regular lines (for linear polarization) or spherical dots for circular polarization; cf. Fig. 3.16) with a feature size in the 100-nm range, well below the diffraction limit;
(2) Wider periodic, parallel lines with a spacing close to the laser wavelength;
(3) An array of larger irregular slabs of several-μm size, separated by deep trenches (similar to a dried-out river bed);
(4) A regular array of conical tips, with spacing and size in the μm-range.

For convenience, the different features have been termed [22, 79] "HSFL" (high spatial frequency LIPSS) resp. "LSFL" (low spatial frequency LIPSS) for types 1 resp. 2 (cf. Fig. 3.15); "groves" (type 3), and "spikes" (type 4) (cf. Fig. 3.17). In general, HSFL and LSFL are oriented perpendicular to each other, a typical example is shown in Fig. 3.15.

An crucial parameter, affecting feature type and size appears to be the total dose of absorbed energy, i.e. the laser fluence and the number of superimposed laser shots

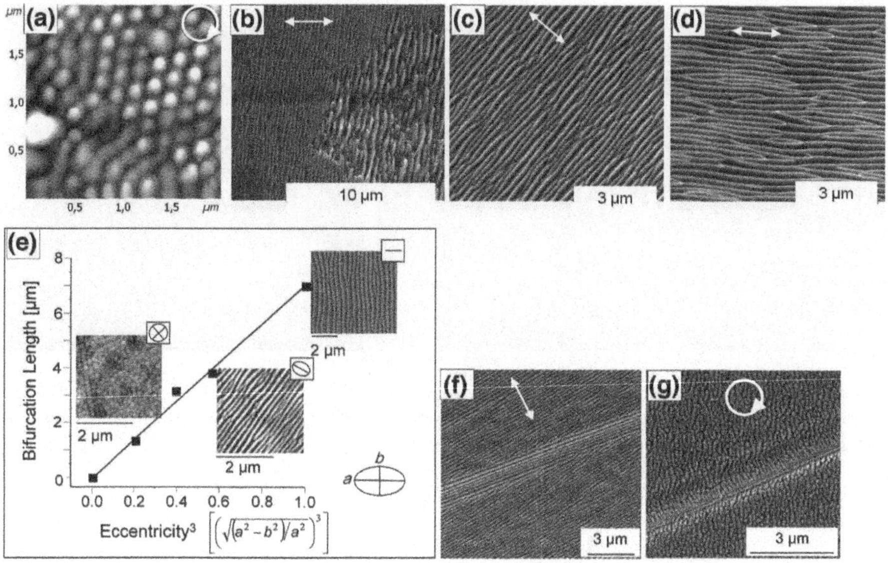

Fig. 3.16 Polarization dependence of LIPSS orientation. The respective polarization, circular in **a** [87] and linear in **b–d** [26] is indicated by the white arrows. **e** shows the effect of elliptical polarization [88]. In **f**, **g**, the influence of macroscopic scratches on the surface is shown, overriding the polarization dependence [89]

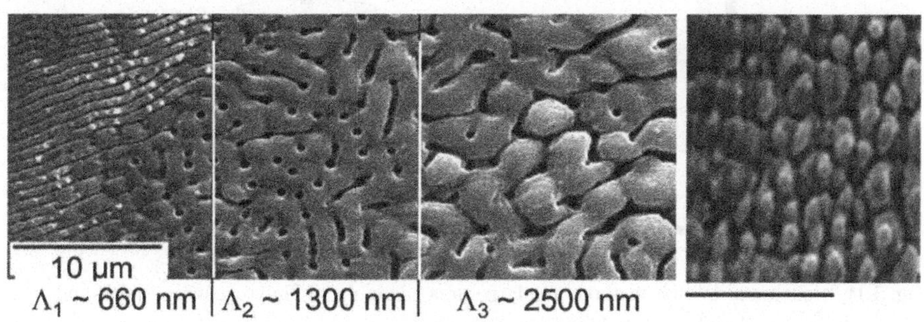

Fig. 3.17 LIPSS on silicon. The three areas on the left [90], showing LSFL and grooves, are from one single spot with Gaussian profile from the edge (left) to the center; the right panel shows spikes [91]. The scale is the same on all panels

[91–93]. This can be seen in Fig. 3.17 where the increasing intensity within the Gaussian spot profile results in a transition from LSFL to grooves

An important step towards an application of LIPSS structures was achieved by the discovery that it is possible to coherently modify larger areas by scanning the laser across the sample surface [94–97] (Fig. 3.18).

In general, there are two scanning methods as shown in Fig. 3.19: (a) Keeping the target fixed in position and scanning the laser spot by means of a pair of orthogonally

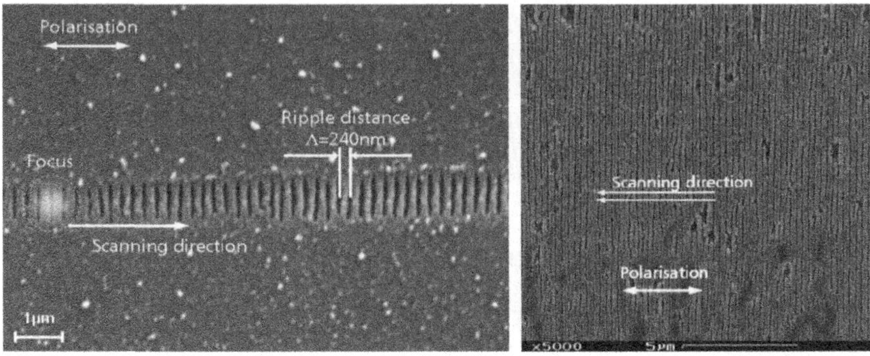

Fig. 3.18 Coherently continued ripples by scanning parallel to the polarization on fused silica. Left: single track: right: coherently continued ripples in 2D by scanning several tracks with an offset of 400 nm (one spot diameter) (from [96])

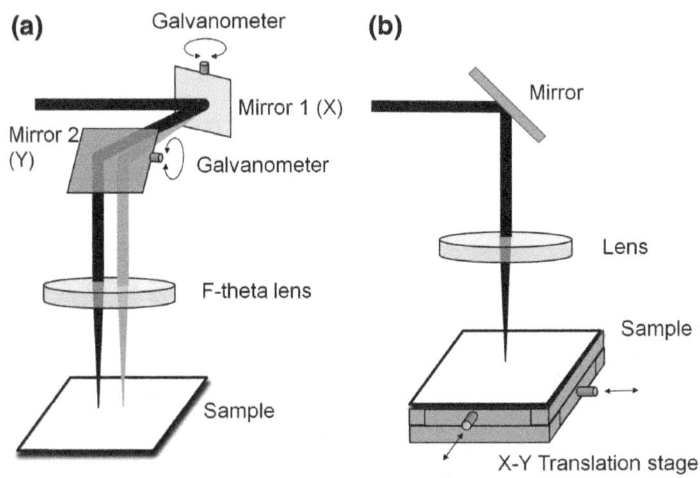

Fig. 3.19 Schematic of two methods for large area scanning. **a** Galvo scanning head and fixed sample, **b** fixed spot and mechanically scanning sample

moving mirrors ("scanning head"); the moving beam is, then, focused onto the target by an F-theta (telecentric) lens. (b) Keeping the laser spot fixed and moving the target, mounted on a set of precision translation stages.

Both techniques bear their individual advantages: procedure (a) allows a very high scanning speed, usually controlling the mirror motion by a galvanometric drive (galvoscanners); further, it allows a very compact and rigid set-up when using a commercial scanning head. Method (b) allows very large target areas to be covered, depending only on the translation stage size and precision; however, it is, generally, much slower and may be less compact; therefore, it is, mostly, used for laboratory application.

Typically, the full surface is covered by first writing lines along one direction (e.g. "*X*") followed by adjacent tracks, displaced in the other direction (e.g. "*Y*"). Then, there are two further options: writing the lines either along or perpendicular to the laser polarization. Generally, it turns out that scanning normal to the ripples direction, i.e., usually, along the polarization, yields more regular patterns than the other sense. Another important parameter for the pattern quality is the scanning speed, or more precisely its ratio to the repetition rate, yielding longitudinal pulse overlap and, thus, the effective number of pulses acting on one spot area. A similar role is played by the scanning pitch, i.e. the separation of adjacent tracks (or lateral pulse overlap). The importance of these parameters becomes evident when considering the dose-dependence of the generated patterns (cf. Figs. 3.15 and 3.17).

3.3.2.1 LIPSS and Functionality

As was already addressed in Sect. 3.2, most of the functional nano-structuring can be, easily, achieved by LIPSS formation. Here, we just want to compile, again, the relevant features. It should be noted that, since LIPSS functionality only involves the surface of the modified material, usually very short laser pulses in the femto- or picosecond range are used to avoid thermal effects in the target volume [98]. A comprehensive review on LIPSS-functionalized surfaces can be found in [22–24, 32].

A Wettability

The modification of surface wettability by LIPSS was first reported in 2006 by the FORTH group on silicon [99] and by Groenendijk and Meijer [100], who patterned a stainless steel surface to become super-hydrophobic (cf. Fig. 3.20). Such surface could, subsequently, be used as a mould for plastic replicas exhibiting similar hydrophobicity.

There have been many approaches since then, showing, e.g., an improvement of hydrophobicity by chemical alkysilane post-treatment [102] or the formation of hierarchical multiscale patterns [103]. Again, it appears that the irradiation dose plays an important role for controlling the functionality [37, 99]. In fact, is not only possible to make the surface (super-) hydrophobic but also hydrophilicity (super-wetting) can be achieved [23, 32, 37], even obtaining a surface where a water film creeps upwards against gravity [104, 105] (Fig. 3.21).

A peculiar effect, so far not fully understood, is the "aging" of processed metal surface [93]: even a multiscale-structured steel surface is not immediately super-hydrophobic but only develops this property after longer dwelling (from hours to days) in ambient atmosphere [37, 93, 103]. The most probable cause is some uncontrolled chemical modification, e.g. oxidation.

B Tribology

Closely related to the topic of modified wettability is the influence of LIPSS on the tribological properties of the treated surface. The underlying idea is two-fold: on the one hand, the depressions and grooves may serve to provide a reservoir for a possible lubricant; on the other hand, the reduced surface area, consistent of smooth, narrow ridges or tips, may reduce the resulting interaction between the surface and another body, rolling or sliding on that surface. A control of friction can have important implications: for a friction clutch in power transmission between a driving and a driven shaft it should be maximized; on the other hand, for wear reduction between moving surfaces, it should be reduced.

Consequently, the first proposal for an application of LIPSS intended to reduce friction and wear in a system of rapidly rotating hard disk and sliding head [106]. Following experiments on diamond like carbon (DLC) [107, 108], using femtosecond LIPSS, showed a reduction of both adhesive and lateral force (Fig. 3.22).

Chen et al. succeeded in reducing the friction coefficient of a SiC seal in water lubricant by about 20% [109] (Fig. 3.23).

Significant wear reduction on silicon [110] and on PEEK surface, a biocompatible thermoplastic for implanted joints, have been reported [111, 112]. Recently, Bonse et al. [113, 114] demonstrated that not only the LIPSS texture but also the contacting materials as well as possible lubricants play an important role.

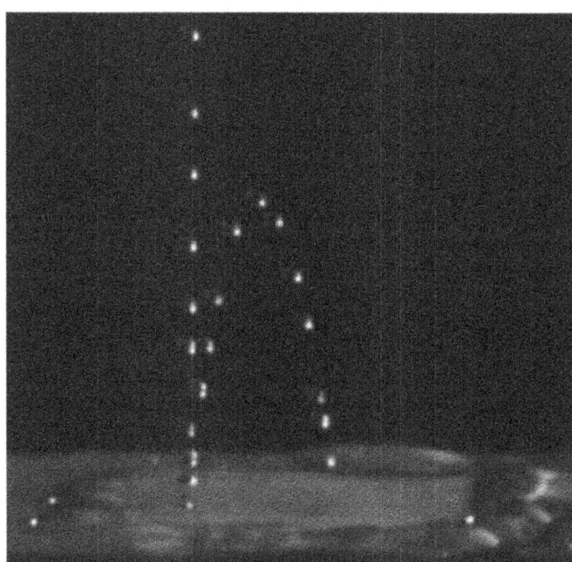

Fig. 3.20 Super-hydrophobicity: falling droplet bounces back from a LIPSS surface [101]

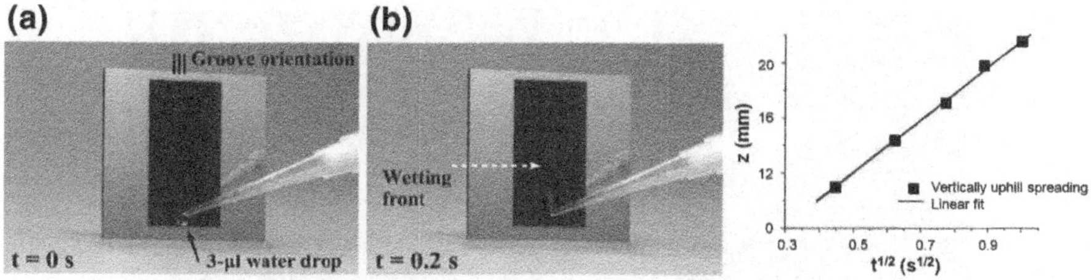

Fig. 3.21 Hydrophilicity: water droplet spreading up-hill (from [105])

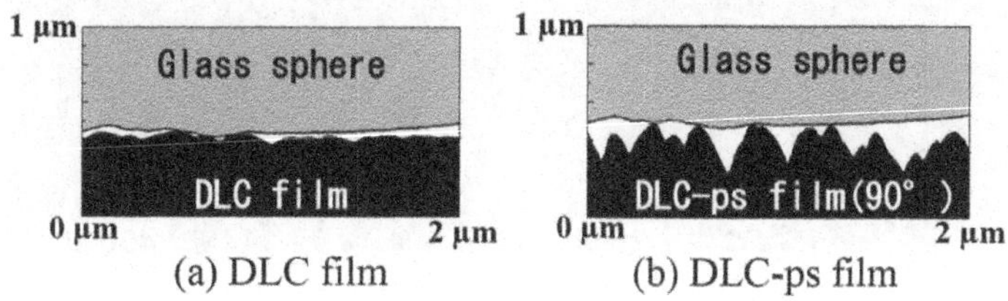

Fig. 3.22 Tribology: influence of LIPSS roughness on contact area (from [107])

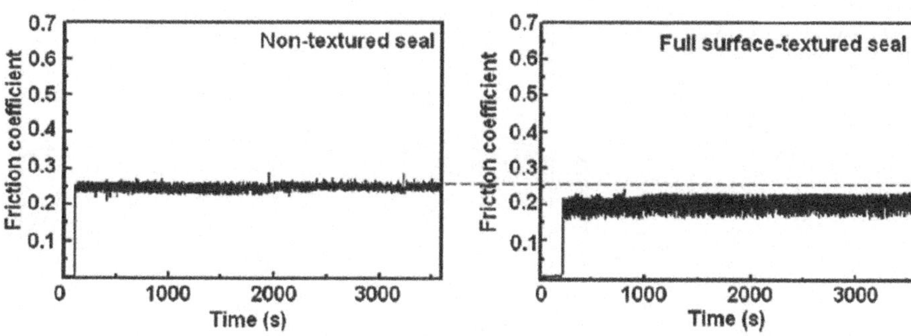

Fig. 3.23 LIPSS-reduction of friction coefficient of SiC in water lubricant (right). For comparison, the coefficient of the non-textured surface (left) is indicated by the dotted line (from [109])

C Color

Already in Sect. 3.2.2, examples are given for selective surface coloring by LIPSS formation (cf. Fig. 3.8; [41]). An impressive overview is presented in [32]. Particular interest appertains to the formation of black silicon, notably in order to optimize spectral absorption for photovoltaic applications. However, this has to be considered with care, since the structural surface modification is associated with the formation of electronic defects, significantly reducing the carrier lifetime [115]. Eric Mazur's group succeeded in overcoming this problem by conducting the laser processing under SF_6 atmosphere [53] and subsequent annealing at 1200 K to remove hierarchical secondary structures [46] (Fig. 3.24).

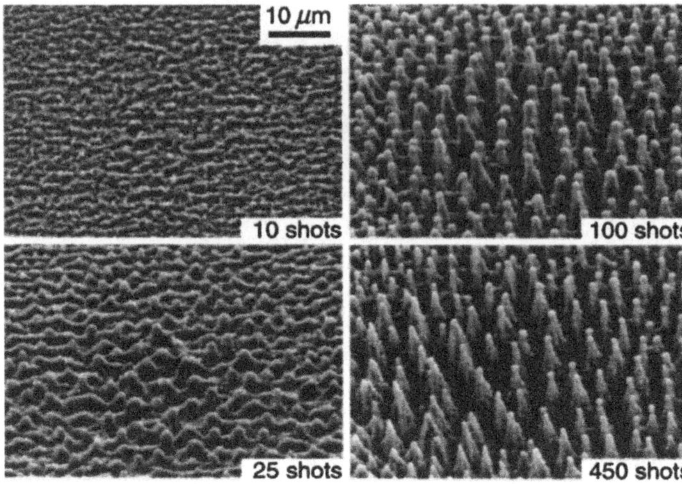

Fig. 3.24 "Black silicon", formed by laser processing (from [46])

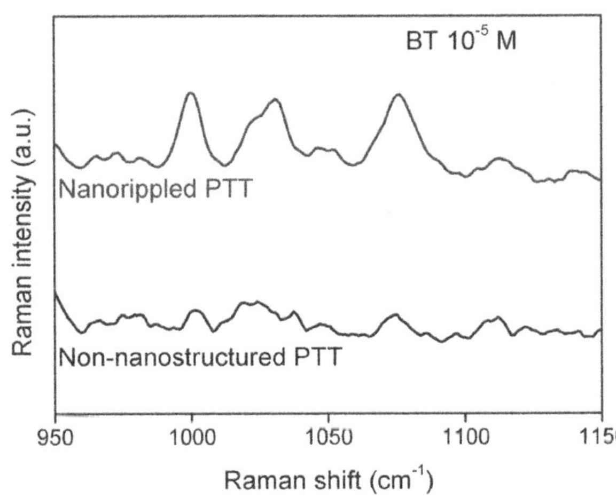

Fig. 3.25 Field enhancement of SERS on a LIPSS-textured surface (from [116])

D Field Enhancement/Surface Enhanced RAMAN Scattering

As indicated before, at sufficiently high irradiation dose the regular line-shaped LIPSS convert into an array of regular, sharp tips, capable to a considerable local electrical field enhancement. This can be exploited by using the array as a template for surface enhanced RAMAN scattering (SERS) [116–118] (Fig. 3.25).

E Templates for Biological and Technological Films

As generally holds for nanostructured surfaces (cf. Sect. 3.2.4), an interesting application of LIPSS patterned surface is the possibility of growing on top biological [119] or technological films [120] with peculiar features.

In the fields of medical and biological functionality, the main application concerns cell adhesion [121, 122]. There can be either a positive or, as well, a negative effect [62], depending on substrate and on biomaterial. A typical application concerns the optimization of titanium implants [82].

Another aspect applies to the LIPSS-template induced self-organized growth of ordered structures, both for biomaterials, e.g. directional cell spreading on titanium [123], and for technological materials, e.g. vertically aligned carbon nanofiber growth [120].

3.3.2.2 Lithographic Model: Laser-Plasmon Interference

The first approach to understand LIPSS formation was triggered by the apparent grating-like structures with a feature size close to the laser wavelength [21, 68–76]: the pattern was assumed to result from modulated ablation [68] following an illumination interference pattern between the incident wave and a surface-scattered wave [69, 72–74]. Keeping the basic idea of modulated ablation due to inhomogeneous energy deposition, the model was refined by ascribing the illumination pattern to interference between the incident wave and surface plasmon-polaritons, excited by the laser field. In order to apply this model also to non-conducting targets, a laser-induced transient high conduction-electron density was considered in a Drude-model for quasi-free electrons [124]. For more details see [22]. Though this model is very successful to predict all details of periodic surface structures of the LSFL type, its validity for other structures like HSFL, grooves, spikes is only very limited. The model is practically purely based on an assumption of particular electromagnetic field distributions occurring in the interaction, an active role of the material in structure formation is neglected, so far.

3.3.2.3 Surface Dynamic Model: Self-organization and Hydrodynamics

An alternative picture of the basic LIPSS formation mechanisms, developed by our group [26] has been triggered by the apparent similarity of LIPSS with patterns produced by self-organized epitaxy [12] or ion beam sputtering [19]. Our model is closely related to the concepts developed for the latter [17, 18] and assumes a much more active role of the material than the lithographic model (Sect. 3.3.2.2). The basic idea of the model postulates that the fast energy input to the target surface generates a state of transient thermodynamic instability (a state of disorder, similar but not equal to a melt) which, then, has to rapidly relax. This could be considered as a competition between surface roughening by erosion and surface smoothing

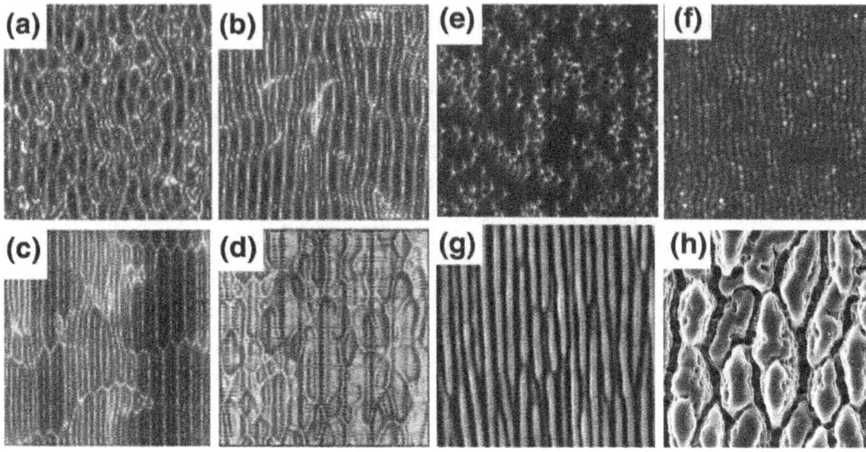

Fig. 3.26 Irradiation dose dependence of LIPSS: Comparison between simulated structures (surface dynamic model [126]; panels (**a**)–(**d**)) and experimental structures (panels (**e**)–(**h**)). The dose is gradually increased from (**a**) to (**d**) respectively from (**e**) to (**h**). Unfortunately, this is only a qualitative comparison, a quantitative comparison is not possible because of unknown material parameters, such as absorptivity, energy dissipation, etc

because of diffusion and surface tension, resulting in self-organization of the surface morphology. The process is described in the framework of nonlinear dynamics and can cover, by introducing several steps of nonlinearity as a function of feedback and degree of instability (as a function of absorbed energy dose), all types of structures observed (HSFL, LSFL, grooves, spikes), as is shown in Fig. 3.26. For more details, see [125].

References

1. K. Liu, L. Jiang, Metallic surfaces with special wettability. Nanoscale **3**, 825 (2011)
2. M. Woodson, J. Liu, Functional nanostructures from surface chemistry patterning. Phys. Chem. Chem. Phys. **9**, 207–225 (2007)
3. A. Zuzuarregui, M.C. Morant-Miñana (eds.), *Research Perspectives on Functional Micro- and Nanoscale Coatings* (IGI Global ACME, Information Science Reference; Hershey, PA, 2016)
4. J.T. Simpson, S.R. Hunter, T. Aytug, Superhydrophobic materials and coatings: a review. Rep. Progr. Phys. **78**, 086501 (2015)
5. M.F. Montemor, Functional and smart coatings for corrosion protection: a review of recent advances. Surf. Coat. Technol. **258**, 17–37 (2014)
6. J.V. Barth, G. Costantini, K. Kern, Engineering atomic and molecular nanostructures at surfaces. Nature **437**, 671–679 (2005)
7. H.-Y. Guo, Q. Li, H.-P. Zhao, K. Zhou, X.-Q. Feng, Functional map of biological and biomimetic materials with hierarchical surface structures. RSC Adv. **5**, 66901 (2015)
8. S. Ji, P.A. Ramahianti, T.-B. Nguyen, W.-D. Kim, H. Lim, Simple fabrication approach for superhydrophobic and superoleophobic Al surface. Microel. Eng. **111**, 404–408 (2013)

9. C. Acikgoz, M.A. Hempenius, J. Huskens, G.J. Vancso, Polymers in conventional and alternative lithography for the fabrication of nanostructures. Eur. Polym. J. **47**, 2033–2052 (2011)
10. S. Choo, H.-J. Choi, H. Lee, Replication of rose-petal surface structure using UV-nanoimprint lithography. Mater. Lett. **121**, 170–173 (2014)
11. E. Brinksmeier, R. Gläbe, L. Schönemann, CIRP J. Manuf. Sci. Technol. **5**, 1–7 (2012)
12. C. Teichert, Self-organized semiconductor surfaces as templates for nanostructured magnetic thin films. Appl. Phys. A **76**, 653–664 (2003)
13. A.A. Tseng, K. Chen, C.D. Chen, K.J. Ma, Electron beam lithography in nanoscale fabrication: recent development. IEEE Trans. Electron. Packag. Manuf. **26**, 141 (2003)
14. F. Watt, A.A. Bettiol, J.A. Van Kan, E.J. Teo, M.B.H. Breese, Ion beam lithography and nanofabrication: a review. Int. J. Nanosci. **4**, 269–286 (2005)
15. Y. Chen, Nanofabrication by electron beam lithography and its applications: a review. Microelectron. Eng. **135**, 57–72 (2015)
16. T. Som, D. Kanjilai (eds.), *Nano fabrication by Ion-Beam Sputtering: Fundamentals and Applications* (CRC Press, Roca Baton, Stanford, 2012)
17. J. Muñoz-García, L. Vázquez, R. Cuerno, J.A. Sánchez-García, M. Castro, R. Gago, *Self-Organized Surface Nanopatterning by Ion Beam Sputtering*, ed. by Z.M. Wang. Toward Functional Nanomaterials (Springer, Dordrecht, Heidelberg, London, New York, 2009), pp. 323–398
18. M. Castro, R. Cuerno, L. Vázquez, R. Gago, Self-Organized Ordering of Nanostructures Produced by Ion-Beam Sputtering, arXiv:cond-mat/0506453v1 [cond-mat.stat-mech] 17 June 2005
19. F. Frost, B. Ziberi, A. Schindler, B. Rauschenbach, Surface engineering with ion beams: from self-organized nanostructures to ultra-smooth surfaces. Appl. Phys. A **91**, 551–559 (2008)
20. A.F. Lasagni, *Exploring the Possibilities of Laser Interference Patterning for the Rapid Fabrication of Periodic Arrays on Macroscopic Areas*, eds. by F.A. Lasagni, A.F. Lasagni. Fabrication and Characterization in the Micro-Nano Range, Advanced Structured Materials **10** (Springer, 2011), pp. 1–27
21. H.M. van Driel, J.E. Sipe, J.F. Young, Laser-induced periodic surface structures on solids: a universal phenomenon. Phys. Rev. Lett. **49**, 1955–1958 (1982)
22. J. Bonse, S. Höhm, S.V. Kirner, A. Rosenfeld, J. Krüger, Laser-induced periodic surface structures—a scientific evergreen. IEEE J. Sel. Top. Quantum Electron. **23**, 9000615 (2017)
23. A. Vorobyev, C. Guo, Multifunctional surfaces produced by femtosecond laser pulses. J. Appl. Phys. **117**, 033103 (2015)
24. F.A. Müller, C. Kunz, S. Gräf, Bio-inspired functional surfaces based on laser-induced periodic surface structures. Materials **9**, 476 (2016)
25. K.C. Phillips, H.H. Gandhi, E. Mazur, S.K. Sundaram, Ultrafast laser processing of materials: a review. Adv. Opt. Photonics **7**, 684–712 (2015)
26. J. Reif, F. Costache, M. Bestehorn, *Self-organized surface nano-structuring by femtosecond laser processing; Chapter 9*, eds. by J. Perriere, E. Millon, E. Fogarassy. Recent Advances in Laser Processing of Materials (Elsevier, 2006) p. 275
27. J. Reif, F. Costache, M. Henyk, S.V. Pandelov, Ripples revisited: non-classical morphology at the bottom of femtosecond laser ablation craters in transparent dielectrics. Appl. Surf. Sci. **197–198**, 891–895 (2002)
28. K.M. Tanvir Ahmmed, C. Grambow, A.M. Kietzig, Fabrication of micro/nano structures on metals by femtosecond laser micromachining. Micromachines **5**, 1219–1253 (2014)
29. R. Blossey, Self-cleaning surfaces—virtual realities. Nat. Mater. **2**, 301–306 (2003)
30. A.I. Hochbaum, J. Aizenberg, Bacteria pattern spontaneously on periodic nanostructure arrays. Nano Lett. **10**, 3717–3721 (2010)
31. S. Kawata, V.M. Shalaev (eds.), *Tip Enhancement*, (Adv. Nano-Optics Nano-Photonics **2**) (Elsevier, 2017)
32. A.Y. Vorobyev, C. Guo, Direct femtosecond laser surface nano/microstructuring and its applications. Laser Photonics Rev. **7**, 385–407 (2013)

33. J. Lin, Y. Cai, X. Wang, B. Ding, J. Yu, M. Wang, Fabrication of biomimetic superhydrophobic surfaces inspired by lotus leaf and silver ragwort leaf. Nanoscale **3**, 1258 (2011)
34. L. Gao, T.J. McCarthy, Teflon is hydrophilic. Comments on definitions of hydrophobic, shear versus tensile hydrophobicity, and wettability characterization. Langmuir **24**, 9183–9188 (2008)
35. T. Young, An essay on the cohesion of fluids. Phil. Trans. R. Soc. London **95**, 65–87 (1805)
36. K. Seoh, M. Kim, D.H. Kim, *Re-derivation of Young's Equation, Wenzel Equation, and Cassie-Baxter Equation Based on Energy Minimization*, ed. by M. Aliofkhazraei. Surface Energy (InTech open, 2015)
37. O. Varlamova, K. Hoefner, M. Ratzke, J. Reif, D. Sarker, Modification of surface properties of solids by femtosecond LIPSS writing: comparative studies on Silicon and Stainless Steel. Appl. Phys. A **123**, 725 (2017)
38. J.F. Padday, A.R. Pitt, Surface and interfacial tensions from the profile of a sessile drop. Proc. R. Soc. London A **329**, 421–431 (1972)
39. R. Tadmor, Line energy and the relation between advancing, receding, and young contact angles. Langmuir **20**, 7659–7664 (2004)
40. B.D. Wilts, B.A. Zubiri, M.A. Klatt, B. Butz, M.G. Fischer, S.T. Kelly, E. Spiecker, U. Steiner, G.E. Schröder-Turk, Butterfly gyroid nanostructures as a time-frozen glimpse of intracellular membrane development. Sci. Adv. **3**, e1603119 (2017)
41. B. Dusser, Z. Sagan, H. Soder, N. Faure, J.P. Colombier, M. Jourlin, E. Audouard, Controlled nanostructrures formation by ultra fast laser pulses for color marking. Opt. Express **18**, 2913–2924 (2010)
42. J.-P. Donnet, T.-K. Wang, Surface microstructure of carbon black: advances in characterization by scanning tunneling microscopy. Macromol. Symp. **108**, 97–109 (1996)
43. W. Niedermeier, H. Raab, J. Stierstorfer, S. Kreitmeier, D. Göritz, The microstructure of carbon black investigated by atomic force microscopy. Kautsch. Gummi Kunstst. **47**, 799–805 (1994)
44. A. Ignatiev, P. O'Neill, G. Zajac, The surface microstructure optical properties relationship in solar absorbers: black chrome. Solar Energy Mater. **1**, 69–79 (1979)
45. Y. Yan, H.-C. Yuan, V.E. Yost, K. Jones, M. Al-Jassim, H.M. Branz, Microstructure and surface chemistry of nanoporous "Black Silicon" for photovoltaics, in *Proc. 35th IEEE Photovoltaic Specialists Conference Rec.*, pp. 2255–2257, 2010
46. C. Wu, C.H. Crouch, L. Zhao, J.E. Carey, R. Younkin, J.A. Levinson, E. Mazur, R.M. Farrell, P. Gothoskar, A. Karger, Near-unity below-band-gap absorption by microstructured silicon. Appl. Phys. Lett. **78**, 1850 (2001)
47. C.A. Spindt, A thin-film field-emission cathode. J. Appl. Phys. **39**, 3504 (1968)
48. S. Tsujino, F. le Pimpec, J. Raabe, M. Buess, M. Dehler, E. Kirk, J. Gobrecht, A. Wrulich, Static and optical field enhancement in metallic nanotips studied by two-photon photoemission microscopy and spectroscopy excited by picosecond laser pulses. Appl. Phys. Lett. **94**, 093508 (2009)
49. A. Mustonen, P. Beaud, E. Kirk, Th Feurer, S. Tsujino, Five picocoulomb electron bunch generation by ultrafast laser-induced field emission from metallic nano-tip arrays. Appl. Phys. Lett. **99**, 103504 (2011)
50. A.H. Zewail, 4D ultrafast electron diffraction, crystallography, and microscopy. Annu. Rev. Phys. Chem. **57**, 65–103 (2006)
51. H.-H. Wang, C.-Y. Liu, S.-B. Wu, N.-W. Liu, C.-Y. Peng, T.-H. Chan, C.-F. Hsu, J.-K. Wang, Y.-L. Wang, Highly Raman-enhancing substrates based on silver nanoparticle arrays with tunable sub-10 nm gaps. Adv. Mater. **18**, 491 (2006)
52. H. Ko, S. Singamaneni, V.V. Tsukruk, Nanostructured surfaces and assemblies as SERS media. Small **4**(10), 1576–1599 (2008)
53. T.-H. Her, R.J. Finlay, C. Wu, S. Deliwala, E. Mazur, Microstructuring of silicon with femtosecond laser pulses. Appl. Phys. Lett. **73**, 1673 (1998)
54. A. Karatas, A.H. Algan, Template synthesis of tubular nanostructures for loading biologically active molecules. Curr. Top. Med. Chem. **17**, 1555–1563 (2017)

55. A.K. Epstein, A.I. Hochbaum, P. Kim, J. Aizenberg, Control of bacterial biofilm growth on surfaces by nanostructural mechanics and geometry. Nanotechnology **22**, 494007 (2011)
56. M.V. Graham, N.C. Cady, Nano and microscale topographies for the prevention of bacterial surface fouling. Coatings **4**, 37–59 (2014)
57. F. Hizal, I. Zhuk, S. Sukhishvili, H.J. Busscher, H.C. van der Mei, C.-H. Choi, Impact of 3D hierarchical nanostructures on the antibacterial efficacy of a bacteria-triggered self-defensive antibiotic coating. ACS Appl. Mater. Interfaces. **7**, 20304–20312 (2015)
58. A. Kritharidou, Z. Georgoussi, C. Tsamis, E. Makarona, Zinc oxide nanostructures as low-cost templates for neuronal circuit. Proc. SPIE **8765**, 87650G (2013)
59. J. Kim, J.W. Grate, P. Wang, Nanostructures for enzyme stabilization. Chem. Eng. Sci. **61**, 1017–1026 (2006)
60. N. Carlsson, H. Gustafsson, C. Thörn, L. Olsson, K. Holmberg, B. Åkerman, Enzymes immobilized in mesoporous silica: a physical–chemical perspective. Adv. Colloid Interface Sci. **205**, 339–360 (2014)
61. W. Yang, P.J. Lopez, G. Rosengarten, Diatoms: self assembled silica nanostructures, and templates for bio/chemical sensors and biomimetic membranes. Analyst **136**, 42–53 (2011)
62. N. Epperlein, F. Menzel, K. Schwibbert, R. Koter, J. Bonse, J. Sameith, J. Krüger, J. Toepel, Influence of femtosecond laser produced nanostructures on biofilm growth on steel. Appl. Surf. Sci. **418**, 420–424 (2017)
63. D. Köhne, T. Fricke-Begemann, R. Weichenhain-Schriever, J. Ihlemann, Large area silica nano grids by homogeneous high resolution laser patterning of SiOx-films. JLMN **10**, 158–161 (2015)
64. D. Huerta-Murillo, A.I. Aguilar-Morales, S. Alamri, J.T. Cardoso, R. Jagdheesh, A.F. Lasagni, J.L. Ocaña, Fabrication of multi-scale periodic surface structures on Ti-6Al-4 V by direct laser writing and direct laser interference patterning for modified wettability applications. Opt. Lasers Eng. **98**, 134–142 (2017)
65. M. Bieda, M. Siebold, A.F. Lasagni, Fabrication of sub-micron surface structures on copper, stainless steel and titanium using picosecond laser interference patterning. Appl. Surf. Sci. **387**, 175–182 (2016)
66. S. Indrišiūnas, B. Voisiat, A. Žukauskas, G. Račiukaitis, Direct laser beam interference patterning technique for fast high aspect ratio surface structuring. Proc. SPIE **9359**, 935003-1 (2015)
67. M. Gedvilas, B. Voisiat, S. Indrišiūnas, G. Račiukaitis, V. Veiko, R. Zakoldaev, D. Sinev, E. Shakhno, Thermo-chemicalmicrostructuring of thinmetal films using multi-beam interference by short (nano- & picosecond) laser pulses. Thin Solid Films **634**, 134–140 (2017)
68. M. Birnbaum, Semiconductor surface damage produced by ruby lasers. J. Appl. Phys. **36**, 3688–3689 (1965)
69. D.C. Emmony, R.P. Howson, L.J. Willis, Laser mirror damage in germanium at 10.6 μm. Appl. Phys. Lett. **23**, 598–600 (1973)
70. M. Siegrist, G. Kaech, F.K. Kneubühl, Formation of a periodic wave structure on the dry surface of a solid by TEA-CO_2 laser pulses. Appl. Phys. **2**, 45–46 (1973)
71. A.K. Jain, V.N. Kulkarni, D.K. Sood, J.S. Uppal, Periodic surface ripples in laser-treated aluminum and their use to determine absorbed power. J. Appl. Phys. **52**, 4882–4884 (1981)
72. F. Keilmann, Y.H. Bai, Periodic surface structures frozen into CO_2 laser-melted quartz. Appl. Phys. A **29**, 9–18 (1982)
73. Z. Guosheng, P.M. Fauchet, A.E. Siegman, Growth of spontaneous periodic surface structures on solids during laser illumination. Phys. Rev. B **26**, 5366–5381 (1982)
74. J.E. Sipe, J.F. Young, J.S. Preston, H.M. van Driel, Laser-induced periodic surface structure. I. Theory. Phys. Rev. B **27**, 1141–1154 (1983)
75. J.F. Young, J.S. Preston, H.M. van Driel, J.E. Sipe, Laser-induced periodic surface structure. II. Experiments on Ge, Si, Al, and brass. Phys. Rev. B **27**, 1155–1172 (1983)
76. J.F. Young, J.E. Sipe, H.M. van Driel, Laser-induced periodic surface structure. III. Fluence regimes, the role of feedback, and details of the induced topography in germanium. Phys. Rev. B **30**, 2001–2015 (1984)

77. M. Bolle, S. Lazare, M. Le Blanc, A. Wilmes, Submicron periodic structures produced on polymer surfaces with polarized excimer laser ultraviolet radiation. Appl. Phys. Lett. **60**(1992), 674–676 (1992)
78. A.M. Ozkan, A.P. Malshe, T.A. Railkar, W.D. Brown, M.D. Shirk, P.A. Molian, Femtosecond laser-induced periodic structure writing on diamond crystals and microclusters. Appl. Phys. Lett. **75**, 3716–3718 (1999)
79. A. Borowiec, H.K. Haugen, Subwavelength ripple formation on the surfaces of compound semiconductors irradiated with femtosecond laser pulses. Appl. Phys. Lett. **82**, 4462–4464 (2003)
80. J. Wang, C. Guo, Ultrafast dynamics of femtosecond laser-induced periodic surface pattern formation on metals. Appl. Phys. Lett. **87**, 251914 (2005)
81. E. Rebollar, J.R. Vazquez de Aldana, J.A. Perez-Hernandez, T.A. Ezquerra, P. Moreno, M. Castillejo, Ultraviolet and infrared femtosecond laser induced periodic surface structures on thin polymer films. Appl. Phys. Lett. **100**, 041106 (2012)
82. A.Y. Vorobyev, Chunlei Guo, Femtosecond laser structuring of titanium implants. Appl. Surf. Sci. **253**, 7272–7280 (2007)
83. P. Gregorčič, M. Sedlaček, B. Podgornik, J. Reif, Formation of laser-induced periodic surface structures (LIPSS) on toolsteel by multiple picosecond laser pulses of different polarizations. Appl. Surf. Sci. **387**, 698–706 (2016)
84. K.R.P. Kafka, D.R. Austin, H. Li, A.Y. Yi, J. Cheng, E.A. Chowdhury, Time-resolved measurement of single pulse femtosecond laser-induced periodic surface structure formation induced by a pre-fabricated surface groove. Opt. Express **23**, 19432–19441 (2015)
85. J. Reif, O. Varlamova, S. Uhlig, S. Varlamov, M. Bestehorn, On the physics of self-organized nanostructure formation upon femtosecond laser ablation. Appl. Phys. A **117**, 179–184 (2014)
86. E.L. Gurevich, Mechanisms of femtosecond LIPSS formation induced by periodic surface temperature modulation. Appl. Surf. Sci. **374**, 56–60 (2016)
87. O. Varlamova, F. Costache, M. Ratzke, J. Reif, Control parameters in pattern formation upon femtosecond laser ablation. Appl. Surf. Sci. **253**, 7932–7936 (2007)
88. J. Reif, *Basic Physics of Femtosecond Laser Ablation, Ch. 2*, eds. by A. Miotello, P. Ossi. Laser-Surface Interactions for New Materials Production; Springer Ser. Mater. Sci. **130** (Springer, Berlin Heidelberg, 2010)
89. J. Reif, O. Varlamova, F. Costache, Femtosecond laser induced nanostructure formation: self-organization control parameters. Appl. Phys. A **92**, 1019–1024 (2008)
90. J. Reif, *Processing with Ultrashort Laser Pulses*, ch. 6, ed. by P. Schaaf. Laser Processing of Materials, Springer Ser. Mater. Sci. vol 139 (Springer, Berlin, Heidelberg, 2010), pp. 113–129
91. O. Varlamova, C. Martens, M. Ratzke, J. Reif, Genesis of femtosecond-induced nanostructures on solid surfaces. Appl. Opt. **53**, I 10-I 15 (2014)
92. O. Varlamova, M. Bounhalli, J. Reif, Influence of irradiation dose on laser-induced surface nanostructures on silicon. Appl. Surf. Sci. **278**, 62–66 (2002)
93. A.-M. Kietzig, M.N. Mirvakili, S. Kamal, P. Englezos, S.G. Hatzikiriakos, Laser-patterned super-hydrophobic pure metallic substrates: Cassie to Wenzel Wetting Transitions. J. Adhes. Sci. Technol. **25**, 2789–2809 (2011)
94. J. Gottmann, R. Wagner, Sub-wavelength ripple formation on dielectric and metallic materials induced by tightly focused femto-second laser radiation. Proc. SPIE **6106**, 61061R (2006)
95. M. Gedvilas, G. Račiukaitis, K. Regelskis, P. Gečys, Formation of gratings by self-organization of the chromium thin film on the glass substrate under irradiation with laser pulses. JLMN **3**, 58–62 (2008)
96. M. Hörstmann-Jungemann, J. Gottmann, D. Wortmann, Nano- and microstructuring of SiO_2 and sapphire with Fs-laser induced selective etching. JLMN **4**, 135–140 (2009)
97. M. Zamfirescu, M. Ulmeanu, F. Jipa, O. Cretu, A. Moldovan, G. Epurescu, M. Dinescu, R. Dabu, Femtosecond laser induced periodic surface structures on ZnO thin films. JLMN **4**, 7–10 (2009)
98. J. Reif, *Material Response to Laser Energy Deposition (Thermal and Hyperthermal Processes)*, ch. 2 eds. by M. Castillejo, P. Ossi, L. Zhigilei. Lasers in Materials Science; Springer Ser. Mater. Sci. vol 191 (Springer, Berlin, Heidelberg, 2014)

99. V. Zorba, L. Persano, D. Pisignano, A. Athanassiou, E. Stratakis, R. Cingolani, P. Tzanetakis, C. Fotakis, Making silicon hydrophobic: wettability control by two-lengthscale simultaneous patterning with femtosecond laser irradiation. Nanotechnology **17**, 3234 (2006)
100. M. Groenendijk, J. Meijer, Microstructuring using femtosecond pulsed laser ablation. J. Laser Appl. **18**, 227–235 (2006)
101. M. Groenendijk, Fabrication of super hydrophobic surfaces by fs laser pulses. Laser Technol. J. **5**, 44–47 (2008)
102. M. Barberoglou, V. Zorba, E. Stratakis, E. Spanakis, P. Tzanetakis, S.H. Anastasiadis, C. Fotakis, Bio-inspired water repellent surfaces produced by ultrafast laser structuring of silicon. Appl. Surf. Sci. **255**, 5425–5429 (2009)
103. P. Bizi-Bandoki, S. Benayoun, S. Valette, B. Beaugiraud, E. Audouard, Modifications of roughness and wettability properties of metals induced by femtosecond laser treatment. Appl. Surf. Sci. **257**, 5213–5218 (2011)
104. A.Y. Vorobyev, C. Guo, Metal pumps liquid uphill. Appl. Phys. Lett. **94**, 224102 (2009)
105. A.Y. Vorobyev, C. Guo, Laser turns silicon superwicking. Opt. Express **18**, 6455–6460 (2010)
106. J.J. Yu, Y.F. Lu, Laser-induced ripple structures on Ni-P substrates. Appl. Surf. Sci. **148**, 248–252 (1999)
107. A. Mizuno, T. Honda, J. Kikuchi, Y. Iwai, N. Yasumaru, K. Miyazaki, Friction properties of the DLC film with periodic structures in nano-scale. Tribol. Online **1**, 44–48 (2006)
108. N. Yasumaru, K. Miyazaki, J. Kiuchi, Control of tribological properties of diamond-like carbon films with femtosecond-laser-induced nanostructuring. Appl. Surf. Sci. **254**, 2364–2368 (2008)
109. C.Y. Chen, C.J. Chung, B.H. Wu, W.L. Li, C.W. Chien, P.H. Wu, C.W. Cheng, Microstructure and lubricating property of ultra-fast laser pulse textured silicon carbide seals. Appl. Phys. A **107**, 345–350 (2012)
110. J. Eichstadt, G.R.B.E. Romer, A.J. Huis in't Veld, Towards friction control using laser-induced periodic surface structures. Phys. Procedia **12**, 7–15 (2011)
111. S. Hammouti, A. Pascale-Hamri, N. Faure, B. Beaugiraud, M. Guibert, C. Mauclair, S. Benayoun, S. Valette, Wear rate control of peek surfaces modified by femtosecond laser. Appl. Surf. Sci. **357**, 1541–1551 (2015)
112. J. Dufils, F. Faverjon, C. Héau, C. Donnet, S. Benayoun, S. Valette, Combination of laser surface texturing and DLC coating on PEEK for enhanced tribological properties. Surf. Coat. Technol. **329**, 29–41 (2017)
113. J. Bonse, R. Koter, M. Hartelt, D. Spaltmann, S. Pentzien, S. Höhm, A. Rosenfeld, J. Krüger, Tribological performance of femtosecond laser-induced periodic surface structures on titanium and a high toughness bearing steel. Appl. Surf. Sci. **336**, 21–27 (2015)
114. J. Bonse, S. Höhm, R. Koter, M. Hartelt, D. Spaltmann, S. Pentzien, A. Rosenfeld, J. Krüger, Tribological performance of sub-100-nm femtosecond laser-induced periodic surface structures on titanium. Appl. Surf. Sci. **374**, 190–196 (2016)
115. J. Reif, O. Varlamova, M. Ratzke, M. Schade, H.S. Leipner, Tz Arguirov, Multipulse feedback in self-organized ripples formation upon femtosecond laser ablation from silicon. Appl. Phys. A **101**, 361–365 (2010)
116. E. Rebollar, M. Sanz, S. Pérez, M. Hernández, I. Martín-Fabiani, D.R. Rueda, T.A. Ezquerra, C. Domingo, M. Castillejo, Gold coatings on polymer laser induced periodic surface structures: assessment as substrates for surface-enhanced Raman scattering. Phys. Chem. Chem. Phys. **14**, 15699–15705 (2012)
117. H. Messaoudi, S. Kumar Das, J. Lange, F. Heinrich, S. Schrader, M. Frohme, R. Grunwald, *Femtosecond-Laser Induced Periodic Surface Structures for Surface Enhanced Raman Spectroscopy of Biomolecules*, eds. by S. Sakabe, C. Lienau, R. Grunwald. Progress in Nonlinear Nano-Optics (Springer, Cham, 2015), pp. 207–2019
118. R. Buividas, P. Stoddart, S. Juodkazis, Laser fabricated ripple substrates for surface-enhanced Raman scattering. Ann. Phys. **524**, L5–L10 (2012)
119. R. Vilar (ed.), Laser Surface Modification of Biomaterials—Techniques and Applications; Woodhead Publ. Ser. Biomat. **111** (Elsevier, Duxford, 2016)

120. Y.F. Guan, A.V. Melechko, A.J. Pedraza, L.R. Baylor, P.D. Rack, Maskless synthesis of Vertically Aligned Carbon Nanofibers (VACNF) field emission arrays on a Laser-Induced Periodic Surface Structures (LIPSS) template, in *Proceeding of 20th International Vacuum Nanoelectronics Conference*, pp. 143–144, 2007
121. A. Ranella, M. Barberoglou, S. Bakogianni, C. Fotakis, E. Stratakis, Tuning cell adhesion by controlling the roughness and wettability of 3D micro/nano silicon structures. Acta Biomater. **6**, 2711–2720 (2010)
122. E.L. Papadopoulou, A. Samara, M. Barberoglou, A. Manousaki, S.N. Pagakis, E. Anastasiadou, C. Fotakis, E. Stratakis, Silicon scaffolds promoting three-dimensional neuronal web of cytoplasmic processes. Tissue Eng. **16**, 497–502 (2010)
123. T. Shinonaga, M. Tsukamoto, T. Kawa, P. Chen, A. Nagai, T. Hanawa, Formation of periodic nanostructures using a femtosecond laser to control cell spreading on titanium. Appl. Phys. B **119**, 493–496 (2015)
124. J. Bonse, A. Rosenfeld, J. Krüger, On the role of surface plasmon polaritons in the formation of laser-induced periodic surface structures upon irradiation of silicon by femtosecond-laser pulses. J. Appl. Phys. **106**, 104910 (2009)
125. O. Varlamova, J. Reif, S. Varlamov, M. Bestehorn, *Self-organized Surface Patterns Originating from Laser-Induced Instability*; ch. 1 eds. by S. Sakabe, C. Lienau, R. Grunwald. *Progress in Nonlinear Nano-Optics*; Springer Ser. Nano-Opt. Nanophot, vol 2 (2015)
126. J. Reif, O. Varlamova, S. Varlamov, M. Bestehorn, 2012, The role of asymmetric excitation in self-organized nanostructure formation upon femtosecond laser ablation. AIP Conf. Proc. **1464**, 428–441 (2012)

Chapter 4
Laser-Inducing Extreme Thermodynamic Conditions in Condensed Matter to Produce Nanomaterials for Catalysis and the Photocatalysis

Alberto Mazzi, Michele Orlandi, Nainesh Patel and Antonio Miotello

Abstract Vaporization from the aluminum target surface, under nanosecond laser irradiation, was evaluated in the framework of the unsteady adiabatic expansion model, while the homogeneous nucleation of vapor bubbles in the metastable liquid (phase explosion) was simulated in the framework of the classical nucleation theory. The size distribution of the liquid nanodroplets produced in the phase explosion process was found to obey a power law in agreement with the few available experimental data when it is assumed that nanoparticles formation comes from solidification of liquid nanodroplets. Some experimental examples are reported to show that pulsed-laser deposition technique is able to synthesize nanoparticles in a single step with the required features for catalysis and photocatalysis applications.

A. Mazzi
Center for Materials and Microsystems, Fondazione Bruno Kessler, 38123 Povo (Trento), Italy
e-mail: mazzi@fbk.eu

M. Orlandi · A. Miotello (✉)
Dipartimento di Fisica, Università degli Studi di Trento, 38123 Povo (Trento), Italy
e-mail: antonio.miotello@unitn.it

M. Orlandi
e-mail: michele.orlandi@unitn.it

N. Patel
Department of Physics and National Centre for Nanosciences and Nanotechnology, University of Mumbai, Vidyanagari, Santacruz (E), Mumbai 400098, India
e-mail: nainesh11@gmail.com

P. M. Ossi (ed.), *Advances in the Application of Lasers in Materials Science*,
Springer Series in Materials Science 274,
https://doi.org/10.1007/978-3-319-96845-2_4

4.1 Introduction

Catalysts in form of nanocluster-assembled coating are under intense investigation in the catalysis community due to their exceptional activity and selective nature in catalytic processes as compared to the corresponding bulk counterpart. Large surface-to-volume atomic ratio, size- and shape-dependent properties, and high concentration of low-coordinated active surface sites are responsible for such remarkable performance. Here we report on how pulsed laser deposition (PLD) technique is able to synthesize nanoparticles (NPs) in a single step with the required relevant features for catalysis application. First of all we discuss a modeling of the phase explosion mechanism induced by the nanosecond laser irradiation of pure metals. Our work is based on a continuum description of the target material, following a thermodynamic study of the metastable liquid metals at temperature close to the thermodynamic critical point, T_c. Finally, an example, recently obtained in our lab, is presented that consists of a PLD produced nanosized Fe(III)-based water oxidation catalyst (WOC) for photoanodes functionalization.

4.2 Mechanisms Involved in PLD to Synthesize NPs

The mechanisms involved in PLD, provided electron-lattice thermalization has occurred in the target, have been widely investigated and clarified along the years as summarized in many published papers and books [1–8]. In previous studies, it was established that the following thermal processes may lead to material loss from a laser-irradiated surface: vaporization, normal boiling, explosive boiling, and spallation. The relevance of these processes depends on the laser pulse duration as well as on the temperature attained in the irradiated zone. However, vaporization is not important for the shortest time-scales (<1 ns), normal boiling is subject to a major kinetic obstacle in the process of gas bubble diffusion, such motion being sufficiently slow that it will simply not occur for $t < 100$ ns. Phase explosion, notwithstanding the unfavorable time-scale (1–100 ns) advocated by Martynyuk [2], was found to be the most efficient mechanism in the ablation process when looking at thermal processes.

An important aspect related to PLD, is the synthesis of NPs. With PLD, NPs synthesis occurs via direct cluster ejection [9, 10] from the target (phase explosion) or collisional sticking and aggregation in the ablated plume flow, favored by inert gas intentionally introduced into the deposition chamber. To deposit NPs in vacuum (to avoid contamination with ambient gas), we have to look at the phase explosion process. In the pressure–temperature (p–T) phase diagram, the boiling temperature is established by the binode line (calculated from the Clausius-Clapeyron equation) where the liquid and vapor phases are in equilibrium. Under rapid heating it is possible to superheat the liquid above the boiling point: however there is a well defined upper limit for the superheating of a liquid, the spinode. Using the Berthelot equation, the spinode function is determined by posing two conditions: (1) the pressure derivative as function of volume at constant T is zero, and (2) the temperature derivative as function of entropy at fixed pressure is zero. This means that the second derivatives of the Gibbs thermodynamic potential are zero. When T approaches T_c, the fluctuations Δv (volume) and Δh (enthalpy) increase sharply. The decrease of density is mainly due to the intensified fluctuation of the specific volume and the increase of the specific heat is mainly due to the increasing fluctuation of Δh. The normal boiling involves heterogeneous nucleation at a temperature only minimally higher than the boiling temperature. On the contrary, in the superheating process, when the temperature lies sufficiently near T_c ($T/T_c \geq 0.80$), the rate of homogeneous nucleation increases dramatically and phase explosion occurs by homogeneous nucleation: the hot region breaks down into vapor plus liquid droplets that are violently expelled out.

4.3 Thermodynamic Modeling of Phase Explosion in the Nanosecond Laser Ablation of Metals

Here we try to connect the theory of homogeneous nucleation to the size distribution of the liquid nanodroplets formed in the phase explosion process [11] and implement the approach by considering a 20 ns full width half maximum (FWHM) laser pulse irradiating an aluminum bulk target [12]. We numerically solved the heat diffusion problem in the target external layers through a finite difference method. Laser ablation was modeled by the progressive removal of discrete layers, considering both normal vaporization and phase explosion. Our simulations describe the formation of liquid nanodroplets through the phase explosion mechanism, with temporal and spatial resolution. The composition of the sputtered material is studied as a function of time and the size distribution of the produced aluminum nanodroplets is reported.

4.3.1 Thermodynamics of Metastable Liquid Metals

In order to develop quantitative simulations of the nanosecond laser ablation of metals, we needed to consider semi-empirical scaling laws available in literature to describe the temperature dependence of the main thermodynamic quantities of liquid metals in the deep metastable regime.

Indeed, under rapid laser heating, the target external layers are melted and they can reach a temperature well above the normal boiling point T_b for short time intervals, without undergoing a phase transition. These states lie in the portion of phase diagram enclosed by the binodal and the spinodal curves [13], which merge at the thermodynamic critical point T_c. The metastable regime corresponds, for metals commonly employed in laser ablation experiments, to a temperature higher than 3000 K (see Table 4.1).

Table 4.1 Phase transition temperatures and critical parameters for a set of metals [11]

Metal	T_m (K)	T_b (K)	T_c (K)	ρ_c (kg/m^3)	P_c (10^8 Pa)
Al	933	2792	6319	634	3.2
Fe	1811	3134	8059	1467	5.4
Co	1768	3200	7710	1350	5.4
Ni	1728	3186	7241	2159	6.5
Cu	1358	2835	5741	2363	4.6
Ag	1235	2435	5851	2718	3.3
Au	1337	3129	7003	5066	3.9

The thermodynamic quantities, in the vicinity of T_c, are usually modeled according to the theory of critical exponents as functions of the expansion parameter $\Delta T = (T_c - T)/T_c$.

The density of the liquid and vapor phase ρ_l and ρ_v in the laser-irradiated target was assumed to follow the coexistence curve as discussed in our recent study [11].

A key quantity that regulates the energetics of the liquid–vapor phase transition is the enthalpy of vaporization Δh_v, which was approximated by the Watson scaling law [14].

The binodal curve, which represents the equilibrium phase change, was determined by integrating numerically the Clausius–Clapeyron equation. In this way, the temperature dependence of the saturated vapor pressure p_s is obtained.

As a semi-empirical model for the surface tension σ , we adopted the Guggenheim formula [15], by using the experimental data collected in a comprehensive review by Keene [16].

Moreover, in order to treat the dynamics of the thermal processes, we needed to estimate the thermal conductivity K of the metastable liquid metals. To this aim, we assumed the thermal conductivity of liquid metals to be entirely due to the electronic contribution. Then we considered the Wiedemann–Franz law, which expresses the electronic contribution to the thermal conductivity as a function of the electrical conductivity and which was proven to be valid for Fermi liquids up to the metal–insulator transition [17]. In turn, we adopted a model for the electrical conductivity of metastable liquid metals proposed by Korobenko et al. [18], based on the Drude—Sommerfeld free electron model.

Finally, the isobaric specific heat capacity of the liquid phase $c_{p,l}$ was expressed as a function of the temperature through a semi-empirical scaling law [19].

The thermophysical quantities discussed here are reported in Fig. 4.1 as function of the expansion parameter ΔT for aluminum.

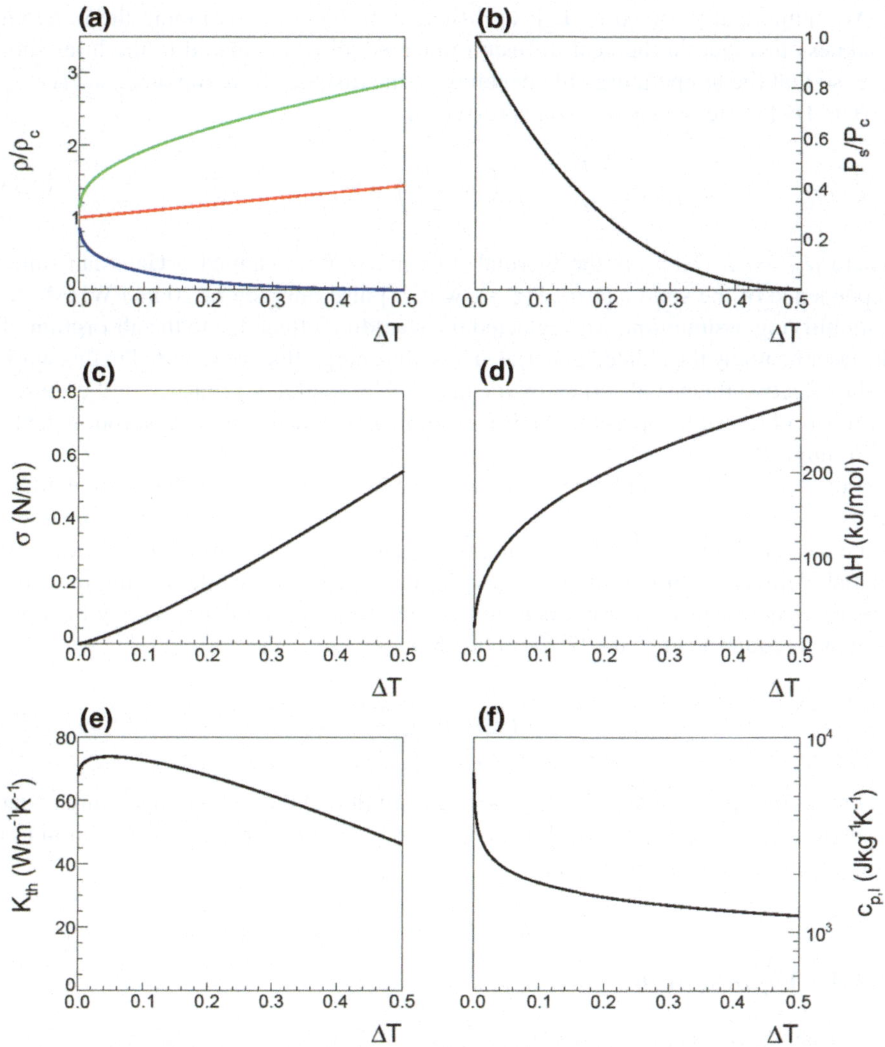

Fig. 4.1 Thermophysical quantities calculated for aluminum as functions of the expansion parameter ΔT. **a** Coexistence curve: ρ_l in green, ρ_v in blue and $\rho_d = (\rho_l + \rho_v)/2$ in red, normalized to the critical density, ρ_c. **b** Saturated vapor pressure normalized to the critical pressure, p_c. Surface tension, enthalpy of vaporization, thermal conductivity and isobaric specific heat capacity are plotted in panels (**c**), (**d**), (**e**) and (**f**), respectively. In particular, the curves in (**a**), (**c**), (**d**) and (**f**) clearly show the peculiar features of the near-critical region

4.3.2 Heat Diffusion Problem

The results reported here were obtained by considering pure aluminum as the irradiated material.

We numerically solved a 1D heat diffusion problem, by assuming the in depth thickness involved in the heat diffusion much small as compared to the laser spot size, so that the temperature only depends on the distance from the outer surface.

The 1D Fourier equation was expressed as:

$$\frac{\partial T}{\partial t} = \frac{\partial}{\partial z}\left(\alpha_{\text{th}}\frac{\partial T}{\partial z}\right) + S(z,t), \tag{4.1}$$

where $\alpha_{\text{th}} = K/\rho_l c_{p,l}$ is the thermal diffusivity. We assumed a Gaussian time-dependence of the source term, $S(z,t)$, with a pulse duration of 20 ns FWHM. As a simplifying assumption, we neglected the shielding effect due to the absorption of the laser beam by the ablated material. Thus, the energy fluence reported in this work is the effective fluence absorbed by the target. The spatial dependence of the source term was obtained by applying the Beer–Lambert law with a fixed absorption depth of 20 nm [20].

Equation (4.1) is a nonlinear partial differential equation, since the thermal diffusivity depends on the temperature. For the numerical solution, we chose a Forward Time Centered Space (FTCS) [21] algorithm using a non-uniform 1D grid with 140 spatial elements with size increasing with the cubic power, while moving inwards. The time increments ($\mathrm{d}t$) were determined step-by-step in order to satisfy the convergence requirement of the FTCS algorithm:

$$\mathrm{d}t \le \min_j \frac{\mathrm{d}z_j^2}{2\alpha_j}, \tag{4.2}$$

where j is the spatial index. The space–time evolution of the target temperature under a single laser pulse is presented in Fig. 4.2 in the case of irradiation of aluminum with an energy fluence of 3 J/cm^2.

4.3.3 Vaporization

Our simulations take into account two of the main laser ablation mechanisms: vaporization and phase explosion. In particular, to calculate the vaporization yield, we adopted the model developed by Kelly [22] which considers the gas dynamics of the vaporized material from the Knudsen layer (KL), where the vapor atoms thermalize to the unsteady adiabatic expansion that leads the sputtered material away from the target. That method assumes the flow velocity at the boundary of the KL to be equal to the speed of sound:

$$u_K \approx \left(\frac{\gamma k_B T_K}{m}\right)^{\frac{1}{2}} = \left(\frac{5 k_B T_K}{3m}\right)^{\frac{1}{2}}, \tag{4.3}$$

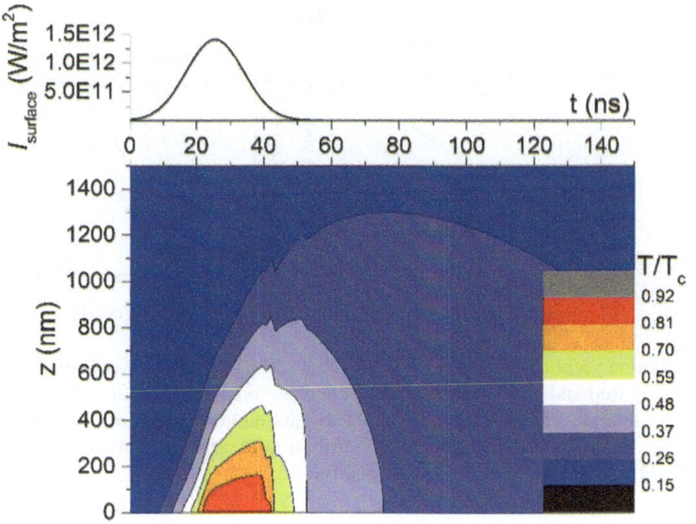

Fig. 4.2 Time dependence of the incoming power density and contourplot of the temperature of liquid Al as a function of time and depth obtained in the case of energy fluence of 3 J/cm^2

where the subscript K refers to the outer boundary of the KL and the heat capacity ratio was taken as $\gamma = 5/3$ as in the case of the ideal gas. As originally obtained by Anisimov [23], the temperature and density ratio across the KL are given by $T_K = 0.669 T_s$ and $\rho_K = 0.308\rho_s$, where T_s is the temperature of the liquid surface, while ρ_s indicates the saturated vapor density at temperature T_s.

The pressure of the liquid surface under the laser ablation process (usually indicated as recoil pressure) was obtained through the scaling law $p_l = 0.55 p_s(T_s)$ that was proven in [23, 24] to be valid in the case of intense vaporization.

Finally, by applying the conservation relation $\rho_l u_l = \rho_K u_K$, we obtained an estimate of the surface recession velocity u_l, which directly depends on the vaporization rate.

The vaporization yield and rate obtained for a single 20 ns shot on aluminum are reported in Fig. 4.3 as a function of both energy fluence and time.

4.3.4 Phase Explosion

A metallic surface, irradiated by a high power nanosecond laser pulse, undergoes an extremely rapid temperature increase with the nanosecond timescale allowing the lattice thermalization. The heating rate of the target external layers, at an energy fluence of some J/cm^2, commonly employed in the nanosecond laser ablation of metals, is usually of about $10^{11} \div 10^{12}$ K/s. At such high heating rates, it has been proven that the normal boiling process due to the heterogeneous nucleation of vapor

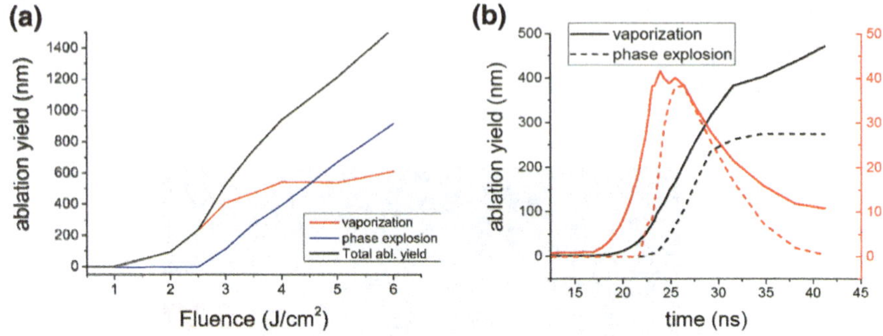

Fig. 4.3 **a** Ablation yield calculated for a single shot (20 ns FWHM) on aluminum as a function of energy fluence. Red and blue line indicate the contribution of vaporization and phase explosion, respectively. **b** Ablation yield (black plot, left axis) and ablation rate (red plot, right axis) as a function of time obtained at 3.5 J/cm^2. The contribution of vaporization and phase explosion is reported with solid lines and dashed lines, respectively

bubbles has a negligible kinetics [25]. Thus, the liquid phase becomes metastable, overcoming the boiling temperature, and it undergoes homogeneous vapor bubble nucleation when the temperature approaches the spinodal line, at approximately $0.9T_c$.

In our phase explosion model, the vapor pressure in the nucleated vapor bubbles p_v was calculated through the expression reported by Debenedetti [26] in the framework of the classical theory of homogeneous nucleation:

$$p_v = p_s \exp\left[(p_l - p_s)\frac{m}{\rho_l k_B T}\right]. \tag{4.4}$$

We also assumed that the vapor bubbles have a spherical shape, and that they are nucleated in labile equilibrium with the metastable liquid. At a given temperature, the initial size of the vapor bubbles is given by the critical radius:

$$r_c = \frac{2\sigma}{p_v - p_l}$$

In the simulations discussed here, we assumed the homogeneous nucleation process to be regulated by the steady-state nucleation rate. This corresponds to neglect the time lag of attainment of the steady nucleation process, τ_{lag}: this simplification, in the case of nanosecond laser heating, is consistent with the results of our recent study [27]. In this context, we can calculate the nucleation rate of near-critical vapor nuclei per unit volume through the expression reported by Skripov [28]:

$$J_s = \frac{\rho_l}{m}\sqrt{\frac{6\sigma}{(3-b)\pi m}}\, e^{-\frac{\Delta h_v}{RT}}\, e^{-\frac{W_c}{k_B T}}$$

where $b = 1 - p_l/p_v$ and $W_c = \frac{4}{3}\pi r_c^2\sigma$ is the free energy of formation of a critical nucleus.

In the simulations discussed below, we adopted the same model reported in [27] to study the single vapor expansion dynamics in the metastable liquid, derived from the original method proposed by Lee et al. [29]. In brief, since the critical radius of vapor bubbles monotonically decreases with the temperature, during the heating process a nucleated vapor bubble spontaneously becomes unstable and starts growing.

As in our previous works, here we assumed that the phase explosion occurs when the volume fraction occupied by vapor bubbles is $\eta_{\max} = V_v/V_{tot} = 0.30$, which provides a closely packed random distribution of bubbles. Finally, phase explosion results in the ejection from the target surface of a mixture of vapor and liquid nanodroplets, which are directly formed in the target.

4.3.5 Computational Framework

Our laser ablation simulations consisted in the solution of the 1D heat diffusion problem (1) through the finite difference method described above, coupled with the estimation of the vaporization rate and a 3D simulation of phase explosion, in a system having square section of 100 nm side and a depth of 10 μm along the z axis, normal to the target surface.

The temperature and the thermodynamic quantities were considered uniform within each discrete layer. The solution of the heat equation proceeded together with the calculation of the vaporization rate from the surface, given by the surface recession velocity as discussed in Sect. 4.3.3. The number of nucleated vapor bubbles was computed in each layer at any time through the steady nucleation model and, when it was not zero in the considered layer volume, the bubbles were randomly generated, initially with the critical size.

When the full depth of a layer was vaporized or when the volume fraction occupied by the vapor bubbles reached the maximum value, $\eta_{\max}$, the layer was removed and the relative amount of vapor and liquid was computed as atomic percentage. As a result, we can estimate the relative ablation rate of vapor and liquid nanodroplets as a function of time, during a laser pulse.

The heat diffusion equation is solved in one dimension even if the appearance of gas bubbles would imply a symmetry breaking. For computational simplicity, we neglect the impact of the vapor bubbles on heat conduction and light absorption.

At the end of the simulations, the distribution of the nucleated vapor bubbles over all the target depth was taken as a whole and the residual liquid inter-volumes were identified through the MC algorithm developed in our previous works [11]. In particular, this method proceeds through the identification of randomly generated spherical volumes in the remaining liquid, followed by the evaluation of the clusters of spheres that permeate each other. These sets of permeated spheres are finally identified as distinct liquid clusters, i.e. nanodroplets.

4.3.6 Results and Discussion

The laser ablation simulations described above were performed by considering 20 ns FWHM irradiation of a pure aluminum target at different energy fluences, ranging from 1.0 to 6.0 J/cm^2.

Homogeneous nucleation was found to be significant at temperature just above $0.90T_c$, as it is widely accepted in literature. We can have an idea of the space–time evolution of the nucleation process by looking at the temperature color map reported in Fig. 4.2.

The nucleation–vaporization simulation described above provided us the ablation yield, that was reported in Fig. 4.3a as a function of the energy density absorbed by the target.

From Fig. 4.3b, we can observe how the vapor–nanoparticle mixture is generated as a function of time in our simulation. Homogeneous nucleation occurs promptly during the laser pulse and, in this way, the liquid NPs are emitted through the phase explosion process.

Our approach, even with some simplifying assumptions, is able to catch some features of the physics of nanosecond laser ablation. The ablation yield starts to become significant at an absorbed energy density of 1.0 J/cm^2, followed by a fluence range in which only vaporization is effective. At 2.5 J/cm^2 we observed the phase explosion threshold, above which liquid nanodroplets are expected in the ablated material. Finally, phase explosion becomes the dominant mechanism at 4.5 J/cm^2.

In the case shown in Fig. 4.3b (3.5 J/cm^2), the nanodroplet formation process continues for several nanoseconds, during flattop surface heating, and in this way some surface layers are sequentially removed. On the other hand, the vaporization process occurs in a significantly broader temperature range, and correspondingly in a larger time interval.

As in our previous works, the size distribution of the liquid nanodroplets formed in the phase explosion process was obtained. Only the liquid layers with fully established nucleation ($\eta \approx \eta_{\max}$) were taken into account in this plot. Here we can see the droplet size histogram in a log–log plot (Fig. 4.4), which shows a power scaling law for particles containing $10^3 \div 10^4$ atoms. The power law distribution in this case can be expressed as $f(N) \propto N^{-1.9}$, where N is the number of atoms per cluster. Interestingly, the exponent seems to be quite independent on the laser fluence, since its value remains almost constant in the range between 3.0 and 6.0 J/cm^2.

This rule is consistent with our recent results obtained in the absence of spatial gradients [11] and also with the results of a molecular dynamics approach that described femtosecond laser ablation of aluminum [30].

To our knowledge, there have been some attempts to find an analogous scaling law for the size distribution of the metal NPs deposited through femtosecond laser ablation, although with poor statistics. The results of Noël et al. [31] are reported in Fig. 4.5: the experimental size distributions were obtained from Cu nanoparticles deposited on mica substrates through PLD. In that work, a Ti:Sapphire laser was used, operating at $\lambda = 800$ nm with pulse duration of 100 fs a

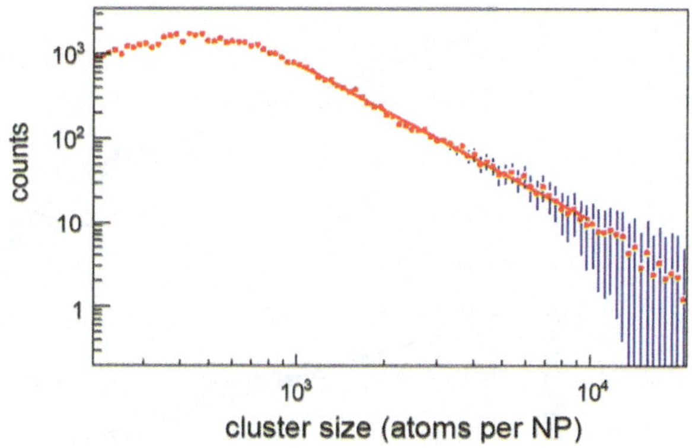

Fig. 4.4 Size distribution of the liquid nanodroplets identified in the simulation at 3.5 J/cm^2. Vertical blue lines are error bars, evaluated according to Poisson statistics

t a repetition rate of 1 kHz. Different values for the exponent of the power function were found, depending on the energy fluence, although with significant experimental uncertainties. On the other hand, these studies are missing in the case of nanosecond lasers.

In the size distribution of aluminum NPs, at an energy fluence between 3.0 and 4.0 J/cm^2, the average number of atoms per cluster was of about $\bar{N} = 450$ atoms, while the average particle size, calculated as the diameter of a spherical equivalent particle at room temperature was of about $\bar{d} = 3.2$ nm.

This average size of the liquid nanodroplets directly ejected from the target is consistent with the experimental observations of pure metal NPs synthesized by PLD in high vacuum using nanosecond laser pulses [32, 33].

In order to graphically show the spatial 3D distribution of the identified liquid nanodroplets, we report a scatter plot in Fig. 4.6, representing the liquid nanodroplets found in the simulation at 3.5 J/cm^2. In our MC method for liquid inter-volume identification, the nanodroplets are reconstructed as clusters of spherical liquid volumes trapped between the nucleated vapor bubbles. Here the spheres belonging to the same cluster are plotted in the same color, which depends on the cluster size, calculated as the diameter of an equivalent spherical nanoparticle.

In the frame reported in Fig. 4.6, the z axis points outwards from the target surface, which was moved in the negative direction during the simulation due to vaporization and phase explosion. In this way the droplets observed at larger z were formed before that at smaller z.

We can see that the larger clusters are formed as non-spherical agglomerates that will eventually evolve to spherical nanodroplets, to minimize the Gibbs free energy, during their flight away from the target or on the substrate.

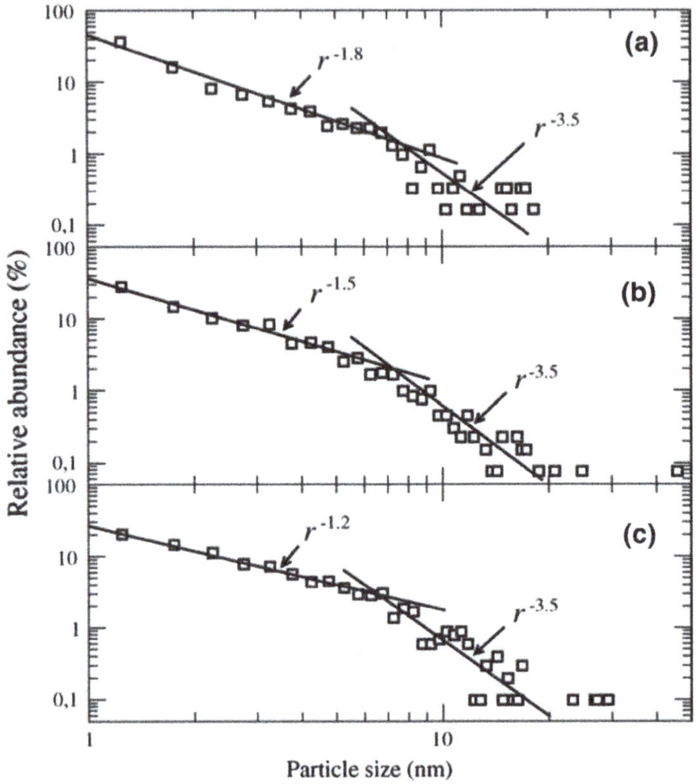

Fig. 4.5 Nanoparticle size distributions in logarithmic scale for laser ablation of copper with laser fluences of 0.8 (**a**), 2.0 (**b**), and 4.0 J/cm^2 [31]. The fitting power function is indicated in the plots as r^{-A}, where r is the nanoparticle radius

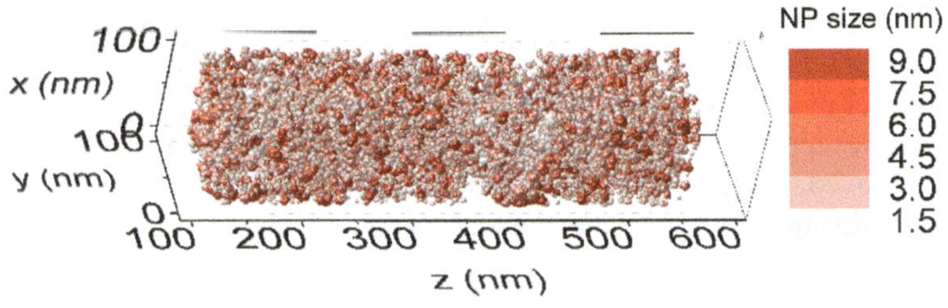

Fig. 4.6 Liquid nanodroplets 3D distribution obtained in the simulation at 3.5 J/cm^2. The nanodroplets are reported as clusters of spheres. The droplet color depends on the cluster size

4.4 Pulsed Laser Deposition of Nanostructured Catalysts: An Application for PEC (Photo-Electrochemical Cell) Technology

PEC devices are considered as prime technology for the production of solar fuels, defined as the conversion and storage of solar energy into the chemical bonds of high energy-density molecular species, such as H_2 from water splitting or CO_2 reduction products (CO, CH_4, CH_3OH, CH_2O_2, CH_2O) [34]. In its most basic implementation, a device of this kind consists of an electrochemical cell where one or both the electrodes are photoactive. The absorption of UV-visible photons at these photoelectrodes leads to the separation of electron-hole pairs which are then employed for driving catalytic fuel-forming redox reactions [35]. Recently, much effort has been dedicated towards designs where the absorption/charge-separation and the catalysis functions are optimized on different materials and then integrated by functionalization of absorbers with redox catalysts. For instance, inorganic semiconductors or molecular chromophores have been employed as absorbers and functionalized by coupling with molecular catalysts or nanostructured metals or metal-oxides [36]. This design allows to take advantage of materials with already optimized optical properties, in particular from photovoltaic applications, but also presents new problems. Of critical importance are the nature of the absorber/catalyst junction and the parasitic light blocking effect [37] by surface-deposited catalysts. Both of these are strongly affected by the catalyst morphology: for the former, dense ion-impermeable structures yield buried junctions with poor PEC performance, while porous ion-permeable architectures can lead to adaptive junctions which are known to enhance charge-separation and photovoltage generation [38]. In the latter case, compact thin films with smooth surfaces generally exhibit strong bulk-like absorption, leading to significant optical losses [39] while nanosized 3D structures can feature lower refractive indexes and/or quantum confinement effects yielding a substantially transparent material [40]. In this context, PLD offers a versatility unparalleled by other fabrication techniques, with the possibility of precisely controlling quantity, morphology and crystal phase of a nanostructured material by tuning the deposition parameters, thus enabling a systematic investigation of how these properties correlate with PEC performance. In the following section, an example recently studied in our lab is presented [41], consisting of a nanosized Fe(III)-based WOC for photoanodes functionalization.

4.4.1 *Porous Versus Compact Catalyst Morphology for Photoanodes Functionalization*

We present here a comparative investigation of an amorphous-Fe_2O_3 WOC fabricated with a morphology consisting of either a porous nanoparticle-assembled coating or a dense compact thin layer. Both were employed as functional coatings on nanocrystalline hematite thin layers (HTL) light absorbers that are easily fabricated as host substrates with acceptable baseline performances in photoelectrochemical water oxidation and good reproducibility [42], marking them as good model systems. Furthermore, the choice of iron oxide as both the absorber material and WOC is motivated by the need to find earth-abundant, cheap and scalable materials to enable the large-scale application of solar water-splitting. Besides satisfying all these requisites [43]. Iron oxide is also non-toxic and environmentally safe.

Both WOC morphologies were fabricated by the PLD technique using a KrF excimer laser (Lambda Physik LP 220i) with an operating wavelength of 248 nm, pulse duration of 25 ns, repetition rate of 20 Hz, and laser fluence of 2.0 J/cm^2. A metallic iron disk was used as target, and substrates were set at 5.5 cm distance. A reactive atmosphere consisting of 45 Pa O_2 was maintained in the deposition chamber during the experiment. The number of pulses was fixed at 2000, to ensure that all samples are realized with the same quantity of material. Substrate temperatures were maintained either at room temperature (RT) or at 300 °C (DEP 300 °C).

Cross-sectional SEM analysis of RT samples on silicon shows a complex morphology with scattered large particles (diameter about 30–300 nm) covered by a porous structure made of smaller particles (average diameter 5 ± 2 nm as a conservative estimate) assembled in irregular wires (Fig. 4.7a). Surface coverage appears to be complete (Fig. 4.7b). Annealing of RT samples at 300 °C leads to aggregation of the smaller particles (AN300, Fig. 4.7c) and partial collapse of the porous structure. Deposition with the substrate at 300 °C (DEP300) yields a dense NPs-assembled layer (Fig. 4.7d, cross-section and Fig. 4.7e, top-view) similar to the structures typically obtained at lower deposition pressures [44]. The effect is ascribed to higher particles mobility when they hit the heated substrate, promoting aggregation to bigger (diameter >10 nm) and partially-fused particles.

The relatively high oxygen pressure employed here, by reducing the kinetic energy of the ablated material, limiting the plume expansion, and promoting collisions between ablated particles, induces the formation of nanoclusters in the plume. These nanoclusters are cooled by the background gas before reaching the substrate, where they are deposited as a porous structure. In the case of lower pressure instead, the plume is less confined, the ablated material has a higher kinetic energy and the higher mobility results in aggregation of the particles after deposition, yielding a more compact microstructure composed of partially fused NPs [45].

UV-Vis absorption spectra on quartz substrates (Fig. 4.8) show that DEP300 has a much higher absorbance than RT or AN300 in all the range.

Given that the quantity of deposited material at a given pressure is essentially set by the number of pulses, which is kept constant for all depositions, the difference between RT and DEP300 is quite remarkable and likely linked to the different morphology. DEP300 can be considered a bidimensional thin-film, although with a nanostructured surface, while a substantial fraction of the RT layer is composed of very small particles (5 ± 2 nm). In this conditions, quantum confinement can dominate the optical properties, as reported for other materials [46] and for other nanosized hematites [47]. A small change in the refractive index of the film could also lead to enhanced transparency. However, the observation of a well defined emission band

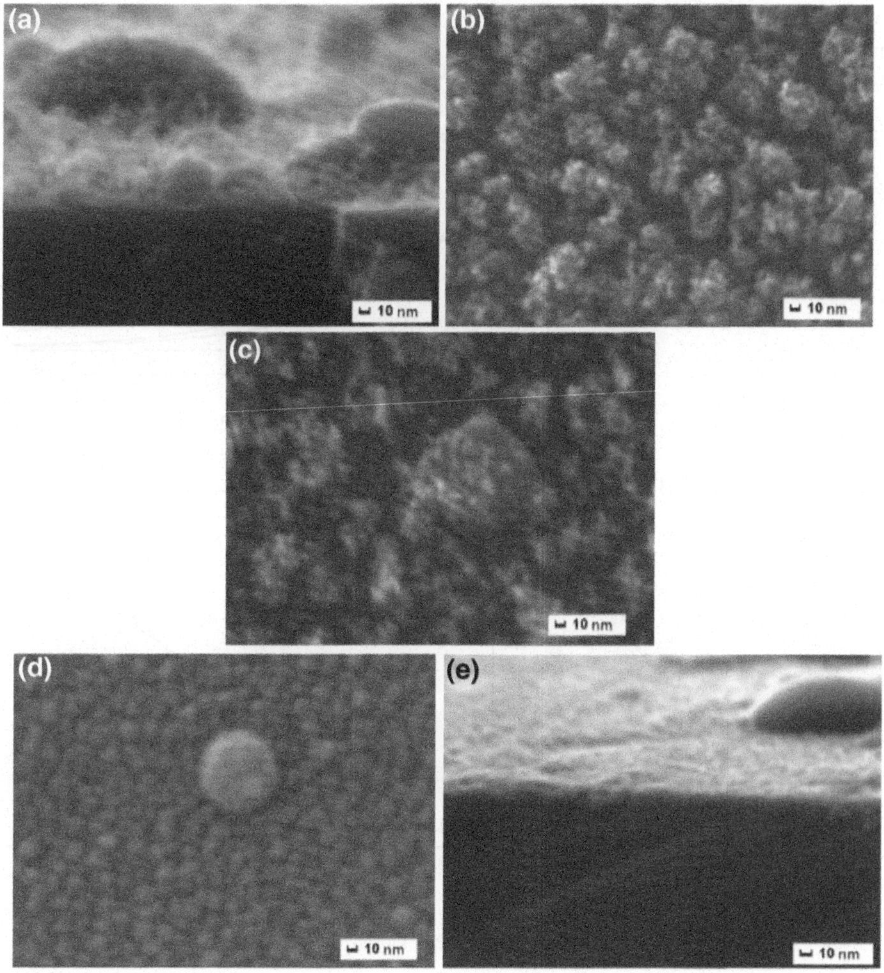

Fig. 4.7 SEM analysis of **a**, **b** RT, **c** AN300, **d**, **e** DEP300. Substrate is silicon

at ca. 450 nm for RT and AN300, consistent with the band gap, is suggestive of the presence and possibly of the co-existence of quantum-optical phenomena. The photoresponse of the integrated absorber-catalyst system, recorded by scanning the photoelectrodes under AM 1.5G illumination (at 100 mW cm^{-2} simulated sunlight) in 0.1 M NaOH (pH 13.3), was found to be deeply affected by the catalyst coating morphology (Fig. 4.9).

With AN300 the photocurrent improves along the whole potential region of interest, reaching 0.4 mA cm^{-2} at 1.23 V versus RHE with respect to ca. 0.08 mA cm^{-2} for the unmodified HTL at the same potential, a five-fold increase. The curve of HTL-AN300 also shows a cathodic shift of the photocurrent onset to 0.9 V versus RHE for HTL-AN300 with respect to 1.1 V versus RHE observed for the bare HTL

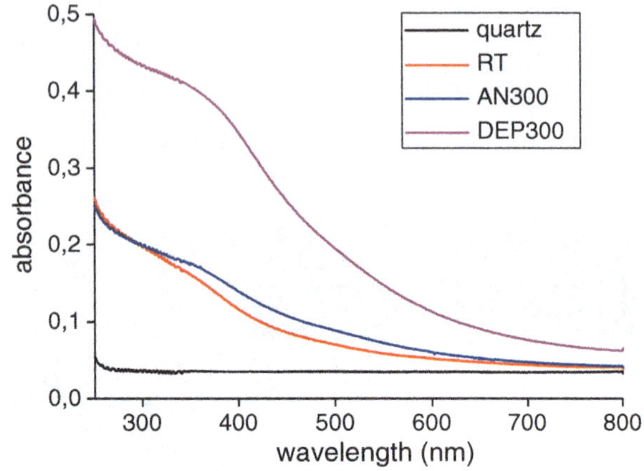

Fig. 4.8 UV-Vis absorption on quartz substrates

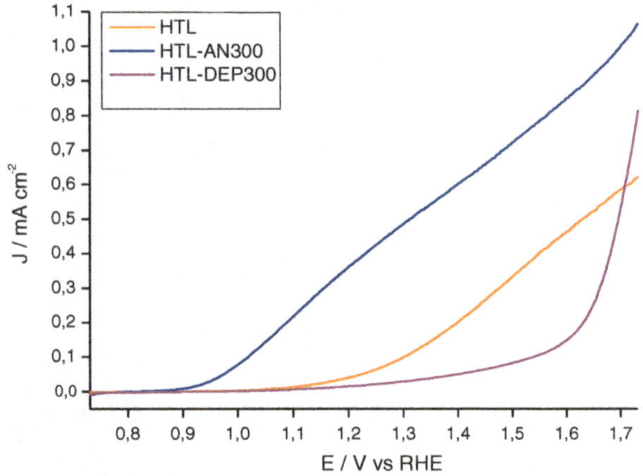

Fig. 4.9 J-V curves of HTL, HTL-AN300 and HTL-DEP300 under light

film, representing one of the most prominent WOC induced decreases in photoanodic overpotential [48]. Differently, for HTL-DEP300 the photocurrent generated at applied bias lower than the dark electrocatalysis onset is largely inferior to that observed with the unmodified HTL, suggesting a significant degree of recombination or the blocking of the hole transfer from the underlying HTL, where photocharge is generated, to the electrolyte.

In conclusion, PLD proves to be an effective method to control morphology at the nanometric scale, thus enabling comparative studies over its effects on the properties of functional materials. The same WOC material, amorphous Fe_2O_3, exhibits here completely different optical and electronic properties depending on morphology: while compact layers lead to bulk-like absorption and inefficient buried junctions, a porous structure results in a transparent material with much improved PEC performance, likely due to the formation of an adaptive junction.

4.5 Conclusions

A thermodynamic study of the metastable liquid metals at temperature close to the thermodynamic critical point was reported. Thereafter, the phase explosion process induced by the nanosecond laser irradiation of pure metals was described, based on a continuum description of the thermal evolution of the target material. Vaporization from the target (aluminum) surface was evaluated in the framework of the unsteady adiabatic expansion model, while the homogeneous nucleation of vapor bubbles in the metastable liquid was simulated in the framework of the classical nucleation theory. The size distribution of the liquid NPs formed in the phase explosion process was found to obey a power law in the range between 10^3 and 10^4 atoms per NP, in agreement with the few available experimental data when it is assumed that NPs formation comes from solidification of liquid nanodroplets. It is proved that PLD technique is able to synthesize NPs in a single step with the required relevant features for catalysis application: favorable surface-to-volume atomic ratio, size- and shape-dependent properties, and high concentration of low-coordinated active surface sites. An example, recently studied in our lab, is finally presented consisting of a PLD produced nanosized Fe(III)-based WOC for photoanodes functionalization.

References

1. D.B. Chrisey, G.K. Hubler (eds.), *Pulsed Laser Deposition of Thin Films* (Wiley, New York, 1994)
2. M.M. Martynyuk, Sov. Phys. Tech. Phys. **21**, 430 (1976)
3. (a) R. Kelly, A. Miotello, Appl. Surf. Sci. **96–98**, 205 (1996); (b) A. Miotello, R. Kelly, B. Braren, C.E. Otis, Appl. Phys. Lett. **61**, 2784 (1992)
4. A. Miotello, R. Kelly, Appl. Phys. A Mater. Sci. Process. **69**, S67 (1999)
5. N.M. Bulgakova, I.M. Bourakov, Appl. Surf. Sci. **197**, 41 (2002)
6. A. Vogel, V. Venugopalan, Chem. Rev. **103**, 577 (2003)
7. R.D. Torres, S.L. Johnson, R.F. Haglund Jr., J. Hwang, P.L. Burn, P.H. Holloway, Crit. Rev. Sol. State and Mat. Sci. **36**, 16 (2011)
8. L.V. Zhigilei, Z. Lin, D.S. Ivanov, J. Phys. Chem. C **113**, 11892 (2009)
9. L.V. Zhigilei, B.J. Garrison, Appl. Phys. Lett. **74**, 1341 (1999)
10. A. Miotello, R. Kelly, Appl. Phys. Lett. **67**, 3535 (1995)
11. A. Mazzi, F. Gorrini, A. Miotello, Phys. Rev. E **92**, 031301(R) (2015)
12. A. Mazzi, A. Miotello, J. Coll. Int. Sci. **489**, 126 (2017)
13. M.M. Martynyuk, Combust. Explosion Shock Waves **13**, 178 (1977)
14. K.M. Watson, Ind. Eng. Chem. **35**, 398 (1943)
15. E.A. Guggenheim, J. Chem. Phys. **13**, 253 (1945)
16. B.J. Keene, Int. Mat. Rev. **38**, 157 (1993)
17. N.H. March, in *Liquid Metals: Concepts and Theory* (Cambridge University Press, Cambridge, 1990), pp. 398
18. V.N. Korobenko, A.D. Rakhel, J. Exp. Theor. Phys. **112**, 649 (2011)
19. G.A. Pinhasi, A. Ullmann, A. Dayan, Int. J. Heat Mass Trans. **50**, 4780 (2007)
20. A. Peterlongo, A. Miotello, R. Kelly, Phys. Rev. E **50**, 4716 (1994)
21. W.H. Press, S.A. Teukolsky, W.T. Vetterling, B.P. Flannery, Camb. Univ. Press **32**, 10013 (2007)
22. R. Kelly, J. Chem. Phys. **92**, 5047 (1990)

23. S.I. Anisimov, Sov. Phys. JETP **27**, 182 (1968)
24. C.J. Knight, AIAA J. **17**, 519 (1979)
25. R. Kelly, A. Miotello, Phys. Rev. E **60**(3), 2616 (1999)
26. P.G. Debenedetti, *Metastable Liquids: Concepts and Principles* (Princeton University Press, 1996)
27. A. Mazzi, F. Gorrini, A. Miotello, Appl. Surf. Sci. **418**, Part B, 601 (2017)
28. V.P. Skripov, *Metastable Liquids* (Wiley, 1974)
29. H.S. Lee, H. Merte, Int. J. Heat Mass Trans. **39**, 2427 (1996)
30. C. Wu, L.V. Zhigilei, Appl. Phys. A **114**, 11 (2014)
31. S. Noël, J. Hermann, T. Itina, Appl. Surf. Sci. **253**, 6310 (2007)
32. P.T. Murray, E. Shin, Mater. Lett. **62**, 4336 (2008)
33. J.C. Alonso, R. Diamant, P. Castillo, M.C. Acosta-García, N. Batina, E. Haro-Poniatowski, Appl. Surf. Sci. **255**, 4933 (2009)
34. (a) P.V. Kamat, J. Bisquert, J. Phys. Chem. C **117**, 14873 (2013); (b) S.N. Habisreutinger, L. Schmidt-Mende, J.K. Stolarczyk, Angew. Chem. Int. Edn. **52**, 7372 (2013)
35. R. van de Krol, in *Photoelectrochemical Hydrogen Production*, ed. by R. van de Krol, M. Grätzel (Springer, Boston, 2012), p. 13
36. (a) W.J. Youngblood, S.-H.A. Lee, Y. Kobayashi, E.A. Hernandez-Pagan, P.G. Hoertz, T.A. Moore, A. L. Moore, D. Gust, T.E. Mallouk, J. Am. Chem. Soc. **131**, 926 (2009); (b) S.Y. Reece, J.A. Hamel, K. Sung, T.D. Jarvi, A.J. Esswein, J.J.H. Pijpers, D.G. Nocera, Science **334**, 645 (2011); (c) V. Cristino, S. Berardi, S. Caramori, R. Argazzi, S. Carli, L. Meda, A. Tacca, C.A. Bignozzi, Phys. Chem. Chem. Phys. **15**, 13083 (2013); (d) M. Orlandi, S. Argazzi, A. Sartorel, M. Carraro, G. Scorrano, M. Bonchio, F. Scandola, Chem. Comm. (Camb) **46**, 3152 (2010); (e) B. Klahr, S. Gimenez, F. Fabregat-Santiago, J. Bisquert, T.W. Hamann, J. Am. Chem. Soc. **134**, 16693 (2012); (f) J. Sun, D.K. Zhong, D. R. Gamelin, Energy Env. Sci. **3**, 1252 (2010)
37. J.R. McKone, N.S. Lewis, H.B. Gray, Chem. Mat. **26**, 407 (2014)
38. F. Lin, S.W. Boettcher, Nat. Mater. **13**, 81 (2014)
39. L. Trotochaud, T.J. Mills, S.W. Boettcher, J. Phys. Chem. Lett. **4**, 931 (2013)
40. M. Fondell, T.J. Jacobsson, M. Boman, T. Edvinsson, J. Mat. Chem. A **2**, 3352 (2014)
41. M. Orlandi, N. Dalle Carbonare, S. Caramori, C.A. Bignozzi, S. Berardi, A. Mazzi, Z. El Koura, N. Bazzanella, N. Patel, A. Miotello, ACS Appl. Mat. Int. **8**, 20003 (2016)
42. N. Dalle Carbonare, S. Carli, R. Argazzi, M. Orlandi, N. Bazzanella, A. Miotello, S. Caramori, C.A. Bignozzi, Phys. Chem. Chem. Phys. **17**, 29661 (2015)
43. P.C.K. Vesborg, T.F. Jaramillo, RSC Adv. **2**, 7933 (2012)
44. M. Orlandi, S. Caramori, F. Ronconi, C.A. Bignozzi, Z. El Koura, N. Bazzanella, L. Meda, A. Miotello, A.C.S. Appl, Mat. & Int. **6**, 6186 (2014)
45. (a) A. Infortuna, A. S. Harvey, L.J. Gauckler, Adv. Funct. Mat. **18**, 127 (2008); (b) I. Petrov, P. B. Barna, L. Hultman, J.E. Greene, J. Vac. Sci. Tech. A **21**, S117 (2003)
46. D. Raymand, T.J. Jacobsson, K. Hermansson, T. Edvinsson, J. Phys. Chem. C **116**, 6893 (2012)
47. L. Vayssieres, C. Sathe, S.M. Butorin, D.K. Shuh, J. Nordgren, J. Guo, Adv. Mat. **17**, 2320 (2005)
48. R.D. Smith, M.S. Prevot, R.D. Fagan, S. Trudel, C.P. Berlinguette, J. Am. Chem. Soc. **135**, 11580 (2013)

Chapter 5
Insights into Laser-Materials Interaction Through Modeling on Atomic and Macroscopic Scales

Maxim V. Shugaev, Miao He, Sergey A. Lizunov, Yoann Levy, Thibault J.-Y. Derrien, Vladimir P. Zhukov, Nadezhda M. Bulgakova and Leonid V. Zhigilei

Abstract Computer simulations and theoretical analysis of laser-materials interactions are playing an increasingly important role in the advancement of modern laser technologies and broadening the range of laser applications. In this chapter, we first provide an overview of the current understanding of the laser coupling and transient variation of optical properties in metals, semiconductors and dielectrics, with the focus on the practical implications on the energy deposition and distribution in the

---

M. V. Shugaev · M. He · L. V. Zhigilei (✉)
Department of Materials Science and Engineering, University of Virginia, 395 McCormick Road, Charlottesville, VA 22904-4745, USA
e-mail: lz2n@virginia.edu

M. V. Shugaev
e-mail: mvs9t@virginia.edu

M. He
e-mail: mh5wz@virginia.edu

S. A. Lizunov · Y. Levy · T. J.-Y. Derrien · V. P. Zhukov · N. M. Bulgakova
HiLASE Centre, Institute of Physics of the Czech Academy of Sciences, Za Radnicí 828, 25241 Dolní Břežany, Czech Republic
e-mail: lizunov@fzu.cz

Y. Levy
e-mail: levy@fzu.cz

T. J.-Y. Derrien
e-mail: derrien@fzu.cz

V. P. Zhukov
e-mail: zukov@ict.nsc.ru

N. M. Bulgakova
e-mail: bulgakova@fzu.cz

S. A. Lizunov
Institute of Thermophysics SB RAS, 1 Lavrentyev Ave, 630090 Novosibirsk, Russia

V. P. Zhukov
Institute of Computational Technologies SB RAS, 6 Lavrentyev Ave, 630090 Novosibirsk, Russia

P. M. Ossi (ed.), *Advances in the Application of Lasers in Materials Science*, Springer Series in Materials Science 274,
https://doi.org/10.1007/978-3-319-96845-2_5

irradiated targets. The continuum-level modeling of the dynamic evolution of laser-induced stresses, nonequilibrium phase transformations, and material redistribution within the laser spot are then discussed, and the need for the physical insights into the mechanisms and kinetics of highly nonequilibrium processes triggered by the laser excitation is highlighted. The physical insights can be provided by atomistic modeling, and several examples are discussed where large-scale molecular dynamics simulations are used for investigation of the mechanisms of the generation of crystal defects (vacancies, interstitials, dislocations, and twin boundaries) and the material redistribution responsible for the formation of laser-induced periodic surface structures in the single-pulse ablative regime. The need for the integrated computational approach fully accounting for the strong coupling between processes occurring at different time- and length-scales is highlighted.

5.1 Introduction

The continuous progress in the advancement of laser applications is increasingly relying on thorough theoretical understanding of laser-induced processes that control the structural and phase transformations responsible for material modification, selective removal and/or transfer. Experimental diagnostics capable of resolving the laser-induced processes on femto-/picosecond time-scales and simultaneously on sub-micrometer length-scales, however, are currently limited to specific systems and irradiation conditions [1–3]. Even with the advancement of existing and the development of new time-resolved diagnostic methods, the experimental probing alone is unlikely to completely uncover the relationships between numerous processes involved in laser-matter interactions, which take place simultaneously or sequentially with overlap in time and space. Hence, numerical simulation of processes occurring in laser-irradiated materials is an invaluable tool for understanding the complex phenomena of laser-materials interactions, especially when materials are driven far out of electronic, thermal, and mechanical equilibrium [4]. A combination of numerical simulation and experimental probing of laser-induced processes performed for the same experimental conditions is likely to be the most efficient way of gaining deep insights into the rapid structural and phase transformations triggered by the laser excitation.

In this chapter, we provide an overview of theoretical treatment and numerical modeling of different stages of laser-materials interactions, from the laser excitation of the electronic sub-system in the material to the thermal and mechanical response to the laser energy deposition leading to material modification, redistribution, or ejection from the irradiated target. The overview is focused on computational predictions that can be directly mapped to experimental observations and used for optimization of laser processing conditions in practical applications. We start with analysis of the transient modification of optical properties due to the laser-induced electronic excitation, as it defines the fraction of the incident laser energy that is absorbed by the irradiated material and its initial distribution in the target. The solid knowledge of the laser energy deposition and redistribution enables a reliable analysis of the material response to the rapid laser heating, which may involve plastic deformation, melting and resolidification, hydrodynamic flow of molten material, photomechanical damage/spallation, explosive boiling of material superheated up to the limit of its thermodynamic stability, and the ejection of material from the target driven by the relaxation of laser-induced stresses, the explosive release of vapor, and/or material ionization and charge separation. Many of the processes involved in laser-materials interactions have been addressed in atomistic or continuum-level simulations, and have been reviewed in chapters published as follow ups on two earlier editions of the School on Lasers in Materials Science [5–8]. In this chapter, we focus our attention on methods capable of addressing the processes occurring on the scale of the whole laser spot as well as on multiscale computational approaches where atomistic simulations inform the larger-scale continuum models on the kinetics and mechanisms of laser-induced structural and phase transformations, and help to design advanced continuum-level models fully accounting for the complexity of short pulse laser-materials interactions.

The variation of optical properties of metals, semiconductors and dielectrics excited by the laser irradiation are discussed next in Sect. 5.2, the continuum-level modeling of thermal and mechanical processes occurring on the scale of the laser spot is reviewed in Sect. 5.3, and several examples of the use of atomistic molecular dynamics method for revealing the processes responsible for material modification by short laser pulses are provided in Sect. 5.4. Finally, in Sect. 5.5, we briefly outline the prospects for the advancement of the computational treatment of the complex phenomena of laser-materials interactions.

5.2 Transient Response of Materials to Ultrafast Laser Excitation: Optical Properties

In this section, we provide a brief overview of several analytical models developed to describe the *optical* response of laser-irradiated materials, which is swiftly changing already during the laser pulse action and continue to evolve after pulse termination. We focus on simplified analytical models, which can be integrated into large-scale continuum or atomistic simulations of ultrashort laser-matter interaction. While for the bandgap materials the existing optical models usually account for the dynamic variation of surface reflectivity due to generation of charge carriers in the conduction band, in the case of metals the majority of simulations reported in literature are based on an assumption of constant reflectivity equal to that of a cold metal at a specific wavelength. This can lead to misinterpretation of experimental data, as the optical properties (both reflectivity and absorption coefficient) of the free electron population in metals may change considerably when the energy density of up to several eV per electron is deposited by the laser pulse. When the surface reflectivity strongly varies upon ultrafast laser excitation of a sample, the actual absorbed energy can be, depending on the electronic structure of the material, either under- or overestimated. While many details of the transient optical response of materials undergoing strong electronic excitation and rapid equilibration are still not fully understood, the simplified models discussed in this section may provide general guidance for interpretation of experimental observations.

5.2.1 Metals: Transient Optical Properties

Theoretical description of the optical response of metals as well as bandgap materials upon their ionization with generation of a dense free-electron plasma is usually based on the theory proposed by Paul Karl Ludwig Drude in 1900 [9]. The Drude theory was developed for explaining the electrical conduction in metals under the classical assumption that free electrons can be represented as particles scattered in collisions with immobile ions. It was refined by Hendrik Antoon Lorentz who proposed that the mean free path of electrons was limited by collisions. This simple theory gives a physical justification for Ohm's law and can predict the electron current in metals under the action of electromagnetic waves. As the frequency-dependent dielectric function of a material is directly related to the oscillating current of free electrons, the Drude theory is widely used for describing the free-carrier contribution to the optical response of solids to laser light (known as the Drude model or sometimes the Drude-Lorentz model) [10, 11].

5.2.1.1 The Drude Model

The Drude model (or free electron model) provides the basic formulation of the free carrier contribution in the material response to electromagnetic waves of a specific angular frequency ω via the expression for the complex dielectric function, which reads in the general form as [11, 12]:

$$\varepsilon = \varepsilon_{\text{core}} + i\frac{n_e e^2 \tau_c}{\omega \varepsilon_0 m_e^* (1 - i\omega\tau_c)}. \tag{5.1}$$

Equation (5.1) can be rewritten as

$$\varepsilon = \varepsilon_1 + i\varepsilon_2 \text{ with}$$

$$\varepsilon_1 = \varepsilon_{\text{core}} - \frac{n_e e^2 \tau_c^2}{\varepsilon_0 m_e^* (1 + \omega^2 \tau_c^2)} \text{ and } \varepsilon_2 = \frac{n_e e^2 \tau_c}{\omega \varepsilon_0 m_e^* (1 + \omega^2 \tau_c^2)}. \tag{5.2}$$

Here n_e and e are the free electron density and elementary charge, $1/\tau_c$ is the collision frequency of electrons. The collisions between electrons and between electrons and lattice atoms/ions provide damping of the optical response of material to laser radiation. When free electrons are moving in a periodic potential of the lattice in a solid medium, they respond to external forces differently than in vacuum. To simplify the description of this effect, the effective mass m_e^* is introduced, which is usually considered to be constant independent of the excitation conditions. $\varepsilon_{\text{core}}$ is a core dielectric function, which is introduced to approximate the interband contributions (absorption of electromagnetic radiation by an electron in an occupied state below the Fermi level with transition to an unoccupied state in a higher band) [11]. This part of the dielectric function is usually considered as independent of frequency. In vacuum $\varepsilon_{\text{core}} = 1$ and often the same is considered for metals [10, 11]. Note that, according to the Drude model, the dielectric permittivity is connected with the electric conductivity $\sigma(\omega)$ in metals as $\varepsilon = 1 + i\sigma(\omega)/\varepsilon_0\omega$.

It is seen from (5.2) that ε_1 becomes negative starting from some frequency $\omega_p = \sqrt{n_e e^2/\varepsilon_0 \varepsilon_{\text{core}} m_e^*}$, taking into account that $(1/\tau_c) \ll \omega_p$. At this characteristic frequency ω_p, called plasma frequency, the response of the material changes from a metallic to a dielectric one. The electromagnetic field with angular frequency below the plasma frequency cannot penetrate the material. This corresponds to the collisionless plasma regime, when the light is totally reflected. In metals, electromagnetic waves can penetrate a very thin surface layer whose size is determined by the electron collision frequency $(1/\tau_c)$ and can be calculated from the dielectric function, see below.

The optical properties of cold metals are reasonably well described by the Drude model in infrared and, partially, visible spectral ranges via adjusting the plasma and collision frequencies, while at higher frequencies the model fails [10]. The reasons for the failure include an increasing contribution of intraband transitions and an enhanced sensitivity of the electromagnetic wave to surface roughness at shorter laser wavelengths. To extend the description for a wider spectral range, several modifications of the Drude model have been proposed.

5.2.1.2 The Combined Drude-Lorentz Model

The Lorentz model was developed for insulating materials, assuming that the electrons are bound to the nuclei in a similar way as small light balls can be bound to large heavy balls by springs [13]. Under the action of an external periodically-varied force (electromagnetic wave), such system has resonant frequencies (harmonic oscillators). Then, the complex dielectric function of a material can be described by accounting for both free and core (bound) electrons through a combination of the Drude and Lorentz models [14, 15]:

$$\varepsilon = \varepsilon_D + \varepsilon_L, \tag{5.3}$$

where ε_D is given by (5.1), and ε_L is the Lorentzian term written in the general form as [13, 14, 16]

$$\varepsilon_L = -\sum_{l=1}^{N} \frac{f_l \Omega_{L,l}^2}{(\omega^2 - \Omega_{L,l}^2) - i\omega\Gamma_{L,l}}. \tag{5.4}$$

Here $\Omega_{L,l}$ is a resonant frequency l ($l = 1, \ldots, N$) with a weighting factor f_l defining the fraction of electrons that are bound to this frequency ($\sum_{l=1}^{N} f_l = 1$), and $\Gamma_{L,l}$ is the damping constant. Adding one or several Lorentzian terms to the Drude model enables expansion of the accessible spectral range toward shorter wavelengths, where the optical properties can be satisfactorily described at a fixed temperature via fitting the model parameters ($\varepsilon_{\text{core}}$, ω_p, τ_c, $\Omega_{L,l}$, $\Gamma_{L,l}$, f_l) [14, 16–18].

It is clear that the above approach gives a rather schematic description of different contributions and parameters affecting the optical response of metals, and its application to simulations of laser-matter interactions must involve a fitting procedure. However, the Drude model and the combined Drude-Lorentz model represent a valuable tool that enables an effective description of the swift changes in the optical properties of solids irradiated by ultrashort powerful laser pulses within the framework of large-scale continuum or atomistic models.

5.2.1.3 High Free-Electron Temperatures: Plasma-like Behavior

When the electron temperature in metals (T_e) approaches the Fermi temperature (T_F), the electron gas starts to behave as a hot dense plasma with associated change in optical response ([19] and references therein) that has to also be accounted for in simulations. The dielectric permittivity at the limit of the hot electron plasma ($T_e \gg T_F$), ε_p, with the Maxwellian distribution function can be expressed in the following form [19–22]:

$$\varepsilon_p = 1 + i\frac{4\omega_p^2}{3\omega(k_B T_e)^{5/2}\sqrt{\pi}} \int_0^\infty \mathrm{d}E_e \frac{E_e^{3/2}\exp(-E_e/k_B T_e)}{-i\omega + 3\sqrt{\pi/2}(k_B T_e/E_e)^{3/2}\nu_{ei}^T G}, \tag{5.5}$$

where E_e and T_e are the electron energy and temperature, respectively; k_B is the Boltzmann constant; $\nu_{ei}^T = 4\sqrt{2\pi} Z n_i e^4 \ln\Lambda / \left(\sqrt{m_e} k_B^{3/2} T_e^{3/2}\right)$ is the electron-ion collision frequency with the Coulomb logarithm, $\ln\Lambda$, which can be approximately taken to be equal to 2 at the electron temperatures of the order of 10 eV [23]; the factor $G = \left(\nu_{ei}^T/\gamma_\sigma + (1-2i)\omega\right)/\left(\nu_{ei}^T + (1-2i)\omega\right)$ with $\gamma_\sigma \approx (0.87 + Z)/(2.2 + Z)$ accounts for the electron-electron collisions [22]; Z is the number of free electrons per ion in the metal, and $n_i = n_e/Z$ is the number density of ions. Note that generally Z can vary upon irradiation of transition metals by ultrashort laser pulses due to excitation of d-electrons [24].

By convention, the description of the optical properties of metals can be divided into three regimes depending on the electron temperature: (1) low electron temperatures, much smaller than T_F; (2) large electron temperatures, above T_F, and (3) intermediate electron temperatures, when both metallic and plasma behavior of the conduction electrons can be manifested. In the examples shown in Fig. 5.1, the following matching between the metallic and plasma ranges of the dielectric function was used:

$$\varepsilon = \begin{cases} \varepsilon_m, & T_e < 0.3T_F \\ \varepsilon_m + \varepsilon_p, & 0.3T_F < T_e < T_F \\ \varepsilon_p, & T_e > T_F \end{cases} \tag{5.6}$$

Here ε_m is calculated according to (5.1) or (5.3). Note that different methods were proposed for bridging the metallic and nondegenerate plasma behaviors via power [19] or linear [25] functions. However, we did not find much difference in the simulation results after applying the different functions. Expression (5.6) gives a sufficiently smooth transition between the two types of metal optical response.

5.2.1.4 Reflection and Absorption Coefficients

As soon as the dielectric function is determined by one of the methods described above, the reflection and absorption coefficients of a material exposed to a short pulse laser heating can be calculated using the Fresnel equations. The refractive index n and the extinction coefficient k are expressed as

$$n = \sqrt{\frac{1}{2}\left(\varepsilon_1 + \sqrt{\varepsilon_1^2 + \varepsilon_2^2}\right)}, \tag{5.7}$$

$$k = \sqrt{\frac{1}{2}\left(-\varepsilon_1 + \sqrt{\varepsilon_1^2 + \varepsilon_2^2}\right)}. \tag{5.8}$$

Then the absorption α and reflection R coefficients are calculated as follows:

$$\alpha = \frac{4\pi k}{\lambda}, \tag{5.9}$$

$$R = \frac{(n-1)^2 + k^2}{(n+1)^2 + k^2}. \tag{5.10}$$

Here λ is the laser wavelength and the expression for R is given for the normal incidence of the laser beam. The absorption length $l_{ab} = 1/\alpha$ characterizes the depth of electromagnetic wave penetration into metallic sample. For an arbitrary angle of incidence, the Fresnel relation for R includes the angular dependence, which is a function of the light polarization direction relative to the beam incidence [26]. Additionally, when the surface layer of an irradiated material is strongly inhomogeneous (either in terms of composition or due to strong gradients of the electron and/or lattice temperatures and the electron density), application of the multilayer

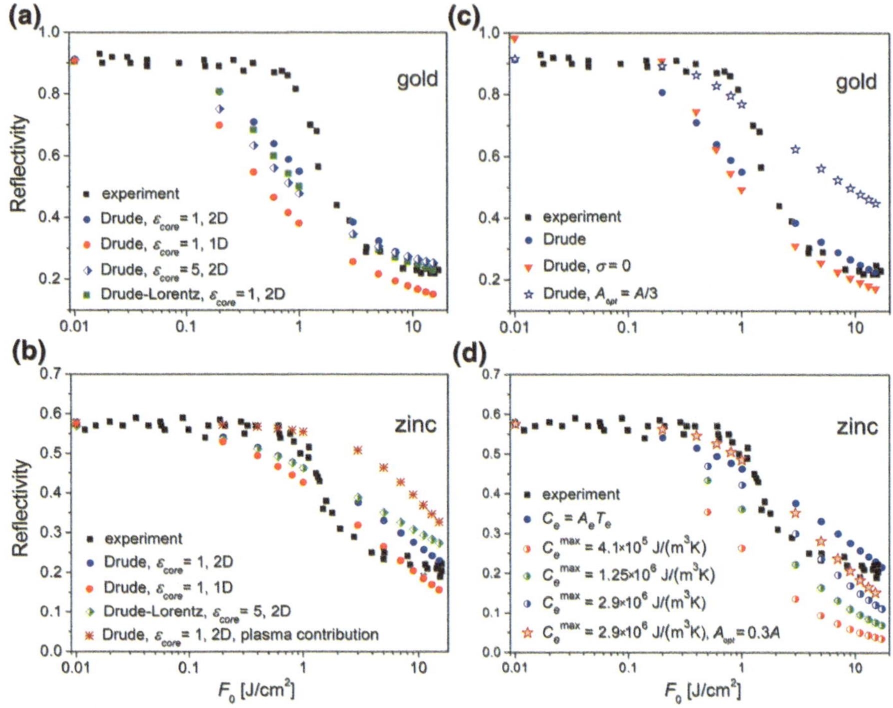

Fig. 5.1 Comparison of the reflectivity of gold and zinc as a function of the average fluence, F_0, obtained in TTM modeling combined with different models of the optical response. The target is irradiated by 66-fs Gaussian laser pulses at 800 nm wavelength. The experimental data were obtained using the hemiellipsoidal metallic reflector technique [34, 36]. Results obtained with the Drude model, (5.1), and the combined Drude-Lorentz model, (5.3), are marked as "Drude" and "Drude-Lorenz" respectively. The simulated reflectivity with integration over the irradiation spot is marked as 2D. 1D corresponds to the simulated reflectivity at the center of the irradiation spot, where $F = 2F_0$. To match the experimental data [36] at low fluences, surface roughness σ was set to 36 and 17 nm for zinc and gold, respectively, unless otherwise is specified. **a** Results for gold. The plasma response model was included in the simulations, but it does not give a noticeable effect in the studied fluence range and is therefore not shown here. **b** Results for zinc without and with plasma contribution, (5.6). **c** The Drude model with $\varepsilon_{core} = 1$ for gold: comparison of 2D simulations with and without accounting for surface roughness and under assumption that only ~30% of free electrons contribute to the optical response. **d** Simulations for zinc (2D-TTM with the Drude model at $\varepsilon_{core} = 1$) with the heat capacity calculated as the linear dependence $C_e = A_e T_e$ over the whole fluence range and with saturation at the values given in the legend. The data obtained under assumption that only ~30% of free electrons contribute to the optical response are shown by star symbols

approach for reflectivity calculations can be useful for achieving a better agreement with experimental data [12, 27, 28].

5.2.1.5 Surface Roughness Effect on Reflectance

The optical properties of material surfaces are dependent on their quality (primarily the surface roughness, oxidation, contamination, etc.) [29]. The handbooks of optical properties provide the data for thoroughly polished metals up to a mirror quality [30]. In many experiments on laser-matter interaction, the reflection coefficients of virgin surfaces can differ (usually are smaller) compared to the handbook data. For rough surfaces with the reflectivity close to the mirror quality ($\sigma \ll \lambda$, where σ stands for the mean-square roughness size), the following expressions can be used to account for the surface roughness in specular and diffuse reflectances, $R_{s,\mathrm{rough}}$ and $R_{d,\mathrm{rough}}$ respectively [31, 32]:

$$R_{s,\mathrm{rough}} = R\exp\left(-\left(\frac{4\pi\sigma}{\lambda}\right)^2\right), \quad R_{d,\mathrm{rough}} = R\left(\frac{4\pi\sigma}{\lambda}\right)^2 \tag{5.11}$$

where R is the reflection coefficient for the ideally polished surface.

5.2.1.6 Two-Dimensional Two-Temperature Modeling (2D-TTM) of Metal Reflectivity Under Conditions of Ultrafast Laser Irradiation

The measurements of the dynamic reflectivity of material samples irradiated with ultrashort laser pulses are usually based on the two major techniques: pump-probe temporally-resolved reflectivity studies [33] and integral measurements of the reflected light using, for example, a hemiellipsoidal metallic reflector technique [34]. Both techniques assume that the light is reflected from a non-uniformly heated laser-irradiation spot that implies a necessity of 2D modeling. Even for the pump-probe technique, the probe pulse focused on a central area of the Gaussian irradiation spot can exhibit averaging over the non-uniformly excited area of the spot. Here, to evaluate the validity of different optical response models, we present the results of simulations based on the two-temperature model (TTM). The TTM is based on the assumption that, during the laser pulse, the laser energy is absorbed by the conduction electrons, which are heated to a high temperature, while the lattice remains cold. As a result, the heat transfer in such nonequilibrium system can be described by two heat flow equations, for electrons and lattice, with the heat exchange between them [35]. We use a "quasi-2D" approach, at which the heat flow simulations were performed only in the direction toward the sample depth, while the reflectivity was integrated over the irradiated spot. Such approach is justified by a small laser-affected depth compared to a typical laser spot size. An implicit numerical scheme was used to solve the TTM equations with integration using the Thomas algorithm. The time and spatial steps were thoroughly checked with respect to the numerical scheme convergence for both the temperature evolution and the integral reflectivity.

Figure 5.1 presents the results of numerical simulations for gold and zinc samples irradiated by single femtosecond laser pulses (66 fs FWHM) of Ti:sapphire laser (800 nm wavelength) with direct comparison to the reflectivity measurements using the hemiellipsoidal metallic reflector technique [36] . The simulations were based on the TTM, similar to that described in [37] but extended to two dimensions to simulate the integral reflected light. All of the optical models discussed above were used in an attempt to find the best agreement between the modeling and experiments.

The laser pulse was assumed to be temporally and spatially Gaussian. Since no results of ab initio calculations of thermophysical parameters of the electronic subsystem based on the electron density of states (DOS) [24, 38] are available for zinc, a simplified dependence of the heat capacity $C_e = A_e T_e$ and a constant electron-lattice coupling factor g [36, 39] were used in the calculations for Zn. The electron thermal conductivity was taken similarly to [24] as $K_e(T_e, T_l) = \upsilon_F^2 C_e(T_e)\tau_c(T_e, T_l)/3$, where υ_F is the Fermi velocity. The collision frequency of free electrons was approximated via the sum of electron-electron and electron-phonon scattering rates as $1/\tau_c = 1/\tau_{e-e} + 1/\tau_{e-ph} = AT_e^2 + BT_l$ [24]. The coefficients A and B were assumed to be constant and were estimated as described in [40], which yielded $A = 1.805 \times 10^6\ \mathrm{K}^{-2}\ \mathrm{s}^{-1}$ and $B = 1.82 \times 10^{12}\ \mathrm{K}^{-1}\ \mathrm{s}^{-1}$ with the plasma frequency $\omega_p = 2.05 \times 10^{16}\ \mathrm{s}^{-1}$.

In the most simulations, the same τ_c values were used in the thermal conductivity expression and in the Drude model, (5.1) . It was found that the estimated B value did not give a reasonable agreement with the measured optical properties of well-polished zinc samples [16]. To fit the room temperature reflectivity, the scaling factor $k_s = 3$ was introduced similarly to [19], which yielded the final $B = 5.46 \times 10^{12}\ \mathrm{K}^{-1}\ \mathrm{s}^{-1}$. As noted in [41, 42], not all electron-electron collisions can contribute to the material optical response. To account for this effect, in a number of simulations the A value was changed in the dielectric function to $A_{\mathrm{opt}} = \eta A$ with introducing a scaling factor $\eta < 1$.

For the simulations with the combined Drude-Lorentz model, the single Lorentzian term was used with parameters that correspond to Fit #1 from [16]. In order to reduce the surface reflectivity reported in literature for highly polished zinc samples [16] to the experimental values of work [36] at low fluences, the surface roughness parameter $\sigma = 36$ nm was chosen, assuming the specular reflection, i.e., the first expression in (5.11).

The same 2D-TTM simulations were performed for gold with parametrization from [43]. The n and k values were taken from [30]. To fit the experimental reflectivity at low fluences reported in [36], the mean-square surface roughness of 17 nm was introduced. In the case of the combined Drude-Lorentz model, five Lorentzian terms were taken into account for gold with the parameters from [17]. The simulations were performed in the wide range of the laser fluence from 0.01 up to 20 J/cm^2 (average fluence over the irradiation spot).

Figure 5.1a, b show the results of the application of models described above to gold and zinc, respectively, and the results are compared with experimental measurements performed with the hemiellipsoidal metallic reflector technique [36] (black squares). For gold, Fig. 5.1a, the simulation data are presented for the Drude model, (5.1) with two values of $\varepsilon_{core} = 1$ (blue dots) and 5 (blue half-filled diamonds), and the combined Drude-Lorentz model, (5.3) with $\varepsilon_{core} = 1$ (green squares). As can be seen, all models give essentially the same results and are not in a good agreement with the experimental data in the fluence range of ~0.2–1.5 J/cm^2, while at higher and lower fluences the simulation results are satisfactory. It should be noted that variation of ε_{core} makes it possible to shift the reflectivity to a higher level at large laser fluences. In a number of studies it is argued that not all free electrons contribute to the optical response of metals and, hence, the effective electron-electron collision frequency in the dielectric function must be reduced compared to that responsible for the electron thermal conductivity [41, 42]. The simulations for gold have shown that, when accounting for this effect, it is possible to achieve a better agreement with the experimental data at laser fluences up to ~1 J/cm^2, while at higher fluences a considerable deviation between the experimental and theoretical data arises (blue-edged stars in Fig. 5.1c, $A_{opt} = A/3$).

The plasma response model was included to the simulations for gold, but it does not give a noticeable effect in the studied fluence range (therefore not shown in Fig. 5.1). It should be noted that the temperature maximum achieved in simulations was only slightly higher than the Fermi temperature of gold, which explains the small plasma contribution. For zinc, where the maximum temperature at the highest simulated fluence was around $2T_F$, the plasma effect on reflectivity is substantial, as shown in Fig. 5.1b (asterisks). The addition of the hot-plasma behavior to the optical response makes it possible to keep the reflectivity at a constant level up to $F_0 \approx 1$ J/cm^2, as observed experimentally. However, at higher fluences the plasma-like optical model yields a much slower decrease of the reflectivity as compared to the experimental measurements. Red dots in Fig. 5.1a, b show the results of 1D simulations for gold and zinc, respectively, performed at laser fluences that correspond to the center of the laser spot. The results demonstrate that 1D modeling can considerably overestimate the total absorbed laser energy, as clearly exemplified by the results shown for gold.

The simulation data obtained for gold with the Drude model and $\sigma = 0$ (ideally smooth surface, red triangles in Fig. 5.1c) overestimate the reflectivity at low fluences and underestimates at high fluences. As seen from Fig. 5.1b, the simulation results for zinc are not satisfactory almost in the whole range of laser fluences. The possible reasons of the discrepancy between the computational predictions and experimental data are discussed below.

5.2.1.7 Importance of the Thermophysical Properties of the "Free Electron Gas" for Description of Optical Response of Metals

As mentioned above, for the free electron population in zinc we used the simple linear dependence of the heat capacity on the electron temperature, $C_e = A_e T_e$ [37, 39], in view of absence of other data. At high free-electron temperatures, close to or above T_F, the heat capacity is expected to saturate at a value of the order of that of non-degenerate gas, $1.5 k_B n_e$, though plasma non-ideality at high densities as well as incomplete equilibrium [44] can influence this saturation value. According to DOS simulations [24, 38], the majority of metals exhibit a tendency of the heat capacity saturation at high electron temperatures (note that the temperature of free electrons is considered here as a measure of their average energy even when the electrons are not fully thermalized immediately after ultrashort laser excitation). Figure 5.1d presents the 2D-TTM simulation data for zinc with the Drude model used to account for the optical response. Blue dots are predicted by assuming the linear electron temperature dependence of the heat capacity ($C_e = A_e T_e$ with $A_e = 81.9$ J/(m^{3}K)) in the whole range of laser fluences. The other half-filled dots show the results of calculations where the linear dependence was used up to a certain electron temperature, above which the value of C_e was fixed (kept constant upon further heating of the electrons). The results are shown for the following three levels of saturation of C_e: $C_e^{\max} = 4.1 \times 10^5$ J/(m^{3}K), 1.25×10^6 J/(m^{3}K), and 2.9×10^6 J/(m^{3}K). We note that $C_e^{\max} = 2.9 \times 10^6$ J/(m^{3}K), which corresponds to $T_e \approx 35{,}000$ K, is close to the values of the electron heat capacity of a number of metals at temperatures where the saturation becomes pronounced [24], and this value gives the best fit to the experimental data for zinc at fluences >1 J/cm^2. Accounting for the limited contribution of the electron-electron collision frequency in the dielectric function leads to even better agreement between simulations and experiment (red-edged stars in Fig. 5.1d, $A_{\text{opt}} = A/3$). To further improve the simulation accuracy at higher fluences, $\varepsilon_{\text{core}} > 1$ can be introduced to account for interband transitions. However, the fluence range where the reflectivity exhibits a sudden drop for both metals is not still properly described that calls for further studies.

The following reasons can be hypothesized to be responsible for the observed sudden drop of the reflectivity and the discrepancy between the experimental and modeling data:

(i) At the first turn, it can be the effect of unsatisfactory description of thermophysical properties of the electron subsystem of the material (note that the electron thermal conductivity of free electrons is also affected by the temperature dependence of the electron heat capacity, see above). As demonstrated in Fig. 5.1d, the choice of the temperature dependence of the heat capacity has a strong effect on the calculated reflectivity. The electron-lattice coupling factor g, which is considered here to be constant for zinc, can also have an impact on the transient optical response of metals, and its temperature dependence [24] should be accounted for in interpretation of the results of pump-probe experiments. However, during irradiation by femtosecond laser pulses, when the electron-lattice coupling is insignificant, the influence of g on the incident pulse reflection should not be pronounced.

(ii) We assume here that the surface roughness behaves in the same manner irrespectively to the electron temperature. However, at high laser fluences, the incident laser wave can be scattered differently than at low fluences. As a result, the light absorption can be significantly altered by the interference between the scattered surface wave and the incident laser wave. For instance, for linearly polarized laser light, the interference effect results in the transient periodic absorption pattern leading to the formation of laser-induced periodic surface structures (LIPSS) at relatively low fluences, near the melting threshold [45, 46], as well as in the regimes of the developed ablation [47] . However, in the ablative regimes the LIPSS can exhibit very high regularity [47], indicating a change in laser-surface coupling.

(iii) The collision frequency of the electrons does not satisfy the simple relation $1/\tau_c = AT_e^2 + BT_l$ at high heating rates/values, and the A and B coefficients can be variable, thus affecting the temperature dependence of the electron thermal conductivity (see above). Additionally, the fraction of free electrons actively participating in the material optical response can also be changing with the electron temperature, thus leading to the variation of A_{opt} [42].

(iv) For the adequate comparison of simulation results with experimental data, the multilayer model must be used for metals at high heating rates to account for a highly inhomogeneous temperature profile that can be formed even within the absorption depth.

(v) The effects of nonequilibrium within the electron subsystem at high-power laser heating are still poorly understood but can have important consequences on the reflectivity of metals.

The simple optical models considered above are needed for large-scale multi-physics simulations of laser interactions with metals, where ab initio models cannot be applied. However, it must be noted that even the existing ab initio models, which require large computational resources, cannot provide an adequate description of the experimental data [48]. Two examples of computational analysis of ultrashort laser interaction with metals, where the dynamic optical response plays a key role, are presented below.

5.2.1.8 Dynamic Optical Response of Metals to Ultrashort Laser Pulses

Surface plasmon-polariton decay length. The origin of LIPSS produced on material surfaces by ultrashort laser pulse irradiation is mainly attributed to the interference of the incident and surface scattered electromagnetic waves [45]. However, the factors affecting the quality of the surface pattern are still debated. Recently, two opposite views on the problem of periodic laser-induced pattern quality were published. Öktem et al. [49] state that, for inscription of a highly regular structure on a metallic surface, the distant points of the irradiation spot must interact via a mutual electric field. Hence, the surface scattered wave has to propagate a distance covering the whole irradiation spot. Limiting the laser-irradiation spot diameter to several wavelengths (12 μm in [49]) can facilitate a better interaction between the surface electromagnetic waves excited in different parts of the irradiation spot, thus enabling better quality of LIPSS. On the contrary, Gnilitskyi et al. [47] have demonstrated experimentally and computationally that metals with a short decay length of the scattered waves enable imprinting of highly regular LIPSS on their surfaces. Although a small irradiation spot size was used in [47] (10.4 μm), it was proven that limiting the interaction between the scattered surface waves (surface plasmon polaritons, or SPP as

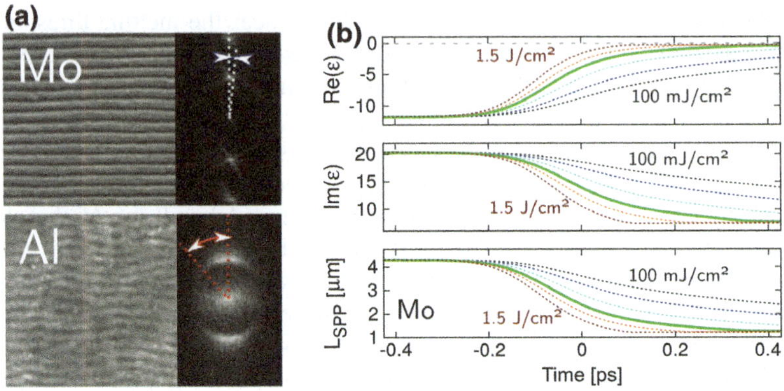

Fig. 5.2 **a** LIPSS imprinted by ultrashort (213 fs) laser pulses at the wavelength of 1030 nm on surfaces of molybdenum (top) with $L_{SPP} \approx 4.3$ μm and aluminum (bottom) with $L_{SPP} \approx 65$ μm under similar irradiation conditions. On the right, 2D Fourier transforms of the images are presented, demonstrating the dispersion of the LIPSS orientation angle (5.3° for Mo and 26.7° for Al). **b** The temporal evolution of the real and imaginary parts of the dielectric function and the SPP decay length, L_{SPP}, for molybdenum under the conditions of (**a**), predicted in TTM simulations supplemented by the Drude model. Colored lines correspond to decreasing fluence from left to right: 1.5, 1, 0.69, 0.4, 0.2, and 0.1 J/cm^2. The results are adapted from [47]

a particular case) allows preserving their coherence with the incident laser light and creation of a regular periodic pattern of light absorption, free of bifurcation points. Figure 5.2a shows LIPSS imprinted on two metals, Mo with the decay length of SPP $L_{SPP} \approx 4.3$ μm (top) and Al with $L_{SPP} \approx 65$ μm (bottom), that clearly demonstrates this concept. On the right from LIPSS images, the estimated dispersion of the LIPSS orientation angle is presented, showing only 5.3° for Mo and 26.7° for Al. This finding is supported by the work of Ruiz de la Cruz et al. [50], where the highly regular periodic surface structures were formed on Cr surfaces with the irradiation spot diameter of 80 μm, much larger than L_{SPP} (4.1 μm [47]). Note that all LIPSS discussed above were produced at near-IR laser wavelength. The plasmon polariton theory proposed in [47] predicts the tendency of L_{SPP} decreasing with wavelength so that the highly regular LIPSS can be imprinted on surfaces of a wider number of metals at visible and near UV spectral range of laser radiation.

The above-cited L_{SPP} values were estimated based on the metal properties under normal conditions. An important question arises on how the SPP decay length evolves upon excitation of metal surfaces by an ultrashort laser pulse. This question was addressed in a series of TTM simulations supplemented by the Drude model, (5.1), (5.7)–(5.10), to account for the transient variation of the optical properties. The simulations have revealed that L_{SPP} decreases upon excitation by ultrashort laser pulses [47] (Fig. 5.2b, the case of molybdenum), thus enhancing the feasibility of obtaining highly regular periodic structures. An important conclusion was made that the tendency of decreasing L_{SPP} upon ultrashort laser irradiation is inherent to all metals. Furthermore, the transient L_{SPP} value is smaller for higher laser fluences,

Fig. 5.2b, thus supporting the mechanism of the efficient LIPSS formation in the ablation regimes.

Evolution of modulated absorption pattern under the conditions of LIPSS formation. Another TTM simulation example, where modeling of the optical response has proven to be of high importance for understanding of the physical processes, is related to the temporal evolution of the modulated temperature profile generated by the interference of the incident and surface scattered electromagnetic waves upon ultrashort laser irradiation [40, 51]. As mentioned above, the formation of periodic surface patterns is believed to be triggered by the periodic absorption of the linearly polarized laser light [45]. The questions on how strong can be the modulation of the lattice temperature produced by the wave interference and how long the temperature modulation can survive on material surface are of high importance for understanding of the post-irradiation evolution of material toward the periodic surface relief. These questions were addressed in [40, 51], where the results of 2D-TTM simulations performed for different materials (titanium, gold, silicon, fused silica) under the assumption that the incident laser intensity is spatially modulated across the beam diameter, as shown in Fig. 5.3a, are reported. It was demonstrated that once free electrons absorbing the laser light experience a spatial modulation of their temperature with the periodicity close to the laser wavelength, they transfer this modulation to the lattice. The periodic lattice temperature profile can survive on the surface up to hundreds of picoseconds even in metals, where high heat conduction effectively smooths the heating inhomogeneity.

Figure 5.3b underlines the importance of taking into account the dynamical change of the optical properties of irradiated materials on the example of gold (800 nm wavelength, 100 fs pulse duration, laser fluence of 1.1 J/cm^2) [51]. It shows the temporal evolution of the surface temperature modulation determined by the parameter $a_l = T_{l,\max} - T_{l,\min}$ (the difference between the maximum and minimum of the lattice temperature in the periodic modulation) called the modulation amplitude. The simulations performed with constant reflection and absorption coefficients taken at their room temperature values (dashed line) yielded only slight modulation of the order of 40 K. The a_l value is first increasing due to lattice heating during the electron-lattice thermalization stage (approximately 10 ps) and then starts to gradually decrease as the result of smoothing the inhomogeneous temperature distribution by heat conductivity. When the dynamic change of the optical properties has been implemented into simulations in the form of the Drude model, the maximum modulation amplitude achieved during material evolution increases by approximately 10 times (solid line). The break in the line corresponds to the stage of the surface melting in both modulation minima and maxima, after which the surface layer experiences further heating. Interestingly, in modeling with the constant optical parameters (dashed line) the melting threshold was not reached (maximum surface temperature ~800 K), which contradicts the experimental observations (the ablation threshold of gold under the studied irradiation conditions is ~0.4 J/cm^2 [52]).

As mentioned above, the simulations with the constant optical parameters do not agree with the experimental value of the damage threshold fluence. The explanation is presented in Fig. 5.3c showing the dynamics of the laser light absorption [51]. As

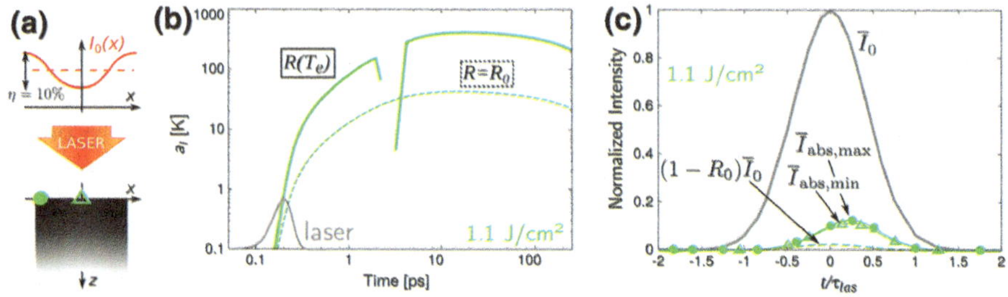

Fig. 5.3 **a** Schematics of the modeled problem on the temporal evolution of the modulated temperature profile generated by the periodically modulated laser intensity across the laser beam (adapted from [40]). It was assumed that the electromagnetic wave at the surface experiences 10% modulation as a result of the interference of the incident laser pulse and the surface scattered wave. **b** The modulation amplitude a_l as a function of time for laser-irradiated gold surface. The laser pulse parameters are: 800 nm wavelength, 100 fs pulse duration, laser fluence of 1.1 J/cm^2. The laser pulse profile is shown by grey line. The results of modeling with the constant and electron temperature dependent optical properties are presented by dashed and solid lines, respectively. **c** The simulated laser energy absorption by gold for the case given in (**b**). The temporal profile of the laser pulse is depicted by a grey solid line. The dashed green line corresponds to a simulation performed with the constant optical properties. Green solid lines are obtained for the temperature modulation maxima and minima with using the Drude model to describe the transient variation of optical properties of gold during and after the laser irradiation (highlighted by dots and triangles, respectively). All lines are normalized by the intensity maximum of the incident laser pulse. Panels (**b**) and (**c**) are adapted from [51]

the reflectivity of gold is very high at 800 nm wavelength (~97.5%), only a very small fraction of the beam energy is absorbed. When the dynamic change of the reflectivity is accounted for in the simulations, a considerably higher fraction of the laser energy is absorbed (green solid lines for the modulation maxima and minima highlighted by dots and triangles, respectively), thus leading to a reasonable agreement with the experimental observations [52]. An interesting feature of absorption can be noticed in Fig. 5.3c: the absorption maximum in gold is shifted toward the tail of the laser pulse, contrary to the case of dielectrics discussed in the next section.

5.2.2 Bandgap Materials

The wide bandgap dielectrics are transparent for light in visible and IR spectral ranges. The ultrashort pulse ablation in these wavelength ranges is achieved via non-linear effects involving multiphoton or tunnelling ionization, which can be followed by collisional multiplication of free electrons effectively absorbing laser light (the so-called avalanche process). The details of numerical models for description of the processes excited in wide bandgap dielectrics can be found in [6, 53, 54] (see also Supplementary Materials for [54]). Here we only briefly focus on the variation of the optical properties of transparent materials under irradiation by powerful ultrashort laser pulses. It is stressed, however, that in order to account for a correct description of ultrashort laser pulse interaction with bandgap materials via continuum-level modeling, these optical properties must be coupled to the description of the local beam intensity in the material. The latter can involve different levels of numerical description from the basic depletion of the laser intensity along its pathway by the different ionization and heating processes [54] to a full Maxwell model [55].

Spatiotemporal dynamics of the optical parameters in transparent dielectrics are well described within the Drude formalism via the complex dielectric function $\varepsilon(n_e)$ that accounts for the contributions from unexcited matter and generated dense plasma of free electrons [12]:

$$\varepsilon = 1 + (\varepsilon_g - 1)\left(1 - \frac{n_e}{n_{\text{val}}}\right) - i\frac{n_e e^2 \tau_c}{\omega \varepsilon_0 m_e^*(1 - i\omega\tau_c)} \tag{5.12}$$

(compare with (5.1)), where n_{val} is the total number of the valence electrons in the unexcited material and ε_g is its dielectric function. From this equation, we can introduce the critical electron density for bandgap materials $n_{cr} = \varepsilon_0 m_e^* R_e(\varepsilon_g)\omega^2/e^2$, at which the oscillation frequency of free electrons (or plasma frequency ω_p, see above) is in resonance with the electromagnetic wave frequency ω. The critical density n_{cr} is one of the fundamental parameters characterizing the interaction of electromagnetic waves with ionisable materials. Note that this value can differ from the critical plasma density in gas plasmas, which is equal to $\varepsilon_0 m_e \omega^2/e^2$. As soon as the condition $n_e > n_{cr}$ is reached for a particular light wavelength, free electron plasma strongly reflects such light. Equation (5.12) together with the Fresnel formulas are usually introduced to the models of laser excitation of bandgap materials based on the rate equations for the generation of free electrons. As noted above, the coefficient of light reflection from an inhomogeneous dense plasma at the sample surface layer should be calculated within a multilayer reflection model.

Figure 5.4 presents the simulation data for fused silica irradiated by 130 fs laser pulses at 800 nm wavelength [54]. It must be underlined that the damage threshold fluence predicted in the simulations (2.27 J/cm^2) is in excellent agreement with experimental measurements. It is observed that, as soon as the critical plasma density is exceeded in the surface layer of fused silica (Fig. 5.4a), the laser light start to be reflected from the metalized material surface (Fig. 5.4b). In this case the reflection is more efficient in the beam tail when the material becomes considerably ionized, contrary to laser excited metals, where light reflection is reducing in the second part of the laser beam (Fig. 5.3c). Interestingly, the mirror-like behavior of the ionized surface layer leads to an effective reduction of the duration of the laser pulse that penetrates into the sample. At the largest simulated fluence, as much as half of the laser beam energy is reflected from the surface. It must be also underlined that a noticeable part of the beam is transmitted through the transparent material even at the highest laser fluences studied in the simulations, mainly in the first half of the pulse duration. The same formalism can also be applied to semiconductor materials as discussed below.

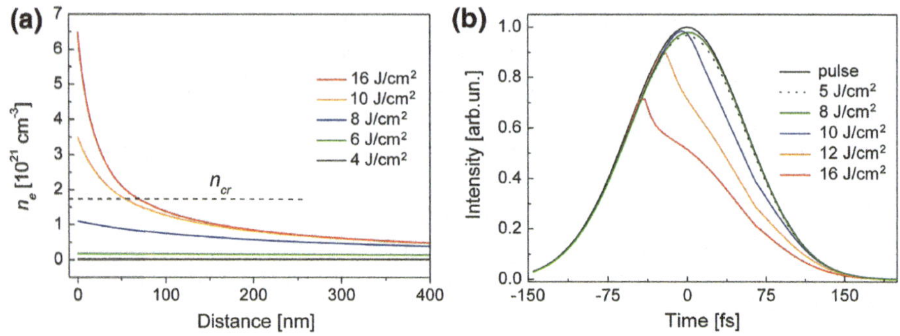

Fig. 5.4 **a** The spatial profiles of the free-electron density at a time of 100 fs after the maximum of the laser pulse with 130 fs duration and 800 nm wavelength. **b** Dynamic changes in the laser intensity of the beam entering the sample after its partial reflection from the laser-excited surface layer, $(1 - R(t)) \times I_0(t)$ where $I_0(t)$ and $R(t)$ are the incident laser intensity (black solid line) and the time-dependent reflection coefficient, respectively. Adapted from [54]

5.2.3 Semiconductors: Non-thermal Melting and Pump-Probe Experiments

In this section, we address the problem of theoretical description of time-resolved reflectivity measurements for silicon irradiated by ultrafast laser pulses. Although silicon is, for historical reasons, one of the most studied materials, the modeling of its nonequilibrium behavior upon intense laser irradiation remains challenging, particularly in the regime of material modification. For picosecond laser pulses of moderate intensity, when the free electron density remains considerably below the critical plasma density, the approach developed by van Driel based on the rate and energy balance equations [56] represents an excellent example of a successful numerical model. As soon as the free electron density produced by laser irradiation approaches the n_{cr} value, the transient optical properties of laser-excited silicon were found to be well described by an analogue of the Drude model, similar to (5.12) [57].

A sensitive method to study the thermodynamic pathways of matter out of equilibrium is to compute and compare the transient reflectivity dynamics obtained for a material in the pump-probe experiments. A pump laser pulse excites the material while a low-energy probe pulse is applied with a short variable delay to measure how the reflectivity evolves. The reflected probe beam contains information on the instantaneous state of the material through its optical properties. For silicon, an adequate computational description of the transient optical properties matching well the experimental data was achieved up to the melting threshold [12].

Well above the melting threshold fluence, close to 0.6 J/cm^2, the quantum-level effects triggered by the high level of excitation of the electron-hole pairs become dominant and substantially complicate the mathematical description of the laser-induced processes at large scale, particularly in 2D and 3D cases. As was found experimentally for semiconductors using time-resolved X-ray diffraction, excitation of ~10% of the total density of the valence electrons to the conduction band is sufficient to induce a bond softening followed by the lattice structural change already at femtosecond time scale, usually with a delay of ~300–400 fs after the maximum of the laser pulse [58–60]. At the atomistic level, the excitation of a large number of electron-hole pairs leads to considerable change of the interatomic potential resulting in lattice disordering, called non-thermal melting. Such ultrafast process is observed primarily in solids with anomalous thermodynamic properties, which exhibit compaction upon melting [61, 62]. It happens in a thin electronically overexcited region at the material surface, up to several dozens of nanometers [60], which, however, is sufficient to strongly affect the transient reflectivity of the irradiated sample. Non-thermally molten material is brought in a highly stretched state (negative pressure) followed by material compaction, which results in the transition from non-thermal to thermal liquid state called liquid-liquid phase transition (LLPT) [62–64]. During this transition stage, which lasts few picoseconds, the free electron subsystem transfers the energy necessary to achieve the thermally molten state to the lattice.

The effects of non-thermal melting in laser-excited materials have extensively been studied for silicon by molecular dynamic simulations [65–70]. The theoretical description of non-thermal melting and associated change in the material reflectivity, suitable for incorporation into continuum-level models describing pump-probe experiments (e.g., [71]), remains a challenging task. So far, only a few simplified analytical models of the non-thermal melting have been proposed, and the description of this process remains rather phenomenological [72, 73]. Moreover, the change of reflectivity of a highly and swiftly excited semiconductor experiencing non-thermal melting includes not only the contributions from free-carrier density variation and state filling [74] but also bandgap shrinking and collapse [60, 75].

During the construction of the model of ultrashort laser excitation of silicon [76], two additional effects have been found to strongly affect the sample reflectivity. (i) When calculating the ambipolar diffusion of the electron-hole pairs, the use of the density-dependent electron-phonon coupling time [77] was found to provoke a transient melting in a sub-surface region that affects the phase detected by the probe pulse and leads to reflectivity oscillations on the picosecond timescale. As this effect was not observed experimentally, the coupling time in the over-dense region was limited to 2 ps, in accordance with time-dependent x-ray scattering experiments [78]. (ii) At relatively low fluences, but above the melting threshold, the transient melting can happen heterogeneously in a rather thick surface layer of silicon. For such a layer, the material refractive index can be described as a simple interpolation between the dielectric permittivity of the excited solid and the molten material permittivity. At high fluences, the transient melting is rather happening via formation of homogeneously distributed melting centers below the surface, as shown by molecular dynamics simulations [79]. The homogeneous melting can be accounted for in the description of the macroscopic optical properties in the framework of the Lorenz-Lorentz model of the effective medium [28], where a fraction of the material is a locally excited solid and another fraction is molten.

By applying the irradiation parameters described in [71] to the model, a good agreement was obtained with the experimental data on the transient pump-probe reflectivity at 1 μm wavelength for the regimes below the melting threshold. For such irradiation parameters, when the reflectivity dynamics is governed by the response of the excited free electrons, the probe signal is reasonably well described by the Drude model in the form given by (5.12). Upon reaching conditions for thermal melting, the molten silicon starts to contribute to the reflectivity signal, and the model predictions deviate from the measurements. Evidently, an adequate model of homogeneous nucleation of the molten phase is required. However, the main feature of this regime, a swift increase of the reflectivity during the laser pulse due the free electron excitation followed by the reflectivity decrease attributed to the electron recombination and a new raise of reflection conditioned by thermal melting of the lattice, is qualitatively reproduced in the modeling. In the regimes of the non-thermal phase transition (at fluences 3–4 times exceeding the thermal melting threshold), the measurements do not reveal the stage of reflectivity decrease due to the electron recombination [71]. The reflectivity increase as a result of the bandgap collapse at the femtosecond time scale [60] has not yet been described in the large-scale numerical simulations. An adequate description of this effect calls for further model developments.

5.3 Continuum-Level Modeling of Thermal and Mechanical Response to Laser Excitation at the Scale of the Laser Spot

The laser excitation of the optically active states in the irradiated target followed by the initial ultrafast non-thermal material response to the electronic excitation and thermalization of the deposited laser energy, discussed in the previous section, set the stage for slower thermal and mechanical processes leading to material modification, redistribution, and removal/ablation. The understanding of these processes is critical for the advancement of laser applications based on material modification, selective removal and/or deposition. A set of computational techniques capable of revealing the information on the heat transfer, elastic and plastic deformation, phase transformations (melting, resolidification, vaporization and volume ablation processes), hydrodynamic flow of transiently melted material, and generation of unusual configurations of crystal defects are briefly reviewed in this section, with the focus on methods capable of addressing the processes occurring on the scale of the whole laser spot.

5.3.1 Thermal Modeling of Laser Melting and Resolidification

At the most basic level, computational description of thermal processes in laser-material interactions should account for the thermal energy redistribution and phase transformations in a region affected by the laser energy deposition. A continuum model based on the numerical solution of the heat diffusion equation complemented by source terms describing the laser energy deposition and the heat of melting/solidification can often serve as the first step in the analysis of the laser-induced processes. The simple thermal model can be expressed in this case by the following equation:

$$\rho C_p \frac{\partial T}{\partial t} = \nabla \cdot (K_{\text{th}} \nabla T) + S_{\text{laser}} - S_m - S_e, \tag{5.13}$$

where ρ, C_p, and K_{th} are the density, heat capacity, and thermal conductivity of the target material, respectively, and the three source terms on the right side of the equation represent the thermal energy sources/sinks due to the laser energy deposition, S_{laser}, the release/absorption of heat at melting/solidification front, S_m, and the energy loss through vaporization from the surface of the irradiated target, S_e. As discussed in Sect. 5.2, the spatial distribution of the laser energy deposition is defined by the optical properties of the material and the energy redistribution during the equilibration of the optically excited states, while the action of the other two source terms is spatially localized in the vicinity of the liquid-crystal interface and the surface of the target.

In the description of melting and resolidification, a simple phase-change model based on an assumption of local equilibrium at the solid-liquid interface (heat-flow limited interface kinetics formulated within the framework of the Stephan problem [80, 81]) is not suitable for the conditions of short pulse laser irradiation, when strong temperature gradients are created and fast thermal energy flow to/from the solid-liquid interface can lead to significant superheating/undercooling of the interface [82–86]. Therefore, a nonequilibrium kinetic description [86–88], in which the instantaneous velocity of the solid-liquid interface is defined by local temperature of the interface predicted by (5.13), is needed for a realistic representation of the movement of the solid-liquid interface. The temperature dependence of the velocity of the melting or solidification front can be described by a kinetic equation formulated as follows [86–88]:

$$V_i(T_i) = V_0(T_i)\left[1 - \exp\left(-\frac{\Delta G(T_i)}{k_B T_i}\right)\right], \tag{5.14}$$

where T_i is the temperature at the solid-liquid interface, $V_0(T_i)$ is a prefactor, and $\Delta G(T_i)$ is the difference in the Gibbs free energy between the liquid and crystalline phases. At small deviations from the equilibrium melting temperature T_m, the free energy difference can be approximated as $\Delta G(T_i) \approx \Delta H_m(T_i) \cdot (1 - T_i/T_m)$, where

$\Delta H_m(T_i)$ is the latent heat of solid-liquid phase transformation. The prefactor $V_0(T_i)$ can be fitted to the results of a series of solid-liquid coexistence atomistic simulations providing the temperature dependence of the interface velocity, as described in Sect. 5.4.1. Note that the decrease of the solidification velocity with increasing undercooling, predicted by (5.14) $V_0(T_i)$ decreasing with increasing undercooling [88], may permit the transformation of the undercooled liquid into a stable amorphous phase, which would stop the advancement of the crystallization front and result in the formation of amorphous regions in the resolidified target. The possibility of phase transformations occurring through the homogeneous nucleation of a new phase should also be included to account for the nucleation of liquid regions inside a crystal superheated up to the limit of its stability against melting [89–92] or homogeneous nucleation of new crystallites in a strongly undercooled molten part of the target [83–85, 93].

As an example of the application of the thermal model described above, we consider the results of a simulation of a silicon target irradiated by 30 ps laser pulse at an incident fluence of 0.3 J/cm^2 and a laser wavelength of 532 nm. The temperature and phase state evolution in the irradiated target is evaluated by solving (5.13 and 5.14) in cylindrical coordinates (r, z) and illustrated by a series of contour plots in Fig. 5.5. Short pulse laser irradiation results in a fast heating and melting of a surface region of the target, with the depth of melting reaching ~60 nm in the center of the laser spot. The laser heating and melting are followed by a rapid cooling and solidification, as can be seen from the upwards movement of the liquid-crystal interface shown in Fig. 5.5 by black lines. The solidification process proceeds through the epitaxial crystal regrowth, and the crystallization kinetics is described by (5.14) with parameters evaluated based the results of atomistic simulations discussed below, in Sect. 5.4.1. At the periphery of the laser spot, a lower energy density deposited by the laser pulse results in a rapid undercooling of the molten region down to temperatures below 0.71 T_m, where the atomic rearrangements become too sluggish for the advancement of the crystallization front (see Sect. 5.4.1). As a result, the advancement of the front at the periphery of the laser spot stops at ~1.5 ns, leading to the formation of stable amorphous phase. At the center of the laser spot, however, the temperature of the solidification front remains above the level required for the transition to the amorphous phase until the front reaches the surface of the target. As a result, a ring-shaped amorphous region surrounding the fully recrystallized central part of the laser spot is formed, as illustrated by Fig. 5.5f. The observation of the ring-shaped amorphous region is consistent with experimental observations for silicon targets irradiated by a single 30 ps laser pulse [94] or by several 100 fs laser pulses [95].

The rapid solidification proceeding under conditions of strong undercooling may produce a high density of crystal defects, such as vacancies, interstitials, dislocations, twin and grain boundaries. In particular, as suggested by the results of atomistic simulations discussed in Sect. 5.4.1, a high concentration of point defects (vacancies and interstitials) can be generated in short pulse laser-induced melting and resolidification of metals and semiconductors. In order to investigate the distribution of point defects on the scale of the laser spot, the predictions of the atomistic simulations can be utilized to design and parametrize a continuum-level description of

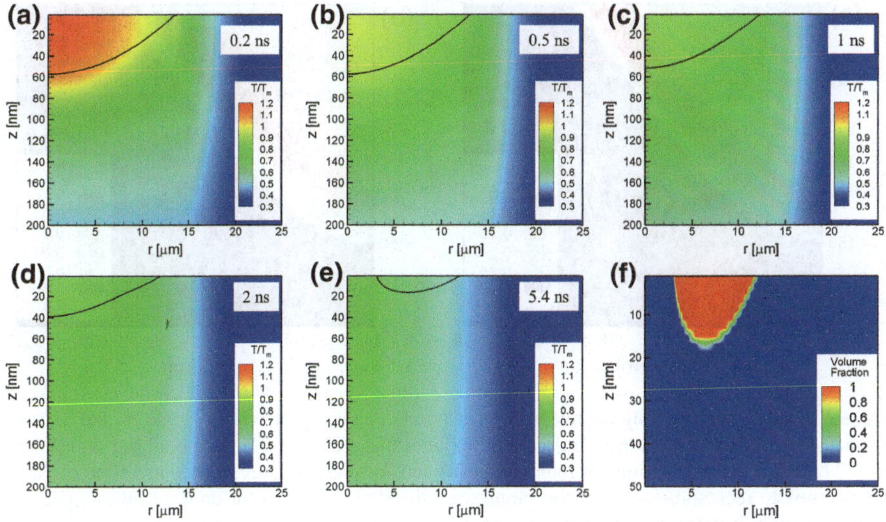

Fig. 5.5 Computational predictions of temperature evolution **a–e**, and formation of a ring-shaped amorphous region **f** in the surface region of a Si target irradiated by a 30 ps laser pulse at an incident fluence of 0.3 J/cm^2 and a laser wavelength of 532 nm. Figures **a–e** show the contour plots of temperature distribution at different time after the laser irradiation. The simulation is performed for 5.4 ns, until the end of the resolidification process. The black lines show the location of the interface between the crystalline and molten/amorphous parts of the target. Figure **f** demonstrates the spatial distribution of a volume fraction of the amorphous phase, formed by 5.4 ns due to fast cooling of the molten region. Note that in (**f**) the scale in z direction is different from (**a**)–(**e**)

the generation of point defects within the general framework of the thermal model discussed above. Equations (5.13 and 5.14) can be complemented with the following equations accounting for the generation, diffusion, and recombination of vacancies and interstitials in the irradiated target [96–99]:

$$\frac{\partial C_I}{\partial t} = \nabla \cdot (D_I \nabla C_I) + K_{IV}\left(C_I^{\mathrm{eq}} C_V^{\mathrm{eq}} - C_I C_V\right) + S_I, \tag{5.15}$$

$$\frac{\partial C_V}{\partial t} = \nabla \cdot (D_V \nabla C_V) + K_{IV}\left(C_I^{\mathrm{eq}} C_V^{\mathrm{eq}} - C_I C_V\right) + S_V, \tag{5.16}$$

where D_I and D_V are the diffusion coefficients of interstitials and vacancies, respectively, K_{IV} is the interstitial-vacancy recombination rate, C_I, C_V, C_I^{eq}, and C_V^{eq} are the current and equilibrium concentrations of interstitials and vacancies, S_I and S_V are the source terms describing the generation of interstitials and vacancies at the propagating solidification front. The parameters of the model can be obtained from atomistic simulations, such as the ones discussed in Sect. 5.4.1.

The ability of the model implementing (5.13–5.16) to predict the distribution of point defects in a laser-processed target is exemplified in Fig. 5.6, which shows concentrations of interstitials and vacancies produced in the silicon target irradiated by a

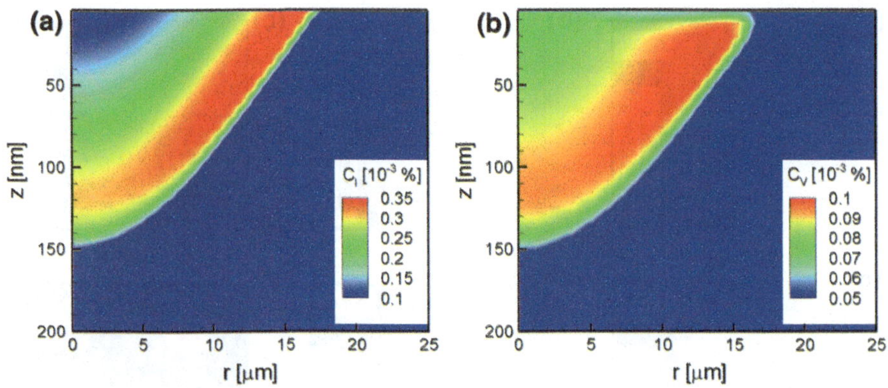

Fig. 5.6 The regions with high concentration of point defects produced in a single crystal Si target irradiated with a single 1 ns laser pulse at an incident fluence of 0.6 J/cm^2 and a laser wavelength of 532 nm. Spatial distributions of interstitials (**a**) and vacancies (**b**) are plotted at a time of 20 ns, after complete recrystallization of the target, and the scale is shown in units of atomic percent. The generation of point defects at the solidification front is parameterized based on the results of atomistic simulations, as discussed in Sect. 5.4.1

nanosecond laser pulse. In contrast to the simulation of picosecond laser irradiation discussed above and illustrated by Fig. 5.5, the irradiation by a longer, 1 ns, laser pulse decreases the temperature gradients and cooling rate, and does not result in the formation of amorphous phase. The substantial undercooling in the vicinity of the solidification front, however, leads to the formation of distinct regions of strong supersaturation of both vacancies and interstitials in the parts of the transiently molten region where the maximum undercooling was achieved. At later stage of the solidification process, the interfacial temperature increases due to release of the latent heat, and, as suggested in Sect. 5.4.1, the temperature increase from 0.91 to 0.95 T_m results in decreasing levels of vacancy and interstitial concentrations.

5.3.2 Thermoelastic Modeling of the Dynamic Evolution of Laser-Induced Stresses

The fast rate of energy deposition in short pulse laser processing applications may result not only in the rapid melting and resolidification of a surface region of an irradiated target but also lead to the generation of strong stresses within the absorption region. The laser-induced stresses are particularly high in the regime of stress confinement [100–103], when the time of the laser heating (defined by the laser pulse duration, τ_p, or the time of the electron–phonon equilibration, $\tau_{e\text{-}ph}$, whichever is longer) is shorter than the time required for the mechanical relaxation (expansion) of the heated volume, $\tau_s \sim L_p/C_s$, where C_s is the speed of sound in the target material and L_p is the effective depth of the laser energy deposition. The relaxation of the initial compressive stresses generated by the laser energy deposition can result in the emergence of an unloading tensile component of the stress wave that is sufficiently intense to cause the formation subsurface voids [83, 104–107] or separation of a surface layer from the target in a process commonly called photomechanical spallation [100–103]. At lower laser fluences, the relaxation of laser-induced stresses in the vicinity of a free surface of the target can result in the co-emission of longitudinal, shear, and surface acoustic waves that can be utilized in many practical applications, including nondestructive evaluation of mechanical properties and surface defects [108–110], acoustic desorption [111–115] , and acoustic activation of surface diffusion [116–118].

The computational description of the dynamic evolution of thermoelastic stresses and the emission of stress waves can be described in the framework of the thermoelasticity model [43, 119–122], which complements the thermal model described above with the thermoelastic wave equation:

$$\rho\frac{\partial^2 u_i}{\partial t^2} = \frac{\partial \sigma_{ij}}{\partial x_j} = C_{ijmn}\left[\frac{\partial^2 u_m}{\partial x_j \partial x_n} - \delta_{mn}\alpha_L\frac{\partial T}{\partial x_j}\right], \tag{5.17}$$

$$\sigma_{ij} = C_{ijmn}\left[\frac{\partial u_m}{\partial x_n} - \delta_{mn}\alpha_L(T - T_0)\right], \tag{5.18}$$

where T, u_i, σ_{ij}, α_L, and C_{ijmn} are temperature, displacement, stress tensor, linear thermal expansion coefficient, and stiffness tensor, respectively. The transient heating/cooling induced by adiabatic compression/expansion of the material during the wave propagation can be accounted for by including an additional source term in the form of $-\alpha_L T C_{iimn}\frac{\partial^2 u_m}{\partial t \partial x_n}$ to the heat diffusion equation, (5.13). In literature, the stress tensor is commonly formulated for the case of isotropic material [120, 123]:

$$\sigma_{ij} = \delta_{ij}\lambda \cdot \left(\frac{\partial u_k}{\partial x_k}\right) + \mu \cdot \left(\frac{\partial u_i}{\partial x_j} + \frac{\partial u_j}{\partial x_i}\right) - 3\delta_{ij}B\alpha_L(T - T_0), \tag{5.19}$$

where $\lambda = B - 2G/3$ and $\mu = G$ are Lamé constants, B and G are bulk and shear moduli of the material. This formulation of the thermoelasticity model is suitable for the description of an irradiation process not only in Cartesian but also in cylindrical coordinates [123]. Equation (5.19), however, is not valid for description of highly anisotropic materials, e.g., single crystal silicon, where the description provided by (5.17 and 5.18) has to be applied.

The capabilities of the thermoelasticity model are illustrated in Fig. 5.7, which provides a snapshot of laser-generated elastic waves in a silica glass plate. The simulation is performed for an absorbed fluence of 10 J/cm^2, a laser spot size of 21 μm, and the characteristic absorption depth assumed to be 10 μm based on the depth of grooves observed in experiments. This simulation is motivated by an intriguing experimental observation of the spatial distribution of laser-induced damage in a silica glass plate undergoing laser cutting [124]. This study demonstrates the generation of damage lines on the rear surface of the plate, which are located not directly

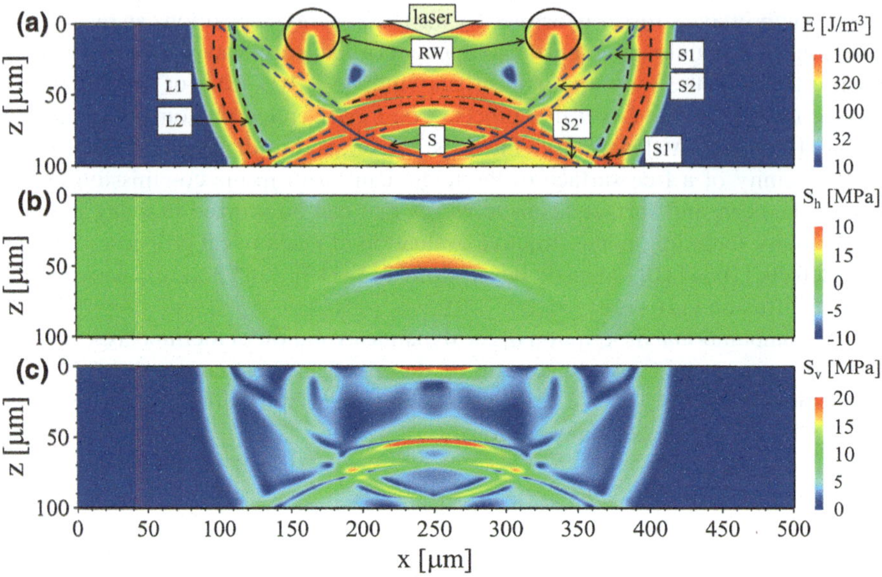

Fig. 5.7 Generation of elastic waves in a silica substrate irradiated by a laser pulse at 10 J/cm^2. The energy density (**a**), hydrostatic stress (**b**), and Von-Misses stress (**c**) are plotted for a time of 25 ns after the laser pulse. In **a**, the longitudinal and shear waves are schematically outlined by black and blue lines, respectively. Rayleigh waves, longitudinal compressive, longitudinal tensile, and shear waves generated by rapid laser heating are marked as RW, L1, L2, S, respectively. Shear waves emitted due to interaction of L1 and L2 waves with top and bottom free surfaces are marked as S1, S2, S1′, S2′. By the time of 25 ns, L1 and L2 waves are already reflected from the bottom free surface and propagate upwards. The thermal stresses generated by laser heating cannot be completely relaxed by the material expansion towards the free surface and, therefore, are still present in the vicinity of the irradiated area

underneath the laser spot but at ~37° angle. The rear surface damage is attributed in [124] to the interaction of a shear wave emitted from the irradiated spot and a surface Rayleigh wave generated due to reflection of a longitudinal wave from the rear surface. To verify this hypothesis, thermoelasticity modeling of the generation of stress waves in a silica glass substrate is performed in Cartesian coordinates for a two-dimensional geometry. Laser-induced heating of the target triggers generation of a variety of acoustic waves that interact with the free surfaces of the silica glass plate and create a complex dynamic pattern of transient stresses within the plate, as depicted in Fig. 5.7. The calculations, however, do not support the original explanation of the rear surface damage suggested in [124], as the reflection of the longitudinal waves from the bottom surface of the plate does not produce Rayleigh waves of any significant magnitude. Therefore, an alternative explanation of the experimental observations is needed.

5.3.3 *Material Redistribution Through Elastoplasticity and Hydrodynamic Flow*

The generation of strong thermoelastic stresses can not only lead to the emission of elastic waves and photomechanical damage but can also result in material redistribution within the laser spot or removal from the target. In particular, the relaxation of laser-induced stresses plays a major role in the laser-induced forward transfer (LIFT) [125–128] and thin film patterning/scribing [93, 129–131], where the initial motion of the film driven by the laser-induced stresses defines to a large extent the final outcome of the target modification and/or parameters of the ejected material. The computational analysis of laser interactions with thin films has been performed with both atomistic [93, 131, 132] and continuum [122, 123, 133] models, and has provided important insights into the complex interplay between the dynamic relaxation of laser-induced stresses, phase transformations, and material redistribution/removal.

As an example of continuum simulation of LIFT, the results of a simulation of the forward transfer of a SnO_2 film deposited on a silica or a polymer substrate and irradiated by a nanosecond laser pulse are illustrated by Fig. 5.8. Two dimensional finite element formulation of the elasticity equation in cylindrical coordinates is applied to study the initial stage of LIFT process involving delamination of the film from the donor substrate and formation of a microbump [122]. The simulation is performed in the irradiation regime where the vaporization of the film material is negligible, and the film dynamics is governed mainly by mechanical processes. The temperature increase predicted by the thermal model is used to calculate the initial distribution of elastic stresses, and the following evolution of the film is described by solving the elasticity equation.

The effect of the substrate elastic properties on the initial stage of LIFT process is assessed by performing simulations for silica and polymer substrates, as shown in Fig. 5.8. Heating the film leads to initiation of a bending wave near the edges of the irradiated spot that propagates in the film and collapses at the center of the spot. According to Fig. 5.8c, three stages of the film detachment can be distinguished in the case of silica substrate: increase of the wave amplitude (1), wave propagation (2), and the final swift uplifting of the central part of the film (3). The simulations reveal a remarkable fact that both the wave amplitude and the velocity component of the film motion at the wave peak are increasing with propagation of the wave toward the center. In the case of the polymer substrate, however, the film has a possibility to expand not only in the direction of the free boundary but also partially toward the donor substrate, compressing the latter. This lowers the stress value at the edges of the spot and, thus, prevents the formation of large velocity gradients normal to the substrate. Therefore, the film is separated uniformly and gently transferred to the acceptor, mitigating damage to the receiver substrate.

For longer laser pulses, outside the regime of stress confinement, the relaxation of thermoelastic stresses takes place during the laser energy deposition, and the processes that control the material response to the laser irradiation are largely of thermal nature and include melting, evaporation from the surface, and, at high laser intensi-

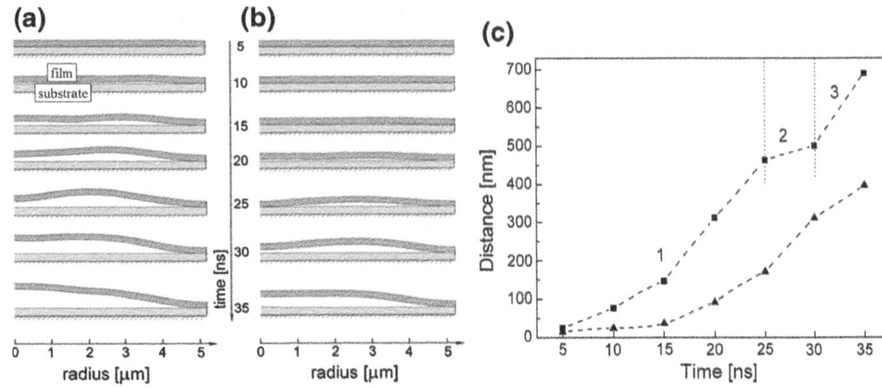

Fig. 5.8 Stages of a 250 nm SnO_2 film deformation demonstrating the initiation of a bending wave at the edge of the irradiation spot followed by detachment of the film from a silica (**a**) and a polymer (**b**) substrates. The maximum distance between the film and the substrate as a function of time (**c**) is shown by squares for the silica and triangles for the polymer substrates. The following stages of the film detachment are identified: increase of the wave amplitude (1), wave propagation (2), and the final swift uplifting of the central part of the film (3). The images are adapted from [122]

ties, a volumetric ablation through phase explosion, or explosive boiling [134–136]. The melt dynamics induced by pressure gradients, however, can still play an important role in laser processing even for continuous wave (CW) laser irradiation. In particular, the melt expulsion from the center of the laser spot is one of the primary mechanisms of CW laser damage and material removal [137–140]. The two main driving forces responsible for melt expulsion are the recoil pressure created by the evaporation process and the Marangoni effect. The melt expulsion by the recoil pressure is caused by the spatial variation of the vapor pressure within the laser spot [141–143], whereas the Marangoni effect is related to the variation of the surface tension along the non-uniformly heated surface of the molten pool [144, 145].

Computationally, the melting, evaporation, and melt flow in the course of material processing by long laser pulses can be described in hydrodynamic simulations based on the solution of Navier-Stokes equations [146]. In particular, the relative contributions of the recoil pressure and Marangoni effects to the CW laser melting and damage of aluminum films have been studied for a broad range of process parameters realized in both laser damage and material processing applications [146]. The simulations, performed with a two-phase hydrodynamic model, reveal only a marginal effect of the Marangoni stresses on the overall picture of melt flow and the melt-through time. The recoil pressure effect, on the contrary, is found to be capable of strongly decreasing the melt-through time in a certain range of laser intensity. Moreover, at high laser intensities, the surface of the molten pool can become unstable with respect to the appearance of waves propagating towards the edge of the molten pool, as illustrated in Fig. 5.9. The formation of waves on the surface of the molten pool and their interaction with the edge of the pool can produce an irregular complex shape of the rim of resolidified material commonly observed around the

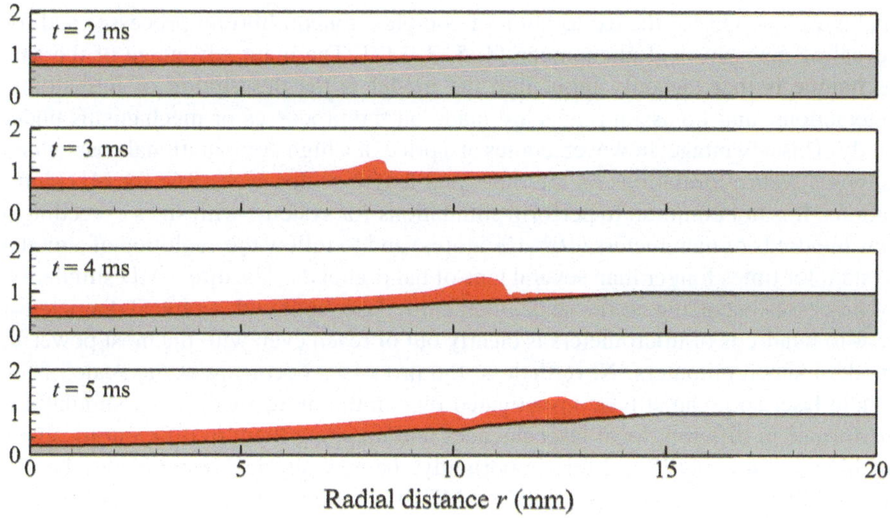

Fig. 5.9 The evolution of the shape of the molten pool (red region) in laser melting of a 1-mm-thick aluminum film irradiated by a CW laser with intensity of 3×10^6 W cm^{-2} and FWHM laser spot radius of 1 cm. The snapshots from the simulation are taken at times t after the start of the laser irradiation shown on the frames. The gradient of the recoil pressure within the laser spot induces fluid flow in the radial direction and leads in the appearance of a wave propagating towards the edge of the molten pool. The melt expulsion has a strong effect on the melt-through time of the free-standing film. The image is adapted from [146]

"crater" generated in the central part of the laser spot by the melt expulsion and active evaporation [138–140].

5.4 Molecular Dynamics Modeling of Laser-Materials Interactions

The brief overview provided in Sect. 5.3 demonstrates the ability of the continuum-level models to provide a computationally-efficient description of a broad range of laser-induced thermal and mechanical processes at experimental time- and length-scales. The predictive power of the continuum methods, however, is limited by the need for a priori knowledge of all processes that take place under given irradiation conditions and the availability of reliable mathematical description of these processes. The complexity and highly nonequilibrium nature of laser-induced processes, particularly in the short-pulse laser irradiation regime, can challenge some of the key assumptions of the continuum models and calls for complementary approaches that may help to design and verify the continuum models.

The atomistic modeling based on the molecular dynamics (MD) simulation technique is capable of providing atomic-level insights into the laser-induced processes and is well suited for the exploration of complex nonequilibrium processes and for revealing new physical phenomena [4, 5, 7, 147]. The main advantage of the MD technique is that the only input into the model is the description of interatomic interactions, and no assumptions are made on the processes or mechanisms under study. This advantage, however, comes at a price of a high computational cost, which imposes severe limitations on the time- and length-scales accessible for MD simulations. It is not common to perform simulations for systems with sizes exceeding a few hundreds of nanometers (10^8–10^9 atoms) and to follow the evolution of a model system for times longer than several tens of nanoseconds. The direct MD simulation of processes occurring on the scale of an entire laser spot with a typical diameter of tens to hundreds of micrometers is clearly out of reach even with the most powerful modern supercomputers. Nevertheless, the processes occurring at the scale of the whole laser spot can still be investigated by combining results of MD simulations performed at different local laser fluences and mapping them to different locations within the laser spot [79]. More importantly, the new physical insights into the fast nonequilibrium processes, provided by MD simulations, can serve as the basis for the design of advanced continuum-level models fully accounting for the complexity of short pulse laser-materials interactions. Several examples illustrating the ability of MD simulations to reveal the processes contributing to material modification by short laser pulses are provided below.

5.4.1 Molecular Dynamics: Generation of Crystal Defects

The ability of short pulse laser irradiation to produce high densities of crystal defects and unusual defect configurations has been demonstrated in a number of large-scale MD simulations [83–86, 148, 149]. In particular, the simulations reveal the mechanisms responsible for surface nanocrystallization [83, 93, 147], generation of unusual metastable phases [149], high densities of dislocations, twin boundaries, and vacancies [83, 84, 86, 148]. Two examples discussed in this section are chosen to illustrate the application of MD simulations for parametrization of continuum-level models and for revealing new mechanisms of microstructure modification activated by short pulse laser irradiation.

As discussed in Sect. 5.3.1, the continuum-level simulations of laser melting and resolidification of silicon substrate can predict the morphology of amorphous phase and the distribution of point defects in the region of a target modified by laser irradiation. These predictions, however, rely on the availability of a kinetic model for the temperature dependence of the solidification front propagation, conditions for the formation of a stable amorphous phase, and the generation of crystal defects in the solidification process. Under conditions when direct experimental probing of the rapid nonequilibrium phase transformations and generation of crystal defects in short pulse laser processing remains challenging, the MD simulations can provide the information needed for the design of continuum models.

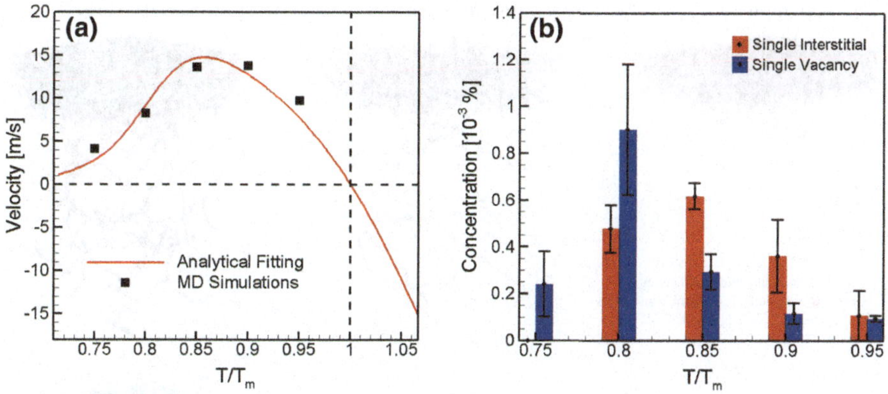

Fig. 5.10 Velocity of the crystal-liquid interface as a function of temperature (**a**), and concentration of point defects generated during solidification (**b**) predicted in MD simulations of a two-phase crystal-liquid Si system with a (100) orientation of the crystal-liquid interface under constant temperature and zero pressure. In **a**, the black squares show the velocity of the solidification front predicted in MD simulations, and the red line depicts fitting of (5.14) to the MD results. In **b**, the red and blue bars represent the concentration of single interstitials and vacancies, respectively, averaged over four simulations performed under the same conditions. The black bars show the standard deviation of the mean calculated based on the four simulations

This application of MD model is illustrated in Fig. 5.10 that shows the temperature dependence of the velocity of the solid-liquid interface and the concentration of point defects generated in silicon undergoing rapid solidification at different levels of undercooling below the equilibrium melting temperature. The simulations start from a solid-liquid coexistence system and are performed under constant undercooling and zero external pressure. The velocity of the epitaxial growth of a (100) crystal-liquid interface is evaluated at different temperatures, and the results are used to define the parameters of the analytical dependence given by (5.14), as shown in Fig. 5.10a. The analysis of the resolidified parts of the system reveals that a substantial number of individual point defects and their clusters are generated in the rapid solidification process. The concentrations of single interstitials and single vacancies are averaged over four simulations performed under the same undercooling conditions, and the results are shown in Fig. 5.10b as a function of temperature. The generation of clusters of point defects is not observed in simulations performed at temperatures higher than 0.9 T_m. The computational predictions on the concentrations of vacancies and interstitials left behind the rapidly advancing solidification front are used for parametrization of the corresponding source terms in (5.15) and (5.16) used in the continuum-level modeling of the laser-induced generation of crystal defects in silicon targets (see Sect. 5.3.1).

The MD method has also been successfully applied to explain recent electron backscatter diffraction measurements [150], which demonstrated strong effect of the crystallographic orientation of grains in polycrystalline metal targets on the generation and accumulation of crystal defects in the surface region of the irradiated

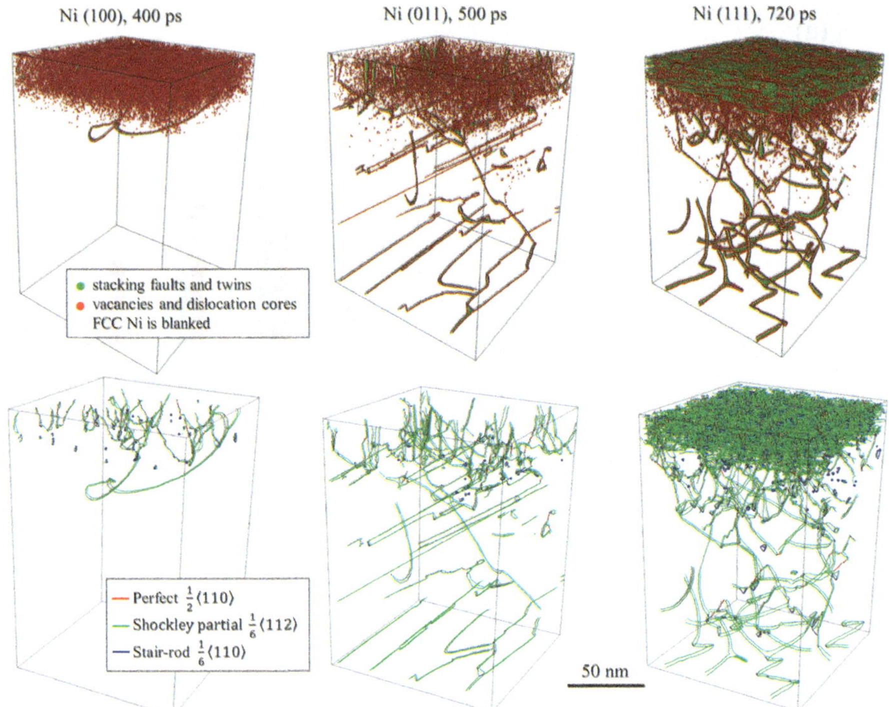

Fig. 5.11 The defect configurations predicted in TTM-MD simulations of (100), (011), and (111) Ni targets irradiated by a 50 fs laser pulse at an absorbed laser fluence of 60 mJ/cm^2. The top row depicts atomic configurations where only the atoms that belong to crystal defects are shown and are colored based on the local structure environment. The bottom row shows the dislocation configurations, with dislocations colored based on their type. The configurations are shown after complete resolidification of the targets

material. In order to adapt the classical MD method for simulation of laser interaction with metals, the description of the laser excitation of the conduction band electrons, electron-phonon coupling, and electronic heat conduction is incorporated into the model by combining MD with TTM discussed above, in Sect. 5.2.1. The results of large-scale atomistic simulations performed with the combined TTM-MD model [151] for Ni targets with (001), (011), and (111) surface orientations confirm the significant influence of the crystal orientation on the formation of sub-surface defects and provide detailed information on the defect configurations generated in each target, as shown in Fig. 5.11.

The results of the simulations also reveal the mechanisms responsible for the generation of dislocation in the three targets. The fast laser heating and the corresponding stresses generated in the surface region of the irradiated targets result in the emission of numerous Shockley partial dislocations in targets with (011) and (111) surface orientations, with the latter exhibiting the highest density of dislocations activated in 3 different {111} slip planes and interacting with each other to produce complex dislocation reactions. In the case of (100) target, however, even though the thermal stresses have approximately the same magnitude as for (011) and (111) surface orientations, the shear stress resolved on existing slip systems does not reach the critical value required for the onset of dislocation emission. In addition to the prompt dislocation emission during the dynamic relaxation of laser-induced stresses, the generation of dislocation loops and other crystal defects takes place during the resolidification of the transiently melted top regions of the irradiated targets. In particular, even though no dislocation emission is observed in the simulation with (100) surface orientation during the initial dynamic relaxation of the laser-induced stresses, a number of dislocations are generated in this system during the following propagation of the solidification front. In all simulations, the rapid advancement of the solidification front results in the generation of a high concentration of vacancies in the resolidified region, which can reach the level of 10^{-3} of the lattice sites, as can be seen from the upper row of snapshots shown in Fig. 5.11.

One notable effect that is only observed in a course of the epitaxial resolidification of the target with (111) surface orientation is the formation of a high density of coherent twin boundaries, Σ3 {111}, at the final stage of the solidification process [86], as shown in Figs. 5.11 and 5.12. The initial stage of the resolidification process proceeds by the epitaxial regrowth of the crystal. The defect structure of the resolidified region, however, changes abruptly when the temperature at the solidification front drops down to about 0.86 T_m, and a large number of coherent twin boundaries appear in the planes parallel to the advancing (111) crystal-liquid interface. The depth of the region affected by the growth twinning as well the thickness of twined domains predicted in atomistic simulation are in a good quantitative agreement with the results of experimental study, as can be seen from the comparison of images shown in Fig. 5.12b, c.

To predict the dependence of the thickness of the surface layer affected by the growth twinning on laser fluence, a series of continuum-level simulations are performed with the TTM enhanced with a description of nonequilibrium melting and solidification [86]. The dependence of the velocity of solidification front calculated in MD simulations and the threshold temperature corresponding to the onset of growth twining are used to parametrize the continuum-level model. The predicted evolution of temperature, melting depth, and depth affected by growth twinning is shown in Fig. 5.12d. This example demonstrates the utility of a multiscale approach that eliminates the need to perform a series of computationally-expensive MD simulations for each set of conditions.

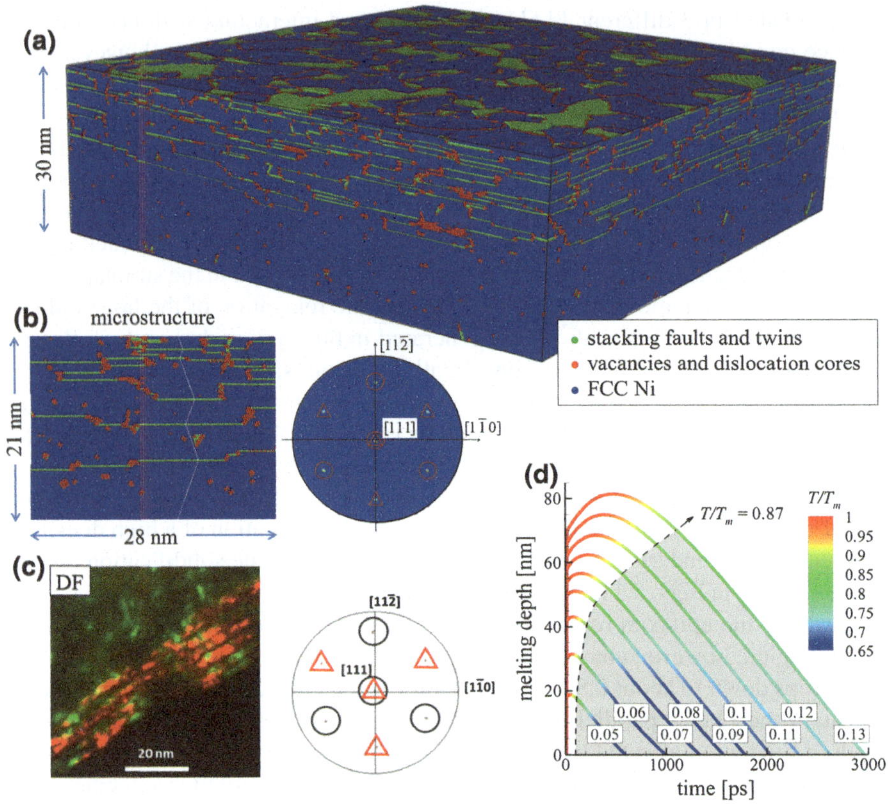

Fig. 5.12 Atomic configuration generated in a TTM-MD simulation of Ni (111) target irradiated by a 50 fs laser pulse at an absorbed laser fluence of 60 mJ/cm^2 (**a**), a representative vertical cross-section of the surface layer of the resolidified target and ⟨111⟩ pole figure calculated for 20 nm top region of the target (**b**), experimental results confirming the presence of twinned domains in the surface region of laser-processed Ni (111) target (**c**), and the evolution of the melting depth and temperature predicted in a series of continuum-level simulations parametrized based on MD results (**d**). In **a**, only a top 30 nm surface region is shown, and atoms are colored according to the local crystal structure environment. In **b**, the dashed white line shows the altering orientation of atomic planes in the twinned domains in the left panel, and the original and new/twinned ⟨111⟩ axes are marked by circles and triangles, respectively, in the pole figure shown in the right panel. In **c**, transmission electron microscopy (TEM) dark field images of the central part of the irradiated laser spot are superimposed to produce a false color image highlighting the spatial distribution of the epilayers (green) and twinned domains (red) in the left panel, while electron backscatter diffraction (EBSD) indexed crystal orientations are plotted in the ⟨111⟩ pole figure shown in the right panel. In **d**, the gray region corresponds to the conditions of strong undercooling ($T < 0.87\ T_m$), when the growth twinning is predicted in the theoretical analysis and atomistic TTM-MD simulations; different curves correspond to the different values of absorbed laser fluence marked in the plot in units of J/cm^2. The images are adapted from [86]

5.4.2 Molecular Dynamics: Ablative Generation of Laser-Induced Periodic Surface Structures

As already discussed above, the direct modeling of processes occurring on the scale of an entire laser spot is not feasible with MD method, and the results of the atomistic modeling can either be used for parametrization of a continuum model (e.g., see Sect. 5.3.1) or mapped to different locations within the laser spot, thus producing a "mosaic" view of processes occurring at different local fluences deposited within the spot by the (typically Gaussian) laser beam [79]. The latter approach, however, cannot account for the long-range lateral (parallel to the surface) redistribution of the molten material within the laser spot and the formation of large-scale surface topography frozen in during resolidification. Thus, in general, the analysis of the lateral flow in the molten pool remains squarely in the domain of continuum-level models, as illustrated in Fig. 5.9 and discussed in Sect. 5.3.3. Only in special cases of strong localization of the laser energy deposition down to submicron dimensions produced, e.g., through tight focusing of the laser beam [152–155], local field enhancement in the vicinity of a tip of a scanning probe microscope [156–159], or interference of multiple laser beams [160, 161], large-scale MD simulations can be directly used for investigation of the formation of frozen surface features in the course of the laser-induced lateral material redistribution [93, 131, 160–162]. One example illustrating the ability of MD simulations to provide a complete picture of the complex dynamic processes leading to the submicron surface patterning is discussed below for the single-pulse generation of LIPSS in the regimes of strong ablation [163] .

As discussed above, in Sect. 5.2.1, the formation of LIPSS is related to the dynamic optical response of the irradiated material to the ultrashort laser excitation, when the interference of the incident laser wave and surface electromagnetic wave produces spatial modulation of the laser energy deposition [45, 164]. Typically, LIPSS are produced by repetitive irradiation of the same area by multiple laser pulses with relatively low laser fluence, in the regime of surface melting and resolidification [164–168]. It is generally believed that an increase of the laser fluence well above the ablation threshold would result in disordering or even complete erasure of the periodic structures by the recoil pressure of the ablation plume. A number of recent studies, however, have demonstrated that the formation of high-quality LIPSS is possible in the single pulse irradiation regime at laser fluences well above the ablation threshold [47, 169–171].

In order to establish the mechanisms of the single pulse ablative LIPSS formation, a large-scale TTM-MD simulation has been performed for a Cr target irradiated by a 200 fs laser pulse [163]. The irradiation is assumed to produce a spatially modulated energy deposition arising from the interference of the incident laser wave and surface electromagnetic wave. The period of the sinusoidal modulation is taken to be 260 nm, and the maximum and minimum of absorbed fluences are 2000 and 3000 J/m^2 respectively, which corresponds to 20% modulation of the laser energy deposition with respect to the average level of 2500 J/m^2. The latter is approximately 70% above the phase explosion threshold for Cr target. The initial dimensions of the

TTM-MD domain are 260 nm × 43 nm × 87 nm, which corresponds to 81 million atoms, and the longest dimension matches the period of the laser energy deposition profile.

The response of the Cr target to the modulated laser energy deposition is illustrated in Fig. 5.13, where a series of snapshots of atomic configurations obtained in the simulation is shown up to 2100 ps after the laser pulse, when the surface of the target completely solidifies. The simulation reveals that the fast heating of Cr target by the laser pulse leads to an explosive decomposition of about 30–40 nm deep surface layer into a mixture of vapor and liquid droplets. Lateral pressure gradients in the plume generated by the spatially modulated laser ablation drive the vapor and liquid droplets to the region located above the minima of the laser energy deposition at the target surface. The material redistribution leads to formation of a high-density region evolving into an elongated liquid wall extending up to ~600 nm above the surface of the target. The upper part of the liquid wall disintegrates into droplets while the base of the wall solidifies on the timescale of ~2 ns, producing a ~100-nm-long frozen surface feature. Interestingly, while the amount of material removed from the trough area of the modified surface corresponds to the ablation depth of ~40 nm, the frozen surface protrusion extends above the level of the original surface of the target marked by dashed rectangles in Fig. 5.13. This computational prediction is consistent with the results of experimental study of the single-pulse LIPSS formation on Au surface [170, 172], where frozen wall-like structures, similar to those shown in Fig. 5.13, are observed. The computational insights into the LIPSS formation mechanisms may help in designing approaches for increasing the processing speed and improving the quality of the laser-patterned periodic surface structures.

5.5 Concluding Remarks

The overview of the diverse range of theoretical and computational approaches to the description of laser-materials interactions provided in this Chapter illustrates the ability of the advanced computational methods to provide a reliable description of different facets of these complex phenomena. In particular, while further advances in the theoretical treatment of the optical response of electronically-excited materials are certainly desired, the current models reviewed in Sect. 5.2 do provide a description of the transient variation of the optical properties of metals, semiconductors, and dielectrics excited by the laser irradiation that is sufficiently accurate for reliable evaluation of the fraction of the incident laser energy absorbed by the target, as well as spatial distribution of the deposited energy within the target. Similarly, the continuum-level models are capable of predicting the temperature evolution, elastic and plastic deformation, generation of stress waves, phase transformations (melting, resolidification, amorphization, vaporization, and volume ablation processes), hydrodynamic flow of transiently melted material, and generation of crystal defects on the macroscopic scale of the whole laser spot, as illustrated by several examples discussed in Sect. 5.3. Finally, the large-scale atomistic modeling has been demon-

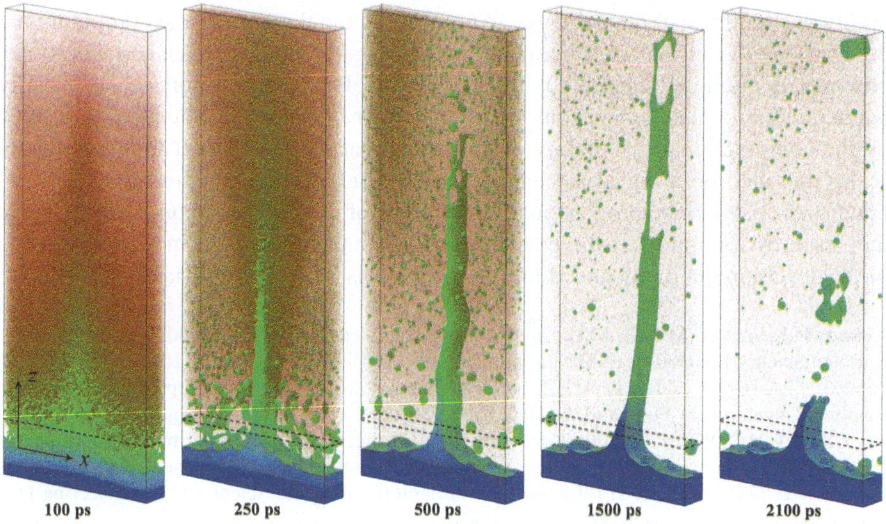

Fig. 5.13 A sequence of atomistic snapshots generated in a TTM-MD simulation of single pulse ablative LIPSS formation on a Cr surface. The snapshots show a part of the system with z coordinate ranging from -80 to 600 nm with respect to the location of the original surface of the target marked by the dashed rectangles. The atoms in the snapshots are colored by their potential energy, with color scale ranging from -3.8 eV (blue) to -1.4 eV (red). With this coloring scheme, dark-blue regions correspond to the solid phase, light-blue and green represent liquid phase and free surfaces, and red atoms belong to the vapor phase. The images are adapted from [163]

strated to yield important insights into the mechanisms and kinetics of complex, highly nonequilibrium processes involved in the laser-materials interactions. The examples provided in Sect. 5.4 illustrate the ability of large-scale MD simulations to reveal the critical information on the kinetics of laser-induced melting and resolidification, generation of crystal defects (vacancies, interstitials, dislocations, and twin boundaries), as well as the rapid material ejection and redistribution leading to the formation of LIPSS.

The remaining challenges include the incorporation of the detailed material-specific information on the electronic structure into the theoretical treatment of the transient optical and transport properties under conditions of strong electronic excitation, design of a reliable description of the metastable states, nonequilibrium phase transformations, and generation of crystal defects in the continuum-level models, as well as the extension of atomistic modeling to complex multi-component materials and multi-phase targets. Even more importantly, the development of integrated computational tools synergistically combining and enhancing the capabilities of different models is clearly needed in the field of modeling of laser-materials interactions. The factors that hamper the emergence of integrated models include the complexity of the task of combining dissimilar physical concepts and computational approaches, as well as the natural tendency of the research groups to focus on the areas of their expertise. Close and long-term interdisciplinary collaborations are highly desired for making a disruptive progress in the theoretical and computational description of laser-materials interactions. The establishment of such collaborations, however, is often incompatible with current models of research funding, especially when international collaborations are involved. An additional challenge is presented by the computational complexity of models operating on different time and length scales and incorporating advanced mathematical algorithms. Optimization of computational codes for heterogeneous architectures of modern supercomputers is a task that requires continuous involvement of experts in the high-performance computing in the development of advanced multiscale models for laser-materials interactions.

Acknowledgements M.V.S., M.H., and L.V.Z. acknowledge financial support provided by the National Science Foundation (NSF) through Grants CMMI-1436775, CMMI-1562929, and DMR-1610936. S.A.L., Y.L., T.J.-Y.D., V.P.Z., and N.M.B. acknowledge financial support of the European Regional Development Fund and the state budget of the Czech Republic (project BIATRI: Grant No. CZ.02.1.01/0.0/0.0/15_003/0000445 and NPU I Project No. LO1602. Computational support enabling large-scale atomistic modeling was provided by the Oak Ridge Leadership Computing Facility (INCITE project MAT130) and NSF through the Extreme Science and Engineering Discovery Environment (project TG-DMR110090).

References

1. E.S. Massaro, A.H. Hill, E.M. Grumstrup, ACS Photon. **3**, 501 (2016)
2. A.M. Lindenberg, S.L. Johnson, D.A. Reis, Annu. Rev. Mater. Res. **47**, 425 (2017)
3. A. Barty, S. Boutet, M.J. Bogan, S. Hau-Riege, S. Marchesini, K. Sokolowski-Tinten, N. Stojanovic, R. Tobey, H. Ehrke, A. Cavalleri, S. Düsterer, M. Frank, S. Bajt, B.W. Woods, M.M. Seibert, J. Hajdu, R. Treusch, H.N. Chapman, Nat. Photon. **2**, 415 (2008)
4. M.V. Shugaev, C. Wu, O. Armbruster, A. Naghilou, N. Brouwer, D.S. Ivanov, T.J.-Y. Derrien, N.M. Bulgakova, W. Kautek, B. Rethfeld, L.V. Zhigilei, MRS Bull. **41**, 960 (2016)
5. C. Wu, E.T. Karim, A.N. Volkov, L.V. Zhigilei, in *Lasers in Materials Science*, ed. by M. Castillejo, P.M. Ossi, L.V. Zhigilei (Springer International Publishing, Switzerland, 2014), pp. 67–100
6. N.M. Bulgakova, V.P. Zhukov, in *Lasers in Materials Science*, ed. by M. Castillejo, P.M. Ossi, L.V. Zhigilei (Springer International Publishing, Switzerland, 2014), pp. 101–124
7. L.V. Zhigilei, Z. Lin, D.S. Ivanov, E. Leveugle, W.H. Duff, D. Thomas, C. Sevilla, S.J. Guy, in *Laser-Surface Interactions for New Materials Production: Tailoring Structure and Properties*, ed. by A. Miotello, P.M. Ossi (Springer, New York, 2010), pp. 43–79
8. N.M. Bulgakova, R. Stoian, A. Rosenfeld, I.V. Hertel, in *Laser-Surface Interactions for New Materials Production: Tailoring Structure and Properties*, ed. by A. Miotello, P.M. Ossi (Springer, New York, 2010), pp. 81–97
9. P. Drude, Ann. Phys. **306**, 566 (1900)
10. M.A. Ordal, R.J. Bell, R.W. Alexander, L.L. Long, M.R. Querry, Appl. Opt. **24**, 4493 (1985)
11. M.S. Dresselhaus, MIT Solid State Physics Course (2001)
12. K. Sokolowski-Tinten, D. von der Linde, Phys. Rev. B **61**, 2643 (2000)
13. F. Wooten, *Optical Properties of Solids* (Academic Press, New-York, 1972)
14. Y. Ren, J.K. Chen, Y. Zhang, J. Huang, Appl. Phys. Lett. **98**, 191105 (2011)
15. S.I. Ashitkov, P.S. Komarov, V.V. Zhakhovsky, Y.V. Petrov, V.A. Khokhlov, A.A. Yurkevich, D.K. Ilnitsky, N.A. Inogamov, M.B. Agranat, J. Phys.: Conf. Ser. **774**, 012097 (2016)
16. J. Zuo, A. Erbe, Phys. Chem. Chem. Phys. **12**, 11467 (2010)
17. A.D. Rakić, A.B. Djurišić, J.M. Elazar, M.L. Majewski, Appl. Opt. **37**, 5271 (1998)

18. H.S. Sehmi, W. Langbein, E.A. Muljarov, Phys. Rev. B **95**, 115444 (2017)
19. S.E. Kirkwood, Y.Y. Tsui, R. Fedosejevs, A.V. Brantov, V.Y. Bychenkov, Phys. Rev. B **79**, 144120 (2009)
20. Y.T. Lee, R.M. More, Phys. Fluids **27**, 1273 (1984)
21. D.V. Knyazev, P.R. Levashov, Phys. Plasmas **21**, 073302 (2014)
22. A.V. Brantov, V.Y. Bychenkov, W. Rozmus, J. Exp. Theor. Phys. **106**, 983 (2008)
23. H. Brysk, P.M. Campbell, P. Hammerling, Plasma Phys. **17**, 473 (1975)
24. Z. Lin, L.V. Zhigilei, V. Celli, Phys. Rev. B **77**, 075133 (2008)
25. M.B. Agranat, N.E. Andreev, S.I. Ashitkov, M.E. Veĭsman, P.R. Levashov, A.V. Ovchinnikov, D.S. Sitnikov, V.E. Fortov, K.V. Khishchenko, JETP Lett. **85**, 271 (2007)
26. E.I. Nefyodov, S.M. Smolskiy, *Understanding of Electrodynamics, Radio Wave Propagation and Antennas* (Scientific Research Publishing, Inc., 2013)
27. S.A. Furman, A.V. Tikhonravov, *Basics of Optics of Multilayer Systems* (Editions Frontières, Gif-Sur-Yvette, France, 1992)
28. O.A. Bulgakova, N.M. Bulgakova, V.P. Zhukov, Appl. Phys. A **101**, 53 (2010)
29. H. Raether, *Surface Plasmons on Smooth and Rough Surfaces and on Gratings* (Springer, Berlin, 1986)
30. E.D. Palik, *Handbook of Optical Constants of Solids* (Academic Press, San Diego, 1998)
31. C.-D. Wen, I. Mudawar, Int. J. Heat Mass Transf. **49**, 4279 (2006)
32. H.E. Bennett, J.O. Porteus, J. Opt. Soc. Am. **51**, 123 (1961)
33. J. Hohlfeld, U. Conrad, J.G. Muller, S.S. Wellerrshoff, E. Matthias, in *Nonlinear Optics in Metals*, ed. by K.H. Bennemann (Clarendon Press, Oxford, 1998), pp. 219–267
34. A.Y. Vorobyev, C. Guo, J. Appl. Phys. **110**, 043102 (2011)
35. S.I. Anisimov, B.L. Kapeliovich, T.L. Perel'man, Sov. Phys.—JETP **39**, 375 (1974)
36. S.A. Lizunov, R. Fang, A.Y. Vorobyev, V.P. Zhukov, N.M. Bulgakova, Ch. Guo, to be published (2018)
37. S.S. Wellershoff, J. Hohlfeld, J. Güdde, E. Matthias, Appl. Phys. A **69**, S99 (1999)
38. E. Bévillon, J.P. Colombier, V. Recoules, R. Stoian, Appl. Surf. Sci. **336**, 79 (2015)
39. O. Matsuda, O.B. Wright, D.H. Hurley, V. Gusev, K. Shimizu, Phys. Rev. B **77**, 224110 (2008)
40. Y. Levy, T.J.-Y. Derrien, N.M. Bulgakova, E.L. Gurevich, T. Mocek, Appl. Surf. Sci. **374**, 157 (2016)
41. N.A. Inogamov, Y.V. Petrov, J. Exp. Theor. Phys. **110**, 446 (2010)
42. J. Winter, S. Rapp, M. Schmidt, H.P. Huber, Appl. Surf. Sci. **417**, 2 (2017)
43. M.V. Shugaev, N.M. Bulgakova, Appl. Phys. A **101**, 103 (2010)
44. B. Rethfeld, A. Kaiser, M. Vicanek, G. Simon, Phys. Rev. B **65**, 214303 (2002)
45. J.E. Sipe, J.F. Young, J.S. Preston, H.M. Van Driel, Phys. Rev. B **27**, 1141 (1983)
46. J. Bonse, J. Krüger, S. Höhm, A. Rosenfeld, J. Laser Appl. **24**, 042006 (2012)
47. I. Gnilitskyi, T.J.-Y. Derrien, Y. Levy, N.M. Bulgakova, T. Mocek, L. Orazi, Sci. Rep. **7**, 8485 (2017)
48. Z. Chen, B. Holst, S.E. Kirkwood, V. Sametoglu, M. Reid, Y.Y. Tsui, V. Recoules, A. Ng, Phys. Rev. Lett. **110**, 135001 (2013)
49. B. Öktem, I. Pavlov, S. Ilday, H. Kalaycıoğlu, A. Rybak, S. Yavaş, M. Erdoğan, F.Ö. Ilday, Nat. Photon. **7**, 897 (2013)
50. A. Ruiz de la Cruz, R. Lahoz, J. Siegel, G.F. de la Fuente, J. Solis, Opt. Lett. **39**, 2491 (2014)
51. Y. Levy, N.M. Bulgakova, T. Mocek, Proc. SPIE **10228**, 102280T (2017)
52. P.P. Pronko, S.K. Dutta, D. Du, R.K. Singh, J. Appl. Phys. **78**, 6233 (1995)
53. N.M. Bulgakova, V.P. Zhukov, I. Mirza, Y.P. Meshcheryakov, J. Tomáštík, V. Michálek, O. Haderka, L. Fekete, A.M. Rubenchik, M.P. Fedoruk, T. Mocek, Proc. SPIE **9735**, 97350N (2016)
54. I. Mirza, N.M. Bulgakova, J. Tomáštík, V. Michálek, O. Haderka, L. Fekete, T. Mocek, Sci. Rep. **6**, 39133 (2016)
55. J.R. Peñano, P. Sprangle, B. Hafizi, W. Manheimer, A. Zigler, Phys. Rev. E **72**, 036412 (2005)
56. H.M. Van Driel, Phys. Rev. B **35**, 8166 (1987)
57. B.E. Sernelius, Phys. Rev. B **40**, 12438 (1989)

58. P. Saeta, J.K. Wang, Y. Siegal, N. Bloembergen, E. Mazur, Phys. Rev. Lett. **67**, 1023 (1991)
59. K. Sokolowski-Tinten, J. Bialkowski, D. Von Der Linde, Phys. Rev. B **51**, 14186 (1995)
60. A. Rousse, C. Rischel, S. Fourmaux, I. Uschmann, S. Sebban, G. Grillon, P. Balcou, E. Förster, J.P. Geindre, P. Audebert, J.C. Gauthier, D. Hulin, Nature **410**, 65 (2001)
61. H. Hu, H. Ding, F. Liu, Sci. Rep. **5**, 8212 (2015)
62. M. Beye, F. Sorgenfrei, W.F. Schlotter, W. Wurth, A. Fohlisch, Proc. Natl. Acad. Sci. **107**, 16772 (2010)
63. S. Sastry, C. Austen, Angell. Nat. Mater. **2**, 739 (2003)
64. V.V. Vasisht, S. Saw, S. Sastry, Nat. Phys. **7**, 549 (2011)
65. P. Stampfli, K.H. Bennemann, Phys. Rev. B **46**, 10686 (1992)
66. P. Silvestrelli, A. Alavi, M. Parrinello, D. Frenkel, Phys. Rev. Lett. **77**, 3149 (1996)
67. A. Gambirasio, M. Bernasconi, L. Colombo, Phys. Rev. B **61**, 8233 (2000)
68. H.O. Jeschke, M.E. Garcia, M. Lenzner, J. Bonse, J. Krüger, W. Kautek, Appl. Surf. Sci. **197–198**, 839 (2002)
69. E.S. Zijlstra, A. Kalitsov, T. Zier, M.E. Garcia, Phys. Rev. X **3**, 011005 (2013)
70. T. Zier, E.S. Zijlstra, A. Kalitsov, I. Theodonis, M.E. Garcia, Struct. Dyn. **2**, 054101 (2015)
71. C.V. Shank, R. Yen, C. Hirlimann, Phys. Rev. Lett. **50**, 454 (1983)
72. M. Combescot, J. Bok, J. Lumin. **30**, 1 (1985)
73. N.M. Bulgakova, I.M. Burakov, Y.P. Meshcheryakov, R. Stoian, A. Rosenfield, I.V. Hertel, J. Laser Micro/Nanoeng. **2**, 76 (2007)
74. R. Gunnella, G. Zgrablic, E. Giangrisostomi, F. D'Amico, E. Principi, C. Masciovecchio, A. Di Cicco, F. Parmigiani, Phys. Rev. B **94**, 155427 (2016)
75. E.N. Glezer, Y. Siegal, L. Huang, E. Mazur, Phys. Rev. B **51**, 6959 (1995)
76. T.J.-Y. Derrien, N.M. Bulgakova, Proc. SPIE **10228**, 102280E (2017)
77. T. Sjodin, H. Petek, H.-L. Dai, Phys. Rev. Lett. **81**, 5664 (1998)
78. M. Harb, R. Ernstorfer, T. Dartigalongue, C.T. Hebeisen, R.E. Jordan, R.J.D. Miller, J. Phys. Chem. B **110**, 25308 (2006)
79. C. Wu, L.V. Zhigilei, Appl. Phys. A **114**, 11 (2014)
80. R.F. Wood, G.E. Giles, Phys. Rev. B **23**, 2923 (1981)
81. K.N. Vonatsos, D.I. Pantelis, Appl. Phys. A **80**, 885 (2005)
82. I. Chowdhury, X. Xu, Numer. Heat Transf. Part A Appl. **219** (2003)
83. C. Wu, M.S. Christensen, J.M. Savolainen, P. Balling, L.V. Zhigilei, Phys. Rev. B **91**, 035413 (2015)
84. C. Wu, L.V. Zhigilei, J. Phys. Chem. C **120**, 4438 (2016)
85. M.V. Shugaev, C.Y. Shih, E.T. Karim, C. Wu, L.V. Zhigilei, Appl. Surf. Sci. **417**, 54 (2017)
86. X. Sedao, M.V. Shugaev, C. Wu, T. Douillard, C. Esnouf, C. Maurice, S. Reynaud, F. Pigeon, F. Garrelie, L.V. Zhigilei, J.-P. Colombier, ACS Nano **10**, 6995 (2016)
87. V. Gupta, H. Song, J. Im, MRS Proc. **397**, 465 (1995)
88. K.A. Jackson, Interface Sci. **10**, 159 (2002)
89. S.-N. Luo, T. Ahrens, T. Çağın, A. Strachan, W. Goddard, D. Swift, Phys. Rev. B **68**, 134206 (2003)
90. D.S. Ivanov, L.V. Zhigilei, Phys. Rev. Lett. **91**, 105701 (2003)
91. Z. Lin, L.V. Zhigilei, Phys. Rev. B **73**, 184113 (2006)
92. Z. Lin, E. Leveugle, E.M. Bringa, L.V. Zhigilei, J. Phys. Chem. C **114**, 5686 (2010)
93. D.S. Ivanov, Z. Lin, B. Rethfeld, G.M. O'Connor, T.J. Glynn, L.V. Zhigilei, J. Appl. Phys. **107**, 013519 (2010)
94. P.L. Liu, R. Yen, N. Bloembergen, R.T. Hodgson, Appl. Phys. Lett. **34**, 864 (1979)
95. Y. Fuentes-Edfuf, M. Garcia-Lechuga, D. Puerto, C. Florian, A. Garcia-Leis, S. Sanchez-Cortes, J. Solis, J. Siegel, Appl. Phys. Lett. **110**, 211602 (2017)
96. A. La Magna, V. Privitera, G. Fortunato, M. Cuscunà, B.G. Svensson, E. Monakhov, K. Kuitunen, J. Slotte, F. Tuomisto, Phys. Rev. B **75**, 235201 (2007)
97. T. Sinno, R.A. Brown, W. von Ammon, E. Dornberger, J. Electrochem. Soc. **145**, 302 (1998)
98. R.A. Brown, D. Maroudas, T. Sinno, J. Cryst. Growth **137**, 12 (1994)

99. E. Dornberger, T. Sinno, J. Esfandyari, J. Vanhellemont, R.A. Brown, W. von Ammon, Proc. Electrochem. Soc. **98**, 170 (1998)
100. L.V. Zhigilei, B.J. Garrison, J. Appl. Phys. **88**, 1281 (2000)
101. E. Leveugle, D.S. Ivanov, L.V. Zhigilei, Appl. Phys. A **79**, 1643 (2004)
102. G. Paltauf, P.E. Dyer, Chem. Rev. **103**, 487 (2003)
103. L.V. Zhigilei, Z. Lin, D.S. Ivanov, J. Phys. Chem. C **113**, 11892 (2009)
104. J.M. Savolainen, M.S. Christensen, P. Balling, Phys. Rev. B **84**, 193410 (2011)
105. A.A. Ionin, S.I. Kudryashov, A.E. Ligachev, S.V. Makarov, L.V. Seleznev, D.V. Sinitsyn, JETP Lett. **94**, 266 (2011)
106. S.I. Ashitkov, N.A. Inogamov, V.V. Zhakhovskii, Y.N. Emirov, M.B. Agranat, I.I. Oleinik, S.I. Anisimov, V.E. Fortov, JETP Lett. **95**, 176 (2012)
107. S.V. Starikov, V.V. Pisarev, J. Appl. Phys. **117**, 135901 (2015)
108. A. Lomonosov, A.P. Mayer, P. Hess, in *Modern Acoustical Techniques for the Measurement of Mechanical Properties*, ed. by M. Levy, H. Bass, R. Stern (Academic Press, San Diego, California, 2001), pp. 65–134
109. P. Hess, Phys. Today **55**, 42 (2002)
110. V.V. Kozhushko, A.M. Lomonosov, P. Hess, Phys. Rev. Lett. **98**, 195505 (2007)
111. C. Krischer, D. Lichtman, Phys. Lett. A **44**, 99 (1973)
112. B. Lindner, U. Seydel, Anal. Chem. **57**, 895 (1985)
113. V.V. Golovlev, S.L. Allman, W.R. Garrett, C.H. Chen, Appl. Phys. Lett. **71**, 852 (1997)
114. A.V. Zinovev, I.V. Veryovkin, J.F. Moore, M.J. Pellin, Anal. Chem. **79**, 8232 (2007)
115. A.M. Dow, A.R. Wittrig, H.I. Kenttämaa, Eur. J. Mass Spectrom. **18**, 77 (2012)
116. A.J. Manzo, H. Helvajian, Proc. SPIE **8969**, 896908 (2014)
117. M.V. Shugaev, A.J. Manzo, C. Wu, V.Y. Zaitsev, H. Helvajian, L.V. Zhigilei, Phys. Rev. B **91**, 235450 (2015)
118. C. Wu, V.Y. Zaitsev, L.V. Zhigilei, J. Phys. Chem. C **117**, 9252 (2013)
119. B.S. Yilbas, J. Phys. D Appl. Phys. **35**, 1210 (2002)
120. D.Y. Tzou, E.J. Pfautsch, J. Eng. Math. **61**, 231 (2008)
121. Y.-M. Lee, T.-W. Tsai, J. Phys. D Appl. Phys. **41**, 045308 (2008)
122. Y.P. Meshcheryakov, M.V. Shugaev, T. Mattle, T. Lippert, N.M. Bulgakova, Appl. Phys. A **113**, 521 (2013)
123. Y.P. Meshcheryakov, N.M. Bulgakova, Appl. Phys. A **82**, 363 (2006)
124. E. Vanagas, J. Kawai, D. Tuzhilin, I. Kudryashov, A. Mizuyama, K.G. Nakamura, K. Kondo, S. Koshihara, M. Takesada, K. Matsuda, S. Juodkazis, V. Jarutis, S. Matsuo, H. Misawa, J. Microlithogr. Microfabr. Microsyst. **3**, 358 (2004)
125. J. Bohandy, B.F. Kim, F.J. Adrian, J. Appl. Phys. **60**, 1538 (1986)
126. I. Zergioti, S. Mailis, N.A. Vainos, C. Fotakis, S. Chen, C.P. Grigoropoulos, Appl. Surf. Sci. **127–129**, 601 (1998)
127. M. Domke, S. Rapp, M. Schmidt, H.P. Huber, Appl. Phys. A **109**, 409 (2012)
128. D. Munoz-Martin, C.F. Brasz, Y. Chen, M. Morales, C.B. Arnold, C. Molpeceres, Appl. Surf. Sci. **366**, 389 (2016)
129. G. Heise, M. Englmaier, C. Hellwig, T. Kuznicki, S. Sarrach, H.P. Huber, Appl. Phys. A **102**, 173 (2011)
130. G. Heise, M. Domke, J. Konrad, S. Sarrach, J. Sotrop, H.P. Huber, J. Phys. D Appl. Phys. **45**, 315303 (2012)
131. D.S. Ivanov, B. Rethfeld, G.M. O'Connor, T.J. Glynn, A.N. Volkov, L.V. Zhigilei, Appl. Phys. A **92**, 791 (2008)
132. N.A. Inogamov, V.V. Zhakhovsky, K.P. Migdal, Appl. Phys. A **122**, 432 (2016)
133. J. Sotrop, A. Kersch, M. Domke, G. Heise, H.P. Huber, Appl. Phys. A **113**, 397 (2013)
134. A. Miotello, R. Kelly, Appl. Phys. Lett. **67**, 3535 (1995)
135. N.M. Bulgakova, A.V. Bulgakov, Appl. Phys. A **73**, 199 (2001)
136. B.J. Garrison, T.E. Itina, L.V. Zhigilei, Phys. Rev. E **68**, 041501 (2003)
137. J. Dowden (ed.), *The Theory of Laser Materials Processing. Heat and Mass Transfer in Modern Technology* (Springer, Dordrecht, 2009)

138. D. Bäuerle, *Laser Processing and Chemistry* (Springer, Berlin, 2000)
139. C. Earl, J.R. Castrejón-Pita, P.A. Hilton, W. O'Neill, J. Manuf. Process. **21**, 214 (2016)
140. W.R. Harp, J.R. Dilwith, J.F. Tu, J. Mater. Process. Technol. **198**, 22 (2008)
141. V. Semak, A. Matsunawa, J. Phys. D Appl. Phys. **30**, 2541 (1997)
142. B.S. Yilbas, S.B. Mansoor, J. Phys. D Appl. Phys. **39**, 3863 (2006)
143. S.P. Harimkar, A.N. Samant, N.B. Dahotre, J. Appl. Phys. **101**, 054911 (2007)
144. Y.L. Yao, H. Chen, W. Zhang, Int. J. Adv. Manuf. Technol. **26**, 598 (2005)
145. V.S. Ajaev, D.A. Willis, Cit. Phys. Fluids **15**, 3144 (2003)
146. A.N. Volkov, L.V. Zhigilei, Int. J. Heat Mass Transf. **112**, 300 (2017)
147. C.-Y. Shih, C. Wu, H. Wu, M.V Shugaev, in *Pulsed Laser Ablation: Advances and Applications in Nanoparticles and Nanostructuring Thin Films*, ed. by I.N. Mihailescu, A.P. Caricato (Pan Stanford, 2018), pp. 421–466
148. Z. Lin, R.A. Johnson, L.V. Zhigilei, Phys. Rev. B **77**, 214108 (2008)
149. C. Wu, D.A. Thomas, Z. Lin, L.V. Zhigilei, Appl. Phys. A **104**, 781 (2011)
150. X. Sedao, C. Maurice, F. Garrelie, J.-P. Colombier, S. Reynaud, R. Quey, F. Pigeon, Appl. Phys. Lett. **104**, 171605 (2014)
151. D.S. Ivanov, L.V. Zhigilei, Phys. Rev. B **68**, 064114 (2003)
152. P.P. Pronko, S.K. Dutta, J. Squier, J.V. Rudd, D. Du, G. Mourou, Opt. Commun. **114**, 106 (1995)
153. J. Koch, F. Korte, T. Bauer, C. Fallnich, A. Ostendorf, B.N. Chichkov, Appl. Phys. A **81**, 325 (2005)
154. A.I. Kuznetsov, J. Koch, B.N. Chichkov, Appl. Phys. A **94**, 221 (2009)
155. U. Zywietz, A.B. Evlyukhin, C. Reinhardt, B.N. Chichkov, Nat. Commun. **5**, 3402 (2014)
156. J. Jersch, F. Demming, J. Hildenhagen, K. Dickmann, Opt. Laser Technol. **29**, 433 (1998)
157. A. Chimmalgi, T.Y. Choi, C.P. Grigoropoulos, K. Komvopoulos, Appl. Phys. Lett. **82**, 1146 (2003)
158. D. Hwang, S.G. Ryu, N. Misra, H. Jeon, C.P. Grigoropoulos, Appl. Phys. A **96**, 289 (2009)
159. C. Huber, A. Trügler, U. Hohenester, Y. Prior, W. Kautek, Phys. Chem. Chem. Phys. **16**, 2289 (2014)
160. Y. Nakata, T. Okada, M. Maeda, Jpn. J. Appl. Phys. **42**, L1452 (2003)
161. D.S. Ivanov, V.P. Lipp, A. Blumenstein, F. Kleinwort, V.P. Veiko, E. Yakovlev, V. Roddatis, M.E. Garcia, B. Rethfeld, J. Ihlemann, P. Simon, Phys. Rev. Appl. **4**, 064006 (2015)
162. X.W. Wang, A.A. Kuchmizhak, X. Li, S. Juodkazis, O.B. Vitrik, Y.N. Kulchin, V.V. Zhakhovsky, P.A. Danilov, A.A. Ionin, S.I. Kudryashov, A.A. Rudenko, N.A. Inogamov, Phys. Rev. Appl. **8**, 044016 (2017)
163. M.V. Shugaev, I. Gnilitskyi, N.M. Bulgakova, L.V. Zhigilei, Phys. Rev. B, **96**, 205429 (2017)
164. J. Bonse, A. Rosenfeld, J. Krüger, J. Appl. Phys. **106**, 104910 (2009)
165. D.L. Andrews, G.D. Scholes, G.P. Wiederrecht, *Comprehensive Nanoscience and Nanotechnology*, 4th ed. (Elsevier, Academic Press, 2011)
166. R. Buividas, M. Mikutis, S. Juodkazis, Prog. Quantum Electron. **38**, 119 (2014)
167. B. Dusser, Z. Sagan, H. Soder, N. Faure, J.P. Colombier, M. Jourlin, E. Audouard, Opt. Express **18**, 2913 (2010)
168. J. Reif, O. Varlamova, F. Costache, Appl. Phys. A **92**, 1019 (2008)
169. M. Gedvilas, J. Mikšys, G. Račiukaitis, RSC Adv. **5**, 75075 (2015)
170. E.L. Gurevich, Appl. Surf. Sci. **374**, 56 (2016)
171. I. Gnilitskyi, V. Gruzdev, N.M. Bulgakova, T. Mocek, L. Orazi, Appl. Phys. Lett. **109**, 143101 (2016)
172. E.L. Gurevich, Y. Levy, S.V. Gurevich, N.M. Bulgakova, Phys. Rev. B **95**, 54305 (2017)

Chapter 6
Ultrafast Laser Micro and Nano Processing of Transparent Materials—From Fundamentals to Applications

Manoj Kumar Bhuyan and Koji Sugioka

Abstract The distinctive features of ultrafast lasers, such as the extremely high peak intensity associated with their ultrashort pulse duration, enable the machining of materials whose band gaps are larger than the photon energies of the laser due to the nonlinear nature of photon absorption by matter. As a result, processing of transparent materials at micro and nanoscale levels can be achieved using ultrafast lasers, which is a matter of particular relevance in the fields of electronics, healthcare, photonics, and energy harvesting. By using ultrafast laser beams that possess a standard Gaussian spatial profile, and by exploiting nonlinear light-matter interaction confinement schemes, fabrication resolutions have declined to just a few tens of nanometers on the surface, and a few hundreds of nanometers inside most transparent materials. However, it is difficult to maintain such fabrication resolutions, especially at nanoscale levels, when forming high aspect ratio (length over diameter) structures inside transparent materials. This issue can be addressed using novel focusing strategies and spatiotemporal beam shaping techniques. In this chapter, we will begin by reviewing the basic concepts of, and experimental methodologies related to, laser machining techniques. Then, machining strategies that have been adopted to facilitate high fabrication resolutions and long fabrication lengths using standard Gaussian laser beams are discussed, along with other competent techniques, thereby highlighting the underlying fundamental mechanisms and their applicability for material processing such as for drilling and cutting. Furthermore, the impact of spatiotemporal beam shaping on material processing is discussed, with a special focus on applications of non-diffractive optical beams for fast and flexible material processing. Finally, we provide a futuristic viewpoint on the potential for achieving super-resolution, yet high-aspect-ratio, material fabrication.

M. K. Bhuyan · K. Sugioka (✉)
RIKEN Center for Advanced Photonics, RIKEN, Wako, Saitama 351-0198, Japan
e-mail: ksugioka@riken.jp

M. K. Bhuyan
e-mail: manoj.femto@gmail.com

P. M. Ossi (ed.), *Advances in the Application of Lasers in Materials Science*, Springer Series in Materials Science 274,
https://doi.org/10.1007/978-3-319-96845-2_6

6.1 Introduction

Ultrafast lasers, defined as lasers with pulse durations shorter than a few picoseconds (ps), are excellent tools for transparent material processing. The ability of such lasers to generate a high peak intensity squeezed into an ultrashort time window makes it possible for them to trigger nonlinear photon absorption processes (described in detail in the following section), leading to optical breakdown of ideally any material which is the basis of ultrafast laser machining technique [1–4]. Furthermore, as will be discussed below, when a laser pulse is focused with an adequate amount of energy, nonlinear photon absorption can occur not only on the surface of most transparent materials, but also inside them, at a very confined region near the focal volume. This absorbed photon energy can induce a range of photo-chemical and photo-physical processes such as cross-linking (polymerization), bond breaking (refractive index alteration), and ablation (material removal in gas, liquid and solid phase), over timescales ranging from femtoseconds to milliseconds. Likewise, material processing down to the nanoscale, and (most importantly) in three-dimensional (3D) space, can be realized using lasers, which has a great deal of relevance in multiple fields including electronics, photonics, healthcare, and energy harvesting. Since there have been numerous research efforts made to accelerate the field of laser-assisted material processing in recent years, this chapter discusses some of the more important approaches, while concentrating primarily on the processing of transparent materials.

As mentioned above, the basic advantages of laser machining techniques lie with their ability to produce localized material breakdown via optical means, thereby causing modifications to material's physical properties. From a fundamental viewpoint, there are three generic aspects associated with ultrafast laser induced optical breakdown of matter.

(i) *The near-instantaneous transfer of energy from a laser pulse to the electronic (sometimes referred to as electron-hole plasma) sub-systems of the material* via *non-linear ionization mechanisms*. Depending on the laser intensity level, the nature of ionization mechanisms can be multi-photon (several photons becoming absorbed simultaneously, thus promoting electrons from the valence to the conduction band of the material), single-photon (conduction band electrons becoming energized by directly absorbing single photons) and scattering (conduction band electrons with energy levels exceeding the material bandgap that can knock additional electrons from the valence band, thereby resulting in an avalanche process). Although, the individual contributions of each ionization mechanism are difficult to calculate (and are still a subject of debate), multi-photon ionization is assumed to play an important role in cases involving dielectric materials possessing wide band gaps, with (ideally) no available free electrons. Note that metals have vast amounts of free electrons and can be energized on a single-photon basis. Additionally, narrow bandgap materials, such as semiconductors, can undergo single photon absorption.

(ii) *Energy transfer from electronic sub-systems to the lattice* via *electron-phonon scattering processes occurring on femtosecond to picosecond timescales*,

(iii) *Lattice re-organization and stabilization on timescales ranging from femtoseconds to microseconds*. Quantitatively, in the case of dielectrics, when the laser intensity exceeds the 10^{12} W/cm^2 level, a laser induced free electron density can be achieved in the range of 10^{18}/cm^3 to the critical density (i.e., nearly 10^{21}/cm^3), which has a strong potential to trigger lattice re-organization during the relaxation cycle of excited matter. The material relaxation can take several pathways [5–8], and can result in various types of material modifications [5, 9–11]. For example, the manifestation can be in the form of a surface crater via ablation, an increased refractive index structure via molecular rearrangement, or a decreased refractive index structure, such as a void formed via localized explosion and stress relaxation inside most transparent materials. More scenarios that consider the experimental findings related to laser structuring are discussed in later sections. In brief, the resultant material modifications depend primarily on the laser assisted energy deposition dynamics. Specifically, the amount of (and the rate at which) optical energy is deposited into or onto the material.

From an experimental viewpoint, the energy deposition dynamics for a given material are controlled via laser beam parameters such as the beam profile, pulse energy and temporal duration, wavelength, polarization state, and focusing geometry. For example, ultrashort laser pulses (those in the femtosecond (fs) to picosecond range) that suppress thermal effects during ablation process, are useful for high-quality material fabrication. In contrast, nanosecond laser pulses are suitable for material processing based on melting and annealing. Laser wavelength is also critical for the machining of biological materials. From an application viewpoint, single-laser-pulse based material processing provides the best option in terms of throughput and fabrication quality. However, since multiple laser pulses can also be employed, especially when the single pulse energy is insufficient to trigger a material phase change, this methodology can provide high-speed and economical material processing using currently available very high repetition rate (up to few hundred MHz) ultrafast lasers with nanojoule energy levels. In multiple-pulse-exposure cases involving pulse energies of just a few microjoules, the resultant material modifications are strongly dependent on the laser polarization state. A spatially homogeneous laser beam profile (which otherwise has a Gaussian profile in the radial direction), such as a flat-topped profile, is also required for uniform energy deposition, especially for large area material processing applications such as sintering, lithography, and pulsed laser deposition [12].

6.2 Direct Fabrication Using Gaussian Laser Beams

This section first introduces Gaussian laser beam-based material processing techniques aimed at the direct fabrication of micro and nanostructures, and then highlights their underlying mechanisms and technical capabilities. Along with surface and bulk processing facilitated by localized (point-like) energy deposition, elongated structuring facilitated by aberration effects and nonlinear beam propagation dynamics are also discussed. Nanoscale structuring techniques based on near-field optical phenomena are then considered, after which a non-laser based (ion beam) machining technique is reviewed to provide a useful comparison.

6.2.1 Standard Fabrication Approach

The methodology of standard laser machining techniques involves focusing a laser beam through a lens (often a microscope objective lens) onto or into a material in order to produce a local change in the physical properties of the treated material. An altered material feature (such as a crater or index change) is mostly referred to as a "structure" throughout this chapter. Figure 6.1a shows a schematic diagram of a *laser based surface structuring technique*. Although the beam (ideal Gaussian profile and wavelength λ) spot size (diameter) S_G at the focal point of an objective lens with a numerical aperture (NA) is $S_G = 1.22\ \lambda/(2\ \text{NA})$, the lateral dimension of the machined structure often becomes smaller than the beam spot size due to the threshold effect over the spatial distribution of the laser intensity. Additionally, various researchers [13–15] have reported that machined structure sizes can be reduced to a few tens of nanometers (which is much smaller than the irradiation wavelength and diffraction spot size), clearly indicating non-linear energy deposition into the material. For instance, Joglekar et al. [13] demonstrated nanoscale structuring of Corning 0211 glass using 600 fs laser pulses at wavelengths of both 527 and 1053 nm, which were focused using an objective lens with an NA of 1.3. Figure 6.1b shows scanning electron microscopy (SEM) images of nanostructures (size down to 30 nm) machined on glass using a single 7.5 nJ energy laser pulse at a 527 nm wavelength. The lateral size of the machined structure was also observed to be in the range from a few tens of nanometers to a micron, as shown in Fig. 6.1c, where 1053 nm wavelength laser pulses are employed. Here, the sub-wavelength scale structuring is attributed to the nonlinear energy confinement mechanisms that lead to ablation events. Note that ablation is a threshold-defined process. This implies that only the material region within which the laser intensity exceeds the threshold value, i.e., 10^{12}–10^{13} W/cm^2, gets ablated. Hence, the lateral size of the ablated zone that defines the surface machining or fabrication resolution can be smaller than the diffraction limit. The deterministic nature of the above-mentioned nanoscale structuring raised concerns over widely accepted nonlinear ionization mechanisms such as multiphoton, tunneling, and avalanche ionization. Joglekar et al. [14] argued for a novel ionization mechanism. Specifically, that Zener ionization could play a role at near-ablation threshold intensity regime.

Although single-pulse structuring techniques involving ultrashort Gaussian beams have been very successful in producing high-quality nanoscale structures (i.e., with almost no heat affected zone around the structure), the techniques suffer from poor machining throughput. Additionally, scaling of structure dimensions (mainly lat-

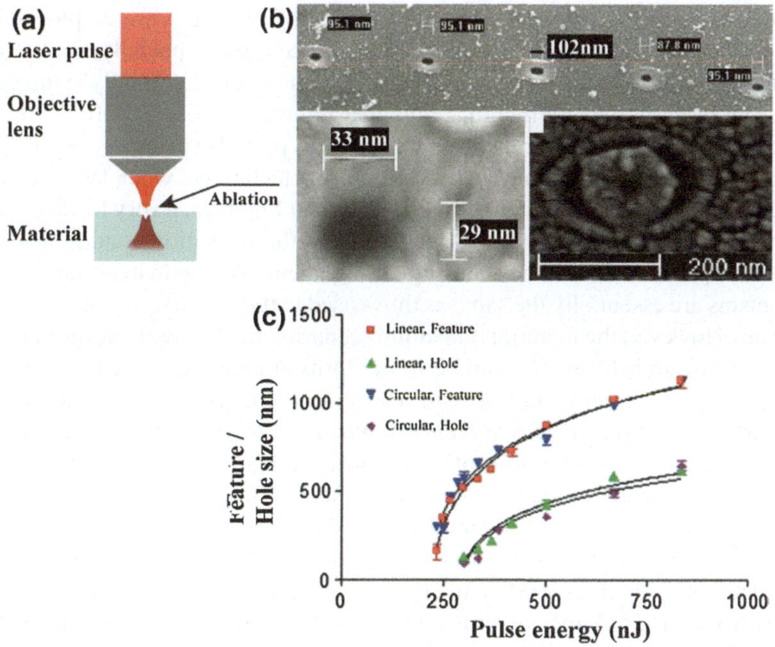

Fig. 6.1 Laser-based surface structuring technique: **a** Typical machining scheme showing a laser pulse focused onto a material through an objective lens. **b** SEM images of single laser pulse (600 fs pulse duration, 527 nm wavelength, energy 7.5 nJ) induced structures on Corning 0211 glass surface. **c** Variation of structure size as a function of laser energy in case of single pulse (600 fs pulse duration, 1053 nm wavelength) excitation of glass [13]. Reproduced with permission from the American Institute of Physics (AIP)

eral) is another issue in such machining schemes. A number of early experiments exploited multiple pulse structuring in order to address the above-mentioned issues. For example, multiple laser pulses (typically less than 10^3) at a repetition rate in the kHz range were used to produce large structures on metals, semiconductors, and dielectrics by invoking incubation effects [16, 17] assisted by laser induced defect formation. By invoking heat accumulation effects [18–21], large numbers of laser pulses (typically in the range of 10^3–10^7) at repetition rates in the kHz to MHz range were also used to produce large (a few microns in diameter) structures, but primarily inside transparent materials. Nevertheless, the machined structure in these cases is always associated with an undesired surrounding heat affected zone.

Very recently, Kerse et al. [22] demonstrated a novel high-throughput ablation concept using a multiple pulse structuring technique. The basis of this idea is to eliminate the heat of the material generated by a laser pulse by removing any ablated materials (debris) with the next pulse, before heat diffusion occurs. This implies that machining with moderate energy pulses, which will eventually excite large volumes of material per pulse (effectively, high ablation rate per pulse), can still be used to process material without creating a heat affected zone, provided suitable pulse trains (i.e., at very high repetition rate in few hundred MHz range) are employed. This technique is thus very useful for high-speed processing, especially for thermally sensitive materials. Indeed, the authors have demonstrated virtually damage-free ablation of brain tissue at a rate of 2 mm^3 per minute using a 432 MHz intra-burst repetition rate, 27 kHz burst repetition rate, and 16 μJ pulse energy.

In contrast to the laser based surface structuring technique, when laser beams are focused inside any transparent material, as shown in Fig. 6.2a, a very localized material phase change may be triggered depending on the excitation conditions defined by the laser parameters and the focusing arrangement. The basic material excitation mechanisms are essentially the same as those used in the case of a surface structuring technique. However, the material relaxation scenarios are different. Nevertheless, the resultant phase transformation can be in the form of either an increase or decrease in the density or refractive index, with respect to the native values of the material. This is the basis of a *laser based bulk structuring technique*. Since the light-matter interaction zone can be flexibly placed anywhere within the material, the above technique is applicable to numerous applications requiring structuring in 3D space. For instance, Glezer et al. [23] demonstrated data storage applications by suitably machining (using single 100 fs laser pulses focused through a 0.6 NA lens) void-like structures as small as 0.008 μm^3 inside several transparent materials including glass, $LiNbO_3$, and polymer, as shown in Fig. 6.2b. Based on the laser induced localized material modification, especially the increased refractive index structure, direct writing of waveguide structures has been demonstrated for emerging technologies including quantum photonics [10, 24]. An illustrative image of waveguide writing is shown in Fig. 6.2c, where either the sample or the focused laser beam can be continuously translated along the desired direction within the material in order to realize waveguides in 3D. A propagation loss down to 0.1 dB/cm has been achieved using laser written waveguides in glass [25–28]. Similarly, localized increased refractive index structures have also been exploited in the fabrication of optical components including couplers and splitters [29, 30], volume Bragg gratings [31], diffractive lenses [32], and waveguide lasers [33]. Note here that above-mentioned demonstrations were based on point-like structures, and were facilitated primarily by either a tight focusing geometry, or by using a low pulse energy close to the corresponding material modification threshold value. Although the dimensions of bulk structures can be down to sub-diffraction sizes, as in the above examples, the scalability of such structures is questionable. For instance, if the laser pulse energy is increased, an axially elongated structure with an almost non-uniform shape may be obtained inside most transparent materials [34]. Such phenomena may be even more pronounced when loose focusing geometries are involved. Figure 6.2d shows a typical elongated microstructure with non-uniform index modification (negative and positive changes in the refractive index at the beginning and the end of the structure, respectively) produced in glass using femtosecond laser pulses with loose focusing (0.5 NA objective lens) geometry. This structure elongation is mainly attributed to the involvement of non-linear propagation dynamics of Gaussian beams along with the optical breakdown processes [35]. For instance, when operating at levels significantly above the material breakdown (laser intensity dependent process) threshold regime,

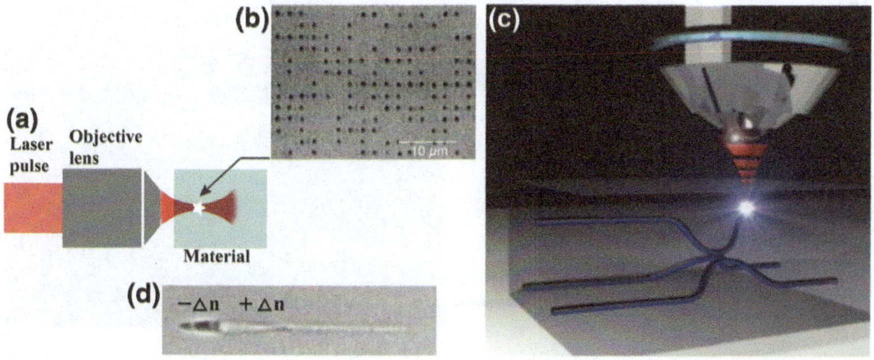

Fig. 6.2 Laser-based bulk structuring technique: **a** Typical machining scheme, showing laser pulse focusing inside a transparent material through an objective lens. **b** Optical micrograph of single pulse induced voids acting as data bits inside fused silica glass for data storage applications [23]. **c** Illustration of laser writing of waveguide structures produced by traversing the sample continuously during irradiation [25]. **d** Optical micrograph of a typical elongated structure inside glass produced using high energy femtosecond laser pulses. The regions with positive and negative changes in the refractive index of the structure are indicated as $+\Delta n$ and $-\Delta n$, respectively [34]. Reproduced with permission from The Optical Society of America (OSA), Laser Focus World, and American Physical Society (APS)

the self-focusing phenomenon (laser power dependent process) may come into the picture, but can be suppressed to some extent by using a tight focusing geometry (such as by using an objective lens with an NA close to one). From a fundamental viewpoint, such self-focusing concentrates light (like a positive lens) due to Kerr effect, that results in plasmas (act like negative lens) that can defocus the beam. This focusing-defocusing cycle continues over some beam propagation distance, producing a self-guided elongated plasma channel in a process referred to as filamentation. Note that the spatio-temporal dynamics associated with the filamentation process is very complex (difficult to control) and hence scalable material processing involving such nonlinear laser beams is limited. Nevertheless, filamentation-based elongated structuring of transparent materials can still be used for potential applications such as deep drilling and cutting [35–40].

By exploiting the filamentation process in loose focusing geometries (using an objective lens of 0.5 NA), and also in multiple pulse illumination schemes, Sudrie et al. [31, 34] first successfully produced extended, but non-uniform, structures (up to 80 μm long and a few μm in diameter) in fused silica glass using femtosecond laser pulses (160 fs pulse duration, 800 nm wavelength, and 7.8 MW power at 200 kHz repetition rate). Ultraviolet (UV) filaments were also exploited to facilitate elongated structuring of $LiNbO_3$ [41] and fused silica [42]. Shah et al. [43] successfully drilled straight hollow channels (≥1 mm long with aspect ratios up to10) in glass in ambient air, and Varel et al. [44] drilled long microchannels in quartz in vacuum conditions. Even in tight focusing conditions where a substantial reduction of nonlinear beam propagation effects is expected, femtosecond laser beams can undergo filamentation process and can induce elongated, yet very narrow structures inside transparent materials. For example, Kydryashov et al. [45] first reported fabricating

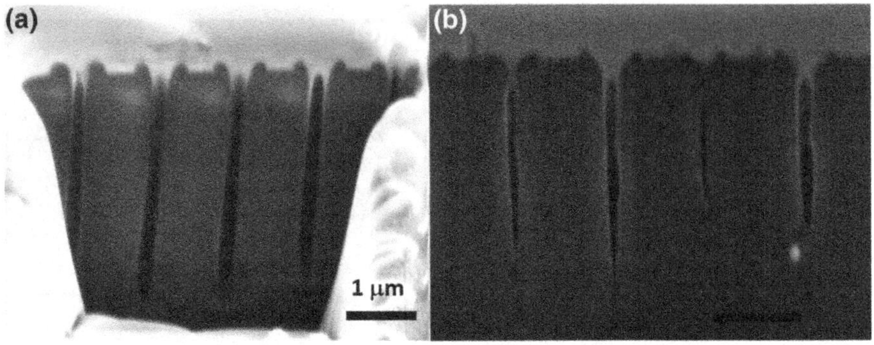

Fig. 6.3 Gaussian laser beam based nanochannel formation in Corning 0211 glass using a single femtosecond laser pulse (600 fs pulse duration, 1053 nm wavelength, and a fluence of 67 J/cm^2). Cross-sectional SEM images of structures machined on **a** reverse surface, and **b** front surface [15]. Reproduced with permission from OSA

1 μm long and 100 nm diameter hollow channel-like structures connecting to the front surface of a sapphire specimen using a single femtosecond laser pulse (600 fs pulse duration, 1053 nm wavelength, and nearly 2 W power). They also reported similar observations inside glass. White et al. [46] produced nanochannels less than 500 nm in diameter and with lengths up to 12 μm connecting to the front surface of fused silica glass materials using a single femtosecond laser pulse (200 fs pulse duration, 800 nm wavelength, and nearly 1–3 μJ pulse energy), whereas Herbstman et al. [15] demonstrated nanochannels between 150 and 240 nm in diameter with lengths up to 8 μm on both the front and reverse surfaces of Corning 0211 glass using a single femtosecond laser pulse (600 fs pulse duration, 1053 nm wavelength, and fluence of 67 J/cm^2), as shown in Fig. 6.3. It was also observed that the structure quality depends on the focusing geometry. For instance, channels with parallel walls were obtained in the case of reverse surface machining (see Fig. 6.3a), and tapered channels were observed for front surface machining (see Fig. 6.3b).

From a fundamental viewpoint, several mechanisms were proposed behind the formation of extended nanochannels in transparent materials using a single femtosecond laser pulse. The first mechanism, proposed by Kudryashov et al. [45], refers to short wavelength Bremsstrahlung radiation from hot dense electron-ion plasma, which can propagate and excite longitudinally beyond the skin depth (typically a few tens of nanometers) inside the material. This can cause melting over a longitudinally extended zone. During the relaxation phase, matter might have ruptured, thus leaving behind an empty zone referred to as a channel, within the material. More realistic mechanisms are based on filamentary propagation of Gaussian beams that cause longitudinal extension of focal zones, and hence extended scale ablation resulting in nanochannels, in the considered transparent materials. Spherical aberration (a linear phenomenon) can also act towards extending the focal zone longitudinally. This is because, compared to the central part of the normally considered Gaussian beams, marginal rays with large incidence angles deviate more at the air-glass interface in the case of a tight focusing geometry. For example, Kanehira et al. [47] produced structures that were nearly 200 μm long embedded with quasi-periodic nanovoids 380 nm in size, and with a pitch of 1.7 μm, in a 900 μm thick Borosilicate glass specimen using 250 laser pulses of 10 μJ energy, whereas Sun et al. [48] and Song et al. [49] reported having observed similar effects. Although the underlying physics are yet to be discovered, Kanehira et al. attributed the mechanism of forming extended structures to both self-focusing and spherical aberration. In addition, those authors also presented some arguments in favor of standing electron plasma wave causing periodic nanovoids. In short, the leading edge of a high intensity laser pulse can, by itself, produce carrier plasmas inside glass that can interact with the trailing part of the pulse, thereby forming an electron plasma wave. When this plasma wave propagates into a region with a large plasma density gradient, it will be reflected and interfere with the forward travelling wave, thereby creating a standing wave pattern. As a result, the free carrier density distribution might be modulated, triggering periodic microexplosions and thus resulted into a periodic nanovoid structure. To summarize, ultrafast Gaussian laser beams have been shown to be capable of machining not only near-spherical (increased refractive index structures and voids), but also elongated sub-micron diameter channel-like structures inside several transparent materials including glass and sapphire on single-pulse and multiple-pulse irradiation basis.

6.2.2 Near-Field Approach

It has been demonstrated that ultrafast lasers associated with far-field focusing arrangements are capable of machining very fine structures down to few tens of nanometers on or inside most technologically relevant materials, as discussed in the previous section. In these cases, the spatial resolution of the machining is primarily dictated by the beam focusing strategy (linear) in conjunction with the applicable nonlinear light confinement processes. Meanwhile, several novel approaches based on near-field optical phenomena have been demonstrated for nanoscale material processing, but (so far) have been strictly limited to the surface of any samples. For instance, tip-based machining approaches [50] combined with either atomic force microscopy (AFM) [51] or near-field scanning optical microscopy (NSOM) [52], have produced spatial machining resolutions down to 30 nm. Here, the strong light localization near the tip causes a change in the thermo-mechanical properties of the material, or even ablation at a very confined spatial scale under the tip. Since these techniques rely on the near-field concept, the working distance is extremely short (a few tens of nanometers). Therefore, an active feedback control system is often employed to maintain the fixed distance between the sample and the tip, and hence to obtain uniform material processing.

Micron size spherical particles (instead of tips) are also exploited for sub-wavelength light localization, and show strong potential for use in fundamental and technological applications including optical nano-sensing [53], nano-surgery [54, 55], and nanoscale material processing [55–60]. Light transmitting through a dielectric microsphere tends to emerge with a high intensity photon flux directly behind the sphere. The birth of the high intensity jet-like beam lies with the fact that a microsphere can be treated as a mesoscopic scattering particle for visible radiation, which means it can scatter light strongly in the forward direction. This scattered light then interferes with the transmitted light refracted through the sphere and produces a highly localized light wave known as "photonic jet". The primary features of photonic jets are as follows: (a) Strong transverse light localization maintained over several wavelengths along the longitudinal direction. For instance, the jet diameter can be as small as just a few hundred nanometers, and the jet length can be extended up to a few micrometers. It has also been shown that the length scale of the photonic jet can be improved further by using a multilayer dielectric microsphere [61, 62], a micro-ellipsoid [63], a liquid filled hollow micro-cylinder [64], a hemispherical cell [65], and a micro-axicon [66]. (b) The spatial distribution and intensity of a photonic jet depend primarily on the sphere diameter and the refractive index contrast between the sphere and its surrounding medium. Most importantly, the peak intensity of the photonic jet can exceed the ablation threshold level for most technologically relevant materials.

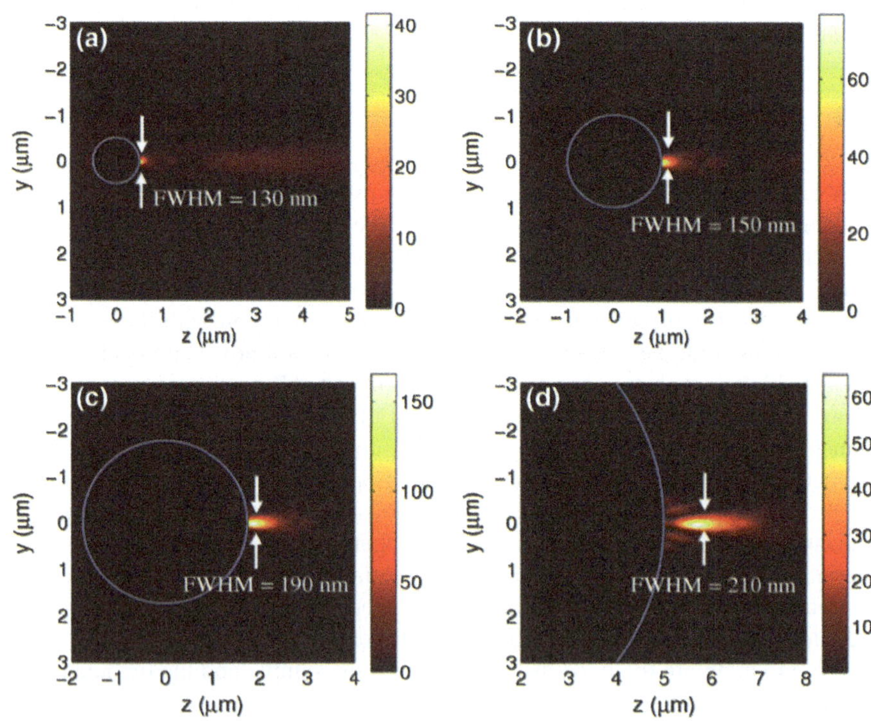

Fig. 6.4 Simulation results: photonic nanojets with a wavelength of 400 nm, produced by microspheres with a refractive index of 1.59 at diameters of **a** 1, **b** 2, **c** 3.5, and **d** 8 μm [53]. Reproduced with permission from OSA

Figure 6.4 shows simulation results for a typical photonic jet, demonstrating its spatial characteristics as a function of sphere diameter [53]. A linearly polarized light wave (polarization direction is perpendicular to the sheet) with a 400 nm wavelength is incident (along the z-direction) on spherical particles that have a refractive index of 1.59 in the ambient environment. The smallest lateral size photonic jet (which eventually dictates the machining spatial resolution), which has a full width at half maximum (FWHM) of 130 nm, was obtained for a 1 μm diameter sphere, as seen in Fig. 6.4a. The jet size increases with increasing sphere diameter. Nevertheless, the possibilities of confining light below the diffraction limit and tuning the beam profile as a function of sphere dimensions, have generated some research interest in the field of nanoscale material processing.

For example, Lu et al. [56] initially reported Si nanostructuring using laser (pulse durations ranging from 150 fs to 23 ns and wavelengths ranging from 248 nm to 800 nm) irradiation through microspheres. Huang et al. [59] demonstrated nanocraters on Ni film using the same technique, whereas Zhou et al. [60] reported nanoscale crater formation on a glass surface. In this last case, a monolayer of 1 μm diameter silica spheres was first deposited on a glass surface, after which a laser pulse (800 nm, 100 fs) was focused (spot size of 5.4 μm) on the microspheres. This resulted in nanocraters just below the spheres.

SEM image of craters and an AFM line profile of a typical crater are shown in Fig. 6.5a, b, respectively. The lateral size (200–400 nm) of the craters were also observed to be linearly scaled with the laser fluence, as illustrated in Fig. 6.5c. These results clearly present novel opportunities for rapid, large-area, and nanoscale mate-

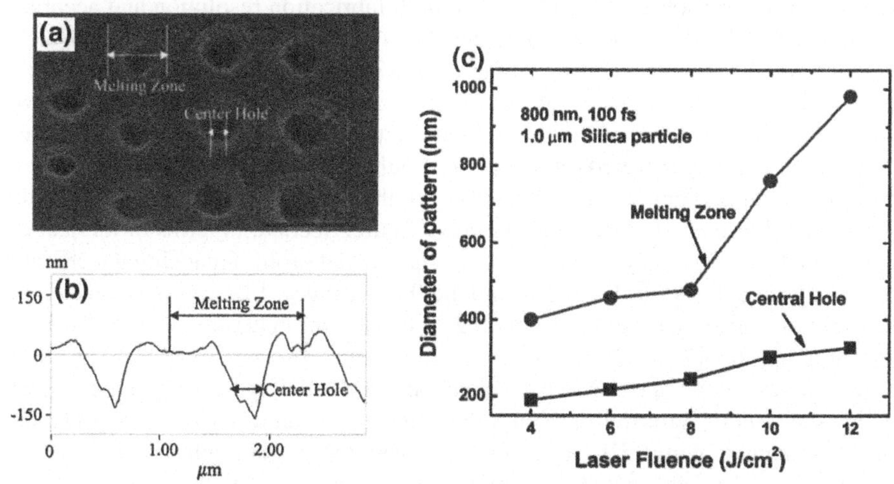

Fig. 6.5 **a** SEM images of photonic nanojet assisted nanocraters on glass surfaces. The laser pulses (wavelength of 800 nm, pulse duration of 100 fs, and fluence of 35 J/cm^2) pass through a monolayer of 1 μm-diameter silica spheres to form surface craters. **b** AFM cross-sectional line profile of a typical crater shown above. **c** Evolution of average crater diameter with respect to the laser fluence [60]. Reproduced with permission from AIP

rial processing. However, the above technique provides little freedom for designing and emplacing the nanostructures, as these elements are primarily dictated by the methods used to deposit the spheres on the relevant materials. To overcome this drawback, McLeod et al. [67] proposed a technique based on an optical trap [68, 69] that holds the microsphere in an aqueous environment. More specifically, zero-order Bessel beams [70, 71] were employed to trap a polystyrene microsphere near the substrate (to be machined), by suitably balancing the scattering force imposed on the sphere by the beam with the counter forces, such as electrostatic repulsion between the sphere and the substrate. The microsphere was therefore self-positioned (without any help of feedback system) at a distance of about 50 nm from the substrate. In this way, the authors demonstrated direct writing of individual as well as continuous structures with lateral sizes down to 100 nm on polyamide films. The self-positioning mechanism also facilitates machining of non-uniform surfaces, and the technique also has the potential to be used for parallel writing using adaptive optics [71–74] such as by trapping several spheres at the same time, followed by laser illumination for nanoprocessing.

6.2.3 Alternative Technology to Laser Machining: Focused Ion Beam (FIB) Machining

With its state of the art fabrication capability down to 10 nm, the focused ion beam based nanofabrication technique [74–77] is mostly utilized for the fabrication (primarily on surface) of plasmonic components, X-ray optics, photonic crystal fabrication, and other such applications where high fabrication resolution and accuracy are important. However, laser machining has the potential to beat the FIB technique in two areas. First, in terms of fabrication resolution, Joglekar et al. [13] have already demonstrated surface craters with sizes down to 30 nm. Since this precision comes from the strength of nonlinear light-matter interactions, it can be predicted that with recently discovered ultrashort laser pulses (such as few cycle laser pulses) and spatiotemporal engineering concepts [78], the fabrication resolution can be further reduced to just a few nanometers in the future. Second, in terms of fabrication depth, the laser (normally considered Gaussian spatial profile) machining technique is already on a par with the FIB technique. For instance, FIB can only machine a few microns deep inside some technologically relevant materials, such as TiO_2, as shown in Fig. 6.6 [75].

The figure shows an SEM cross-sectional image of a typical photonic crystal, which consists of an array of 50 nm diameter nanochannels with an aspect ratio of 10, machined using FIB fabrication, for visible light manipulation. In contrast, machining of channels that are just a few tens of nanometers wide and up to 10 μm long has been demonstrated in glass using Gaussian laser beams. Furthermore, there is a significant increase in fabrication depth [up to a millimeter (mm)] when using zero-order Bessel laser beams, which will be discussed in the final section of this chapter. Note that even though the recent machining demonstration using the FIB

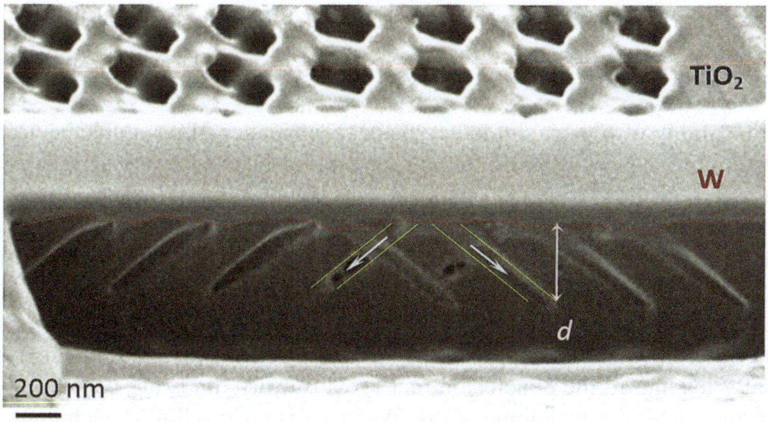

Fig. 6.6 FIB fabrication results: SEM cross-sectional image of a 3D photonic crystal in TiO_2 produced using FIB. The photonic crystal consists of an array of nanochannels (FIB fabricated) of diameter of 50 nm, and an aspect ratio of 10 [75]. Reproduced with permission from OSA

technique represents the current state of the art, the fabricated structures still cannot be integrated completely in the volume. In contrast, it can be foreseen that a fabrication technique involving a laser beam in ambient air (instead of the vacuum environment required for FIB) will emerge in the future, and the final section of this chapter on laser machining using non-diffractive beams provides a potential pathway to the realization of this goal.

6.3 Hybrid Approach

Direct fabrication techniques involving ultrafast Gaussian laser beams can achieve super-resolution structures, but the scalability of the structure diameter is questionable, as discussed in the previous section. In particular, since the direct fabrication of large diameter (more than few microns) structures (especially hollow-like) is extremely difficult, hybrid material processing approaches, such as those discussed below, have been developed. The first hybrid approach relates to single-step processing i.e. laser ablation of materials that are essentially transparent to incident light, in the presence of an aqueous environment in order to facilitate removal of ablated particles. The second approach relates to two- and three-step processing i.e. laser induced material modification followed by chemical etching and as-needed heat treatment. The underlying mechanisms and technical capabilities of such techniques are also discussed below.

6.3.1 Single-Step Processing: Laser Machining in Suitable Environment

Laser machining of transparent materials in the presence of a suitable environment using the multiple laser pulse irradiation scheme has been shown to be very effective for fabricating hollow channels, without any need of post-fabrication processes. Long straight microchannels were machined in transparent materials such as quartz in a vacuum environment, which facilitated the removal of ablated materials [44]. The use of a liquid environment (which is in contact with the machining surface) offers more flexible microchannel fabrication in terms of length and shapes. Different liquids such as water, methanol, ethanol, and isopropanol were evaluated for this process, eventually leading to the conclusion that low viscosity fluids are better candidates for hybrid machining techniques [79]. Water assisted machining techniques have been shown to have great potential for producing arbitrarily shaped hollow channels in transparent materials. The methodology of the machining technique shown in Fig. 6.7a, can be understood as follows: a femtosecond laser pulse is first focused (through the material) at the reverse surface of a transparent material placed in contact with water, after which the laser focal point can be translated inside the material along the chosen trajectory in order to obtain arbitrarily shaped 3D hollow channels.

For example, Li et al. [80], the inventors of the water-assisted machining technique, successfully fabricated 4 μm diameter microchannels with aspect ratios of up to 50 in silica glass at an effective sample translation speed of 0.3 μm/s using an objective lens of 0.55 NA. Figure 6.7b shows an optical micrograph of a typical continuous square-shaped hollow microstructure machined in this way. Using a similar technique and focusing geometry, Iga et al. [81] drilled straight 2-μm-diameter microchannels with aspect ratios of up to 325. Later, Hwang et al. [79] enhanced this method by adding ultrasonic wave agitation in order to machine straight and curved 3D channels with diameters on the order of tens of microns (see Fig. 6.7c, d), and aspect ratios of around 40 in glass at a translation speed of up to 30 μm/s using an objective lens of 0.42 NA. In addition, Ke et al. [82] successfully fabricated very long channels (>200 μm) of very small diameter (<700 nm) and aspect ratios of up to 1000.

Although technical exploitation of the liquid assisted machining technique has been very successful, the underlying physical mechanisms involving laser induced plasma generation, bubble generation and expansion, and the generation and expansion of shock/pressure waves are very complex, and not yet fully understood. However, the essential physics of this technique can be elucidated as follows: the laser pulse focused on the reverse surface of the sample in contact with a liquid can trigger an optical breakdown of the sample, thereby causing matter ablation. Simultaneously, part of the laser pulse energy can also be absorbed by the liquid such as water. This causes rapid heating (typically at the rate of 10^{10} K/s [82]) and water vaporization, thereby creating bubbles inside the water. Such bubbles create a liquid flow, assisted by the thermal convection processes, that carries away the debris. This process is followed by the immediate inflow of the surrounding water into the ablation site. As

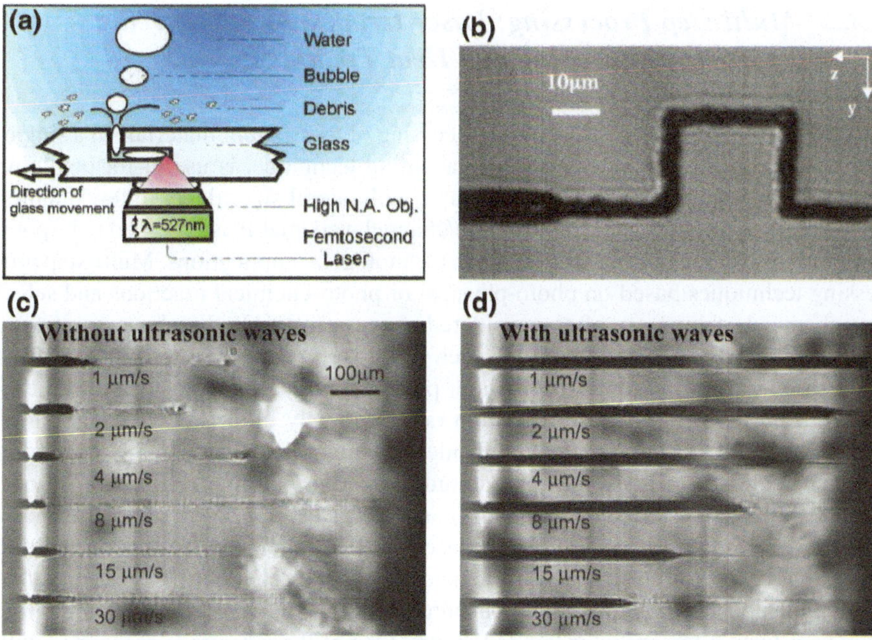

Fig. 6.7 Water-assisted micromachining results in glass: **a** Scheme of laser machining in liquid environment [82], **b** optical micrograph of a continuous, square-shaped hollow microstructure fabricated in glass using this machining technique [80]. Optical micrographs of microchannels machined without (**c**) and with (**d**) the assistance of ultrasonic waves [79]. Reproduced with permission from The American Chemical Society (ACS), OSA, and Springer

the laser focal point moves inside the sample to illuminate the fresh ablation site, the "sample ablation-bubble generation—bubble and debris evacuation" processes continue. Gradually, a hollow channel shaped structure appears.

In a broader sense, the large number of bubbles generated and their fast entrainment through the channels leads to rapid debris evacuation inside the channels. This necessitates the proper selection of laser parameters and liquids for the machining of high quality channels. The rate of bubble expansion and collapse needs to be balanced with the laser repetition rate, which has been normally limited to a range of just a few kHz, while remembering that the bubble lifetime in such ablation conditions is in the order of a millisecond. Thus, any minor fluctuation in laser energy would lead to different bubble dynamics that would interrupt the channel evolution process. In addition, laser focal point scanning has to be maintained precisely at the ablation site (which moves with each pulse) in order to avoid channel clogging. Note here that the water flow inside micron and sub-micron size channels, which is essential for realizing long channels, is assisted by capillary forces that impose no obvious limitations on long hollow structure machining.

6.3.2 Multi-step Processing: Laser Irradiation, Followed by Chemical Etching and Heat Treatment

As discussed in the previous section, processing of transparent materials in a single-step, such as by using water-assisted machining technique, is useful for obtaining long, arbitrary shapes and hollow structures with lateral sizes down to few hundred nanometers. However, since the structure/channel wall quality is observed to be poor, reservations exist regarding their usage in technological applications. Multi-step processing techniques based on photo-physical or photo-chemical reactions and selective chemical etching are often considered as ways to tackle the above-mentioned issue [83, 84]. The methodology of such multi-step processing technique can be understood as follows: femtosecond laser pulses (tens of nanojoules to a few microjoules of pulse energy at high repetition rates, typically in the kHz to MHz range) are first focused at a certain position inside the material, after which the focal point is scanned along a pre-designed path in order to induce soft material modifications. Note that the laser irradiation path must connect at least one of the sample surfaces. After laser irradiation, the sample is placed inside a chemical compound mixture consisting of hydrofluoric and sulfuric acids in order to etch out the laser modified material. Depending on the molar concentration of the acids used, the etching time may vary from a few minutes to a few hours. One of the advantage associated with this technique lies with the use of very low (by at least one order of magnitude) energy pulses compared to the single-step, ablation-based machining methods. Marcinkevicius et al. [85] fabricated an H-shaped 3D structure of interconnected 12 μm wide hollow channels with arbitrary angles of interconnection inside silica glass using this technique. Combining focal spot engineering with a two-step machining technique, Cheng et al. [86] varied the beam focal depth to machine hollow channels with arbitrary cross-sections in glass.

With an aim towards enhancing the potential of such two-step processing techniques, special photosensitive materials, such as Foturan glass, are being considered for technological applications. For such glasses, the etching rate of the laser-exposed region is significantly enhanced compared to the unirradiated region. However, the processing technique requires multiple steps. After laser irradiation along the desired path, the photosensitive material is annealed (up to few hundred centigrade), after which the sample can be treated with chemicals to etch out the laser-modified region. Following this approach, Masuda et al. [87] and Sugioka et al. [88] fabricated a microfluidic device based on Foturan glass that contains a movable plate inside the channels for controlling the flow direction of certain reagents, as shown in Fig. 6.8.

However, there are several undesired effects that are associated with above-mentioned multi-step processing techniques. These include (i) Channel tapering: since the exit point of a laser modified zone is more exposed to chemical etching than the interior parts, tapering (longitudinally) occurs in the resultant channels. The tapering effect is stronger as the laser modified zone decreases in lateral size (below a few tens of microns). Nevertheless, this effect can be neglected when fabricating channels that are just a few hundred microns in diameter, as are often used for

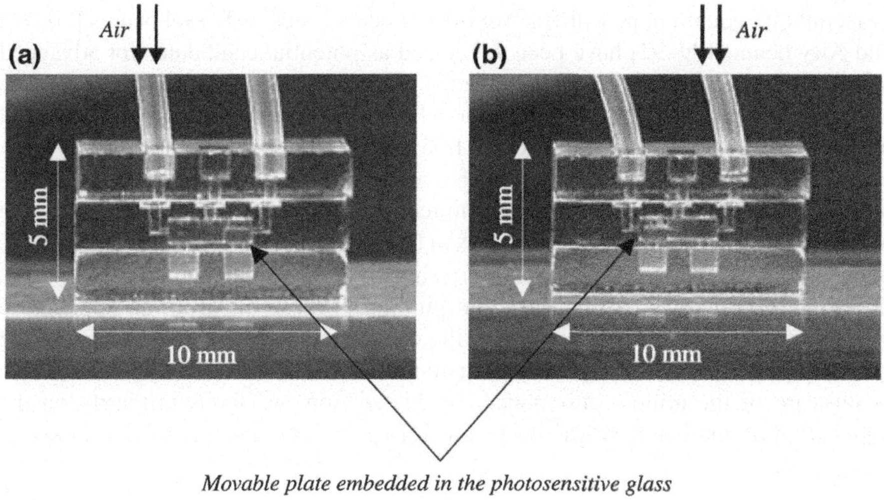

Fig. 6.8 Microstructures fabricated inside photosensitive glass using multi-step processing techniques. A microfluidic device prototype with a freely movable microplate that can switch the flow direction of reagents is shown. The movable plate inside the glass was realized by laser structuring followed by heat treatment and chemical etching. **a** By infusing compressed air from the left opening of the top part, the microplate moves to the right side. **b** As compressed air is infused from the right opening of the top part, the microplate moves to the left side [88]. Reproduced with permission from Springer

lab-on-a-chip or bio-chemical applications [86, 88]. A novel possibility for overcoming the tapering effect was explored by Chaitanya et al. [89]. In this technique, the laser modification pattern was designed like double cones facing each other so that after the chemical etching treatment, the resultant channel becomes cylindrically shaped. (ii) Complex integration: chemical etching processes that rely on the chemical selectivity of the laser modified material do not differentiate well between soft material modification (basis of waveguide structure) and strong material modification. Therefore, simultaneously writing both waveguides and hollow channels for opto-fluidic applications is very challenging. In fact, when integrating microfluidics with optical waveguides, the waveguides are always written after the fabrication of the microfluidic components.

6.4 Non-diffractive Approach for Flexible Fabrication

The contemporary approach to material processing involves fabrication using optical beams that do not diffract while propagating from one point to the other, and which can attain both super-resolution (lateral sizes down to 100 nm) as well as longitudinally extended length scales (up to 1 mm). Being propagation invariant solutions of Helmholtz equation, non-diffractive optical beams such as Bessel beams [70, 71] and Airy beams [89–92] have been advocated as potential candidates for advanced material processing. Such beams can ideally overcome diffraction and hence require an infinite amount of energy, which obviously means these beams cannot be realized in practice. However, quasi-diffraction free beams have been experimentally generated that retain their inherent beam characteristics (i.e., non-diffractive and self-reconstruction behavior [93, 94]) over limited propagation distance. Such beams are now being used in numerous applications in fields such as nonlinear optics [94–100], plasma physics [101, 102], material processing [103, 104], optical trapping [69, 105, 106], microscopy [107, 108], and atom guiding [109, 110]. From an application viewpoint, physical parameters such as the central lobe size and the corresponding intensity (specifically, the highest among all the surrounded lobes) of non-diffractive beams, are of the utmost importance. In this section, we focus primarily on the application of quasi-non-diffractive beams for fast and flexible material processing.

6.4.1 Zero-Order Bessel Beams

Durnin et al. [70] first studied solutions of Helmholtz equation and found Bessel type solutions that have electric field structures independent of the propagation distance. In a cylindrical coordinate system, the electric field structure of such a wave can be described as

$$E(r, \Phi, z) = E_0 J_n(k_r r) \mathrm{e}^{ik_z z} \mathrm{e}^{\pm i n \Phi},$$

where E_0 is a constant, r, ϕ and z are the radial, azimuthal, and longitudinal axes, respectively, k is the wave vector relating its radial component, $k_r = k \sin \theta$, and the longitudinal component, $k_z = k \cos \theta$, J_n is the nth order Bessel function of the first kind, and θ is the half-angle of the cone containing all the wave vectors (more explanation of this parameter is given in the following paragraph) that superimpose to form Bessel beams. When $n = 0$, the above equation is reduced to

$$E(r, z) = E_0 J_0(k_r r) \exp(i\, k_z z),$$

which defines the field structure of zero-order Bessel beams. Therefore, its corresponding intensity is proportional to $J_0^2(k_r r)$, clearly indicating concentric ring patterns with a central bright spot. Importantly, the central spot of this Bessel beam, which is equivalent to the focal spot of a Gaussian beam, does not diffract over the beam propagation distance. This unique property has made the beam of particular interest for high-aspect-ratio micro and nanostructuring applications.

For the generation of zero-order Bessel beams, several methods have been proposed using passive components such as annular disk [70], axicon lens [110–113], static hologram [114] and dynamically reconfigurable hologram generated via spatial light modulator (SLM) [114–119]. Among these methods, a generation approach

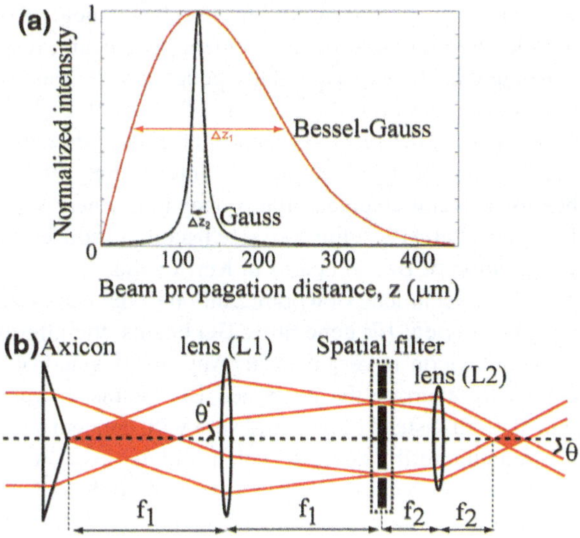

Fig. 6.9 Zero-order Bessel beam characteristics and generation method: **a** Peak intensity variation along the longitudinal direction corresponding to Gaussian and BG beams that have central spot sizes of 1.8 μm. **b** Experimental arrangement for axicon lens-based Bessel beam generation along with a magnification system comprising two lenses (L1 and L2) and a spatial filter to produce high-quality and high cone angle zero-order Bessel beams

based on an axicon lens (a cone-shaped optical element) has attracted significant amounts of attention due to its ability to produce high throughput (nearly 100%) beam transformation (from Gaussian to Bessel beams) in an economical manner. In contrast, diffraction-based optical elements, such as SLM, suffer from typically below 50% beam transformation efficiency. When collimated monochromatic Gaussian laser beams with flat phase front are incident normally on an axicon with n_{ax} refractive index and α wedge angle (sometimes referred to as base angle), the beams deviate towards the main axis (along the beam propagation direction) with an angle θ' and overlap. This produces an interference effect that can be characterized as zero-order Bessel beams (sometimes referred to as Bessel-Gauss (BG) beams) of conical half-angle θ'. With pure geometrical optics considerations, Snell's law at the inclined face of an axicon gives rise to the following expression:

$$\theta' = \arcsin(n_{ax} \sin \alpha) - \alpha.$$

A quantitative comparison of the on-axis (central axis) intensities of Gaussian and BG beams with identical 1.8 μm spot sizes is shown in Fig. 6.9a. As evident from the figure, the longitudinal distance (non-diffracting length) for BG beams is nearly one order of magnitude longer than that for Gaussian beams. The mathematical description of BG beams generated by an axicon lens is usually based on Fresnel diffraction integral evaluated under stationary phase approximations [120]. The on-axis intensity $I(z)$ of BG beams thus produced at a distance z from the beam onset point (the axicon tip) can be expressed as follows [121]:

$$I(z) = 8\pi P_0 z \sin^2\theta' / \lambda_0 w_0^2 \,.\, \exp\left[-2\left(z \sin\theta' / w_0\right)^2\right],$$

where P_0 is the peak power of the laser beam incident on the axicon, λ_0 and w_0 are the wavelength and waist of the incident laser beam, respectively, and θ' is the conical half-angle of the experimentally generated BG beams. Note here that, compared to the BG beams in air, the conical half-angle θ' of BG beams in glass is reduced by a factor of n, the refractive index of glass, whereas the non-diffracting length Z_B gets lengthened by a factor of n. Interestingly, the size of the central core of BG beams does not change while propagating linearly in glass. However, according to Polesana et al. [122], the core size becomes reduced in cases involving the nonlinear propagation of Bessel beams in Kerr media.

From a technical viewpoint, although the use of an axicon has been observed to be very efficient for generating BG beams, the alignment of the illuminating beam over the axicon is very critical. Any off-axis and oblique-axis axicon illumination can easily aberrate the generated BG beams, thereby resulting in a checkerboard type pattern instead of a central spot with concentric rings. Additionally, typical axicon manufacturing defects, such as oblate tips, often modulate the on-axis intensity profile of BG beams. Of course, a methodology based on Fourier filtering can be adopted to obtain near-ideal BG beams. Another important issue, especially within the framework of Bessel beam-based material processing (which typically requires high-angle BG beams [121, 123]) concerns the direct applicability of an axicon lens alone. In fact, large base angle axicon lens are not suitable (and often commercially unavailable) for high cone angle BG beam generation due to large scale aberrations and pulse broadening effects that result from the large thickness of the axicon lens itself. Therefore, telescopic systems are normally employed in association with an axicon lens to produce BG beams of micron and sub-micron central core sizes, which eventually determines the fabrication resolution. A typical axicon lens based Bessel beam generation scheme is depicted in Fig. 6.9b, where a telescope consisting of lens L1 (of focal length f_1) and lens L2 (of focal length f_2) with a de-magnification factor of f_2/f_1 is employed. For example, when $\alpha = 0.5°$, $n_{ax} = 1.45$, $f_1 = 1$ m, and $f_2 = 20$ mm, a BG beam of conical half-angle $\theta = 12°$ can be obtained in air, with its onset position lying at the focal plane of the objective lens, L2. A spatial filtering arrangement [124] at the focal plane of lens L1 followed by the axicon lens was also employed to produce near-ideal BG beams. The spatial filter is mainly composed of a stopper that blocks undesired wave vectors lying inside the ring and an aperture that blocks undesired wave vectors lying outside the ring. This arrangement only allows passage of the wave vectors passing through the stopper-aperture based annulus region, and thereby ensures near-ideal BG beams at the focal plane of lens L2. Interestingly, recently, Dudutis et al. [125] discovered an application of modulated BG beams (i.e., with no spatial filtering) generated by an imperfect axicon lens, for initiating preferential cracks, leading to cutting of transparent materials.

Surface processing:

Exploiting their inherent properties, zero-order BG beams are shown to be useful for large-area, nanoscale surface processing. For the sake of comparison, a smaller spot (sub-micron size is needed for better fabrication resolution) can be obtained using a tight focusing geometry in the case of Gaussian beams. However, tight focusing of

Gaussian beams also results in shrinking the corresponding longitudinal depth to a few microns. Now, when laser processing over a large area (~cm^2) is required, maintaining a fixed distance between the sample and the laser focal point is a tedious job that would normally require an adaptive positioning control system, thereby adding complexity and expense to the manufacturing process. However, BG beams offer potential solutions to this problem. In ablation-threshold regime, the light-matter interaction zone involving primarily the BG central core can be defined and positioned with respect to the material to be processed in a more efficient way compared to comparable Gaussian beams. Note that the central core size of a BG beam is a function of its cone angle (a high cone angle corresponds to a small core size), while the longitudinal depth (non-diffracting length) of a BG beam is a function of the beam apodization (a large input beam size corresponds to a large depth of focus). Most importantly, since both the central core and the longitudinal depth of focus are essentially decoupled from each other, they can be independently controlled, which is an unusual characteristic when compared to Gaussian beams. These unique characteristics remove the stringent criteria of sample-beam positioning that is normally required for large-area processing.

Courvoisier et al. [126] demonstrated that the size and morphology of the structures machined on a glass surface using high cone angle Bessel beams ($\theta \sim 24°$) do not vary much over long distances (up to a few tens of microns), and Sahin et al. [127] successfully machined slits (see Fig. 6.10a) with widths down to 125 nm on a 25 nm thick gold film over glass using femtosecond BG beams ($\theta \sim 20°$). They also reported a prominent plasmonic resonance effect at wavelengths around 900 nm when using nanoslit arrays with 860 nm periodicity, as shown in Fig. 6.10b.

To summarize, large-area surface processing, especially of opaque materials in the ablation threshold regime, can be very productive when using high cone angle BG beams. While this has opened a potential route to machining plasmonic, metasurface structures, such a technique would face scaling restraints in terms of the machined

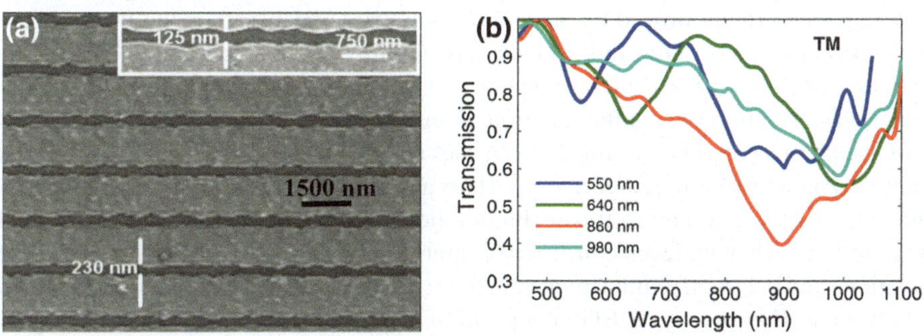

Fig. 6.10 **a** SEM image of nanoslits machined on 25 nm thick gold coated over glass using zero-order BG beam. The darker regions of the image are ablated stripes. The corresponding plasmonic response is shown in (**b**), depicting the transverse magnetic transmission spectra corresponding to four nanoslit arrays with periodicities of 550, 640, 860 and 980 nm cases [127]. Reproduced with permission from AIP

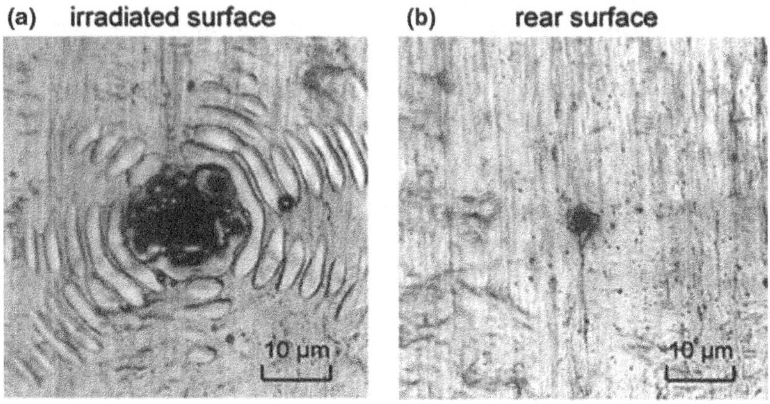

Fig. 6.11 SEM surface profile images of a hole drilled into a 20 μm thick stainless steel sheet using a BG beam (100 pulses of 20 J/cm^2 fluence). **a** Front surface profile, and **b** rear surface profile [128]. Reproduced with permission from Springer

structure size. For instance, when high energy Bessel pulses are employed in order to obtain large structures, a signature mark of concentric ring-shaped structures is obtained, as shown in Fig. 6.11. In fact, these results (as depicted in Fig. 6.11) represent one of the earliest studies by Matsuoka et al. [128] on BG beam machining. SEM surface profile images of a hole drilled into a 20 μm thick stainless steel sheet using a BG beam (100 pulses of 20 J/cm^2 fluence) clearly show that ring patterns are only visible on the front surface of the material. This is because material ablation occurs over the whole region exposed by several outer rings of BG beams, exceeding the ablation threshold fluence. Nevertheless, this ring (instead of point) structure is, in general, an undesired effect that can degrade the quality of high-density structures.

He et al. [129] proposed a technique based on tailored BG beams to address the issue of supplementary, yet unwanted, ring-type surface structuring. The beam tailoring technique followed by the authors was primarily based on the Bessel beam generation method of Durnin [70] using an annulus ring coupled with a lens. In the present case, a binary phase ($0 - \pi$) plate consisting of numerous concentric annular zones was used to replicate the function of annulus rings. Each annular zone creates a BG beam of a certain cone angle that causes all the generated BG beams (by several annular zones) to lie together and interfere at the same point, which is defined by the axicon lens. The key to this technique lies with the phase tuning of individual annular zones that achieves constructive interference only at the Bessel central core positions and destructive interference in the region of outer lobes of all the generated BG beams. This permits a BG beam with a super-intense central core and highly suppressed outer lobes to be generated. When employed for material processing, such a beam can be expected to produce high-quality structures.

He et al. [129] produced such tailored BG beams using a custom designed phase plate for processing of Si using 65 fs laser pulses at a wavelength of 1.5 μm (to which Si is transparent) at a 1 kHz repetition rate. Longitudinal cross-sectional images of

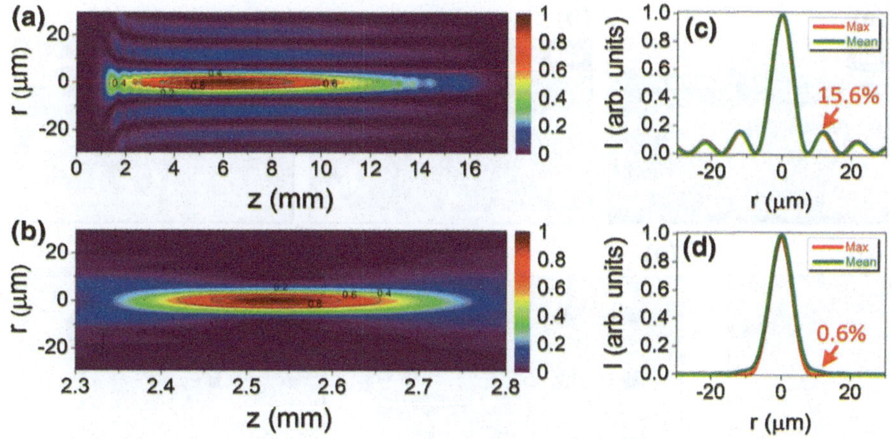

Fig. 6.12 Spatial characteristics of conventional and tailored BG beams. Longitudinal cross-sectional profiles for conventional (**a**) and tailored (**b**) BG beams are shown. The radial intensity profiles corresponding to conventional (**c**) and tailored (**d**) BG beams are also shown [129]. Reproduced with permission from Nature Publishing Group (NPG)

both conventional and tailored BG beams are shown in Fig. 6.12a, b. The radial peak intensity profile for the beams, as shown in Fig. 6.12c, d, clearly indicate that the side-lobe intensity for the tailored BG beam is suppressed to merely 0.6% compared to 15.6% for the conventional BG beam. Note that in both cases, the central core size was 6 μm. The potential use of beam tailoring for the machining of near-parallel channels in 50–100 μm thick Si, with no unwanted features around the channel opening, was demonstrated using a multiple pulse irradiation scheme (limited to a few hundred laser pulses per channel), as shown in Fig. 6.13a–f. From a quantitative view point, the Gaussian beams produced tapered vias/channels with front surface diameters of 20 μm and aspect ratios of less than 3, as can be seen from the SEM cross-sectional image of the channel shown in Fig. 6.13a. The conventional BG beams produced 5 μm diameter near taper-free vias with aspect ratios around 10 (see Fig. 6.13b). However, the applications of such vias are limited due to undesired surface damage caused by the characteristic concentric lobes of BG beams. In contrast, tailored BG beams produced a clean structure (with no additional damage) with characteristics similar to that of a conventional BG beam (see Fig. 6.13c). Figure 6.13d–f show the SEM cross-sectional, front, and rear surface 2D array images of through holes produced in 100 μm thick Si substrates by tailored BG beams.

Volume processing:

The potential application of zero-order BG beams to the machining of sub-micron scale surface features is clearly evident from the earlier paragraphs. Here, we discuss how BG beams are useful for controlled volume fabrication. In the context of BG beam volume interactions with transparent materials such as glass, it is important to note here that (analogous to Gaussian beams), BG beams also suffer from nonlinear

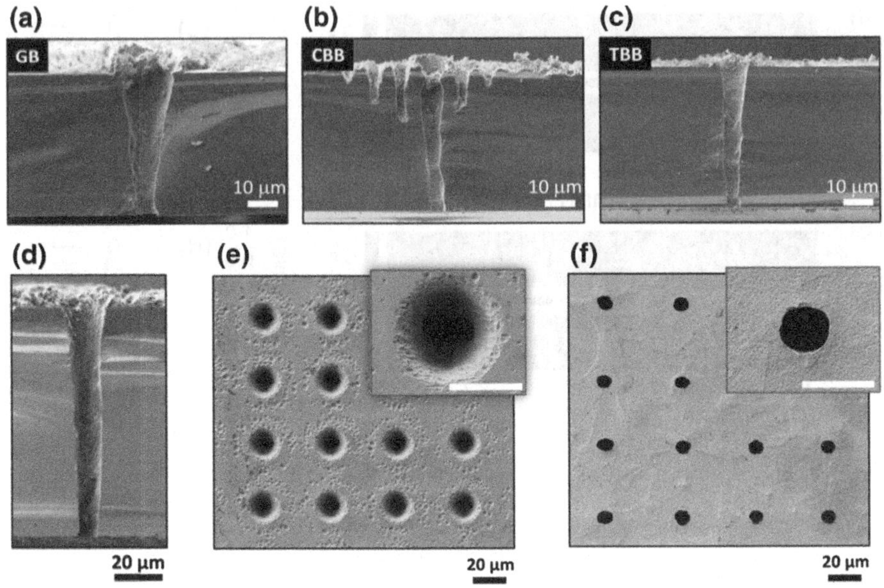

Fig. 6.13 Experimental results for through Si hole fabrication using laser beams with several spatial profiles. SEM cross-sectional profiles of through holes drilled in a 50 μm thick Si substrate using **a** Gaussian, **b** conventional BG, and **c** tailored BG beams are shown. **d–f** SEM cross-sectional, front, and rear surface images of two-dimensional (2D) array of through holes produced in 100 μm thick Si substrates by the tailored BG beam, respectively. The scale bars in the insets of **e** and **f** are 5 μm [129]. Reproduced with permission from NPG

effects, including the Kerr effect and the plasma defocusing effect, while they are propagating in the medium. Yet, it has been observed that BG beams are more robust to nonlinear distortion than Gaussian beams over a broad range of experimental conditions [97, 123]. Nevertheless, three different Bessel beam propagation regimes in Kerr media have been observed: weakly nonlinear Bessel filamentation, unsteady Bessel filamentation, and steady Bessel filamentation [97]. The most important case in the context of material processing refers to steady Bessel filamentation, where an extended plasma column of near-uniform density can be achieved, and where a significant level of energy absorption may occur inside the medium, especially when the plasma number density is above $10^{18}/cm^3$. Experimentally, such a regime can be achieved in a certain parametric window of laser energy, pulse duration, and Bessel cone angle. In a broader sense, high cone angle BG beams normally produce high electron densities (close to the corresponding critical density) and therefore cause a higher level of energy deposition inside the medium, when compared to low cone angle BG beams. In a medium such as glass, even with low-level excitation (employing a low cone angle BG beam), soft material modification (in terms of positive changes in the refractive index) can be obtained, as reported by Marcinkevicius et al. [130], who were the first to demonstrate wave-guiding with low loss and birefringence effects in BG beam induced structures in glass. However, structure size and uniformity could not be maintained when the laser energy increased. In fact, one such (multiple pulse irradiation regime) case led to periodic damage structures, as observed by Gaizauskas et al. [98], who attributed BG beam induced structures to an unsteady Bessel filamentation process. Bhuyan et al. [131] also observed similar behavior when machining glass using femtosecond BG beams ($\theta = 7°$).

In brief, low cone angle BG beams normally induce a soft positive refractive index change (~10^{-4}–10^{-3}), even though the structure size is not scalable solely on the basis of laser pulse energy. A typical example of femtosecond (60 fs) laser-driven positive index structures in fused silica glass using single pulse and multiple pulse (100 pulses at 1 kHz repetition rate) irradiation schemes is shown in Fig. 6.14a, b. A novel approach to single pulse machining involving low cone angle BG beams and exploiting the temporal engineering concept [78] demonstrated by Bhuyan et al. [132] provides additional control over the structural characteristics of material modification. Figure 6.14c shows a phase contrast microscopy (PCM) image of a single picosecond (4.7 ps) pulse induced negative index structure, such as a void inside fused silica glass. The impact of pulse duration (from 60 fs to 5.7 ps) on material processing was therefore studied and a gradual change in refractive index from positive to negative is clearly evident from the PCM images of single pulse induced structures in glass using BG beams ($\theta = 8.3°$, pulse energy of 4.1 μJ), as shown in Fig. 6.15a. A longitudinal cross-sectional profile of the BG beam is also shown in the top of this figure. A parametric window in the energy-pulse duration framework is shown in Fig. 6.15b. Here, the highlighted regions correspond to structures that show unique morphology. For instance, Regimes 1, 2, 3 and 4 represents positive index modified structures, uniform negative index modified structures (such as voids), fragmented negative (soft) index modified structures, and fragmented negative (strong) index modified structures, respectively.

A few important points should be made when BG beams interact with a condensed transparent medium such as glass, thereby resulting in a permanent material modification. First, BG beams are more robust to Kerr nonlinearities than Gaussian beams. For instance, it has been shown that the "self-focusing effects" that play a decisive role (obstructive) in energy deposition inside the medium (which dictates the type of material modification) can be substantially suppressed using high cone angle BG beams, and over a wide variety of energetic conditions, primarily due to the conical nature of the energy flow. In contrast, control of high-energy Gaussian beams is very challenging. Second, when femtosecond and picosecond BG beam are compared, the latter is more stable in the presence of nonlinearity of the medium. This is mainly because the plasma defocusing effect is substantially suppressed due to delayed ionization mechanisms triggered by the picosecond laser beam. The leading part of the femtosecond laser pulse can swiftly generate free carriers in the medium, which have the potential to scatter the remainder of the pulse. However, generally speaking, a picosecond laser pulse of the same energy produces a smaller number of carriers at the leading edge of the pulse due to its low peak intensity (in comparison to a femtosecond pulse), so a low carrier scattering effect can be observed. Additionally, the avalanche effect is more pronounced in the case of picosecond BG beams, which

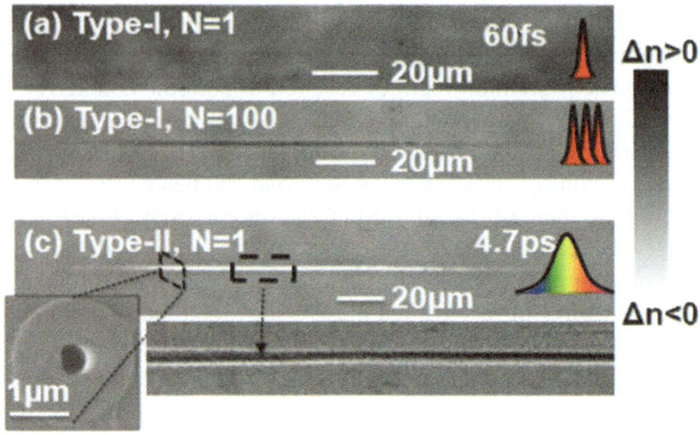

Fig. 6.14 Phase contrast microscopy (PCM) images of BG beam (θ = 8.3°) induced high aspect ratio structures at two different pulse durations. **a** Single pulse based Type-I positive refractive index structure machined using 1.7 μJ pulse energy and 60 fs. **b** Multiple pulse (pulse number, N = 100) based smooth Type-I refractive index structure machined using 1 μJ, 60 fs. **c** Single pulse based Type-II uniform void structure machined using 7 μJ, 4.7 ps. Typical SEM cross-sectional profiles for the fabricated structures at picosecond pulse durations demonstrating submicron lateral confinement are also shown (inset) [133]. Reproduced with permission from Wiley

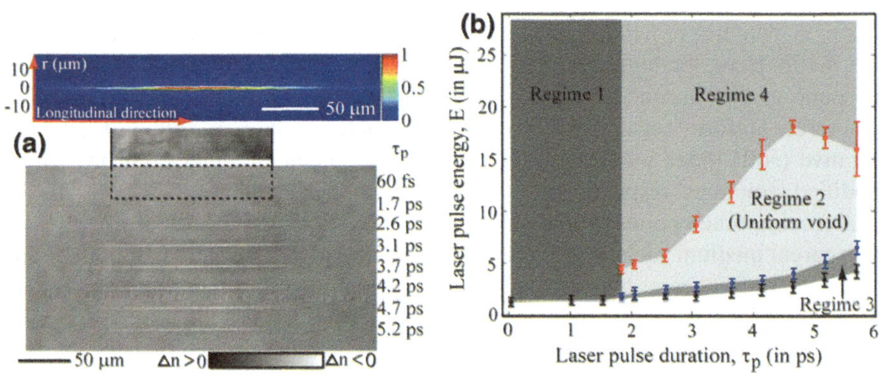

Fig. 6.15 BG beam based bulk structuring results: **a** Pulse duration dependent morphological characteristics of single pulse induced long structures in fused silica glass, showing the gradual evolution from positive to negative refractive index changed structures, **b** energetic conditions in the framework of pulse energy and temporal duration for the machining of structures with different morphologies [132]. Reproduced with permission from AIP

ultimately boosts the carrier number density. This is the reason why, compared with their femtosecond counterparts, picosecond laser pulses are generally more efficient in terms of energy deposition inside a medium.

Indeed, through numerical simulation studies, Bhuyan et al. [132] demonstrated that energy deposition approximately an order of magnitude higher can be obtained inside glass when it is excited by a picosecond (such as 4.7 ps) BG beam with a conical half-angle of 8.3° and a pulse energy of 4.1 μJ, instead of a femtosecond (such as 60 fs) BG beam. Nevertheless, these results pertaining to the nonlinear Bessel beam propagation dynamics in glass indicate the achievement of a high deposited energy density (a few kJ/cm^3, sufficient to trigger a phase change of material) only in the picosecond case. As an extreme case, the temporal engineering concept was applied to a loose focusing geometry, and remarkably long voids with lengths of up to 1 mm and diameters of down to 100 nm were obtained on single-pulse basis, using zero-order BG beams with a conical half-angle of 4.2° (in glass), 1 ps pulse duration, and an energy of just a few microjoules [133].

In brief, void-like structures with potential applications in the field of photonics and fluidics can be realized using BG beams of low and moderate conical half-angles ($\theta <$ nearly 10°) only when a temporal engineering concept (that produces ps pulses) is employed. However, Bhuyan et al. [121] were the first to demonstrate that a single femtosecond BG laser pulse in a tight focusing geometry can drastically alter the material properties, and thus produce voids inside transparent materials. This technique has been explored for different transparent materials such as borosilicate glass [121, 134], fused silica glass [132], sapphire [135], polymers [135–138] and others. Figure 6.16a shows an SEM cross-sectional image of a 21 μm long and 300 nm diameter nanochannel in borosilicate glass that was machined using a single femtosecond BG pulse ($\theta = 17°$ in glass) of 0.85 μJ energy. The channel characteristics have also been observed to be scalable [121]. For instance, the region of the material over which the BG central core intensity exceeds approximately 10^{12-13} W/cm^2 [2] primarily dictates the channel length. This implies that higher is the pulse energy, longer is the channel length, until reaching a saturation value. In contrast, the channel diameter estimation is difficult as the underlying mechanism (which is still being debated) is not yet fully understood. Nevertheless, Bhuyan et al. [121] experimentally obtained a near-linear scaling of channel diameter with respect to the laser pulse energy. One of the important features of single-pulse BG beam machining is that by keeping the pulse energy fixed, the channel length can be varied straightforwardly, without affecting the channel diameter, simply by suitably placing the BG beam within the medium. Note that this unique feature of de-coupling channel length-diameter is not possible with Gaussian beam machining due to its complex and unpredictable beam propagation dynamics.

From a technological viewpoint, BG beam induced extended nanochannels are suitable for numerous applications. For instance, the deep drilling (up to a millimeter) capability of BG beams is unique and far exceeds other state-of-the-art techniques [74–77]. Bhuyan et al. [132] demonstrated water flow through BG beam induced nanochannels in fused silica glass, thereby opening novel possibilities for designing nanoscale devices, such as by injecting suitable nanoparticles inside channels. With an aim towards replicating the design of a photonic crystal, Bhuyan et al. [121] demonstrated a 2D nanochannel array comprised of machined channels (230 nm in diameter and a few tens of microns long), as shown in Fig. 6.16b. The other application that BG beam machining technology aims at is high throughput slicing and/or cutting of thick transparent materials. The methodology of such cutting technique is

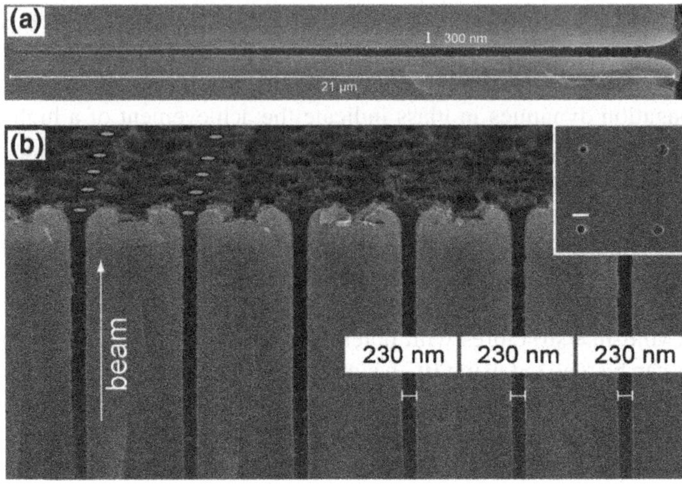

Fig. 6.16 BG beam ($\theta = 17°$) based nanostructuring results. **a** SEM cross-sectional image of a typical nanochannel produced in Corning 0211 glass using a single Bessel pulse of 0.85 μJ energy, and **b** SEM cross-sectional image of a nanochannel array obtained using almost the same machining conditions [121]. Reproduced with permission from AIP

as follows: (a) Choose the laser energy needed to machine a uniform channel-like structure (although strong non-uniform material modification can be used if compromised with the cut surface quality) inside the material using a single laser pulse. Although the length of the channel is not especially critical, the best cutting results are obtained in cases where the channels extend over the sample thickness (especially for inhomogeneous materials such as tempered glass [139]). (b) Optimize the distance between consecutive channels (pitch), which depends on the channel morphology and the lateral dimensions. The pitch values used for cutting experiments thus far are reported to lie between 0.5 and 10 μm. (c) Select the effective number of laser pulses per spot. Ideally, the best case would be achieved using a single laser pulse per spot. However, if single pulse induced material modification is not sufficiently strong, multiple laser pulses per spot will be required. (d) Apply mechanical pressure on either side of the plane containing channel-like structures to achieve cutting of the material.

Several research efforts [138–141] have been made in this direction. For instance, Tsai et al. [140] first exploited femtosecond (120 fs) zero-order BG beams (2 μm core diameter) to realize cutting of 100 μm thick fused silica glass (homogeneous along the sample thickness) at a speed of 1 μm/s (laser repetition rate of 1 kHz, and a pitch of 1 μm), whereas Bhuyan et al. [139] demonstrated cutting of homogenous glass and inhomogeneous tempered glass with thicknesses of up to 1 mm, at a speed up to 270 mm/s (laser repetition rate of 60 kHz), using picosecond BG beams with 2.5 μm core diameter. Those authors also reported that pitch is an important factor for high-quality cutting, as shown in Fig. 6.17a. For instance, an optimum pitch of

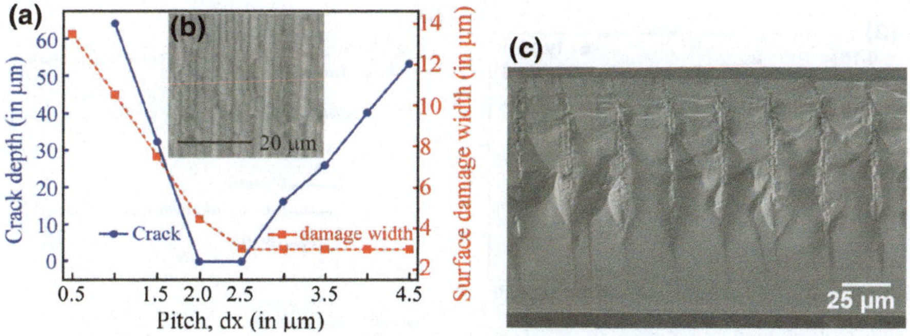

Fig. 6.17 Experimental results of BG beam-based cutting of transparent materials. **a** Sample crack evaluation with respect to the pitch, i.e., distance between consecutive BG beam induced material modifications inside a 700 μm thick tempered glass, **b** typical cut surface of a tempered glass [139]. **c** SEM cross-sectional image of a cut surface of sapphire sample machined with a pitch of 25 μm [141]. Reproduced with permission from Springer and OSA

2 ± 0.5 μm was observed for sample cleaving with minimum surface damage. Rapp et al. [141] also demonstrated sapphire plate cutting using a zero-order BG beam with an exceptionally high pitch (25 μm), as illustrated in Fig. 6.17c. The possibility of cutting with a large pitch was facilitated by preferential crack propagation along the crystal orientation axis. In short, the deep drilling capability of BG beams can be highly beneficial for high-speed cutting of thick materials, irrespective of sample homogeneity along the sample thickness direction.

To better understand and control material processing using BG beams, it is necessary to study the dynamics of material modification. First, since it was clarified that nanoscale yet extended material modifications could be obtained either completely inside the material or with a structure opening on the sample surfaces, the question of whether the dynamics of material modification in each separate case are the same requires consideration. Second, since various structural modification types, such as positive and negative refractive index changes, can be obtained using BG beams with the same focusing geometry and the same pulse energy, but different pulse durations, this again opens up the question of whether the dynamics of both structure types are the same, or if they follow the same route. Although the exact mechanism is yet to be clarified, a number of mechanisms have been proposed based on the experimental findings obtained thus far. The ultrafast dynamics of extended structures machined in glass using zero-order BG beams were first revealed by Velpula et al. [133, 142] using time-resolved optical microscopy techniques. In these experiments, the optical transmission mode was employed to study the behavior of laser induced plasmas, i.e., free carriers inside glass through their scattering and/or absorption natures. PCM mode was employed to track the relative phase change of the excited matrix. The carrier density was then estimated from the absorption characteristics of the excited regions at a fixed time delay with respect to the laser excitation, based on the Drude model [6, 133, 142]. Ideally, laser-excited matter dynamics would be independent of

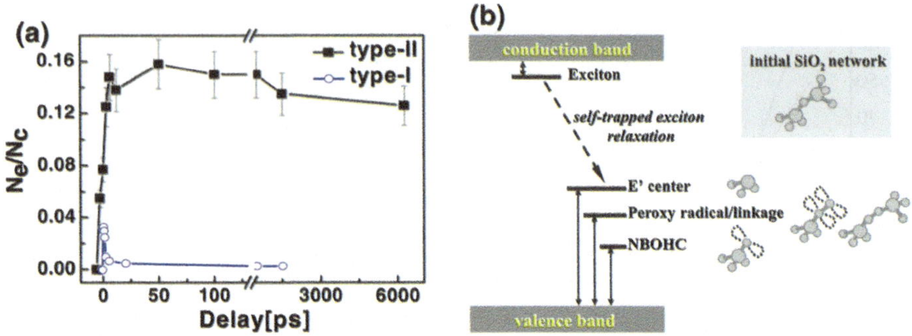

Fig. 6.18 Ultrafast dynamics of fused silica glass excited by BG beams: **a** variation of peak carrier density with respect to the delay time in cases of material modification with both increased (Type-I) and decreased (Type-II) refractive indices [133], **b** schematic illustration of exciton and intrinsic defect energy levels in SiO_2 [6]. Reproduced with permission from Wiley and Springer

the spatial profiles (such as Gaussian, Bessel) of the incident laser beam. However, given the advantage of extended scale matter excitation, which makes the visualization of transient features clearer, the study of matter excited with BG beams is more appealing.

The carrier dynamics corresponding to increased (Type-I) and decreased (Type-II) refractive index structures machined using 8.3° BG beams in fused silica glass, as shown in Fig. 6.18a, are observed to be different from each other. For the Type-I structure, which can be obtained primarily using a single femtosecond BG pulse, the carrier number density swiftly increases, reaching a maximum value that is about an order of magnitude less than the critical value in vacuum ($N_{cr} = 1.74 \times 10^{21}/cm^3$ for 800 nm light), and decreases to almost zero within a few picoseconds after laser excitation. This behavior is often associated with carrier trapping (lifetime of 150 fs in SiO_2 matrix [6]).

The carriers trapped in their own deformed matrix then result in self-trapped exciton formation. Such excitons are very transient (see Fig. 6.18b), and upon relaxation decay into electronic defects known as E′ centers and non-bridging oxygen-hole center (NBOHC). Earlier reports on Gaussian beam excited glass dynamics have revealed that NBOHC defects form smaller Si-O rings, and hence create a compact matrix. Therefore, increased refractive index structures are believed to be formed via electronic defects. In contrast, for the Type-II structure, which can be obtained primarily using a single picosecond BG pulse, the carrier number density increased to a near-critical value and remained almost constant until 17.5 ns, without showing any observable carrier trapping signature. Earlier reports [132, 133] have clarified, through numerical simulations, that the temperature rise in this case can indeed be close to (or above) the fused silica matrix softening point, which is about 1000 °C. Therefore, at such high temperatures, carrier trapping is not expected. For these machining conditions, the pressure waves [133, 143] can be visualized following 1 ns of laser excitation, thereby indicating the existence of high-level matter excitation, which can lead to a high temperature and pressure state. As a result, a micro-explosion may occur over the excited zone, compressing the material along the outward direction and leaving behind a rarified zone at the excitation epicenter. The existence of a high temperature and pressure state was also argued based on the fact that molten material can be obtained at the exit opening of such Type II structures (channels/voids) when they are machined in an appropriate way (see inset to Fig. 6.14c). It was then argued that the hot excited matter must have been ejected from the bulk and deposited around the channel. Therefore, the decreased refractive index structures are believed to be formed due to the high temperature and pressure driven thermo-mechanical expansion of glass. Very recently, Bhuyan et al. [144] reported that the timescale of emergence of such void-like structures (resulting from stress relaxation) can be up to a few hundred nanoseconds, which is otherwise believed to be on the ultrafast timescale [7]. However, more advanced imaging and microscopy tools will be required in order to examine the unique light-matter interaction processes involving non-diffracting zero-order BG beams.

6.4.2 Vortex Beams

Optical vortex beams with zero intensity at the center, and concentric high intensity lobes, have attracted significant attention in fields such as nonlinear optics and plasma physics [102, 144–151], optical tweezers [105, 152], and material processing [151, 152–157]. Femtosecond optical vortices have been generated using spiral phase plates [157–161], holograms [151, 155], and uniaxial birefringent crystals [153, 162]. Interestingly, however, when such high-intensity femtosecond vortex beams are employed for surface processing, an unusual light-matter interaction geometry is facilitated, resulting in the creation of a dense plasma zone in a toroidal region of the material. As a result, ablation within the ring (down to micrometer size) may occur even when using a single laser pulse. Hnatovsky et al. [153] first exploited such a concept and machined highly reproducible microstructures on glass, as shown in Fig. 6.19a, b. One of the more striking observations from that figure is that, upon increasing the laser pulse energy, deep cavities started to form at the center of the annulus ring-type structure. This was attributed to shock wave (moving towards the ring center) driven material compression and ejection. This technique can be potentially useful for the direct fabrication of isolated nanoparticles of different materials.

Non-diffractive higher-order Bessel beams (J_m), consisting of concentric cylindrical regions with gradual diminishing intensity from the inner to outer regions, and Laguerre-Gaussian beams, are also representative of the vortex beam class, and have been exploited for material processing. Like zero-order Bessel beams, ideal higher-order Bessel beams also require infinite levels of energy. However, finite-energy versions of such higher-order Bessel beams can be realized using Gaussian beams passing through a spiral phase plate followed by an axicon lens, or a spatial light modulator masked with a suitable phase. In any case, the vortex charge or helicity (integer values) determines the size of the intense inner ring of the corresponding

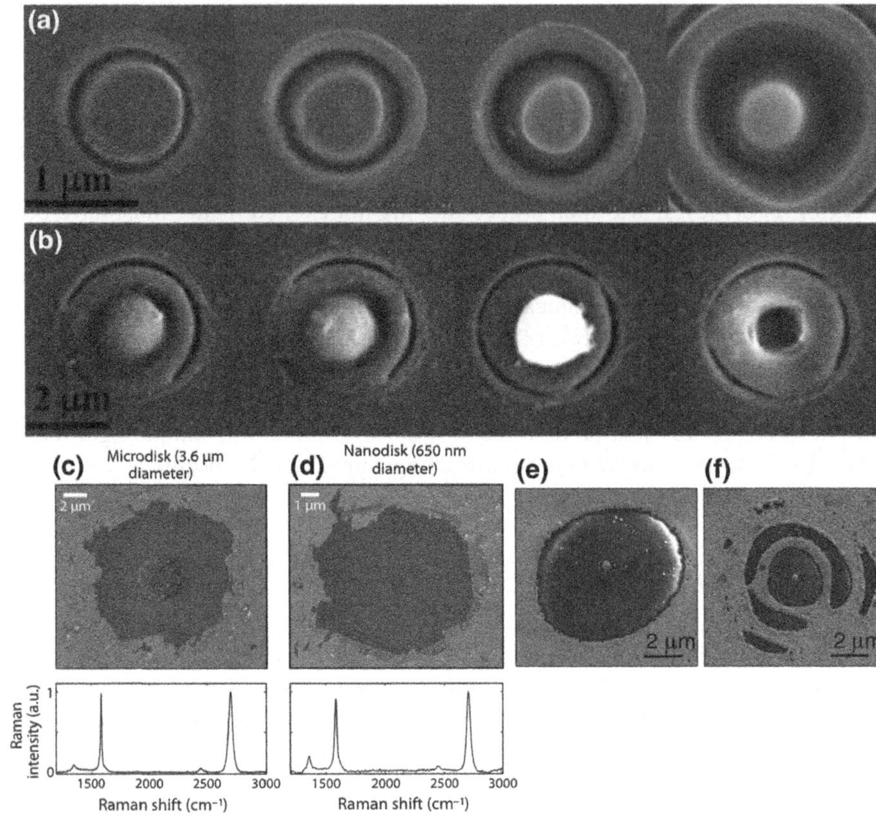

Fig. 6.19 Vortex beam-based surface structuring results: **a**, **b** single pulse ablation of soda-lime glass with double-charge femtosecond vortex pulses using a focusing lens of 0.9 NA [153], **c**, **d** micro- and nano-disks machined on graphene using higher-order Bessel beams in multi-pulse exposure scheme [154], **e**, **f** nano-disks machined on gold film using femtosecond laser first-order Bessel beams [156]. Reproduced with permission of OSA and AIP

Bessel beam. This opens up the novel possibility of designing a light-matter interaction geometry in the form of a cylinder with an arbitrary radius. For instance, using femtosecond laser vortex beams, Wetzel et al. [154] fabricated micro- and nano-disks with sizes as small as 650 nm on some of the most technologically demanding materials available, such as graphene, with clear boundaries confirmed via Raman spectroscopy (see Fig. 6.19c, d). Similarly, Sahin et al. [156] fabricated nano-disks as small as 200 nm on a gold thin film using a femtosecond first-order Bessel beam, as shown in Fig. 6.19e. They also pointed out that concentric rings formed around the central intense ring may produce unwanted structures for higher laser fluences (see Fig. 6.19f).

Vortex beams such as higher-order Bessel beams with an extended depth of focus have also been demonstrated for volume fabrication (i.e., internal material modification) [151, 157]. The underlying process in such cases is almost the same as that for a zero-order Bessel beam. For instance, the nonlinear propagation stability (in the presence of nonlinearity of the medium and multiphoton loss) of first-order Bessel beams [147–150] in a Kerr medium such as glass, is obviously an important factor for uniform energy deposition inside the medium. This is mainly because the multiphoton absorption loss in the high-intensity regions (innermost cylinder) is readily compensated for by the inward energy flow associated with the conical nature of first-order Bessel beams. Similar to the case for zero-order Bessel beams, several researchers [149, 150] have discovered parametric windows that provide stability criteria for nonlinear first-order Bessel beams. Apart from parameters such as the laser pulse energy and the pulse duration, the stability of Bessel beams is extremely sensitive to the cone angle. In other words, higher is the cone angle, better is the nonlinear stability of Bessel beams. For instance, since self-phase modulation and four-wave mixing effects modify the spatial spectrum, modulations to the axial intensity distribution (primarily on the inner cylinder of first-order Bessel beams) are very dominant in the case of low cone angle Bessel beams. Generally speaking, axial intensity modulation results in non-uniformity of the processed structure. In addition, the maximum intensity, and hence the plasma density in the cylindrical (innermost) region of higher order Bessel beams, can be obtained with a lower vortex charge value (smaller inner cylinder diameter).

Therefore, for material processing applications, an optimum balance between the vortex charge and the cone angle for higher-order Bessel beams is highly important, depending on the material to be processed and the type of structure to be obtained. For instance, Bessel beams that have cone angles (in glass) larger than 5° can trigger peak electron densities of more than 10^{19} cm^{-3}, thus facilitating sufficient energy deposition to induce a permanent material modification, as shown in Fig. 6.20a [151]. Xie et al. [151] first exploited vortex Bessel beams (higher order) for volume fabrication of Corning glass and obtained a tube-like refractive index modified structure for potential waveguide applications, as illustrated in Fig. 6.20b. A single pulse (35 μJ energy and 1 ps temporal duration) was shown to be capable of machining a 150 μm long cylindrical positive index modified zone with a 5 μm diameter using J_3 vortex beams. The refractive index change was estimated to be around 10^{-4} from the light (633 nm) guiding experiments (see Fig. 6.20d for the guided mode profile). Figure 6.20e schematically shows the directions of inwardly and outwardly propagating cylindrical matter waves generated by the tubular plasma. These are shock, pressure, and heat waves. Jedrkiewcz et al. [157] demonstrated cylindrical refractive index modifications (wall thickness and diameter of approximately 1 and 8 μm, respectively) in 100 μm thick BK7 glass using a J_8 vortex beam (37 μJ pulse energy, 1 ps temporal duration) with a conical half-angle of 11° (in the medium). Note that a hollow structure fabricated by a single vortex pulse can also be fabricated using a hollow Gaussian beam in a multi-pulse illumination regime with continuous sample translation. However, vortex beams clearly show superiority in terms of fabrication speed and structural uniformity.

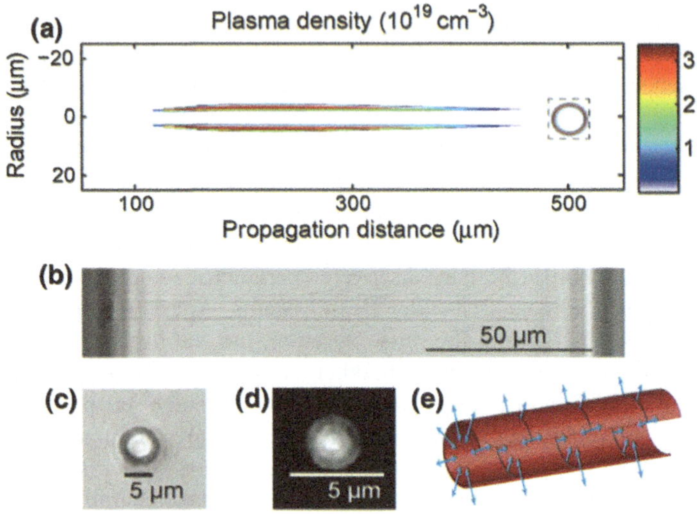

Fig. 6.20 Vortex beam (J_3 beam, θ =6.8°) based micro-structuring results. **a** Electron density distribution inside 150 μm thick glass excited with 5 μJ pulse energy, obtained by numerical simulations. **b** Optical micrograph (longitudinal view) of a typical tubular structure machined (35 μJ pulse energy, 1 ps pulse duration) in glass on single-shot basis. **c** Transverse section of the damage observed under identical machining conditions, but using 20 μJ pulse energy. **d** Image of near-field output guided light in the structure shown in (**c**), at an input wavelength of 632 nm. **e** Schematic view of the propagation direction of mechanical and thermal waves expanding outward and inward (arrows) from the excited tubular sheet volume (circle) [151]. Reproduced with permission from NPG

6.4.3 Curved Beams

An ideal Airy beam, discovered by Berry et al. [90], is basically a solution of the paraxial wave propagation equation, which is non-diffractive in nature. Similar to zero-order Bessel beams, such beams also possess a central intense spot, and propagate over an infinite distance with no diffraction, and uniquely along a parabolic (i.e., curved) trajectory. Like other non-diffractive beams, Airy beams also exhibit self-reconstruction characteristics [94]. Notably, Airy beams require an infinite level of energy in order to manifest their diffraction-free propagation characteristics in free space. Siviloglou et al. [91] first derived a finite energy Airy beam solution by modulating an ideal Airy beam with an exponential decay function. In this case, the field in the $\zeta = 0$ plane is assumed to have form of $\psi\,(s, \zeta = 0) = \mathrm{Ai}(s)\exp(as)$, where Ai denotes the Airy function, and a is a real positive constant. $s = x/x_0$ and $\zeta = z/kx_0^2$ are the dimensionless transverse and longitudinal coordinates respectively, x_0 is the spatial spot size and $k = 2\pi/\lambda_0$ is the wavenumber of the optical wave. Upon substituting this initial field ($\zeta = 0$ plane) profile in the 2D paraxial envelope equation ($i\partial\zeta\psi + 1/2\partial^2 s\psi = 0$), one can obtain the generalized Airy-Exponential field solution as follows:

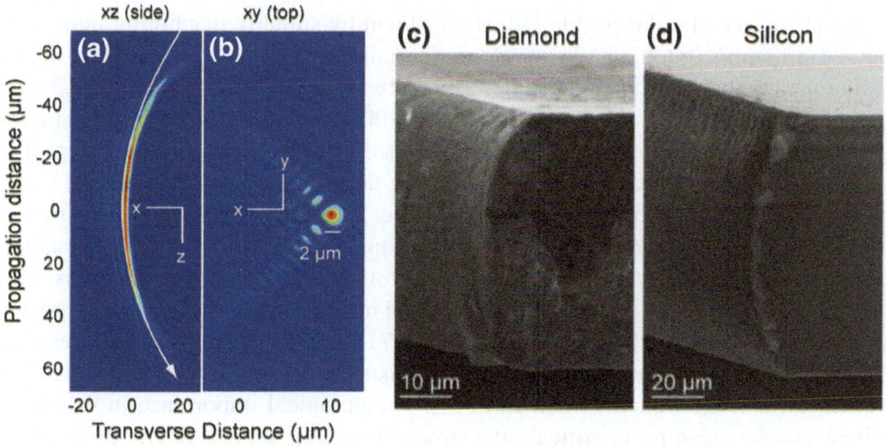

Fig. 6.21 Airy beam-based cutting results: Longitudinal (**a**) and transverse (**b**) cross-sectional images of a typical airy beam generated using a spatial light modulator. Cross-sectional SEM images of cut surfaces of (**c**) diamond and (**d**) Si [170]. Reproduced with permission from AIP

$$\psi(s,\zeta) = \mathrm{Ai}\left[s - (\zeta/2)^2 + ia\zeta\right]\exp\left[as - \left(a\zeta^2/2\right) - i\left(\zeta^3/12\right) + i\left(a^2\zeta/2\right) + i(s\zeta/2)\right].$$

From an experimental viewpoint, finite energy (usually by beam apodization) Airy beams can be generated using a phase mask [100, 163, 164], SLM [164–167], or an optically induced refractive index gradient [168]. Airy beams with central lobe sizes down to the sub-micron scale, which are useful in many important applications, have been generated [99, 100, 106, 108, 169]. These applications include flexible material processing [165, 170, 171], where the existence of stable nonlinear Airy beams is beneficial [172, 173]. Note here that the size (d_a, FWHM) of the central lobe of the Airy beams can be related to the curvature radius over which the beam bends, as $d_a = 1.63\ [R/2k^2]^{1/3}$, where R is the trajectory curvature radius [166]. This implies that the higher the curvature radius, the lower the Airy central lobe size. Figure 6.21a, b show typical longitudinal and transverse cross-sectional images of experimentally generated (via SLM) Airy beams with an arc-shaped (curvature radius of 120 μm) central lobe size of 2 μm extended longitudinally over several tens of microns, as reported by Mathis et al. [170].

When such a beam is used for bulk structuring of materials, the intense curved central lobe is expected to ablate and/or modify the material in a way similar to Bessel beams, but along a curved line if operating in the above-ablation threshold regime. This unique curved machining feature of Airy beams has been exploited for cutting or dicing of technologically important materials such as diamond and Si by following the same methodology used for Bessel beams and filamentation based cutting [36, 139, 140]. For instance, using a pulsed laser source with 800 nm wavelength, 100 fs pulse duration, and 11 μJ pulse energy, Mathis et al. [170] demonstrated cutting (laser material modification followed by cleaving) of 50 μm thick diamond slides with cut surface circular profile radius of 70 μm by suitably machining individual curved structures (pitch of 0.3 μm) along a certain direction, as shown in Fig. 6.21c. Using parameters similar to those described above, except with a pitch of 0.5 μm, the same authors also demonstrated cutting 100 μm thick Si (see Fig. 6.21d for a cross-sectional SEM image of the cut surface) with cut surface circular profile radius of 120 μm. Note that the above cutting demonstrations were realized using a two-pass configuration (effectively, two laser pulses at each spot). Airy beams thus provide an additional degree of control along with mechanical sample rotation or beam steering that can be used to obtain round-cornered, thick materials with curved surface profiles. Besides cutting, curved trench machining using Airy beams was also demonstrated by the same authors [171]. One of the striking observations that they made was the asymmetric debris distribution (present only one side of trench) around the trench, which can be of technological importance in situations where sample cleaning is critical after processing. This unique debris distribution was attributed to the asymmetric intensity profile, as shown in Fig. 6.21a, b.

6.5 Conclusions

In the 30 years since the first ultrafast laser processing experiments were demonstrated by Srinivasan et al. [174] and Küper et al. [175] in 1987, research into this field has become increasingly more active. This is because, due to distinct features such as ultrashort pulse width and extremely high peak intensity, ultrafast lasers offer novel schemes for processing materials that are not accessible using other types of lasers. In particular, the clean ablation ability of ultrafast lasers and their associated multiphoton absorption processes surprised numerous researchers, and opened a new door for high-quality processing of transparent materials [176]. Breakthroughs in terms of internal modification or volume fabrication of transparent materials, such as glass, further accelerated the growth of this field [10, 11, 23]. It was also demonstrated that the multiphoton absorption processes in the near-threshold ablation regime can even lead to fabrication resolution down to a few tens of nanometers, which is far beyond the diffraction limit [13, 14]. Therefore, material fabrication at micro and nanoscale levels is currently one of the major activities in the field of ultrafast laser processing, and has numerous potential applications in the fields of electronics, healthcare, photonics, and energy harvesting. One of the most successful commercial applications of the ultrafast laser processing of transparent materials is scribing and dicing very thin and hard glass in order to produce high-quality edges, flexible geometry, and crack-free components for use in the mass production of displays for cell phones and tablet computers. Meanwhile, from the viewpoint of recent fundamental research, attention is being focused on nanoprocessing based on the nonlinear light confinement processes in order to produce structures of lateral sizes down to few nanometers. Near-field optical phenomena can also be used to improve the fabrication resolution. Further challenges lie in facilitating the deep machining of transparent materials while maintaining micro and nanoscale lateral sizes (high aspect ratio machining).

Non-diffractive optical beams represented by zero-order Bessel beams possessing a long (along the longitudinal direction), narrow and intense region, have become promising tools for high aspect ratio material processing. For example, voids with lengths up to a millimeter and diameters down to 100 nm have been fabricated inside fused silica glass on a single-pulse basis using temporally engineered zero-order Bessel beams [133]. Other non-diffractive optical beams, such as vortex and Airy beams, enable us to imprint new types of structures that cannot be achieved by other techniques. For example, the former can produce tube-like structures, while the latter can produce longitudinally curved structures.

Ultrafast laser processing is already used for commercial and industrial applications, and this trend is expected to be further accelerated because of the distinct advantages ultrafast laser processing has over conventional laser processing or other existing competing techniques. To this end, current research activities are not only focusing on developing advanced processing techniques, but are also aimed at elucidating the detailed mechanisms of ultrafast laser-matter interactions. It is now believed that spatiotemporal beam shaping, including the generation of non-diffractive optical beams, will be one of the key technologies driving future improvements to the fabrication resolution, quality, and efficiency of ultrafast laser based manufacturing.

References

1. S. Nolte, G. Kamlage, F. Korte, T. Bauer, T. Wagner, A. Ostendorf, C. Fallnich, H. Welling, Adv. Eng. Mat. **2**, 23 (2000)
2. R.R. Gattass, E. Mazur, Nat. Photonics **2**, 219 (2008)
3. G.D. Valle, R. Osellame, P. Laporta, J. Opt. A: Pure Appl. Opt. **11**, 013001 (2008)
4. K. Sugioka, Y. Cheng, Light: Sci. Appl. **3**, e149 (2014)
5. Y. Shimotsuma, P.G. Kazansky, J. Qiu, K. Hirao, Phys. Rev. Lett. **91**, 247405 (2003)
6. S. Mao, F. Quéré, S. Guizard, X. Mao, R. Russo, G. Petite, P. Martin, Appl. Phys. A **79**, 1695 (2004)
7. S. Juodkazis, K. Nishimura, S. Tanaka, H. Misawa, E.G. Gamaly, B. Luther-Davies, L. Hallo, P. Nicolai, V.T. Tikhonchuk, Phys. Rev. Lett. **96**, 166101 (2006)
8. A. Mermillod Blondin, J. Bonse, A. Rosenfeld, I.V. Hertel, YuP Meshcheryakov, N.M. Bulgakova, E. Audouard, R. Stoian, Appl. Phys. Lett. **94**, 041911 (2009)
9. D. Du, X. Liu, G. Korn, J. Squier, G. Mourou, Appl. Phys. Lett. **64**, 3071 (1994)
10. K.M. Davis, K. Miura, N. Sugimoto, K. Hirao, Opt. Lett. **21**, 1729 (1996)
11. E.N. Glezer, E. Mazur, Appl. Phys. Lett. **71**, 882 (1997)
12. H. Helvajian, *Direct-Write Technologies for Rapid Prototyping Applications*, ed. by A. Piqué, D.B. Chrisey (Academic Press, San Diego, CA, 2002), Chapter 14
13. A.P. Joglekar, H. Liu, G.J. Spooner, E. Meyhofer, G. Mourou, A.J. Hunt, Appl. Phys. B Lasers Opt. **77**, 25 (2003)
14. A.P. Joglekar, H. Liu, E. Meyhofer, G. Mourou, A.J. Hunt, PNAS **101**, 5856 (2004)
15. J.F. Herbstman, A.J. Hunt, Opt. Express **18**, 16840 (2010)
16. A. Rosenfeld, M. Lorenz, R. Stoian, D. Ashkenasi, Appl. Phys. A **69**, S373 (1999)
17. D. Ashkenasi, M. Lorenz, R. Stoian, A. Rosenfeld, Appl. Surf. Sci. **150**, 101 (1999)
18. C.B. Schaffer, J.F. Garcia, E. Mazur, Appl. Phys. A **76**, 351 (2003)
19. R. Osellame, N. Chiodo, G. Della Valle, S. Taccheo, R. Ramponi, G. Cerullo, Opt. Lett. **29**, 1900 (2004)

20. S.M. Eaton, H. Zhang, P.R. Herman, F. Yoshino, L. Shah, J. Bovatsek, A.Y. Arai, Opt. Express **13**, 4708 (2005)
21. M. Shimizu, M. Sakakura, M. Ohnishi, Y. Shimotsuma, T. Nakaya, K. Miura, K. Hirao, J. Appl. Phys. **108**, 073533 (2010)
22. C. Kerse, H. Kalaycioglu, P. Elahi, B. Cetin, D.K. Kesim, O. Akcaalan, S. Yavas, M.D. Asik, B. Oktem, H. Hoogland, R. Holzwarth, F.O. Ilday, Nature **537**, 84 (2016)
23. E.N. Glezer, M. Milosavljevic, L. Huang, R.J. Finlay, T.-H. Her, J.P. Callan, E. Mazur, Opt. Lett. **21**, 2023 (1996)
24. G.D. Marshall, A. Politi, J.C.F. Matthews, P. Dekker, M. Ams, M.J. Withford, J.L. O'Brien, Opt. Express **17**, 12546 (2009)
25. T. Meany, Laser Focus World **50**, 25 (2014).
26. Y. Nasu, M. Kohtoku, Y. Hibino, Opt. Lett. **30**, 723 (2005)
27. J. Lapointe, Y. Ledemi, S. Loranger, V.L. Iezzi, E.S.L. Filho, F. Parent, S. Morency, Y. Messaddeq, R. Kashyap, Opt. Lett. **41**, 203 (2016)
28. J. Guan, X. Liu, P.S. Salter, M.J. Booth, Opt. Express **25**, 4845 (2017)
29. W. Watanabe, T. Asano, K. Yamada, K. Itoh, J. Nishii, Opt. Lett. **28**, 2491 (2003)
30. C. Florea, K.A. Winick, J. Lightwave Technol. **21**, 246 (2003)
31. L. Sudrie, M. Franco, B. Prade, A. Mysyrowicz, Opt. Commun. **191**, 333 (2001)
32. E. Bricchi, J.D. Mills, P.G. Kazansky, B.G. Klappauf, J.J. Baumberg, Opt. Lett. **27**, 2200 (2002)
33. G.D. Valle, S. Taccheo, R. Osellame, A. Festa, G. Cerullo, P. Laporta, Opt. Express **15**, 3190 (2007)
34. L. Sudrie, A. Couairon, M. Franco, B. Lamouroux, B. Prade, S. Tzortzakis, A. Mysyrowicz, Phys. Rev. Lett. **89**, 4135 (2002)
35. A. Couairon, A. Mysyrowicz, Phys. Rep. **441**, 47 (2007)
36. C.-H. Tsai, C.-S. Liou, J. Manuf. Sci. Eng. **125**, 519 (2003)
37. F. Ahmed, M. Lee, H. Sekita, T. Sumiyoshi, M. Kamata, Appl. Phys. A **93**, 189 (2008)
38. S. Nisar, L. Li, M.A. Sheikh, J. Laser Appl. **25**, 042010 (2013)
39. S. Butkus, D. Paipulas, R. Sirutkaitis, E. Gaizauskas, V. Sirutkaitis, J. Laser Micro/Nano Eng. **9**, 213 (2014)
40. K. Mishchik, C.J. Leger, O.D. Caulier, S. Skupin, B. Chimier, C. Honninger, R. Kling, G. Duchateau, J. Lopez, J. Laser Micro/Nano Eng. **11**, 66 (2016)
41. J. Schwarz, J.-C. Diels, Appl. Phys. A **77**, 185 (2003)
42. D.G. Papazoglou, I. Zergioti, S. Tzortzakis, Opt. Lett. **32**, 2055 (2007)
43. L. Shah, J. Tawney, M. Richardson, K. Richardson, Appl. Surf. Sci. **183**, 151 (2001)
44. H. Varel, D. Ashkenasi, A. Rosenfeld, M. Wahmer, E.E.B. Campbell, Appl. Phys. A **65**, 367 (1997)
45. S.I. Kudryashov, G. Mourou, A. Joglekar, J.F. Herbstman, A.J. Hunt, Appl. Phys. Lett. **91**, 141111 (2007)
46. Y.V. White, X. Li, Z. Sikorski, L.M. Davis, W. Hofmeister, Opt. Express **16**, 14411 (2008)
47. S. Kanehira, J. Si, J. Qiu, K. Fujita, K. Hirao, Nano Lett. **5**, 1591 (2005)
48. H. Sun, J. Song, C. Li, J. Xu, X. Wang, Y. Cheng, Z. Xu, J. Qiu, T. Jia, Appl. Phys. A **88**, 285 (2007)
49. J. Song, X. Wan, X. Hu, Y. Dai, J. Qiu, Y. Cheng, Z. Xu, Appl. Phys. Lett. **92**, 092904 (2008)
50. H. Hu, H.J. Kim, S. Somnath, Micromachines **8**, 90 (2017)
51. A. Kirsanov, A. Kiselev, A. Stepanov, N. Polushkin, J. Appl. Phys. **94**, 6822 (2003)
52. D.J. Hwang, A. Chimmalgi, C.P. Grigoropoulos, J. Appl. Phys. **99**, 044905 (2006)
53. X. Li, Z. Chen, A. Taflove, V. Backman, Opt. Express **13**, 526 (2005)
54. A. Darafsheh, A. Fardad, N.M. Fried, A.N. Antoszyk, H.S. Ying, V.N. Astratov, Opt. Express **19**, 3440 (2011)
55. T.C. Hutchens, A. Darafsheh, A. Fardad, A.N. Antoszyk, H.S. Ying, V.N. Astratov, N.M. Fried, J. Biomed. Opt. **17**, 068004 (2012)
56. Y.F. Lu, L. Zhang, W.D. Song, Y.W. Zheng, B.S. Lukyanchuk, JETP Lett. **72**, 457 (2000)

57. H.-J. Munzer, M. Mosbacher, M. Bertsch, J. Zimmermann, P. Leiderer, J. Boneberg, J. Microsc. **202**, 129 (2001)
58. K. Piglmayer, R. Denk, D. Bauerle, Appl. Phys. Lett. **80**, 4693 (2002)
59. S.M. Huang, M.H. Hong, B.S. Lukyanchuk, T.C. Chong, Appl. Phys. A **77**, 293 (2003)
60. Y. Zhou, M.H. Hong, J.Y.H. Fuh, L. Lu, B.S. Lukyanchuk, Z.B. Wang, L.P. Shi, T.C. Chong, Appl. Phys. Lett. **88**, 023110 (2006)
61. S.C. Kong, A. Taflove, V. Backman, Opt. Express **17**, 3722 (2009)
62. Y. Shen, L.V. Wang, J.T. Shen, Opt. Lett. **39**, 4120 (2014)
63. C. Liu, Prog. Electromagn. Res. Lett. **37**, 153 (2013)
64. G. Gu, R. Zhou, Z. Chen, H. Xu, G. Cai, Z. Cai, M. Hong, Opt. Lett. **40**, 625 (2015)
65. Z. Hengyu, C. Zaichun, C.T. Chong, H. Minghui, Opt. Express **23**, 6626 (2015)
66. Y.E. Geints, A.A. Zemlyanov, E.K. Panina, J. Opt. Soc. Am. B **32**, 1570 (2015)
67. E. McLeod, C.B. Arnold, Nat. Nanotechnol. **3**, 413 (2008)
68. A. Ashkin, J.M. Dziedzic, J.E. Bjorkholm, S. Chu, Opt. Lett. **11**, 288 (1986)
69. V. Garces-Chavez, D. McGloin, H. Melville, W. Sibbett, K. Dholakia, Nature **419**, 145 (2002)
70. J. Durnin, J.J. Miceli, J.H. Eberly, Phys. Rev. Lett. **58**, 1499 (1987)
71. D. McGloin, K. Dholakia, Contemp. Phys. **46**, 15 (2005)
72. E. McLeod, A.B. Hopkins, C.B. Arnold, Opt. Lett. **31**, 3155 (2006)
73. S. Hasegawa, Y. Hayasaki, N. Nishida, Opt. Lett. **31**, 1705 (2006)
74. S. Hasegawa, H. Ito, H. Toyoda, Y. Hayasaki, Opt. Express **24**, 18513 (2016)
75. S. Juodkazis, L. Rosa, S. Bauerdick, L. Peto, R.E. Ganainy, S. John, Opt. Express **19**, 5802 (2011)
76. J. Gierak, Nanofabrication **1**, 35 (2014)
77. M.S. Alias, H.-Y. Liao, T.K. Ng, B.S. Ooi, J. Vac. Sci. Technol. B **33**, 06F701 (2015)
78. A.M. Weiner, Rev. Sci. Instrum. **71**, 1929 (2000)
79. D.J. Hwang, T.Y. Choi, C.P. Grigoropoulos, Appl. Phys. A **79**, 605 (2004)
80. Y. Li, K. Itoh, W. Watanabe, K. Yamada, D. Kuroda, J. Nishii, Y. Jiang, Opt. Lett. **26**, 1912 (2001)
81. Y. Iga, T. Ishizuka, W. Watanabe, K. Itoh, Y. Li, J. Nishii, Jpn. J. Appl. Phys. **43**, 4207 (2004)
82. K. Ke, E.F. Hasselbrink Jr., A.J. Hunt, Anal. Chem. **77**, 5083 (2005)
83. Y. Kondo, J. Qiu, T. Mitsuyu, K. Hirao, T. Yoko, Jpn. J. Appl. Phys. **38**, L1146 (1999)
84. S. Kiyama, S. Matsuo, S. Hashimoto, Y. Morihira, J. Phys. Chem. C **113**, 11560 (2009)
85. A. Marcinkevicius, S. Juodkazis, M. Watanabe, M. Miwa, S. Matsuo, H. Misawa, J. Nishii, Opt. Lett. **26**, 277 (2001)
86. Y. Cheng, K. Sugioka, K. Midorikawa, M. Masuda, K. Toyoda, M. Kawachi, K. Shihoyama, Opt. Lett. **28**, 55 (2003)
87. M. Masuda, K. Sugioka, Y. Cheng, N. Aoki, M. Kawachi, K. Shihoyama, K. Toyoda, H. Helvajian, K. Midorikawa, Appl. Phys. A **76**, 857 (2003)
88. K. Sugioka, Y. Cheng, K. Midorikawa, Appl. Phys. A **81**, 1 (2005)
89. K.C. Vishnubhatla, N. Bellini, R. Ramponi, G. Cerullo, R. Osellame, Opt. Express **17**, 8685 (2009)
90. M.V. Berry, Am. J. Phys. **47**, 264 (1979)
91. G.A. Siviloglou, D.N. Christodoulides, Opt. Lett. **32**, 979 (2007)
92. Y. Hu, G.A. Siviloglou, P. Zhang, N.K. Efremidis, D.N. Christodoulide, Z. Chen, *Nonlinear Photonics and Novel Optical Phenomena*. Springer series in optical sciences, vol. 170, ed. by Z. Chen, R. Morandotti (2012), p. 1
93. Z. Bouchal, J. Wagner, M. Chlup, Opt. Commun. **151**, 207 (1998)
94. J. Broky, G.A. Siviloglou, A. Dogariu, D.N. Christodoulides, Opt. Express **16**, 12880 (2008)
95. M.A. Porras, A. Parola, D. Faccio, A. Dubietis, P. Di Trapani, Phys. Rev. Lett. **93**, 153902 (2004)
96. D.E. Roskey, M. Kolesik, J.V. Moloney, E.M. Wright, Opt. Express **15**, 9893 (2007)
97. P. Polesana, M. Franco, A. Couairon, D. Faccio, P.D. Trapani, Phys. Rev. A **77**, 043814 (2008)
98. E. Gaizauskas, E. Vanagas, V. Jarutis, S. Juodkazis, V. Mizeikis, H. Misawa, Opt. Lett. **31**, 80 (2006)

99. I. Kaminer, M. Segev, D.N. Christodoulides, Phys. Rev. Lett. **106**, 213903 (2011)
100. P. Polynkin, M. Kolesik, J.V. Moloney, G.A. Siviloglou, D.N. Christodoulides, Science **324**, 229 (2009)
101. B. Hafizi, E. Esarey, P. Sprangle, Phys. Rev. E **55**, 3539 (1997)
102. J. Fan, E. Parra, I. Alexeev, K.Y. Kim, H.M. Milchberg, L.Y. Margolin, L.N. Pyatnitskii, Phys. Rev. E **62**, R7603 (2000)
103. M. Duocastella, C. Arnold, Laser Photonics Rev. **6**, 607 (2012)
104. F. Courvoisier, R. Stoian, A. Couairon, Opt. Laser Technol. **80**, 125 (2016)
105. V.G. Shvedov, A.V. Rode, Y.V. Izdebskaya, A.S. Desyatnikov, W. Krolikowski, Y.S. Kivshar, Phys. Rev. Lett. **105**, 118103 (2010)
106. J. Baumgartl, M. Mazilu, K. Dholakia, Nat. Photonics **2**, 675 (2008)
107. T.A. Planchon, L. Gao, D.E. Milkie, M.W. Davidson, J.A. Galbraith, C.G. Galbraith, E. Betzig, Nat. Methods **8**, 417 (2011)
108. T. Vettenburg, H.I.C. Dalgarno, J. Nylk, C. Coll-Llado, D.E.K. Ferrier, T. Cizmar, F.J. Gunn-Moore, K. Dholakia, Nat. Methods **11**, 541 (2014)
109. J. Arlt, T. Hitomi, K. Dholakia, Appl. Phys. B **71**, 549 (2000)
110. M. Mazilu, D.J. Stevenson, F. Gunn-Moore, K. Dholakia, Laser Photonics Rev. **4**, 529 (2009)
111. J.H. McLeod, J. Opt. Soc. Am. **44**, 592 (1954)
112. G. Roy, R. Tremblay, Opt. Commun. **34**, 1 (1980)
113. T. Grosjean, S.S. Saleh, M.A. Suarez, I.A. Ibrahim, V. Piquerey, D. Charraut, P. Sandoz, Appl. Opt. **46**, 8061 (2007)
114. J. Turunen, A. Vasara, A.T. Friberg, Appl. Opt. **27**, 3959 (1988)
115. A. Vasara, J. Turunen, A.T. Friberg, J. Opt. Soc. Am. A **6**, 1748 (1989)
116. J.A. Davis, J. Guertin, D.M. Cottrell, Appl. Opt. **32**, 6368 (1993)
117. N. Chattrapiban, E.A. Rogers, D. Cofield, W.T. Hill, R. Roy, Opt. Lett. **28**, 2183 (2003)
118. T. Cizmar, V. Kollarova, X. Tsampoula, F. Gunn-Moore, W. Sibbett, Z. Bouchal, K. Dholakia, Opt. Express **16**, 14024 (2008)
119. C. Maurer, A. Jesacher, S. Bernet, M. Ritsch-Marte, Laser Photonics Rev. **5**, 81 (2011)
120. V. Jarutis, R. Paskauskas, A. Stabinis, Opt. Commun. **184**, 105 (2000)
121. M.K. Bhuyan, F. Courvoisier, P.A. Lacourt, M. Jacquot, R. Salut, L. Furfaro, J.M. Dudley, Appl. Phys. Lett. **97**, 081102 (2010)
122. P. Polesana, A. Dubietis, M.A. Porras, E. Kucinskas, D. Faccio, A. Couairon, P. Di Trapani, Phys. Rev. E **73**, 056612 (2006)
123. D. Faccio, E. Rubino, A. Lotti, A. Couairon, A. Dubietis, G. Tamosauskas, D.G. Papazoglou, S. Tzortzakis, Phys. Rev. A **85**, 033829 (2012)
124. O. Brzobohaty, T. Cizmar, P. Zemanek, Opt. Express **16**, 12688 (2008)
125. J. Dudutis, P. Gecys, G. Raciukaitis, Opt. Express **24**, 28433 (2016)
126. F. Courvoisier, P.-A. Lacourt, M. Jacquot, M.K. Bhuyan, L. Furfaro, J.M. Dudley, Opt. Lett. **34**, 3163 (2009)
127. R. Sahin, Y. Morova, E. Simsek, S. Akturk, Appl. Phys. Lett. **102**, 193106 (2013)
128. Y. Matsuoka, Y. Kizuka, T. Inoue, Appl. Phys. A **84**, 423 (2006)
129. F. He, J. Yu, Y. Tan, W. Chu, C. Zhou, Y. Cheng, K. Sugioka, Sci. Rep. **7**, 40785 (2017)
130. A. Marcinkevicius, S. Juodkazis, S. Matsuo, V. Mizeikis, H. Misawa, Jpn. J. Appl. Phys. **40**, L1197 (2001)
131. M.K. Bhuyan, F. Courvoisier, P.-A. Lacourt, M. Jacquot, L. Furfaro, M.J. Withford, J.M. Dudley, Opt. Express **18**, 566 (2010)
132. M.K. Bhuyan, P.K. Velpula, J.P. Colombier, T. Olivier, N. Faure, R. Stoian, Appl. Phys. Lett. **104**, 021107 (2014)
133. P.K. Velpula, M.K. Bhuyan, F. Courvoisier, H. Zhang, J.-P. Colombier, R. Stoian, Laser Photonics Rev. **10**, 230 (2016)
134. M.K. Bhuyan, F. Courvoisier, H.S. Phing, O. Jedrkiewicz, S. Recchia, P. Di Trapani, J.M. Dudley, Eur. Phys. J. Spec. Top. **199**, 101 (2011)
135. L. Rapp, R. Meyer, R. Giust, L. Furfaro, M. Jacquot, P.A. Lacourt, J.M. Dudley, F. Courvoisier, Sci. Rep. **6**, 34286 (2016)

136. Y. Yu, L. Jiang, Q. Cao, B. Xia, Q. Wang, Y. Lu, Opt. Express **23**, 32728 (2015)
137. Q. Xie, X. Li, L. Jiang, B. Xia, X. Yan, W. Zhao, Y. Lu, Appl. Phys. A **122**, 136 (2016)
138. G. Wang, Y. Yu, L. Jiang, X. Li, Q. Xie, Y. Lu, Appl. Phys. Lett. **110**, 161907 (2017)
139. M.K. Bhuyan, O. Jedrkiewicz, V. Sabonis, M. Mikutis, S. Recchia, A. Aprea, M. Bollani, P. Di Trapani, Appl. Phys. A **120**, 443 (2015)
140. W.-J. Tsai, C.-J. Gu, C.-W. Cheng, J.-B. Horng, Opt. Eng. **53**, 051503 (2013)
141. L. Rapp, R. Meyer, L. Furfaro, C. Billet, R. Giust, F. Courvoisier, Opt. Express **25**, 9312 (2017)
142. P.K. Velpula, M.K. Bhuyan, C. Mauclair, J.-P. Colombier, R. Stoian, Opt. Eng. **53**, 076108 (2014)
143. M. Sakakura, M. Terazima, Y. Shimotsuma, K. Miura, K. Hirao, Opt. Express **15**, 5674 (2007)
144. M.K. Bhuyan, M. Somayaji, A. Mermillod-Blondin, F. Bourquard, J.P. Colombier, R. Stoian, Optica **4**, 951 (2017)
145. A. Vinçotte, L. Berge, Phys. Rev. Lett. **95**, 193901 (2005)
146. S. Shiffler, P. Polynkin, J. Moloney, Opt. Lett. **36**, 3834 (2011)
147. P. Polynkin, C. Ament, J.V. Moloney, Phys. Rev. Lett. **111**, 023901 (2013)
148. M.A. Porras, C. Ruiz-Jiménez, J. Opt. Soc. Am. B **31**, 2657 (2014)
149. V. Jukna, C. Milián, C. Xie, T. Itina, J. Dudley, F. Courvoisier, A. Couairon, Opt. Express **22**, 25410 (2014)
150. C.L. Arnold, S. Akturk, A. Mysyrowicz, V. Jukna, A. Couairon, T. Itina, R. Stoian, C. Xie, J.M. Dudley, F. Courvoisier, S. Bonanomi, O. Jedrkiewicz, P. Di Trapani, J. Phys. B **48**, 094006 (2015)
151. C. Xie, V. Jukna, C. Milián, R. Giust, I. Ouadghiri-Idrissi, T. Itina, J.M. Dudley, A. Couairon, F. Courvoisier, Sci. Rep. **5**, 8914 (2015)
152. K.T. Gahagan, G.A. Swartzlander, Opt. Lett. **21**, 827 (1996)
153. C. Hnatovsky, V.G. Shvedov, W. Krolikowski, A.V. Rode, Opt. Lett. **35**, 3417 (2010)
154. B. Wetzel, C. Xie, P.-A. Lacourt, J.M. Dudley, F. Courvoisier, Appl. Phys. Lett. **103**, 241111 (2013)
155. W. Cheng, P. Polynkin, J. Opt. Soc. Am. B **31**, C48 (2014)
156. R. Sahin, T. Ersoy, S. Akturk, Appl. Phys. A **118**, 125 (2015)
157. O. Jedrkiewicz, S. Bonanomi, M. Selva, P. Di Trapani, Appl. Phys. A **120**, 385 (2015)
158. T. Watanabe, M. Fujii, Y. Watanabe, N. Toyama, Y. Iketaki, Rev. Sci. Instrum. **75**, 5131 (2004)
159. V.V. Kotlyar, A.A. Almazov, S.N. Khonina, V.A. Soifer, J. Opt. Soc. Am. A **22**, 849 (2005)
160. J. Xin, K. Dai, L. Zhong, Q. Na, C. Gao, Opt. Lett. **39**, 1984 (2014)
161. M. Massari, G. Ruffato, M. Gintoli, F. Ricci, F. Romanato, Appl. Opt. **54**, 4077 (2105)
162. A. Volyar, V. Shvedov, T. Fadeyeva, A.S. Desyatnikov, D.N. Neshev, W. Krolikowski, Y.S. Kivshar, Opt. Express **14**, 3724 (2006)
163. W.C. Cheong, W.M. Lee, X.-C. Yuan, L.-S. Zhang, Appl. Phys. Lett. **85**, 5784 (2004)
164. R. Cao, Y. Yang, J. Wang, J. Bu, M. Wang, X.-C. Yuan, Appl. Phys. Lett. **99**, 261106 (2011)
165. L. Froehly, F. Courvoisier, A. Mathis, M. Jacquot, L. Furfaro, R. Giust, P.A. Lacourt, J.M. Dudley, Opt. Express **19**, 16455 (2011)
166. F. Courvoisier, A. Mathis, L. Froehly, R. Giust, L. Furfaro, P.A. Lacourt, M. Jacquot, J.M. Dudley, Opt. Lett. **37**, 1736 (2012)
167. E. Greenfield, M. Segev, W. Walasik, O. Raz, Phys. Rev. Lett. **106**, 213902 (2011)
168. M. Gecevicius, M. Beresna, R. Drevinskas, P.G. Kazansky, Opt. Lett. **39**, 6791 (2014)
169. A. Bahabad, M.M. Murnane, H.C. Kapteyn, Phys. Rev. A **84**, 033819 (2011)
170. A. Mathis, F. Courvoisier, L. Froehly, L. Furfaro, M. Jacquot, P.A. Lacourt, J.M. Dudley, Appl. Phys. Lett. **101**, 071110 (2012)
171. A. Mathis, L. Froehly, L. Furfaro, M. Jacquot, J.M. Dudley, F. Courvoisier, J. Eur. Opt. Soc. **8**, 13019 (2013)
172. A. Lotti, D. Faccio, A. Couairon, D.G. Papazoglou, P. Panagiotopoulos, D. Abdollahpour, S. Tzortzakis, Phys. Rev. A **84**, 021807 (2011)
173. D. Abdollahpour, S. Suntsov, D.G. Papazoglou, S. Tzortzakis, Phys. Rev. Lett. **105**, 253901 (2010)

174. R. Srinivasan, E. Sutcliffe, B. Braren, Appl. Phys. Lett. **51**, 1285 (1987)
175. S. Kuper, M. Stuke, Appl. Phys. B **44**, 199 (1987)
176. S. Kuper, M. Stuke, Appl. Phys. Lett. **54**, 4 (1989)

Chapter 7
Molecular Orbital Tomography Based on High-Order Harmonic Generation: Principles and Perspectives

Anna Gabriella Ciriolo, Michele Devetta, Davide Faccialà, Prabhash Prasannan Geetha, Aditya Pusala, Caterina Vozzi and Salvatore Stagira

Abstract High-order harmonic generation is a nonlinear effect appearing when atoms or molecules are exposed to intense laser pulses. It consists in the emission of ultrashort bursts of extreme ultraviolet and soft X radiation, giving rise to harmonics in the spectral domain. This emission encodes a wealth of information about the intimate structure of the targeted species, allowing the reconstruction of the outermost distribution of electrons. We will review here the most important aspects of high-order harmonic generation and we will show how this process can be applied to the reconstruction of molecular orbitals. A critical analysis will illustrate benefits, weaknesses and perspectives of this technique.

A. G. Ciriolo · P. P. Geetha · A. Pusala · S. Stagira (✉)
Dipartimento di Fisica, Politecnico di Milano,
Piazza L. da Vinci 32, 20133 Milano, Italy
e-mail: salvatore.stagira@polimi.it

A. G. Ciriolo
e-mail: annagabriella.ciriolo@polimi.it

P. P. Geetha
e-mail: prabhash.prasannan@polimi.it

A. Pusala
e-mail: aditya.pusala@polimi.it

M. Devetta · D. Faccialà · P. P. Geetha · C. Vozzi · S. Stagira
Istituto di Fotonica e Nanotecnologie - CNR,
Piazza L. da Vinci 32, 20133 Milano, Italy
e-mail: michele.devetta@gmail.com

D. Faccialà
e-mail: davide.facciala@polimi.it

C. Vozzi
e-mail: caterina.vozzi@ifn.cnr.it

P.M. Ossi (ed.), *Advances in the Application of Lasers in Materials Science*, Springer Series in Materials Science 274, https://doi.org/10.1007/978-3-319-96845-2_7

7.1 Introduction

When high intensity ($I \approx 10^{13} - 10^{16}\,\mathrm{W/cm^2}$) ultrashort laser pulses interact with a collection of atoms or molecules, a highly non-linear optical process occurs, namely the *High-order Harmonic Generation* (HHG) process. HHG is associated to the emission of coherent radiation extending up to the extreme ultraviolet (XUV) and the X-ray spectral range [1]. This radiation has unprecedented temporal characteristics: the XUV bursts are emitted with duration in the sub-femtosecond domain. Harmonic emission was observed for the first time in 1987 by McPherson et al. [2] in a gas jet of Neon atoms excited with intense ultraviolet pulses and in 1988 by Ferray et al. in noble atom gases excited with a Nd:YAG laser radiation [3]. Since then a lot of effort has been devoted to the study of harmonic generation process and to the development of attosecond pump-probe experiments based on HHG [4]. In particular, the mechanism of HHG in atoms is quite well understood and the possibility to exploit it for the reconstruction of the properties of the generation media, which is the target of HHG spectroscopy, has lead to impressive results. HHG processes in molecules are different from their counterpart in atomic gases, as was demonstrated at the beginning of the millennium by theoretical calculations on harmonic generation as a function of the molecular orientation [5] and by experimental studies on HHG driven by elliptically polarized laser pulses in molecules [6]. The availability of techniques for field-free molecular alignment [7] allowed the study of photoionization and harmonic generation in molecules undergoing impulsive excitation of rotational wavepackets [8–10]; among the numerous applications of this approach, the pioneering work from Itatani and coworkers [11] demonstrated the idea of molecular orbital tomography enabled by HHG in aligned molecules (HHG-MOT). The tomographic reconstruction of molecular orbitals based on HHG resembles somehow the computed tomography of a human body (TC): the tridimensional structure of the body can be retrieved by a suitable elaboration of a large number of X-ray images of the body acquired at different angles. Differently from the usual TC, the orbital tomography is based on the acquisition of several high-harmonic spectra of the target molecules for several polarization directions of the laser pulse in the molecular frame, as sketched in Fig. 7.1; a suitable elaboration of this collection of HHG spectra allows to retrieve the shape of the highest occupied molecular orbital (HOMO).

HHG-MOT belongs to a larger family of molecular investigation techniques based on the self-probing of the target species by the outermost electron [12], that is ionized and driven back towards the parent ion by an intense laser field. Among them we may cite the Laser Induced Electron Diffraction (LIED) approach [13, 14] based on the detection of the electrons scattered by the molecules, and the manipulation of the ionization-recollision process by two-color laser fields [15, 16] driving HHG. Tomographic imaging not based on the recollision approach can be also accomplished by photoemission [17] or STM imaging [18] of molecules deposited on substrates, by measuring the molecular-frame photoelectron angular distribution driven by a laser pulse [19] as well as by X-ray or photoelectron diffraction [20] driven by a Free Electron Laser [21].

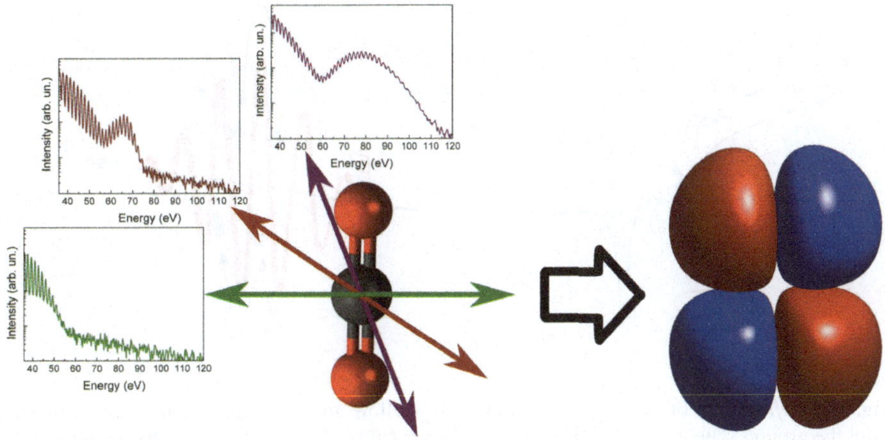

Fig. 7.1 Pictorial view of HHG-based molecular orbital tomography. Left: several HHG spectra are acquired for different directions of the laser polarization in the molecular frame; right: a suitable elaboration of the measurements allows to retrieve the outermost molecular orbital

The seminal paper by Itatani et al., demonstrating HHG-MOT in nitrogen with multicycle near-IR laser pulses, drove several subsequent works about the extension of HHG spectroscopy to polyatomic molecules [22–25], the characterization of the polarization state of harmonics generated in molecules [26, 27], the application of HHG-MOT to multielectronic systems [28] as well as the extension of harmonic spectroscopy [29] and tomography [30] to mid-IR driving pulses. The availability of molecular orientation techniques, able to control the head-tail direction and the 3D molecular alignment [31–34], opened the way to the study of asymmetric molecules [35–38] paving the way to tomographic imaging of asymmetric orbitals [39].

In this chapter we first describe the basic concepts of HHG process (Sect. 7.2) and the recollision spectroscopy in general (Sect. 7.3). Then we give a detailed description of HHG-MOT (Sect. 7.4), presenting also the open problems and possible perspectives.

7.2 High-Order Harmonic Generation

A first semi-classical interpretation of the HHG process was proposed independently in 1993 by Corkum et al. [40] and Kulander et al. [41]. The model they provided describes the HHG process as a temporal sequence of three steps (hence it is referred to as *Three Step Model*) and is described in the left panel of Fig. 7.2.

In the first step (yellow arrow), the generic oscillating electric field $E(t) = E_0 \cos(\omega_0 t)$ has a probability to ionize the atom. Under the extreme intensity regime here considered, the Coulomb potential of atoms is strongly distorted by the intense quasi-static electromagnetic field. So ionization occurs mainly via tunnel ionization.

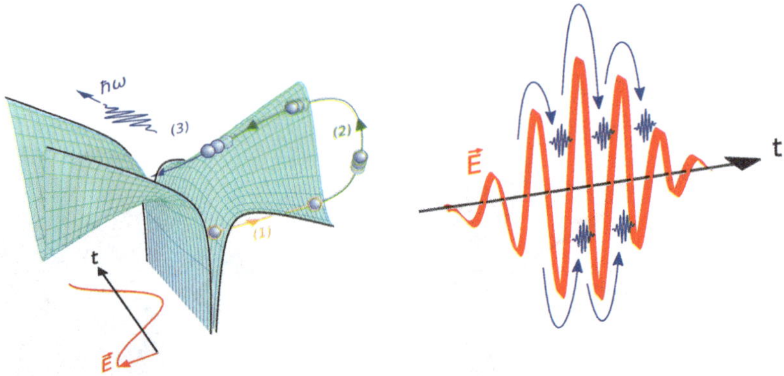

Fig. 7.2 *Left*: sketch of the three step model. (1—yellow arrow) The electron is tunnel ionized from the ground state. (2—green arrow) It is accelerated by the laser field and driven back to the initial position. (3—blue arrow) It recombines emitting a photon of energy $\hbar\omega$. *Right*: the three step process (blue arrow) repeats every half-cycle of the laser field (red line), leading to the emission of a burst of photons (blue line) every half-cycle

The probability to tunnel ionize the target at a given time t_i, depends on the strength of the laser field at that time, and it will reach its maximum when $E(t_i) = E_0$. In this model we assume that the ionized electron has an initial kinetic energy equal to zero. In the second step (green arrow in Fig. 7.2), the freed electron is accelerated by the laser field. By neglecting the Coulomb field of the ionized atom, the motion of the electron can be directly calculated from the Newton dynamics of a negative charge e of mass m inside the oscillating sinusoidal field $E(t)$. Under the effect of the laser oscillating field, the electron can be driven back to the parent ion, and recombination can occur at a certain time $t_r > t_i$. Recombination is associated to the emission of high energy photon $\hbar\omega$. The energy of the photon emitted at t_r is given by the sum of the atom ionization potential I_p plus the kinetic energy K gained by the traveling electron under exposure to the laser field, $\hbar\omega(t_r) = I_p + K(t_r)$. Recombination provides the third step of interaction.

These three steps take place every half cycle of the fundamental laser field (right panel in Fig. 7.2). During each occurrence, emission is temporally confined within a sub-cycle of the laser frequency, so ultrashort bursts of high energy photons are generated. In centrosymmetric targets, a constructive interference occurs for the *odd* harmonic components of the fundamental laser frequency, leading to the emission of the typical comb-shaped spectrum. It is possible to find a collection of (ω, t_i, t_r) values, each associated to a different solution of the classical equation of motion of an electron into an electromagnetic field. Those different solutions provides the whole spectrum of HHG frequencies. For each value of ω, two possible couple of (t_i, t_r) values can be found. Among the two, the solution with the highest excursion time $\tau = t_r - t_i$ is referred to as *long trajectory*, while the shorter one is referred to as *short trajectory*. The (t_r, t_i) solution that maximizes the photon energy $\omega(t_r)$ corresponds to the cutoff of the HHG spectrum. The resulting cutoff energy is given by:

$$\hbar\omega_{\text{cutoff}} = I_p + 3.17U_p, \tag{7.1}$$

where U_p is the ponderomotive energy of the electron, defined as $U_p = e^2E_0^2/4m\omega^2$.

Gray lines in Fig. 7.3 show some of the results obtained by using this model. For each photon energy $\hbar\omega$, the classical ionization times (panel a) and recombination times (panel b) are reported. Long and short trajectory solutions are represented respectively by solid and dashed lines. It can be seen that, at the cutoff, the two trajectories collapse into a single trajectory, called *cutoff trajectory*.

The HHG spectra as predicted by the classical model exhibit a good qualitative agreement with the HHG experimental spectra. In particular, the main success of the classical model is the capability to predict the extension of the generated spectrum. However, this model does not provide quantitative predictions on the intensity of the harmonics generated. In order to do so, we need to move to a quantum model.

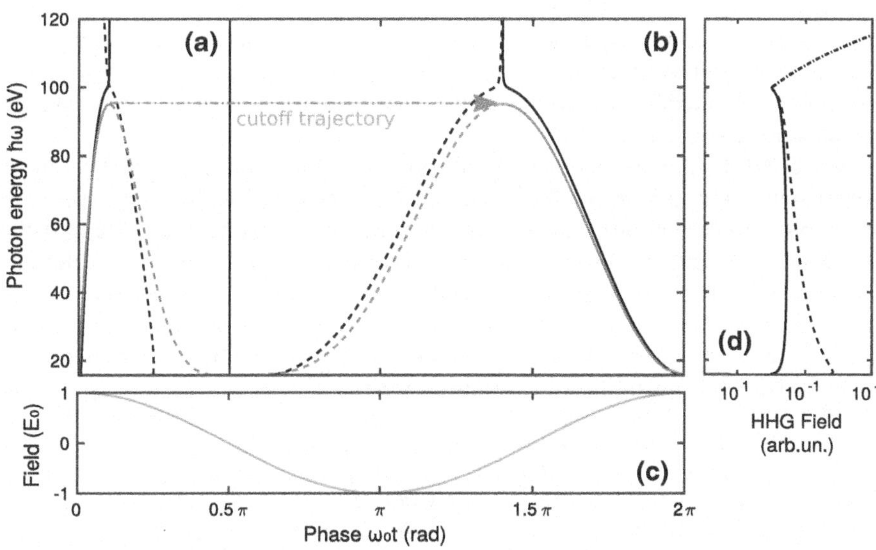

Fig. 7.3 **a, b**—**Gray lines**: Emitted photon energy $\hbar\omega$ as a function of (**a**) the semi-classical ionization phase $\omega_0 t_i$ and (**b**) the recombination phase $\omega_0 t_r$. **a, b**—**Black lines**: Emitted photon energy as a function of (a) the saddle point ionization phase $\omega_0\Re\{t_s'\}$ and (**b**) the recombination phase $\omega_0\Re\{t_s\}$ (see Sect. 7.2.2 for details). Solid and dashed lines represent the results obtained for the long and short trajectories respectively. **c** Laser field evolution. **d** Component a_s of the wavepacket spectrum, as defined in (7.13) and (7.14), for long (solid line), short (dashed line), and cutoff (dashed-dotted line) trajectories. Parameters: ionization potential $I_p = 15.759$ eV (argon), fundamental wavelength $\lambda_0 = 1400$ nm, intensity $I = 1.3662 \times 10^{14}$ W/cm^2

7.2.1 Lewenstein Model

In 1994 Lewenstein [42] derived an approximated quantum method for describing the HHG process. Within this model, the basic assumptions made in the classical three-step model were opportunely adapted to a quantum description of the process.

In this framework we first consider the interaction of the laser field with an isolated atom in the single active electron approximation (SAE). The electron, under the action of the Coulomb field $V(\mathbf{r})$ and the external electric field $\mathbf{E}(t)$, can be described by the wave-function that solves the Schrödinger equation in the length gauge:

$$i\hbar\frac{\partial\Psi(\mathbf{r},t)}{\partial t} = \left[-\frac{\hbar^2}{2m}\nabla^2_{\mathbf{r}} + V(\mathbf{r}) + e\mathbf{r}\cdot\mathbf{E}(t)\right]\Psi(\mathbf{r},t). \tag{7.2}$$

Direct numerical solution of equation (7.2) is time consuming and can be performed only for very simple systems. However, by neglecting the atomic potential $V(\mathbf{r})$ with respect to the external field potential energy $e\mathbf{r}\cdot\mathbf{E}(t)$ during the electron propagation an integral form for the temporal evolution of the electric dipole $\mathbf{D}(t)$ can be derived. The HHG spectrum can then be directly retrieved from the electric dipole. The above mentioned approximation is called *Strong Field Approximation (SFA)*. We will provide a brief overview on the Lewenstein model; a three step scheme inspired to the classical one discussed in the previous paragraph will be used for explaining this quantum approach.

First of all, before the interaction with the laser pulse, the electron stays in its ground state, $\Psi(\mathbf{r},t) = \phi_0(\mathbf{r})e^{iI_pt/\hbar}$, that solves equation (7.2) for $E(t) = 0$. After ionization the electron propagates in the continuum states under the effect of the electric field. During the electron propagation SFA is applied. In this case, solutions of equation (7.2) are continuum plane-wave states $|\mathbf{k}\rangle = e^{i\mathbf{k}\cdot\mathbf{r}}$ of wave-vector $\mathbf{k}$. A general solution can be represented by the superposition of the ground state solution and a collection of plane waves. By assuming that the ground state depletion due to tunnel ionization is negligible, this solutions reads:

$$\psi(\mathbf{r},t) = e^{i\frac{I_p}{\hbar}t}\left[\phi_0(\mathbf{r}) + \frac{1}{\sqrt{(2\pi)^3}}\int b(\mathbf{k},t)e^{i\mathbf{k}\cdot\mathbf{r}}d\mathbf{k}\right]. \tag{7.3}$$

At t', ionization occurs. The probability of ionizing an electron from the ground state of energy $-I_p$ to the continuum state $|\mathbf{k}'\rangle = e^{i\mathbf{k}'\cdot\mathbf{r}}$ depends on the scalar product of the dipole matrix element $\mathbf{d}'$ with the field:

$$\mathbf{E}(t')\cdot\mathbf{d}'(\mathbf{k}') = \mathbf{E}(t')\cdot\langle\mathbf{k}'|\mathbf{r}|\phi_0(\mathbf{r})\rangle \qquad \mathbf{k}' = \mathbf{k}(t') \tag{7.4}$$

During the propagation SFA holds, and, under the external vector potential $\mathbf{A}(t)$, the canonical momentum $\mathbf{p} = \hbar\mathbf{k}(t) - e\mathbf{A}(t)$ is conserved. We introduce the semi-classical action $S(\mathbf{p},t,t')$ that represents the relative phase (in units of $\hbar^{-1}$) accumulated by the electron with respect to the corresponding hole, from the initial time t' to the generic time t, which is defined as:

$$S(\mathbf{p}, t, t') = \int_{t'}^{t} \frac{\left[\mathbf{p} + e\mathbf{A}(t'')\right]^2}{2m} + I_p \ \mathrm{d}t''. \tag{7.5}$$

Propagation and acceleration bring the electron from the initial state $|\mathbf{k}'\rangle = |\mathbf{k}(t')\rangle$ to the final state $|\mathbf{k}\rangle = |\mathbf{k}(t)\rangle$. Eventually the electron recombines from the continuum state $|\mathbf{k}\rangle$ to the ground state. The probability of recombination depends on the associated dipole matrix element:

$$\mathbf{d}(\mathbf{k}) = \langle \phi_0(\mathbf{r})|\mathbf{r}|\mathbf{k}\rangle \qquad \mathbf{k} = \mathbf{k}(t). \tag{7.6}$$

The complex valued electric dipole $\mathbf{D}(t)$ at time t can thus be calculated as an integral over all the possible initial times $t' \leq t$ and canonical momenta $\mathbf{p} \in \mathbb{R}^3$ of three terms related to the ionization, propagation and recombination processes:

$$\mathbf{D}(t) \propto \int_{-\infty}^{t} \int_{\mathbb{R}^3} d_{\mathrm{ion}}[\mathbf{p} + e\mathbf{A}(t')]\mathbf{d}[\mathbf{p} + e\mathbf{A}(t)]e^{-\frac{i}{\hbar}S(\mathbf{p},t,t')}\mathrm{d}\mathbf{p}\mathrm{d}t', \tag{7.7}$$

where $d_{\mathrm{ion}}[\mathbf{p} + e\mathbf{A}(t')] = \mathbf{E}(t') \cdot \mathbf{d}'[\mathbf{p} + e\mathbf{A}(t')]$. The HHG field can be directly computed, except for a constant factor, from the *Fourier transform* of the electron acceleration, which is the second derivative of $\mathbf{D}$:

$$\begin{aligned}\mathbf{E}_H(\omega) &\propto \int_{\mathbb{R}} \ddot{\mathbf{D}}(t)e^{i\omega t}\mathrm{d}t \propto \omega^2 \int_{\mathbb{R}} \mathbf{D}(t)e^{i\omega t}\mathrm{d}t \\ &\propto \int_{\mathbb{R}} \int_{-\infty}^{t} \int_{\mathbb{R}^3} d_{\mathrm{ion}}[\mathbf{p} + e\mathbf{A}(t')]\mathbf{d}_f[\mathbf{p} + e\mathbf{A}(t)]e^{-\frac{i}{\hbar}\tilde{S}(\omega,\mathbf{p},t,t')}\mathrm{d}\mathbf{p}\mathrm{d}t'\mathrm{d}t,\end{aligned} \tag{7.8}$$

where $\tilde{S}(\omega, \mathbf{p}, t, t') = S(\mathbf{p}, t, t') - \hbar\omega$ and $\mathbf{d}_f = \omega^2\mathbf{d}$. The integral in (7.8) is called *Lewenstein integral*, and it basically comes from an extension to the HHG process of the SFA procedure used by Keldysh for modeling the photo-ionization of atoms in a strong laser field [43]. Since the dipole in (7.6) has been expressed in the length form, (7.8) represents the associated Lewenstein integral in the length form. It is also possible to directly compute the momentum matrix element of the electron from:

$$\mathbf{v}(\mathbf{k}) = \langle \phi_0(\mathbf{r})|-i\nabla|\mathbf{k}\rangle \qquad \mathbf{k} = \mathbf{k}(t). \tag{7.9}$$

In this case, the Lewenstein integral in the velocity form is obtained by substituting in (7.8) the term $\mathbf{d}_f = \omega\mathbf{v}$. An approximated solution of the Lewenstein integral can be found by applying the *method of steepest descent*, or *saddle point approximation* (SPA), which allows one to find the complex quantum electron trajectories that give the highest contribution to the integral.

7.2.2 Saddle Point Approximation

In (7.8), the exponential term $e^{-\frac{i}{\hbar}\tilde{S}(\mathbf{p},t,t')}$ is rapidly oscillating with respect to the other terms. As a consequence, the points of the $(\mathbf{p}, t, t')$ domain that give the highest contribution to the Lewenstein integral are those in correspondence of which the gradient of $\tilde{S}$ is nearly zero, namely those points in the surrounding of a saddle point $\mathbf{Q}_s = (\mathbf{p}_s, t_s, t'_s)$. The SPA method consists in expanding $\tilde{S}$ in Taylor series around $\mathbf{Q}_s$ up to the second order: $\tilde{S}(\mathbf{Q}) \approx \tilde{S}(\mathbf{Q}_s) + (\mathbf{Q} - \mathbf{Q}_s) \cdot [H(\mathbf{Q}_s) \cdot (\mathbf{Q} - \mathbf{Q}_s)]/2$, where H is the Hessian matrix of $\tilde{S}$. The Lewenstein integral (7.8) can thus be written as:

$$\mathbf{E}_H(\omega) \approx \sum_s C(s) d_{\mathrm{ion}}[\mathbf{p}_s + e\mathbf{A}(t'_s)]\mathbf{d}_f[\mathbf{p}_s + e\mathbf{A}(t_s)]e^{-\frac{i}{\hbar}\tilde{S}(\omega,\mathbf{Q}_s)}, \quad (7.10)$$

where the sum runs over all the relevant saddle points, and the coefficient $C(s)$ reads:

$$C(s) = K \int_{\mathbb{R}^5} e^{-\frac{i}{2\hbar}(\mathbf{Q}-\mathbf{Q}_s)\cdot[H(\mathbf{Q}_s)\cdot(\mathbf{Q}-\mathbf{Q}_s)]} d\mathbf{Q} = K\sqrt{\frac{(-i2\pi\hbar)^5}{\det[H(\mathbf{Q}_s)]}}. \quad (7.11)$$

A physical interpretation can be attributed to $C(s)$, which accounts for the spreading of the electron wavepacket during the propagation in the continuum. Each saddle point $\mathbf{Q}_s$ is associated to a specific trajectory, representing an electron ionized at t'_s, recombining at t_s and having a canonical momentum $\mathbf{p}_s$. Thus, in the SPA framework, a description based on three steps is recovered. Within this description, $\mathbf{k}^s_{\mathrm{ion}} = [\mathbf{p}_s + e\mathbf{A}(t'_s)]/\hbar$ is the wave-vector of the electron at the ionization time t'_s and $\mathbf{k}^s_{\mathrm{rec}} = [\mathbf{p}_s + e\mathbf{A}(t_s)]/\hbar$ the one at the recombination time t_s. In order to find the saddle points, the equation $\nabla_{\mathbf{Q}}\tilde{S}(\mathbf{Q})\big|_{\mathbf{Q}_s} = 0$ must be solved. A system of three coupled equations is thus obtained:

$$\left.\frac{\partial\tilde{S}(\mathbf{p},t,t')}{\partial t'}\right|_{\mathbf{Q}_s} = 0 = \frac{{\mathbf{k}^s_{\mathrm{ion}}}^2}{2m} + I_p, \quad (7.12a)$$

$$\left.\nabla_{\mathbf{p}}\tilde{S}(\mathbf{p},t,t')\right|_{\mathbf{Q}_s} = 0 = \int_{t'_s}^{t_s} \frac{\mathbf{k}(t)}{m} dt, \quad (7.12b)$$

$$\left.\frac{\partial\tilde{S}(\mathbf{p},t,t')}{\partial t}\right|_{\mathbf{Q}_s} = 0 = \frac{{\mathbf{k}^s_{\mathrm{rec}}}^2}{2m} + I_p - \hbar\omega_s. \quad (7.12c)$$

From (7.12a), the kinetic energy at the ionization time t'_s can be derived. It is equal to $-I_p$. Thus, the electron tunnels out from the ion with an initial imaginary velocity. For $I_p = 0$ the initial velocity is zero and the assumption made in the classical three-step model is recovered. Equation (7.12b) states that the integral of the velocity of the electron from the initial time t'_s to the recombination time t_s is zero. This means that the total spatial displacement between t'_s and t_s is zero. According to this equation, only those electrons that come back to the parent ion give a major contribution to the HHG process. Finally, (7.12c) provides the energy conservation condition. It states that the emitted photon energy $\hbar\omega$ is equal to the sum of the kinetic energy of the recombining electron and of the ionization potential I_p.

It can be observed that, in contrast to the semi-classical model, the solutions obtained are in general complex-valued solutions. In Figs. 7.3a, b the real parts of the quantum solutions (t_s, t_s') as a function of the photon energy $\hbar\omega$ are compared with the corresponding classical values (t_r, t_i). While the real parts are associated to respectively ionization times and recombination times, the imaginary parts contribute to introduce an imaginary term in the phase $\tilde{S}$. This imaginary phase is strictly related to the quantum nature of the process and leads to a modulation of the amplitude of the emitted HHG radiation [44].

With these tools in hands, it is possible to predict how the propagation from the initial state $\mathbf{k}(t_s')$ affects the amplitude of the electronic wavepacket recombining with the ion. From (7.10), the wave-packet spectrum associated to a single trajectory is here defined by:

$$w_s(\omega) = d_{\mathrm{s}}^{\mathrm{i}}(\omega)a_s(\omega) \tag{7.13}$$

where

$$d_{\mathrm{s}}^{\mathrm{i}}(\omega) = d_{\mathrm{ion}}(\mathbf{k}_{\mathrm{ion}}^{s}) \quad , \quad a_s(\omega) = C(s)e^{-\frac{i}{\hbar}\tilde{S}(\omega,\mathbf{Q}_s)}. \tag{7.14}$$

It should be observed that $a_s(\omega)$ takes into account not only the propagation of the electron in the continuum states but also the exponential nature of the tunnel ionization phenomenon, which is field driven. However, it is insensitive to how the Coulomb field affects this probability, which is partially taken into account by the $d_s^i(\omega)$ dipole term. Dashed and solid line of Fig. 7.3d show $a_s(\omega)$ respectively for the short and long trajectories generated from one half-cycle of the driving field. In the cutoff energy region only one of the two solutions survives, which gives the result shown. With the factorization provided, the total field is given by:

$$\mathbf{E}_H(\omega) \approx \sum_s \mathbf{d}_f(\mathbf{k}_{\mathrm{rec}}^{s})d_{\mathrm{s}}^{\mathrm{i}}(\omega)a_s(\omega). \tag{7.15}$$

The interference between the two contributions increases the complexity of the phenomenon. We will see later that there exist favourable conditions where one of the two contributions is suppressed, which allows for an easier interpretation of the results.

7.2.3 *Macroscopic Effects*

HHG experiments are usually performed in a gas jet, where a number of atoms or molecules are simultaneously excited. The harmonic emission is thus the result of a coherent superposition of all the single-atom contributions coming from the macroscopic sample. As a consequence, the propagation of the harmonic radiation in the atomic gas has to be taken into account in order to extract the actual single atom response. For maximizing the qth harmonic generation efficiency, phase-matching between the qth order non-linear polarization of the medium and the harmonic radiation must be achieved.

Being $\Phi_p^{(q)}(\mathbf{r})$ and $\Phi_q(\mathbf{r})$ respectively the qth order polarization phase and the qth harmonic phase evaluated at the spatial coordinate $\mathbf{r}$, two wave-vectors, $\mathbf{k}_p^{(q)}$ and $\mathbf{k}_q$ can be introduced, defined as $\mathbf{k}_p^{(q)} = \nabla_{\mathbf{r}}\Phi_p^{(q)}(\mathbf{r})$ and $\mathbf{k}_q = \nabla_{\mathbf{r}}\Phi_q(\mathbf{r})$. The conversion efficiency of the qth harmonic is increased when the phase mis-match vector

$$\Delta\mathbf{k}_q(\mathbf{r}) = \mathbf{k}_p^{(q)} - \mathbf{k}_q \tag{7.16}$$

is reduced [45]. Several contributions must be taken into account in order to model the two wave-vectors $\mathbf{k}_p^{(q)}$ and $\mathbf{k}_q$. The polarization phase is q times the driving field phase $\Phi(\mathbf{r})$ plus an additional contribution $\Phi_q^{\text{dip}}(\mathbf{r})$ which represents the phase accumulated by the electron in the continuum. This so called dipole phase depends on both the harmonic order and the actual trajectory followed by the electron in the continuum. Moreover, in a focusing geometry, $\nabla_{\mathbf{r}}\Phi(\mathbf{r})$ includes also a geometrical phase factor, named *Gouy* phase, that strongly changes through the focus of the beam.

Due to all these contributions, perfect on-axis phase matching can be achieved, in principle, only at a given position $\mathbf{r}_{\text{opt}}$ that satisfies $\Delta\mathbf{k}_q(\mathbf{r}_{\text{opt}}) = 0$ with $\mathbf{k}_q$ parallel to the propagation axis. Practically, good phase-matching is achieved even for $\Delta\mathbf{k}_q(\mathbf{r}_{\text{opt}}) \neq 0$ when the target is confined within a region Δz much smaller than the coherence length $L_q = \pi/|\Delta\mathbf{k}_q|$. As it has been shown by Salieres et al. [46], it's possible to find experimental conditions in which phase-matching is ruled mainly by the *Gouy* and dipole phase. In particular, on axis phase matching of short trajectories is achieved when the laser is focused behind the atomic gas-jet, while long trajectories are suppressed. In this case, the HHG field in (7.10) becomes:

$$\mathbf{E}_H(\omega) \approx \mathbf{d}_f(\mathbf{k}_{\text{rec}})d^{\text{i}}(\omega)a(\omega), \tag{7.17}$$

where the s subscript is omitted, since we have only one contribution.

7.3 HHG for Atomic and Molecular Spectroscopy

The physics underlying the HHG process is extremely complicated, since it aims at describing the highly nonlinear response of matter to a high-intensity laser excitation. As described in the previous sections, the saddle point procedure leads to a significant simplification in the process description, providing an easy framework for the interpretation based on the disentanglement of three electronic contributions: ionization, laser acceleration and recombination.

The availability of a reliable and simple physical model makes it easier to access the connection between the experimentally measurable quantities and the processes going on into the target system. As discussed in the previous section, the main result of the semi-classical HHG theory is that, within the SFA, a factorization of the HHG amplitude is possible. Indeed, the complex-valued harmonic field can be expressed as a product of two factors:

$$\mathbf{E}_H(\omega) \propto w(\omega)\mathbf{d}_f(\mathbf{k}_{\text{rec}}), \tag{7.18}$$

where $w(\omega)$ is an amplitude factor accounting for the propagation of the laser-driven electron wave packet and for the tunnel ionization probability, while $\mathbf{d}_f(\mathbf{k}_{\text{rec}})$ is related to the dipole recombination matrix element from the continuum wave function to the ground state orbital. By studying the properties of the high-order harmonic spectral emission, information on the radiating process can be directly retrieved concerning the ionization yield, the electron wave packet interaction with the driving external fields and the recombination dipole moment.

Among the three contributions, the tunnel ionization contribution and the propagation contribution exhibit a strong dependence on the driving laser waveform. The recombination dipole element $\mathbf{d}_f(\mathbf{k}_{\text{rec}})$ is instead strictly related to the electronic structure of the target system. Thus, the main information on the processes going on into the system is contained in the recombination factor. The recombination dipole can in principle be extracted from an HHG measurement, provided that the wavepacket factor $w(\omega)$ is known. A practical way for achieving dipole reconstruction consists in calibrating the harmonic spectrum on that of a well-known reference system [11], which is typically provided by a noble gas with similar I_p as that of the system under investigation. For the same driving field, species with the same ionization energy are ionized through potential barriers of the same width, which means at the same instant during the laser cycle. Then, the ionization probability of target system can be assumed to be equal to that of the reference. Moreover, the electron wavepacket propagation is essentially the same as the one observed in the reference system exposed to identical external field conditions. A reference amplitude can be used $w_{\text{ref}}(\omega)$, which can be experimentally determined or numerically computed.

The calibration procedure is reliable only when a number of assumptions are satisfied: (i) electron wave packet acceleration takes place in purely laser-driven continuum regime, (ii) a single orbital with a well-defined I_p is involved in the ionization-recombination process (single active electron approximation—SAE), (iii) the macroscopic response of the system perfectly reproduces the single atom response. Deviations from these assumptions mean respectively that the SFA and SAE approximations are no longer valid and that phase matching is playing a remarkable role in shaping the harmonic spectrum upon propagation within the sample.

By going beyond the above mentioned ideal conditions, a significant complication in the process modeling and in the data interpretation is introduced. HHG measurements have indeed the potentials for exploring multiple aspects of the laser-system interaction which are not included in the semiclassical model. Consequently, refinements and integrations of theory are required to explain the whole informational content of a full spectrum of XUV frequencies. As an example, the influence of the ionic core can induce a perturbation in the laser-driven continuum propagation.

The portion of the HHG spectrum which is mainly affected by the ionic potential is the low-frequency one. The low-order harmonics are indeed produced by those low-energy electrons which travels along trajectories very close to the parent ion before undergoing recombination. In this case, the scattering wave packet may no longer be approximated by plane waves but a treatment based on Coulomb waves has to be used [47].

An additional complication arises in molecules, due to the fact that the outermost electrons are separated by a relatively low energy gap with respect to the case of atoms. Indeed, while in atoms the energy difference among the external shells ranges typically from tens to hundreds of eV, in molecules valence orbitals are only a few eV apart. Since the probability of tunnel ionization decreases exponentially with the binding energy, in atoms only the outermost shell is likely to be ionized; in molecules, instead, several orbitals can be simultaneously involved in the emission process. The possibility to disentangle the contributions from multiple channels in an HHG spectrum has been studied [48]. As an outcome, the disentanglement is practicable only if the different channels are decoupled and electron-electron correlation dynamics among the electron levels involved do not take place. Electron correlation effects can be observed in HHG experiments due to the fact that the driving laser field may induce a perturbation in the electronic configuration of the ionized atoms or molecules, resulting in the activation of multi-electron processes and in the appearance of multiple recombination channels. Multi-electron processes are dynamical processes usually evolving on an ultrafast time scale [49].

The possibility to resolve electron dynamics lies on the ultrafast nature of the XUV emission process. In fact, electron wave packet generated by ionization undergoes continuum propagation and recombination within one optical cycle of the laser field. The recombination alone is thus confined within a temporal window which has a sub-optical cycle duration. Laser systems typically used for HHG applications deliver pulses with an optical cycle of the order of 1–10 fs, which means that processes occurring on the femtosecond time scale can be probed with a resolution down to the attosecond time scale. In this sense, electron correlation dynamics can be studied using the HHG self-probing scheme with an unprecedented temporal resolution.

7.4 Molecular Orbital Tomography Based on HHG

In this section we will introduce and review the principles behind molecular orbital tomography based on high-order harmonic generation and we will then present some recent experimental results obtained with this technique. We will then discuss the limits of molecular tomography and provide some hints about possible solutions to those problems. Eventually the perspectives of this technique will be discussed.

7.4.1 Impulsive Molecular Alignment

Since the tomographic reconstruction of molecular orbitals requires to fix the orientation of the molecular frame (i.e. of the molecular axes) in the laboratory frame, we will first discuss how this is usually achieved.

An intense femtosecond laser pulse is able to induce an impulsive rotational response in a molecular gas, leading to field-free alignment echoes in the excited molecules at suitable delays from the laser pulse arrival. During these echoes, known as rotational revivals, a large fraction of the molecules are aligned within a small solid angle. Hence the acquisition of HHG spectra required by tomographic imaging will be performed in correspondence of one of these revivals, by focusing in the aligned molecular gas a second laser pulse which drives HHG.

The impulsive rotational excitation requires a quantum description according to the Schrödinger equation:

$$\left(\hat{H}+\hat{U}\right)|\Psi\rangle = i\hbar\frac{\partial|\Psi\rangle}{\partial t}. \tag{7.19}$$

In the general case and assuming the molecule as a rigid body, the rotational Hamiltonian operator is given by [50]:

$$\hat{H} = \frac{\hbar^2}{2}\left(\frac{\hat{J}_\xi^{\,2}}{I_A}+\frac{\hat{J}_\eta^{\,2}}{I_B}+\frac{\hat{J}_\zeta^{\,2}}{I_C}\right), \tag{7.20}$$

where $\xi,\ \eta,\ \zeta$ is a system of coordinates in the molecular frame directed along the three principal axes of inertia of the molecule, $\hat{J}_\xi,\ \hat{J}_\eta,\ \hat{J}_\zeta$ are the operators corresponding to the projection of the molecular angular momentum along such axes and $I_A,\ I_B,\ I_C$ are the principal moments of inertia of the molecule.

The operator $\hat{U}$ describes the interaction between the molecule and the laser field. This is mainly governed by the interplay between the electric field component of the laser pulse and the molecular dipole induced by the field itself. In the general case the operator can be expanded in a series of terms like [51]:

$$\hat{U} = -\left(\sum_i p_i E_i + \frac{1}{2}\sum_{ij}\alpha_{ij}E_iE_j + \frac{1}{6}\sum_{ijk}\beta_{ijk}E_iE_jE_k + \cdots\right), \tag{7.21}$$

where all the indexes in the summations run over the three values ξ,η,ζ; $\mathbf{E}$ is the electric field component of the laser pulse; $\mathbf{p}$ is the molecular permanent dipole; $\bar{\alpha}$ is the molecular polarizability tensor; $\bar{\beta}$ is the first molecular hyperpolarizability tensor and so on. It is worth noting that all the molecular quantities are referred to the molecular frame, whereas the laser electric field is defined in the laboratory one; the projection of $\mathbf{E}$ on the molecular axes is thus obtained by a suitable rotation matrix that links the two frames [50].

We will limit here the discussion to linear molecules for the sake of simplicity; a general treatment on rotational phenomena in molecules can be found in [50, 51], whereas an overview on impulsive molecular alignment is presented in [7]. Let's assume that the molecular axis is directed along ζ and that the molecular gas interacts with a laser pulse linearly polarized along the laboratory axis z given by $\mathbf{E}(t) = a(t)\cos(\omega_0 t)\mathbf{u}_z$, with $a(t)$ a slowly-varying envelope and ω_0 the optical carrier frequency that is assumed far from any resonance. For linear molecules the non-vanishing terms of the polarizability are $\alpha_{\zeta\zeta} = \alpha_\parallel$ and $\alpha_{\xi\xi} = \alpha_{\eta\eta} = \alpha_\perp$, whereas the relevant moment of inertia will be indicated as I. In such a case the eigenfunctions of the Hamiltonian in (7.20) are the spherical harmonic functions $|\Phi_{jm}\rangle = Y_j^m(\theta, \psi)$, where θ is the angle between the ζ and z axes and ψ describes the precession of ζ around z; note that the eigenfunctions do not depend on the third Euler angle φ describing the rotation of the molecule around its axis. The rotational eigenenergies are given by $\mathscr{E}_j = \hbar^2 j(j+1)/(2I)$ where j is an integer number and $-j \leq m \leq j$ [50]; hence for every j there are $2j+1$ degenerate states with the same energy but different m. By taking into account the frame transformation, the interaction operator in (7.21) reduces to:

$$\hat{U} = -\frac{a^2(t)}{4}\left[\left(\alpha_\parallel - \alpha_\perp\right)\cos^2\theta + \alpha_\perp\right], \tag{7.22}$$

where we have neglected the contribution of the hyperpolarizability and we averaged the interaction energy over an optical cycle of the laser pulse since the molecular response is much slower than the optical period. One can see that the permanent dipole of the molecule, if present, is anyway not contributing to the rotational response in this approximation.

Equation (7.19) can then be solved numerically by projecting the rotational wavefunction $|\Psi(t)\rangle$ on the Hamiltonian eigenfunctions. One will end up with a system of differential equations in the coefficients of this projection, i.e.:

$$\frac{\partial c_{j'm'}}{\partial t} = \frac{1}{i\hbar}\sum_{jm} c_{jm}(t)\exp\left[-\frac{i}{\hbar}\left(\mathscr{E}_j - \mathscr{E}_{j'}\right)t\right]\langle\Phi_{j'm'}|\hat{U}|\Phi_{jm}\rangle, \tag{7.23}$$

where

$$|\Psi(t)\rangle = \sum_{jm} c_{jm}(t)\exp\left[-\frac{i}{\hbar}\mathscr{E}_j t\right]|\Phi_{jm}\rangle. \tag{7.24}$$

If we could assume that before the interaction with the aligning laser pulse all the molecules in the gas target were in the same rotational quantum state $|\Phi_{JM}\rangle$, the evolution of the system would be readily obtained by solving (7.23)–(7.24) with a suitable initial condition at $t = 0$, i.e. $c_{jm}(0) = \delta_{jJ}\delta_{mM}$ where δ is the Kronecker delta. However this is not the case, since a molecular gas cannot be described by a pure quantum state [7]; indeed the rotational evolution must be determined in the framework of the density matrix formalism [7, 50]. In practice, the gas is described at $t = 0$ by a "collection" of pure initial quantum states $|\Phi_{JM}\rangle$ that are populated with a statistical weight $W(J) = g(J)\exp\left(-\frac{\mathscr{E}_J}{k_B T}\right)$ which depends on the gas temperature T according to the Boltzmann statistics through the rotational energy of the pure state $|\Phi_{JM}\rangle$. The factor $g(J)$ takes into account a normalization term including the energy degeneracy and additional terms that depend on the specific properties of the molecule:

$$g(J) = \frac{f(J)}{\sum_k f(k)(2k+1)\exp\left(-\frac{\mathscr{E}_k}{k_B T}\right)}, \tag{7.25}$$

where $f(J)$ is related to selection rules dictated by the molecular symmetry and by the nuclear spin of the atoms included in the molecule [7, 50, 51]. Hence for each initial quantum state $|\Phi_{JM}\rangle$ one will determine the corresponding evolution of the wavefunction $|\Psi^{(JM)}(t)\rangle$. The expectation value of any operator $\hat{X}$ related to the gas will then be calculated as a weighted sum of the quantum expectation values determined for each initial pure quantum state, i.e.:

$$\langle X\rangle(t) = \sum_{JM} W(J)\langle\Psi^{(JM)}(t)|\hat{X}|\Psi^{(JM)}(t)\rangle, \tag{7.26}$$

whereas the probability dP of finding a molecule at a given angular position (θ, ψ) within $(d\theta, d\psi)$ will be calculated according to:

$$F(\theta, \psi, t) = \frac{dP(\theta, \psi, t)}{d\theta d\psi} = \sum_{JM} W(J)|\Psi^{(JM)}(\theta, \psi, t)|^2 \sin(\theta). \tag{7.27}$$

A concise description of the gas evolution is obtained by replacing $\hat{X} = \cos^2\theta$ in (7.26); in this case the resulting expectation value $A(t) = \langle\cos^2\theta\rangle(t)$ is known as the *alignment factor*. A complete alignment of the molecules along the z axis corresponds to $A = 1$, whereas a complete antialignment (with all the molecules perpendicular to the z axis) leads to $A = 0$. A random distribution of molecules corresponds to $A = 1/3$. Figure 7.4 reports the evolution of the alignment factor calculated for several gases of linear molecules excited at room temperature with a 100-fs laser pulse; the temporal scale is different for each molecule in order to show the complete rotational response in each case. In all the cases one finds that the alignment factor shows a periodic behavior with a rotational period $T_R = 2\pi I/\hbar$. A *full revival* of the alignment occurs at integer multiples of T_R; a *half revival* occurs between two full revivals; in some molecules *quarter revivals* might be observed according to the specific molecular properties. Table 7.1 reports the rotational periods of a few linear molecules with some additional properties; note that α $(\mathrm{F}\cdot\mathrm{m}^2) = \alpha$ $(\text{Å}^3)\cdot 4\pi\varepsilon_0 \times 10^{-30}$.

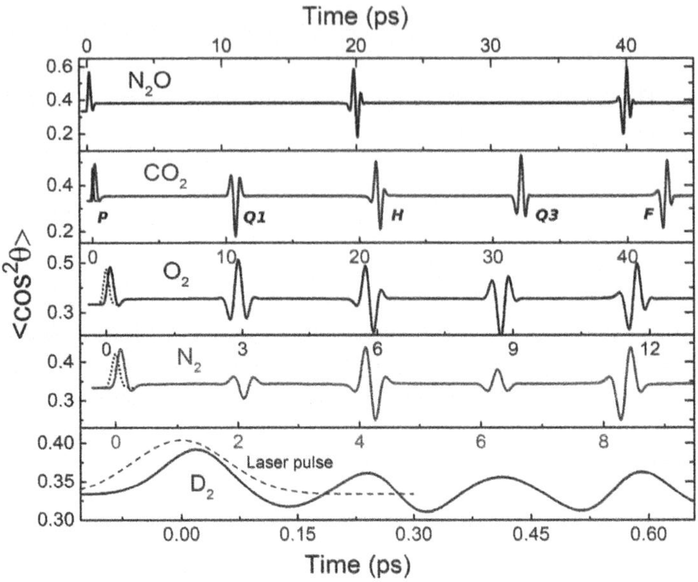

Fig. 7.4 Calculated evolution of the alignment factor $A(t)$ for several molecular gases at 300 K excited by a linearly polarized laser pulse with 100-fs duration and a peak intensity of 2.5×10^{13} W/cm^2. Note that the temporal scale is different for each molecule; the envelope of the laser pulse, that peaks at $t = 0$, is reported in each trace as dashed line for comparison. The alignment echoes are labeled as prompt alignment (P), first quarter (Q1), half revival (H), third quarter (Q3) and full revival (F)

Table 7.1 Properties of some linear molecules [52]

Molecule	I_p (eV)	Companion atom	T_R (ps)	$\alpha_\parallel$ (Å^3)	$\alpha_\perp$ (Å^3)
H_2	15.43	Ar	0.27	0.963	0.087
D_2	15.47	Ar	0.55	0.963	0.087
N_2	15.58	Ar	8.34	2.092	1.289
O_2	12.07	Xe	11.60	1.582	0.883
C_2H_2	11.40	Xe	14.16	4.334	2.017
N_2O	12.89	Xe	39.80	4.772	1.647
CO_2	13.78	Kr	42.73	3.927	1.674

7.4.2 Theory of HHG-based Molecular Orbital Tomography

As already discussed in Sect. 7.2, in the framework of the Single Active Electron and of the Strong Field Approximations the HHG field emitted by a single molecule is given by (7.18), which is reported here again for the sake of clarity:

$$\mathbf{E}_H(\omega) \propto \omega \mathbf{v}[\mathbf{k}_{\text{rec}}(\omega)]w(\omega), \tag{7.28}$$

where

$$\mathbf{v}(\mathbf{k}_{\text{rec}}) = \langle \Psi | -i\nabla | \mathbf{k}_{\text{rec}} \rangle = \mathbf{k}_{\text{rec}} \mathscr{F}_{\mathbf{k}_{\text{rec}}}(\Psi) \tag{7.29}$$

with $|\Psi\rangle$ the highest occupied molecular orbital (HOMO) of the molecule, $|\mathbf{k}\rangle = (2\pi)^{-3/2} \exp[i\mathbf{k} \cdot \mathbf{r}]$ the free-particle wavefunction with electron momentum $\hbar\mathbf{k} = \mathbf{p} + e\mathbf{A}$, $\mathbf{p}$ the canonical momentum which is assumed constant during the electron motion in the continuum, $\mathbf{k}_{\text{rec}}$ the electron wavevector at recollision, $\mathbf{A}(t)$ and $\mathbf{E}(t)$ the vector potential and the electric field of the laser pulse which drives HHG respectively, $w(\omega)$ the electron wavepacket described by (7.13–7.14) and $\mathscr{F}_{\mathbf{k}}(\cdot)$ is the spatial Fourier transform operator. In (7.28) we considered the dipole velocity term $\mathbf{v}$ instead of the usual dipole term $\mathbf{d}$ since the former provides a better agreement with the experimental data [30, 53], although it is not suited for describing the non-linear polarization state of harmonic radiation coming from molecules [26]. It is worth noting that the structural information about the target molecule is encoded in the $\mathbf{v}(\mathbf{k}_{\text{rec}})$ term, which is proportional to the spatial Fourier transform of the HOMO (assumed to be a real function).

For multicycle laser pulses with carrier frequency ω_0 the phase of the wavepacket term $w(\omega)$ in (7.28) gives rise to two features: (a) a frequency-dependent spectral phase known as *attochirp* [54], not contributing to the retrieval of the structural information because orbital tomography is usually based on the acquisition of HHG intensity spectra; (b) the spectral interference giving rise to the harmonic comb structure at odd frequencies $(2n + 1)\omega_0$ that modulates the single-recollision spectrum and that can be removed by extracting the spectral envelope. Hence these phase terms will not be considered hereafter. The amplitude of $w(\omega)$, depending on the ionization yield and the electron wavepacket spreading, contributes to the modulation of the HHG spectrum and cannot be neglected in the tomographic applications [24, 55].

The experimental HHG spectrum emerging from a molecular gas is however the coherent superposition of the emissions from all the molecules, hence it is also affected by macroscopic phase matching effects and depends on the molecular angular distribution. The former effect can be controlled by a suitable interaction geometry, for instance by focusing the laser beam before the gas jet position [45] whereas the angular distribution of molecules can be calculated as shown in Sect. 7.4.1. In the end, the macroscopic harmonic intensity spectrum is reduced to:

$$\begin{aligned} I_{\text{gas}}(\omega) &\propto \left| \int \mathbf{E}_H(\omega, \theta, \psi, \varphi) F(\theta, \psi, \varphi) d\theta d\psi d\varphi \right|^2, \\ \mathbf{E}_H(\omega, \theta, \psi, \varphi) &= \omega\, w(\omega)\, \mathbf{k}_{\text{rec}}\, \mathscr{F}_{\mathbf{k}_{\text{rec}}} \left[\Psi \left(\xi, \eta, \zeta \right) \right], \end{aligned} \tag{7.30}$$

where we emphasized that the molecular angular distribution F may depend on all the angular coordinates of the molecules.

In the majority of cases the HHG spectrum is driven by a laser pulse linearly polarized along the laboratory z axis; according to the saddle-point approximation and in the plateau region of the harmonic spectrum, one can then assume that the recollision electron momentum is real and given by $\mathbf{k}_{\text{rec}} = k_{\text{rec}}(\omega, I_p)\mathbf{u}_z \approx \mathbf{u}_z\sqrt{2m(\hbar\omega - I_p)}/\hbar$.

If the angular distribution F is known, one can retrieve the single-molecule emitted field $\mathbf{E}_H$ from the measured $I_{\text{gas}}(\omega)$ by suitable techniques applied to (7.30); then the HOMO wavefunction Ψ will be obtained by inverse Fourier transform.

We conclude this section with some test cases by approximating the HOMO of simple linear molecules using the Linear Combination of Atomic Orbitals (LCAO). Real-world tomographic applications will instead be shown in Sect. 7.4.3. We will use atomic units hereafter.

For a molecule directed along ζ we will consider three typical types of molecular orbital: (1) a σ_g orbital, similar to the HOMO in N_2, that will be modeled as the overlap between two p_ζ atomic orbitals; (2) a π_g orbital, similar to the HOMO in O_2 and CO_2, that will be modeled as the overlap between two p_η atomic orbitals with opposite signs; (3) a π_u orbital, similar to the HOMO in C_2H_2, that will be modeled as the overlap between two p_η atomic orbitals with the same sign. The p orbitals will be expressed as $p_\zeta = \zeta \exp(-\rho)$ and $p_\eta = \eta \exp(-\rho)$, with $\rho = \sqrt{\xi^2 + \eta^2 + \zeta^2}$. The two "atomic centers" simulating the molecule are placed along the ζ axis in the positions $\zeta = \pm d$. Table 7.2 shows these three model cases with the corresponding Fourier transform in the molecular frame, where $k_\zeta = k\cos\theta$, $k_\eta = k\sin\theta$, $k = k_{\text{rec}}$ and θ is the angle formed by the molecular axis with the z laboratory axis; irrelevant constants are omitted in the reported expressions. Note that the Fourier transform of σ_g and π_g orbitals are real, whereas the π_u one is imaginary.

Figure 7.5 shows the quantity $|\mathbf{E}_H(\omega, \theta)|/[\omega\,|w(\omega)|]$ calculated in the three model orbitals as a function of θ and of the kinetic energy of the recolliding electron, which is given by $k_{\text{rec}}^2/2 = \omega - I_p$, where ω is the frequency of the emitted harmonic field and I_p the molecular ionization potential. In all the cases we assumed a distance $d = 1$ atomic unit between the two centers. One can see that each orbital shows peculiar structures; in particular there are minima of harmonic emission that appear at well defined photon energies and molecular orientations. These minima correspond to a

Table 7.2 Simple molecular orbitals and their spatial Fourier transform

Orbital	LCAO	Wavefunction	Fourier transform
σ_g	$p_\zeta(\zeta - d) - p_\zeta(\zeta + d)$		$\dfrac{k_\zeta \sin(k_\zeta d)}{(1+k^2)^3}$
π_g	$p_\eta(\zeta - d) - p_\eta(\zeta + d)$		$\dfrac{k_\eta \sin(k_\zeta d)}{(1+k^2)^3}$
π_u	$p_\eta(\zeta - d) + p_\eta(\zeta + d)$		$i\dfrac{k_\eta \cos(k_\zeta d)}{(1+k^2)^3}$

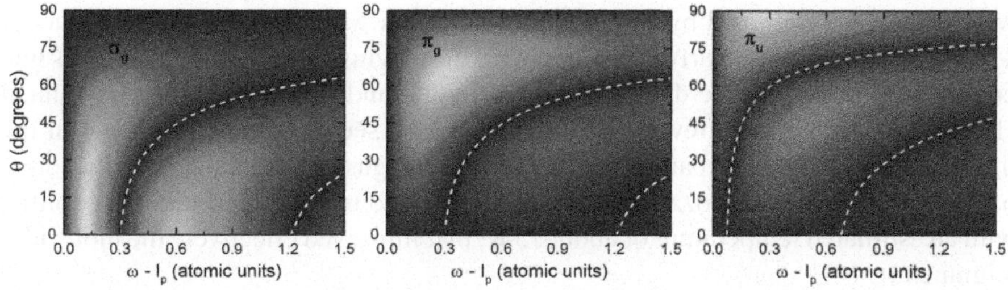

Fig. 7.5 $|\mathbf{E}_H(\omega, \theta)|/[\omega\ |w(\omega)|]$ calculated for σ_g (left), π_g (center) and π_u (right) orbitals as a function of θ and $\omega - I_p$ assuming $d = 1$. The white dashed lines show the minima of harmonic emission calculated according to the two-center interference model (see text)

π phase jump in the emitted field and have been also explained in the framework of a molecular two-center model [8, 9] that predicts destructive interference of harmonic emission in σ_g and π_g orbitals when $R \cos\theta = n\lambda_B$, where R is the distance between the two centers, n is an integer and $\lambda_B = 2\pi/k_{\text{rec}}$ the electron de Broglie wavelength. In π_u orbitals this condition becomes $R \cos\theta = (2n + 1)\lambda_B/2$. White dashed lines in Fig. 7.5 show where these conditions are met; the model indeed predicts very well the position of the minimum harmonic emission.

7.4.3 Experimental Molecular Tomography

We will now discuss an experimental case in order to show how molecular orbital tomography can be achieved in practice. Figure 7.6 shows a sequence of harmonic spectra acquired in a pulsed gas jet of aligned carbon dioxide molecules [30]. The harmonic emission was driven by 1.45-μm, 18-fs laser pulses with a peak intensity of $1.7 \times 10^{14}\ \text{W/cm}^2$ produced by a mid-IR optical parametric amplifier. The molecules

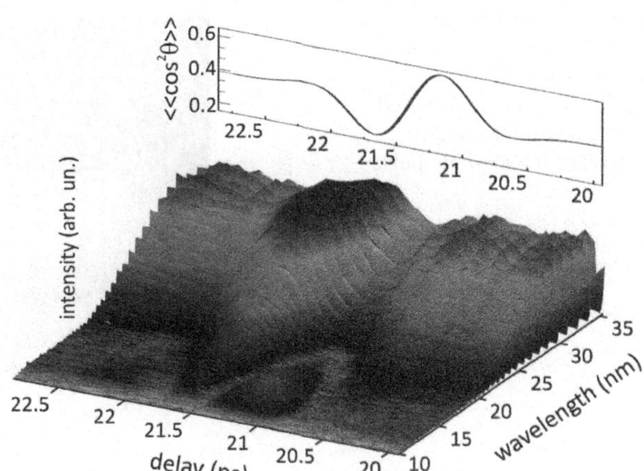

Fig. 7.6 HHG spectra I_{CO_2} acquired in aligned CO_2 molecules as a function of the delay τ between the aligning and the driving laser pulses. The upper panel shows the evolution of the calculated alignment factor during the delay scan

were impulsively aligned by 800-nm, 100-fs pulses with a peak intensity of about 4×10^{13} W/cm^2, collinearly combined with the driving pulses and having the same polarization direction; the delay between aligning and driving pulses was scanned along the half rotational revival of CO_2, as can be seen by the upper panel in the figure. The supersonic expansion in vacuum of the gas exiting the nozzle (absolute backing pressure 4 bar, nozzle diameter 500 μm) led to a strong rotational cooling, with an estimated temperature of about 75 K, that improved effectively the molecular alignment.

Since the angular distribution of molecules changes with the delay, a noticeable delay-dependent modulation of the harmonic spectrum is observed. This modulation is the result of the coherent sum of emissions from molecules having different orientations in the macroscopic molecular ensemble. Figure 7.7 shows how the calculated normalized angular distribution of the molecules evolves in time; one can see that in correspondence of the maximum alignment almost all the molecules are concentrated in a small angular region between 15° and 30°, whereas at antialignment they are almost at 90°.

It is worth noting that the HOMO of carbon dioxide is doubly degenerate, but only the orbital lying in the (z, ζ) plane, i.e. the plane formed by the alignment axis and the molecular axis, contributes to the generation of harmonics [30]. Hence the only significant coordinate in this process is the angle θ between the two axes. From this it comes out that the components of the wavevector $\mathbf{k}$ of the recolliding electron projected on the axes (ξ, ζ) of the molecular frame are the only ones needed for a reconstruction of the molecular orbital, whereas the coordinate η, assumed perpendicular to the (z, ζ) plane, will not contribute to the tomographic imaging. In this sense HHG-MOT is in practice a bidimensional imaging technique.

From (7.30), one can model the measurements shown in Fig. 7.6 according to:

$$I_{CO_2}(\omega, \tau) \propto \left| \int E_H(\omega, \theta) F(\theta, \tau) d\theta \right|^2 . \tag{7.31}$$

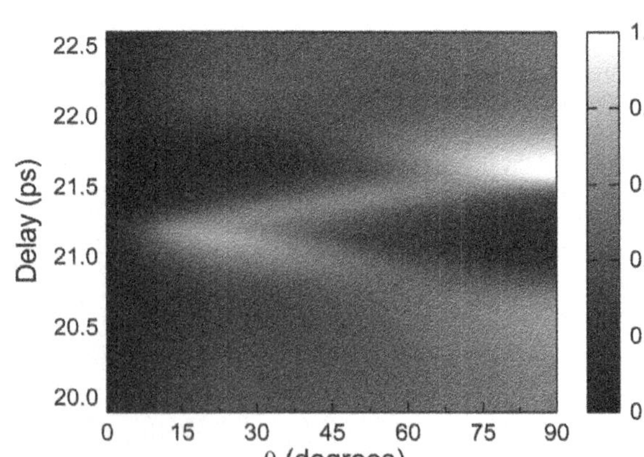

Fig. 7.7 Normalized angular distribution of CO_2 molecules $F(\theta, \tau)$ as a function of the angle θ and of the delay τ from the arrival of the aligning pulse

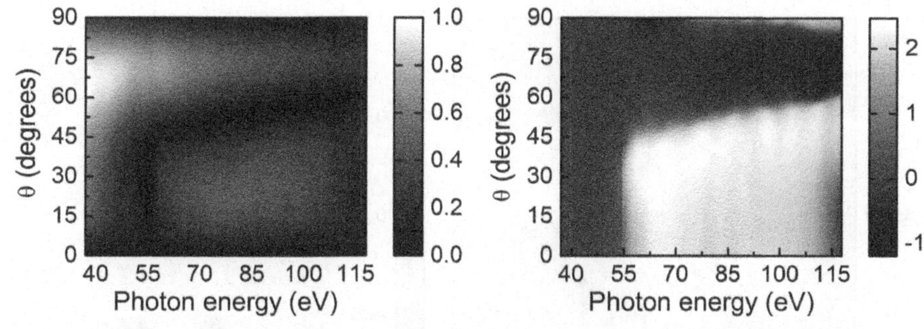

Fig. 7.8 Retrieved amplitude (left) and phase (right) of the harmonic field $E_H(\omega, \theta)$ emitted by CO_2 molecules as a function of photon energy and angle θ formed by the molecules with the HHG driving laser field

$E_H(\omega, \theta)$, required for the tomographic reconstruction of the molecular orbital, is a complex-valued function; for this reason a suitable retrieval procedure for both amplitude and phase of E_H from the measured $I_{CO_2}(\omega, \tau)$ is needed. The details of this procedure can be found elsewhere [30]; here we directly show the retrieved results in Fig. 7.8. One can see that the retrieved amplitude (left panel) shows a minimum that resembles the behavior predicted by the two-center interference model for π_g orbitals (see central panel in Fig. 7.5). The minimum corresponds to a phase jump, clearly visible in the retrieved phase (right panel in Fig. 7.8). However this jump does not amount to π as expected by the LCAO approximation, but only to about 2.2 rad [30]. This discrepancy is related to the limitations of the Strong Field approximation with respect to the real cases; such limitations will be analyzed in Sect. 7.4.4.

Once the $E_H(\omega, \theta)$ quantity is retrieved, one can exploit it for the molecular orbital retrieval. By taking into account the influence of the molecular Coulomb potential at a first approximation level, one finds that [30]:

$$\mathscr{F}_{\mathbf{k}}(\Psi) \approx \frac{\mathrm{Re}\left[E_H(\omega, \theta)\right]}{w(k)\left(\dfrac{k^2}{2} + I_p\right)\left(\dfrac{3k}{2} + \dfrac{I_p}{k}\right)}, \tag{7.32}$$

where $\omega = k^2/2 + I_p$ (in atomic units). The wavepacket amplitude $w(k)$ was in this case evaluated theoretically by an SFA model considering an hydrogenoid atom with the same ionization potential of the investigated molecule [30]. The Fourier transform of the orbital was then calculated according to (7.32) and is shown in the left panel of Fig. 7.9 in the (k_ξ, k_ζ) plane. It is worth noting that the harmonic field E_H is retrieved only in the region of positive wavevector components, hence it is analytically extended to the whole **k** plane by taking into account the known symmetry of the CO_2 molecular orbital, which must satisfy the relations $\mathscr{F}_{-\mathbf{k}}(\Psi) = \mathscr{F}_{\mathbf{k}}(\Psi)$ and $\mathscr{F}_{k_\xi, -k_\zeta}(\Psi) = -\mathscr{F}_{k_\xi, k_\zeta}(\Psi)$ [30]. Furthermore the limited spectral region in which the harmonics were detected corresponds to "empty" circular regions in the spatial Fourier transform of the spectrum, as can be easily seen in the figure.

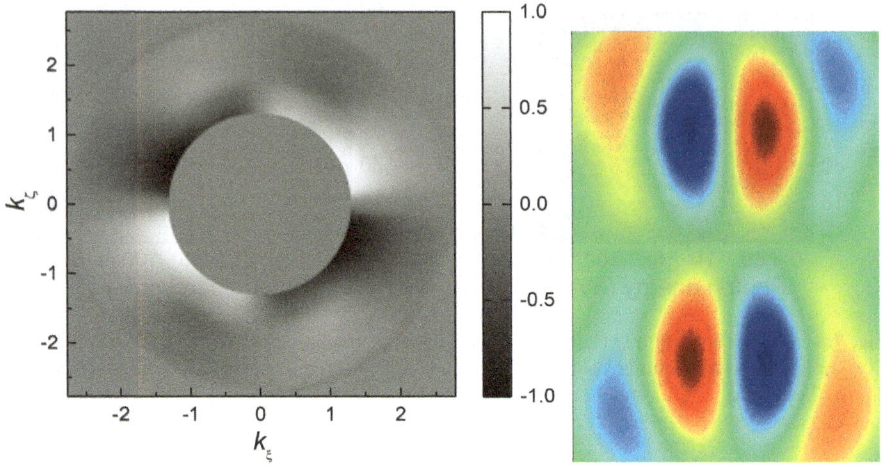

Fig. 7.9 *Left:* normalized Fourier transform of the CO_2 molecular orbital in the (k_ξ, k_ζ) plane, as obtained by (7.32). The wavevectors are expressed in atomic units; the empty central and external circular regions correspond to portions of the HHG spectrum not acquired in the experiment. *Right:* HOMO of carbon dioxide retrieved by HHG-MOT. The lobes surrounding the orbital are related to the incomplete determination of the Fourier transform; extending the detected harmonic spectrum would improve the reconstruction

Anti-transforming $\mathscr{F}_{\mathbf{k}}(\Psi)$ back to the spatial domain gave rise to the retrieved orbital, which is reported in the right panel of Fig. 7.9. One can recognize the typical π_g shape of the CO_2 HOMO, although surrounded by additional lobes that are related to the incomplete determination of the Fourier transform described above. Extending the detection of high harmonics to low photon energies would improve substantially the reconstruction [30, 56].

7.4.4 Open Issues and Possible Solutions

Several studies on HHG in molecules revealed that the Strong Field and the Single Active Electron Approximations, on which the original tomographic approach is based, fail at least when near-IR driving lasers are exploited [57]. Indeed multiple orbital contributions [47, 48, 58] are known to appear in HHG from some molecules, thus obscuring the retrieval of the pure outermost molecular orbital and giving rise to multielectron effects in harmonic generation. However recent experimental evidences [30, 56, 59] demonstrated that this issue is strongly alleviated when mid-IR laser sources are used. Moreover these sources are beneficial to HHG-MOT owing to the favorable scaling of the spectral cutoff with fundamental frequency [60], allowing to extend the harmonic emission towards high photon energies without exposing the target molecules to unsustainable laser intensities [23, 30]. The distinction between contributions from different orbitals can be also achieved by HHG from bichromatic and orthogonally polarized pulses [61].

Several issues come from the nature of the electron recollision process and how this process influences HHG-MOT. The first issue appears when multicycle laser pulses are exploited for HHG; in this case the electron wavepacket collides with the parent ion from opposite directions, hindering the possibility of a complete reconstruction of generic-shape orbitals [62] since suitable assumptions on the orbital symmetry are required in order to replace the missing information [30, 53]. In this case the exploitation of single-cycle driving pulses with controlled Carrier-Envelope Phase, combined to a full control of the molecular orientation, would overcome the problem [62] since single recollision on only one side of the molecule would occur.

A second issue in HHG-MOT comes from the assumption that the electron wavepacket can be modeled as a plane wave, thus allowing to express the recombination matrix element in terms of a spatial Fourier transform. However, the Coulomb interaction between the incoming electron and the parent ion leads to a distortion of the wavepacket [63, 64] which hinders a simple orbital retrieval since it leads to a non trivial harmonic spectral phase as a function of the photon energy and of the molecular orientation [30, 65]. Furthermore the dispersion relation between the harmonic photon energy and the colliding electron momentum, which is essential for the calibration of the spatial frequencies in the HHG-MOT procedure, is not straightforward any more [66]. This problem cannot be easily solved; it can be faced by extending the tomographic approach to Coulomb waves [67] although this would require knowing in advance the shape of the potential well in which the wavepacket propagates [30].

A third issue is related to the calibration of the electron wavepacket amplitude, required for a correct tomographic retrieval. In practice, one often calibrates the wavepacket amplitude only by measuring the HHG spectra in a *companion gas*, i.e. in a noble gas having an ionization potential similar to the molecular one [11] (the companion atoms of a few linear molecules are listed in Table 7.1). However in a correct procedure one should calibrate the wavepacket amplitude by measuring the ionization yield in the molecular frame [68], since it depends on both the ionization potential and the orientation of the molecule with respect to the laser field direction.

A general problem in HHG-MOT is related to the detection of the spectral phase of the emitted harmonics. A standard HHG setup is only able to detect the spectral intensity of the harmonic emission; hence the spectral phase, which is required for a tomographic reconstruction, must be included on the basis of a reasonable assumption [11] or on the basis of a reconstruction from an experimental set of measurements [30] supported by a theoretical model. Recent efforts led to the LAPIN approach that combines interferometric HHG and the exploitation of mixtures of the investigated species with a reference gas; this technique demonstrated a complete determination of the emitted harmonic field in amplitude and phase [69].

7.4.5 Conclusions and Perspectives

The well-developed HHG field has led to a number of important advances both from the fundamental as well as the technological point of view. These advances have resulted in imaging of molecular orbitals [11, 30], mapping the evolution of chemical reactions [70], studying multi-electron dynamics in molecules [48] and harnessing light at the timescales of attosecond. Indeed the understanding and control of the mechanism underlying HHG has opened up the possibility of studying dynamics with attosecond temporal resolution in gases, solids and biomolecules [71].

Another important development in the field was made in 2011, when Ghimire and co-workers successfully generated HHG from the bulk of a ZnO crystal, a semiconductor with a direct band gap of 3.2 eV [72]. This was a great breakthrough since it answered the question if these process can be extended to the solids. Since then different groups have reported the generation of HHG from dielectrics and smaller band-gap semiconductors. This latest development along with the continuous improvement in ultrafast laser technologies pave the way to the intriguing extension of HHG-based tomographic techniques to the investigation of a wide range of solid-state materials with extreme temporal resolution.

Acknowledgements We acknowledge support from the European Research Council Starting Research Grant UDYNI (grant agreement n. 307964), from Laserlab-Europe (EU-H2020 654148), from the Italian Ministry of Research and Education (ELI project—ESFRI Roadmap), from Regione Lombardia through the project FEMTOTERA (ID: CONCERT2014-008) and from Fondazione Cariplo through the project GREENS (n. 2013-0656).

References

1. T. Popmintchev, M.C. Chen, D. Popmintchev, P. Arpin, S. Brown, S. Ališauskas, G. Andriukaitis, T. Balčiunas, O.D. Mücke, A. Pugzlys et al., Science **336**, 1287 (2012)
2. A. McPherson, G. Gibson, H. Jara, U. Johann, T.S. Luk, I.A. McIntyre, K. Boyer, C.K. Rhodes, J. Opt. Soc. Am. B **4**, 595 (1987)
3. M. Ferray, A. L'Huillier, X.F. Li, L.A. Lompré, G. Mainfray, C. Manus, J. Phys. B **21**, L31 (1988)
4. F. Calegari, G. Sansone, S. Stagira, C. Vozzi, M. Nisoli, J. Phys. B At. Mol. Opt. Phys. **49**, 062001 (2016)
5. M. Lein, N. Hay, R. Velotta, J.P. Marangos, P.L. Knight, Phys. Rev. Lett. **88**, 183903 (2002)
6. B. Shan, S. Ghimire, Z. Chang, Phys. Rev. A **69**, 021404 (2004)
7. H. Stapelfeldt, T. Seideman, Rev. Mod. Phys. **75**, 543 (2003)
8. C. Vozzi, F. Calegari, E. Benedetti, J.P. Caumes, G. Sansone, S. Stagira, M. Nisoli, R. Torres, E. Heesel, N. Kajumba, J.P. Marangos, C. Altucci, R. Velotta, Phys. Rev. Lett. **95**, 153902 (2005)
9. T. Kanai, S. Minemoto, H. Sakai, Nature **435**, 470 (2005)
10. J. Itatani, D. Zeidler, J. Levesque, M. Spanner, D.M. Villeneuve, P.B. Corkum, Phys. Rev. Lett. **94**, 123902 (2005)
11. J. Itatani, J. Levesque, D. Zeidler, H. Niikura, H. Pepin, J.C. Kieffer, P.B. Corkum, D.M. Villeneuve, Nature **432**, 867 (2004)
12. S. Haessler, J. Caillat, P. Salières, J. Phys. B At. Mol. Opt. Phys. **44**, 203001 (2011)

13. M. Meckel, D. Comtois, D. Zeidler, A. Staudte, D. Pavičić, H.C. Bandulet, H. Pépin, J.C. Kieffer, R. Dörner, D.M. Villeneuve, P.B. Corkum, Science **320**, 1478 (2008)
14. M.G. Pullen, B. Wolter, A.T. Le, M. Baudisch, M. Hemmer, A. Senftleben, C.D. Schröter, J. Ullrich, R. Moshammer, C.D. Lin, J. Biegert, Nat. Commun. **6**, 7262 (2015)
15. D. Shafir, Y. Mairesse, D. Villeneuve, P. Corkum, N. Dudovich, Nat. Phys. **5**, 412 (2009)
16. H. Niikura, N. Dudovich, D. Villeneuve, P. Corkum, Phys. Rev. Lett. **105**, 053003 (2010)
17. P. Puschnig, S. Berkebile, A.J. Fleming, G. Koller, K. Emtsev, T. Seyller, J.D. Riley, C. Ambrosch-Draxl, F.P. Netzer, M.G. Ramsey, Science **326**, 702 (2009)
18. C.J. Villagomez, T. Zambelli, S. Gauthier, A. Gourdon, S. Stojkovic, C. Joachim, Surf. Sci. **603**, 1526 (2009)
19. P. Hockett, C.Z. Bisgaard, O.J. Clarkin, A. Stolow, Nat. Phys. **7**, 612 (2011)
20. N. Kyo, T. Takahiro, A. Hiroshi, F. Takashi, M. Takuya, M. Shinichirou, O. Kanade, S. Hirofumi, T. Tadashi, T. Kensuke, T. Shota, W. Ken, Y. Makina, Y. Akira, Sci. Rep. **5**, 14065 (2015)
21. J. Küpper, S. Stern, L. Holmegaard, F. Filsinger, A. Rouzée, A. Rudenko, P. Johnsson, A.V. Martin, M. Adolph, A. Aquila, S.C.V. Bajt, A. Barty, C. Bostedt, J. Bozek, C. Caleman, R. Coffee, N. Coppola, T. Delmas, S. Epp, B. Erk, L. Foucar, T. Gorkhover, L. Gumprecht, A. Hartmann, R. Hartmann, G. Hauser, P. Holl, A. Hömke, N. Kimmel, F. Krasniqi, K.U. Kühnel, J. Maurer, M. Messerschmidt, R. Moshammer, C. Reich, B. Rudek, R. Santra, I. Schlichting, C. Schmidt, S. Schorb, J. Schulz, H. Soltau, J.C.H. Spence, D. Starodub, L. Strüder, J. Thøgersen, M.J.J. Vrakking, G. Weidenspointner, T.A. White, C. Wunderer, G. Meijer, J. Ullrich, H. Stapelfeldt, D. Rolles, H.N. Chapman, Phys. Rev. Lett. **112**, 083002 (2014)
22. R. Torres, N. Kajumba, J.G. Underwood, J.S. Robinson, S. Baker, J.W.G. Tisch, R. de Nalda, W.A. Bryan, R. Velotta, C. Altucci, I.C.E. Turcu, J.P. Marangos, Phys. Rev. Lett. **98**, 203007 (2007)
23. C. Vozzi, R. Torres, M. Negro, L. Brugnera, T. Siegel, C. Altucci, R. Velotta, F. Frassetto, L. Poletto, P. Villoresi, S. De Silvestri, S. Stagira, J.P. Marangos, Appl. Phys. Lett. **97**, 241103 (2010)
24. A.T. Le, R. Lucchese, M. Lee, C. Lin, Phys. Rev. Lett. **102**, 203001 (2009)
25. T. Kanai, E.J. Takahashi, Y. Nabekawa, K. Midorikawa, Phys. Rev. A **77**, 041402 (2008)
26. J. Levesque, Y. Mairesse, N. Dudovich, H. Pépin, J.C. Kieffer, P.B. Corkum, D.M. Villeneuve, Phys. Rev. Lett. **99**, 243001 (2007)
27. O. Smirnova, S. Patchkovskii, Y. Mairesse, N. Dudovich, D. Villeneuve, P. Corkum, M.Y. Ivanov, Phys. Rev. Lett. **102**, 063601 (2009)
28. S. Patchkovskii, Z. Zhao, T. Brabec, D.M. Villeneuve, Phys. Rev. Lett. **97**, 123003 (2006)
29. R. Torres, T. Siegel, L. Brugnera, I. Procino, J.G. Underwood, C. Altucci, R. Velotta, E. Springate, C. Froud, I.C.E. Turcu, S. Patchkovskii, M.Y. Ivanov, O. Smirnova, J.P. Marangos, Phys. Rev. A **81**, 051802(R) (2010)
30. C. Vozzi, M. Negro, F. Calegari, G. Sansone, M. Nisoli, S. De Silvestri, S. Stagira, Nat. Phys. **7**, 822 (2011)
31. K. Oda, M. Hita, S. Minemoto, H. Sakai, Phys. Rev. Lett. **104**, 213901 (2010)
32. F. Filsinger, J. Küpper, G. Meijer, L. Holmegaard, J.H. Nielsen, I. Nevo, J.L. Hansen, H. Stapelfeldt, J. Chem. Phys. **131**, 064309 (2009)
33. S. Fleischer, Y. Zhou, R.W. Field, K.A. Nelson, Phys. Rev. Lett. **107**, 163603 (2011)
34. E. Frumker, C. Hebeisen, N. Kajumba, J. Bertrand, H.J. Wörner, M. Spanner, D. Villeneuve, A. Naumov, P. Corkum, Phys. Rev. Lett. **109**, 113901 (2012)
35. P.M. Kraus, A. Rupenyan, H. Wörner, Phys. Rev. Lett. **109**, 233903 (2012)
36. P.M. Kraus, D. Baykusheva, H.J. Wörner, Phys. Rev. Lett. **113**, 023001 (2014)
37. L.S. Spector, M. Artamonov, S. Miyabe, T. Martinez, T. Seideman, M. Guehr, P.H. Bucksbaum, Nat. Commun. **5**, 3190 (2014)
38. P.M. Kraus, O.I. Tolstikhin, D. Baykusheva, A. Rupenyan, J. Schneider, C.Z. Bisgaard, T. Morishita, F. Jensen, L.B. Madsen, H.J. Wörner, Nat. Commun. **6**, 7039 (2015)
39. M. Qin, X. Zhu, Q. Zhang, P. Lu, Opt. Lett. **37**(24), 5208 (2012)
40. P.B. Corkum, Phys. Rev. Lett. **71**, 1994 (1993)

41. K.C. Kulander, K.J. Schafer, J.L. Krause, B. Piraux, *Proceedings of the Workshop, Super Intense Laser Atom Physics (SILAP) III* (Plenum Press, New York, 1993)
42. M. Lewenstein, P. Balcou, M.Y. Ivanov, A. L'Huillier, P.B. Corkum, Phys. Rev. A **49**, 2117 (1994)
43. L. Keldysh et al., Sov. Phys. JETP **20**, 1307 (1965)
44. P. Salières, P. Carrè, L.L. Dèroff, F. Grasbon, G.G. Paulus, H. Walther, R.K.A.W. Becker, D.B. Milošević, A.S.A.M. Lewenstein, Science **292**, 902 (2001)
45. P. Balcou, P. Salières, A. L'Huillier, M. Lewenstein, Phys. Rev. A **55**, 3204 (1997)
46. P. Salières, A. L'Huillier, M. Lewenstein, Phys. Rev. Lett. **74**, 3776 (1995)
47. S. Haessler, J. Caillat, W. Boutu, C. Giovanetti-Teixeira, T. Ruchon, T. Auguste, Z. Diveki, P. Breger, A. Maquet, B. Carré, Nat. Phys. **6**, 200 (2010)
48. O. Smirnova, Y. Mairesse, S. Patchkovskii, N. Dudovich, D. Villeneuve, P. Corkum, M.Y. Ivanov, Nature **460**, 972 (2009)
49. D. Faccialà, S. Pabst, B.D. Bruner, A.G. Ciriolo, S. De Silvestri, M. Devetta, M. Negro, H. Soifer, S. Stagira, N. Dudovich, C. Vozzi, Phys. Rev. Lett. **117**, 093902 (2016)
50. L.D. Landau, E. Lifshitz, *Quantum Mechanics: Non-Relativistic Theory*, 3rd edn. (Elsevier, 1981)
51. H.W. Kroto, *Molecular Rotation Spectra* (Dover Publications, 2003)
52. NIST Computational Chemistry Comparison and Benchmark Database, NIST Standard Reference Database Number 101. http://cccbdb.nist.gov
53. G. Gibson, J. Biegert, Phys. Rev. A **78**, 033423 (2008)
54. G. Doumy, J. Wheeler, C. Roedig, R. Chirla, P. Agostini, L.F. DiMauro, Phys. Rev. Lett. **102**, 093002 (2009)
55. A.T. Le, R.R. Lucchese, S. Tonzani, T. Morishita, C.D. Lin, Phys. Rev. A **80**, 013401 (2009)
56. M. Negro, M. Devetta, D. Facciala, S. De Silvestri, C. Vozzi, S. Stagira, Faraday Discuss. **171**, 133 (2014)
57. P. Salieres, A. Maquet, S. Haessler, J. Caillat, R. Taïeb, Rep. Prog. Phys. **75**, 062401 (2012)
58. B.K. McFarland, J.P. Farrell, P.H. Bucksbaum, M. Gühr, Science **322**, 1232 (2008)
59. A. Rupenyan, P.M. Kraus, J. Schneider, H.J. Wörner, Phys. Rev. A **87**, 031401 (2013)
60. B. Shan, Z. Chang, Phys. Rev. A **65**, 011804 (2001)
61. H. Yun, K.M. Lee, J.H. Sung, K.T. Kim, H.T. Kim, C.H. Nam, Phys. Rev. Lett. **114**, 153901 (2015)
62. E.V. van der Zwan, C.C. Chirilă, M. Lein, Phys. Rev. A **78**, 033410 (2008)
63. H.J. Wörner, H. Niikura, J.B. Bertrand, P. Corkum, D. Villeneuve, Phys. Rev. Lett. **102**, 103901 (2009)
64. M. Meckel, A. Staudte, S. Patchkovskii, D.M. Villeneuve, P.B. Corkum, R. Dorner, M. Spanner, Nat. Phys. **10**(8), 594 (2014)
65. W. Boutu, S. Haessler, H. Merdji, P. Breger, G. Waters, M. Stankiewicz, L.J. Frasinski, R. Taieb, J. Caillat, A. Maquet, P. Monchicourt, B. Carre, P. Salierès, Nat. Phys. **4**, 545 (2008)
66. X. Zhou, R. Lock, W. Li, N. Wagner, M.M. Murnane, H.C. Kapteyn, Phys. Rev. Lett. **100**, 073902 (2008)
67. C. Zhai, L. He, P. Lan, X. Zhu, Y. Li, F. Wang, W. Shi, Q. Zhang, P. Lu, Sci. Rep. **6**, 23236 (2016)
68. D. Pavičić, K.F. Lee, D.M. Rayner, P.B. Corkum, D.M. Villeneuve, Phys. Rev. Lett. **98**, 243001 (2007)
69. J. Bertrand, H. Wörner, P. Salières, D. Villeneuve, P. Corkum, Nat. Phys. **9**, 174 (2013)
70. H.J. Worner, J.B. Bertrand, D.V. Kartashov, P.B. Corkum, D.M. Villeneuve, Nature **466**, 604 (2010)
71. F. Calegari, D. Ayuso, A. Trabattoni, L. Belshaw, S. De Camillis, S. Anumula, F. Frassetto, L. Poletto, A. Palacios, P. Decleva, J.B. Greenwood, F. Martín, M. Nisoli, Science **346**, 336 (2014)
72. S. Ghimire, A.D. DiChiara, E. Sistrunk, P. Agostini, L.F. DiMauro, D.A. Reis, Nat. Phys. **7**, 138 (2011)

Chapter 8
Laser Ablation Propulsion and Its Applications in Space

Claude R. Phipps

Abstract Where lasers shine is in propelling a remote object using a space-based mother ship with an onboard laser. In some cases, there is no other reasonable choice. These cases include small low Earth orbit (LEO) debris reentry, large LEO debris nudging to avoid collisions, direct launch to LEO of small payloads at low cost and raising large geosynchronous (GEO) objects to graveyard orbits. We introduce the new, exciting idea of the laser rocket, in which a "burst mode" laser accelerates a 25-kg spherical probe surrounded by a discardable ablator layer to 3.6 km/s in minutes.

8.1 What Is Laser Ablation Propulsion and What Use Is It?

People often ask me what I'm working on and their eyes glaze over when I say "laser ablation." Ablation is easy to understand. If a sufficiently intense pulsed beam hits a surface, it makes a high-temperature jet. Momentum is mass times velocity, and if the jet velocity is high, only 8 nm of material are removed each laser shot. But that small mass times the tremendous jet velocity (60 km/s) gives a lot of momentum to the surface. If a 100 ps-pulse laser beam illuminates a spot 10 cm in diameter at 1 kHz, the applied force is about 10 N on aluminum, enough to lift 1 kg.

The "Why?" for this technology is that it offers propulsion of remote objects thousands of km distant, at the speed of light. Nothing else can do that. Also, we have the option of instantly varying exhaust velocity during flight, only a matter of changing laser beam intensity. Other advantages are (a) physical separation of the energy source from the fuel, avoiding engine erosion, (b) very small minimum impulse (1nN-s), (c) instant on-off operation and (d) jet velocity which can be high compared to chemical rockets because of the higher temperatures possible in intense

C. R. Phipps (✉)
Stanford University, Stanford, CA, USA
e-mail: crphipps@aol.com

C. R. Phipps
Photonic Associates, LLC, Santa Fe, NM, USA

P. M. Ossi (ed.), *Advances in the Application of Lasers in Materials Science*,
Springer Series in Materials Science 274,
https://doi.org/10.1007/978-3-319-96845-2_8

pulsed laser absorption in surface atoms. Applications include launching small probes into Low Earth Orbit (LEO) or on interstellar journeys from LEO, and space debris re-entry or repositioning.

8.2 Photon Beam Propulsion

The laser impulse coupling coefficient C_m is the ratio of momentum delivered to a target to the incident beam fluence (energy per unit area) for a laser pulse, or of surface pressure to incident intensity,

$$C_m = m_T \delta v_T / W = \delta \mu_E v_E / \Phi = p/I. \tag{8.1}$$

with dimensions N-s/J or N/W. C_m for pure-photon pressure is minute: the "momentum coupling coefficient" for photons reflecting off a polished surface is

$$C_m = 2/c = 6.7 \text{ nN/W}. \tag{8.2}$$

A 10-kW laser reflecting perfectly would produce a thrust of only 67 μN.

The other important parameter for any type of photon propulsion is propellant exit velocity, v_E, simply c for light, but $\sqrt{2kT/m_i} \ll c$ for laser ablation propulsion.

Conservation of energy requires that the efficiency of the whole process is

$$\eta = \psi C_m v_E / 2. \tag{8.3}$$

The parameter $\psi = 1$ for photons and may be larger for ablation propulsion, as we will discuss later.

The history of photon propulsion begins 70 years ago with Tsander [1], Tsiolkovsky [2] and Oberth [3], leading to today's "solar sails." In 1953, Saenger published his concept for photon rockets [4] well before the invention of lasers.

For very long trips, where time is available, solar sails represent a practical use of pure photon propulsion. At 1 kW/m^2 at our distance from the Sun, a 10-km diameter reflective sail will generate 540 N thrust. Using this thrust, a 3 μm, 250 ton Al reflective film with this diameter could accelerate to 3 km/s in 17 days. The main problem is how to deploy such a film. Despite decades of work, the largest sail yet deployed (JAXA IKAROS [5], 2010) is 14 × 14 m.

8.3 Laser Ablation Propulsion

For usefully large forces, for example, enough to counteract gravity or accelerate a several-kg object to orbital speeds in a reasonable time, photon propulsion alone is too weak. In 1972, Kantrowitz realized [6] that the phenomenon of laser ablation could generate coupling coefficients of order 100 N/MW to 10 kN/MW, four to six orders of magnitude larger than that for photons alone. Bunkin and Prokhorov [7] gave theory for the process in 1976.

Variable v_E is achieved by adjusting laser intensity on target—by changing focal spot area, laser pulse duration and energy—which causes exhaust velocity to vary across the range from chemical reactions (approximately 5 km/s) to much higher values easily reaching 50 km/s. This is because $v_E = (2kT_i/m_i)^{0.5}$ and 10,000 K ion temperatures T_i are readily created by a laser pulse. In short, v_E is only a matter of intensity [8]. Thrust can be varied independently of v_E by changing the laser pulse repetition rate.

After a new initiative of the Air Force Office of Scientific Research (AFOSR) and NASA, the first free flight of a 1997 laser rocket design by Myrabo [9] resulted in a flight altitude of 72 m in 2000 [10]. Delrin® (POM, a polymer) was utilized as the ablating propellant and still remains a most promising monopropellant, as we will see later. Similar investigations were reported from other countries (Russia in 1998 [11], Germany in 1999 [12], and Japan in 2001 [13].

8.4 Pulsed Laser Ablation Propulsion Details

Laser ablation propulsion operates, ideally in vacuum, by inducing a jet of vapor and plasma from a target using a high intensity laser pulse. Terminology is defined in our review of the field [14].

Ablation efficiency is defined as in (8.3), where

$$\psi = <v_x^2>/<v_x>^2 \tag{8.4}$$

It can be shown [15] that high intensity ablation plumes correspond to $\psi \leq 1.15$, so we will assume $\psi = 1$ for simplicity in discussing efficiency.

In (8.1), m_T is target mass, δv_T is the change in target velocity, W is pulse energy (J), p is surface pressure at the target, I is intensity (W/m^2), $\Phi = I\tau$ is fluence on target (J/m^2), v_E is exhaust velocity of the laser ablation jet and $\delta\mu_E$ is areal mass density (kg/m^2) in the ablation jet column created by one pulse. The change in velocity of the propelled target from a single pulse is (Fig. 8.1)

$$\delta\, v_T = C_m\, \Phi/\mu_T \tag{8.5}$$

and

$$\delta\, v_{T\|} = \eta_c C_m \Phi/\mu_T \tag{8.6}$$

In (8.5–8.6), μ_T is the target's areal mass density, η_c is an average geometrical efficiency factor taking account of the shape of the target and the fact that the ablation jet will be normal to each facet of its surface, not necessarily antiparallel to the laser beam. The quantity $\delta v_{T\|}$ is the change in target velocity in the beam direction.

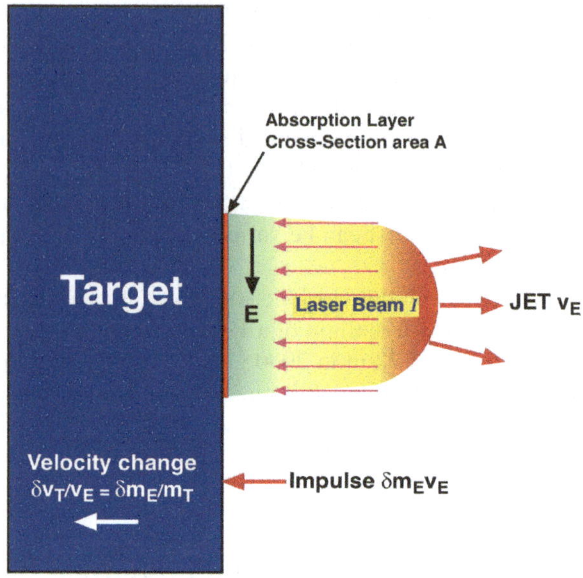

Fig. 8.1 Laser ablation impulse generation

Equation (8.6) is a numerically convenient formulation for space applications because we can deliver a fluence Φ to a region with diameter d_s and be sure that any object within that circle having mass density μ_T and the same η_c will gain the same velocity increment from that pulse. This is valid because space debris tend to exist in families with similar μ_T. For direct comparison to electric propulsion engines, the thrust to electrical power ratio is

$$C_{me} = \eta_{eo} C_m \tag{8.7}$$

Laser electrical-to-optical efficiency η_{eo} can range from 25 to 80%, depending on the laser type. Exhaust velocity can be determined from the product of the easily measured quantities C_m and Q (J/kg ablated) as follows. Where

$$Q = W/\delta\, m_T = \Phi/\delta\, \mu_T \tag{8.8}$$

and because $\delta\mu_T = \delta\mu_E$ by mass conservation, it can be seen dimensionally that the product $C_m Q$ must be exhaust velocity:

$$v_E = C_m Q. \tag{8.9}$$

Equation (8.3) can be extended to show that *ablation* efficiency is given by

$$\eta_{AB} = C_m v_E/2 = \delta\mu_E v_E^2/(2\,\Phi) = C_m I_{sp} g_o/2, \tag{8.10}$$

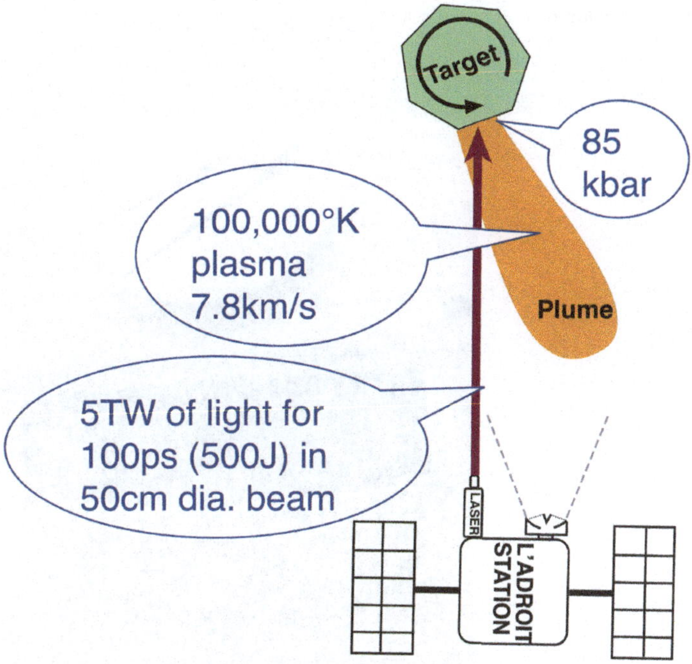

Fig. 8.2 The jet plume is always perpendicular to the surface facet. This figure gives typical parameters for the laser beam and its interaction with the target material

where g_o is the acceleration of gravity and I_{sp} is the so-called specific impulse. C_m and I_{sp} are a constant product in which I_{sp} varies inversely with C_m for engines with the same efficiency. The units of I_{sp} are seconds. The I_{sp} for light is 3.06E7 s. Another constant product is

$$C_m^2 Q/2 = \eta_{AB} \tag{8.11}$$

Because $\delta\mu_T = \rho_T \delta x$, the thickness of the target layer ablated in one pulse is (Fig. 8.2)

$$\delta x = C_m^2\, \Phi/(2\, \rho_T \eta_{AB}) \tag{8.12}$$

For example, with an aluminum target (density ρ_T =2700 kg/m^3), if C_m = 70 N/MW, Φ=35 kJ/m^2 and η_{AB} =1, δx=32 nm. At laser repetition frequency f=50 Hz, even in one minute operation, total ablation depth is just 95 μm. Here we assume a perfectly uniform beam on target, such as is achievable with modern methods of apodization.

In order to determine C_m, we need laser energy W on target and impulse J (N-s) delivered to it. J can be measured using deflection of a pendulum (Fig. 8.3),

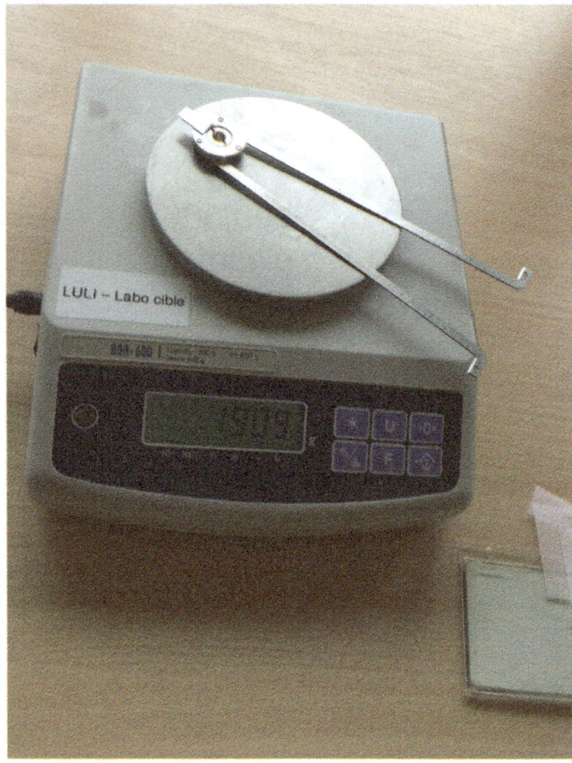

Fig. 8.3 Impulse pendulum

$$J = m_{\text{eff}}\{2g_o L[(1-\cos(\beta/2)]\}^{1/2} \tag{8.13}$$

In (8.13), L is the distance from the pendulum fulcrum to the point where laser impulse is generated. β is the deflection angle of a probe beam reflected from a mirror attached to the pendulum, twice the deflection angle θ. The period of a pendulum depends only on g_o and L, not on the mass, so that cannot be used to get impulse J.

Alternatively, one can use powerful laser interferometric techniques to get velocity directly. In either case the effective mass m_{eff} of the target-plus-pendulum must be known,

$$m_{eff} = \sum m_i L_i / L. \tag{8.14}$$

Finally, fuel usage rate is

$$\dot{m} = P C_m^2 / (2\eta_{AB}). \tag{8.15}$$

The parameter Q(J/kg) is equally critical to determining the ablation efficiency η_{AB}, which governs the effectiveness of a particular laser and laser ablation fuel.

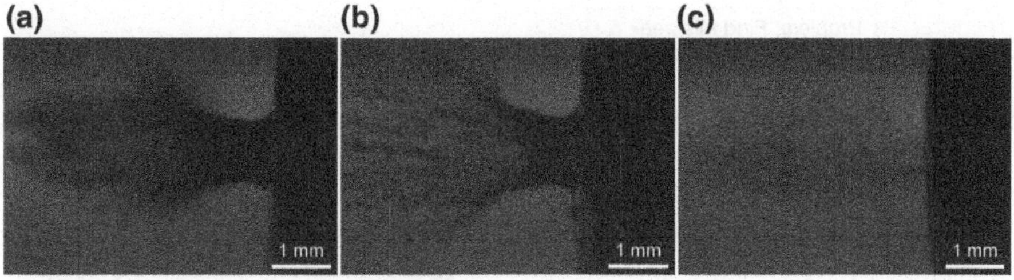

Fig. 8.4 Shadowgraph data from the Lippert group at Paul Scherer Institute shows splashing being inhibited as target viscosity increases from (**a**) to (**c**)

In principle, one may measure v_E with streak photography or Faraday probes to determine $Q = v_E/C_m$, but it is easy to miss a large mass fraction moving at very low velocity (splashing, Fig. 8.4) [16] with such techniques. Because it is difficult to measure ablated material mass with microgram accuracy from before-and-after target mass measurements, an alternative is to determine Q is from

$$Q = \Phi/(\rho_T\, \delta x) = 2\eta_{AB}/C_m^2 \tag{8.16}$$

by measuring the average depth δx of the ablation crater with profilometry or a similar technique. This doesn't work if the underlying material swells due to the laser pulse, a good reason to measure mass loss directly with a sensitive microbalance or by using repetitive pulses.

8.5 Optima

A first optimum is that fluence which gives maximum C_m. Figure 8.5 shows [17, 18] experimental and notional plots of C_m values versus incident fluence Φ to illustrate this optimum. In previous work, we called this maximum C_m value and the fluence at which it occurs C_{mopt} and Φ_{opt}.

For each mission, there is another kind of optimum which gives minimum energy cost to complete the mission. For example, for an Earth to LEO mission, this optimum is 100–300 N/MW [19] (Fig. 8.6). From the figure, it is clear that $C_m = 1000$ N/MW has an infinite cost for a 200 s flight (black dot at the top). In other words, the craft never reaches orbit!

In other cases, optimum performance means using an absolute minimum of ablation fuel mass, where high laser power is not a problem. This is illustrated in Fig. 8.7, from [19]. When mission duration is at a premium and laser energy is not, we have shown [20] that C_m as low as 70 N/MW is best for getting from LEO to Mars.

Our work in ablation propulsion is to find lasers and materials which achieve a desired optimum. Figure 8.6 from [19] shows how well this simple theory tracks simulations.

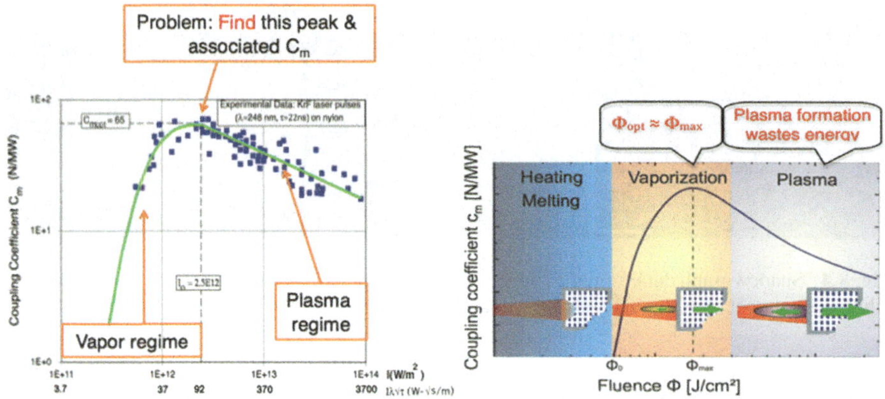

Fig. 8.5 Illustrating C_{mopt} and $\Phi_{opt} = I_{opt}\tau$. Clearly both vapor and plasma regime must be considered in theory to find the optimum

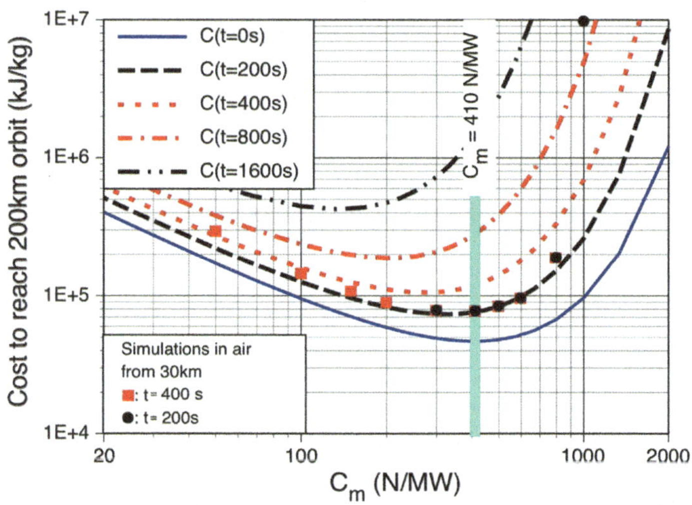

Fig. 8.6 Each mission has an optimum-cost impulse coupling coefficient. In the case studied, laser launch from 35 km to LEO, C_{mopt} is 300–500 N/MW for a 200 s flight. Lines are theory, dots are simulations for a real atmosphere. Flight time depends on laser time-average power

8.6 Why not CW?

For high efficiency in laser ablation propulsion, the laser must consist of repetitive, high intensity pulses [e.g., 20 kJ, 10 ps, 50 Hz] rather then being continuous (CW). There are several reasons for this recommendation [14]. First, high I_{sp} has not been demonstrated for CW lasers in vacuum. Second, our calculations show that intensities needed to achieve even $I_{sp} = 130$ s with CW ablation (about 1 GW/m^2) require a very

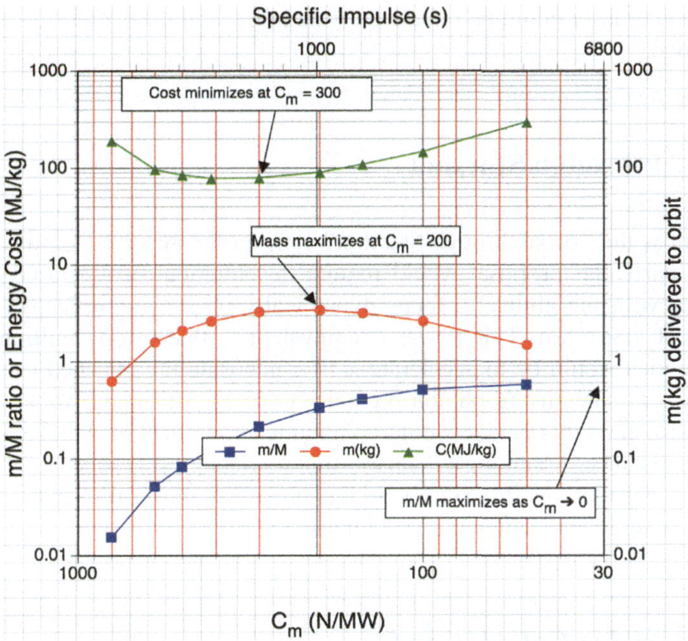

Fig. 8.7 Results of many simulations confirm the analysis, clearly showing that mass, mass ratio and cost optimize at different values of the coupling coefficient C_m. In particular, the mass ratio m/M of delivered mass m to launch mass M maximizes at $C_m = 0$ [$v_E = \infty$], mass delivered to LEO maximizes at $C_m = 200$ N/MW [$v_E = 1$ km/s] in the case studied, and cost minimizes at $C_m = 300$–400 N/MW. Here we assume $\eta_{AB} = 1$, so the exit velocity can be obtained as $v_E = 2/C_m$

small focal spot, e.g., 3.5 cm for a 1 MW laser, difficult at 200 km range or through atmospheric scintillations. In any case, CW ablation has a "welding torch" quality, generating lots of low-velocity splash [Fig. 8.5] which kills I_{sp} quickly compared to a 10 ps pulse stream, and CW laser thermal coupling will be disastrous. We will talk about this parameter later. Last, repetitive pulses give a chance for plasma clearing between shots so that refraction in the plasma jet doesn't interfere with propagation to the target. We do not emphasize CW beams for ablation propulsion in this chapter. When energy is cheap, it could play an important role.

However, nothing prevents using CW lasers to propel objects with photon pressure alone.

Because of fiber laser advances, CW lasers at 1060 nm are now readily available with 10 s of kW output power and reasonable beam quality.

Mason, et al. have proposed [21] using a 10 kW laser together with existing high-altitude telescopes (AMOS, PLATO, Mt. Stromlo) to produce displacements of a few 100 m/day for objects with $\mu \approx 10$ kg/m^2. It is worth noting, however, that the intensity delivered to the target in their proposal was only a few times the solar flux.

A higher power laser could be effective for nudging, and *laser ablation nudging* is very effective.

8.7 Breakthrough Starshot

Most recently, the "Breakthrough Starshot" concept has received venture funding. [22]. This particular proposal would require development of a 100 GW average power laser with the ability to focus dynamically on a 1 m^2 reflective sail over a distance of 4.8E6 km. This power is equivalent to that of one million 100 kW lasers. The idea would be to accelerate a 10 g nanosatellite attached to the sail to 0.27c in two minutes using photon pressure, to reach α-Centauri in 9 years. Left unanswered is how this device would communicate pictures or data back to Earth from such a distance. Assuming a 1 μm transmit wavelength, to exceed the microwave background, calculations show this would require a 1 diameter transmit antenna and a 1 km-diameter receive antenna [23] (Table 8.1).

8.8 Theory for Calculating C_{mopt}

Because of the Fig. 8.4 behavior, in which two completely different physical regimes—vapor and plasma—meet at the optimum fluence where C_m is maximum, two distinct calculations are necessary. In the vapor regime, increasing fluence produces increasing coupling which increases without limit, while in the plasma regime, the reverse is true. Then these are combined to give the value for C_{mopt} that we want.

8.9 Plasma Regime Theory for Ablation Propulsion

Earlier work [24] resulted in a first-principle theory for C_{mp} in the plasma regime, for laser pulses longer than 100 ps and common laser wavelengths.

$$C_{mp} = p/I = 1.84E-4\left[\frac{A^{7/16}}{2^{9/16}Z^{3/8}(Z+1)^{3/16}(I\lambda\sqrt{\tau})^{1/4}}\right]N-s/J, \text{ and} \tag{8.17}$$

$$I_{spp} = 442\frac{2^{9/16}Z^{3/8}(Z+1)^{3/16}(I\lambda\sqrt{\tau})^{1/4}}{A^{7/16}}s. \tag{8.18}$$

$$T_{ep} = 2.56\frac{A^{1/8}Z^{3/4}}{(Z+1)^{5/8}}\left(I\lambda\sqrt{\tau}\right)^{1/2} \tag{8.19}$$

In the limits of this theory, $C_{mp}I_{spp}$ =0.08, and η_{AB} =0.40. Z is the average ion charge state in the plasma plume,

Table 8.1 Breakthrough Starshot parameters

C_m (reflection)	$=2/c=6.7\text{E}{-}9$
Sail area A	1 m^2
Power on sail	1E11 W
F on sail	670 N
Speed	v = 0.27c
Total mass	0.001 kg
Acceleration	68,000G
Accel. Interval	120 s
Distance at cutoff	4.8E6 km (0.032 AU)
Speed at cutoff	0.27c
Gamma	1.03
Laser location	On Earth
Laser wavelength	1 μm
Required aperture	5.8 km
Assumed sail mass	0.0005 kg
Thickness if Al	185 nm (650 atoms)
Al tensile strength	3E8 N/m^2
Assumed sail radius of curvature	5 m
Sail stress under pressure	9.1E9 N/m^2
Ratio of sail stress to Al tensile strength	30
Allowable heat absorption	1000 W/m^2
Needed absorptivity	1.0E-8
Goal	α Centauri
Arrival	9 years
Assumed receive aperture d_R	1.0 km
Assumed data wavelength	1 μm
Assumed data BW	0.2 μm
Assumed transmit power	1 W
Transmit dia. to match d_R	1 m
CMB noise into receiver	2.7 $\mu W/m^2/\mu m$/sterrad
Signal	9.0E-17W
S/N (signal to noise ratio)	40f

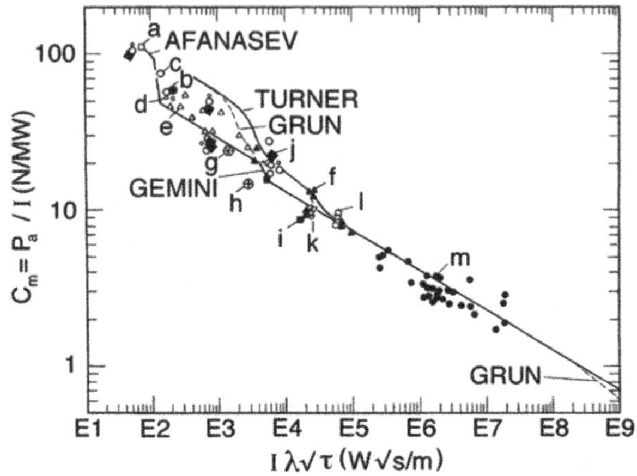

Fig. 8.8 Plasma regime data agreed with theory

$$Z = n_e / \sum_{j=1}^{j\max} (n_{ij}) \gg 1. \tag{8.20}$$

Electron density is determined by the Saha equation,

$$\frac{n_e n_j}{n_{j-1}} = \frac{2u_j}{u_{j-1}} \left(\frac{2\pi\, m_e k T_{ep}}{h^2} \right)^{3/2} \exp\left(-W_{j,j-1}/kT_{ep}\right) \tag{8.21}$$

Z is not the same as the ionization fraction $\eta_i = n_i/(n_o + n_i) \le 1$, a function of neutral density n_{o} and ion density n_{i}.

Data from 47 datasets from the plasma regime available at that time agreed well with the plasma regime theory (Fig. 8.8).

The trick was to plot the data versus the strange parameter $I\lambda\sqrt{\tau}$, based on the functional aspects of the theory, a parameter "halfway" in its character between an intensity and a fluence. Plotting versus other variables gives a scattered result. Bumps in the figure are actually due to changing Z for simulations of each material.

8.10 Vapor Regime Theory

There are two approaches to modeling the vapor regime. The first uses tabulated pairs of pressure and temperature (p, T) from SESAME tables for some elements [25]. By equating laser intensity to energy sinks in the vapor regime, we obtain

$$I = \frac{pv}{a}\left(\frac{\gamma}{\gamma - 1}\right)\left[1 - \frac{T_o}{T} + \frac{q}{C_p T} + \frac{\gamma - 1}{2}\right] + \frac{\sigma\varepsilon}{a} T^4 + f(\tau) \tag{8.22}$$

where

$$f(\tau) = \frac{1}{a}\left[\phi(T, x_h) + \frac{x_h \rho_s C_v (T - T_o)}{\tau}\right]. \tag{8.23}$$

In (8.22) and (8.23), a is total absorption fraction of the target (not a coefficient), and ϕ is a flux limiter from inertial confinement fusion theory. We can relate the quantity p in (8.16) to T by using the Riedel equation [26] in conjunction with the SESAME equation-of-state database (e.g., for Al) maintained at Los Alamos National Laboratory for $T \leq 7890$ K, its triple point.

Equations (8.21) and (8.22) are wavelength-dependent only as λ affects the surface absorptivity a. For the infrared to ultraviolet range studied here, we have used $0.05 \leq a \leq 0.24$ for modeling aluminum [27]. Absorptivity is different from the room temperature value.

We now have a numerical solution which relates p_v and v to I over the range corresponding to our $p(T)$ data, and then compute the vapor regime coupling coefficient as $C_{mv} = p_v/I$.

Another approach is used where ablation threshold Φ_o is well-known but the (p,T) pairs are not available [28]. In this case, where $\xi = \Phi/\Phi_o$, and α [absorption coefficient (m^{-1})], is different from a in [21]. We obtained

$$C_{mv} = \sqrt{\frac{2\rho\, C^2(\xi - 1)\ln \xi}{\alpha\, \Phi_o \xi^2}} \quad \text{and} \tag{8.24}$$

$$I_{spv} = \sqrt{\frac{2\alpha \Phi_o(\xi - 1)}{\rho\, g_o^2 \ln \xi}} \tag{8.25}$$

C is a fit parameter derived by matching ablated mass density data to the expression

$$\mu = (\rho/\alpha)\ln(C\,\xi) \quad \text{kg/m}^2 \tag{8.26}$$

It is interesting to note that the $C_{mv}I_{spv}$ product from (8.19) and (8.20) give $\eta_{AB} = (g_o/2)C_{mv}I_{spv} = (2C/g_o)(1-1/\xi)$, a function which approaches $2C/g_o$ as $\xi \to \infty$. The coupling coefficient in (8.24) maximizes at $\Phi_{opt} = 4.2\ \Phi_o$, 4.2 times the threshold.

It is important to note the two quite different temperatures used in this analysis. T in, e.g., (8.22)–(8.23) is target temperature, of order 1000 K, while T_{ep}, e.g., in (8.19) and (8.21) is the plasma plume temperature and can be 100,000 K.

8.11 Combined Theory

As the caption to Fig. 8.4 points out, both vapor and plasma theory must somehow be combined to find the maximum C_{mopt} and its associated fluence. This is because

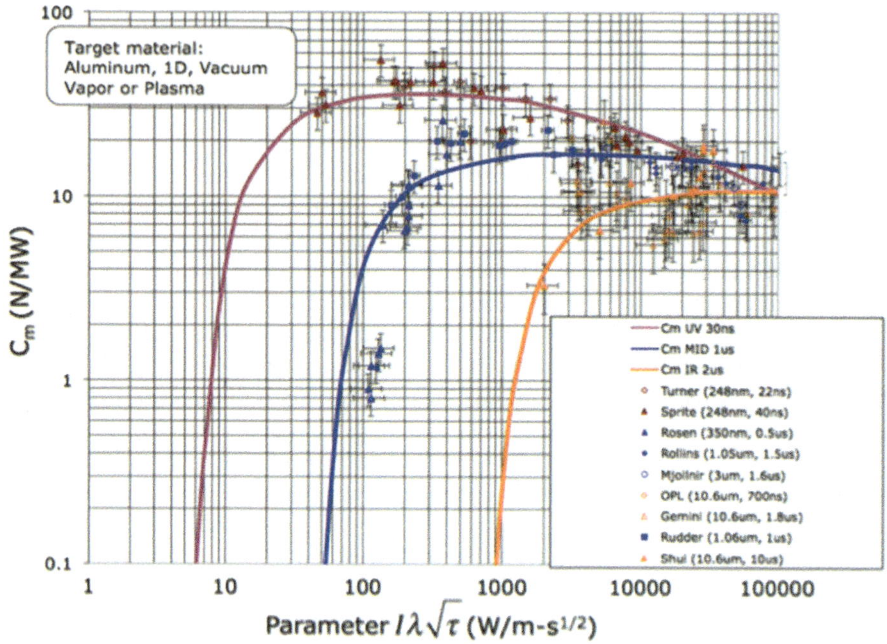

Fig. 8.9 Fitting coupling data with combined theory. Legend refers to data identified in [13]

the plasma regime C_m prediction (8.17) is infinite for zero intensity. We chose to do this with the (8.27) heuristic:

$$C_m = p/I = [(1-\eta_i)p_v + \eta_i\, p_p]/I = (1-\eta_i)\, C_{mv} + \eta_i\, C_{mp}. \tag{8.27}$$

This combination yielded a smooth transition between the two models by attenuating the vapor contribution and emphasizing the plasma contribution as ionization becomes complete using the ionization fraction η_i as a weighting function.

Combined theory specific impulse can be obtained in the same way. The combination yielded reasonable fitting of actual coupling data [29], including the C_{mopt} peak (Fig. 8.9).

It is tedious to use this method [30], because of the number of iterations required to get Z for each $I\lambda\sqrt{\tau}$. Z is a function of T through the Saha equation [31], which determines the relative population of atomic excited states. The procedure is explained in detail [25] in our 2012 paper.

An alternate approach which has less accuracy but permits rapid system design is given by Phipps and Bonnal [32]. Combining (8.17) with the trend $I\sqrt{\tau} = \mathrm{B}$ which applies for $\tau > 100$ ps gives a practical expression:

$$C_{mopt} \approx \frac{f(A,Z)}{I\lambda\sqrt{\tau}^{1/4}} = \frac{f(A,Z)}{B^{1/4}\lambda^{1/4}} = \frac{C_{exp}(A,Z)}{\lambda_{\mu m}^{1/4}} N/W. \tag{8.28}$$

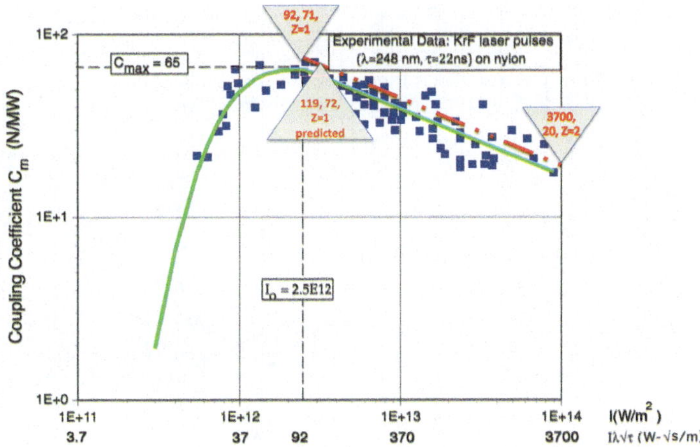

Fig. 8.10 The (8.23) approach gives the value of peak C_m *accurately*, for 22 ns pulses and its $I\lambda\sqrt{\tau}$ coordinate within a few percent

dependent only on A, Z and λ. Equation (8.28) gives a good first estimate of what C_m will be. Using $B = 4.8E8\ W\text{-}s^{0.5}/m^2$ at optimum fluence,

$$\Phi_{opt} = 4.8E8\sqrt{\tau}\ J/m^2. \tag{8.29}$$

and the function C_{exp} is given by

$$C_{exp}(A, Z) = 27\ A^{7/16}\ Z^{-3/8}(Z + 1)^{-3/16}\quad N/MW. \tag{8.30}$$

Equation (8.29) results from the fact that optimum fluence lies just above the optical breakdown threshold of the material, which varies with $\tau^{0.5}$ in 1-D thermal transfer problems [15]. Z (usually 1 or 2) can be estimated from experimental data and ionization state temperatures for a target material. This approach can give surprisingly good results (Fig. 8.10). In this case, for a 248 nm pulsed laser on nylon [33]. Because nylon has the formula $(C_{12}H_{22}N_2O_2)_n$, the correct value for the average value of A for the nylon example [in (8.30)] is 5.94.

8.12 Ultrashort Pulses

The theory on which the preceding sections are based is not valid for pulses shorter than 100 ps. To a greater and greater extent, applications we are considering use exactly that regime. Until recently, very little data existed for ultrashort pulses. For pulses in the 130–500 fs at 800 nm wavelength on Al, simulations by Fournier [34] for 530 nm and 20 ps gave $C_{mopt} = 155$ N/MW. On the other hand, other results [35,

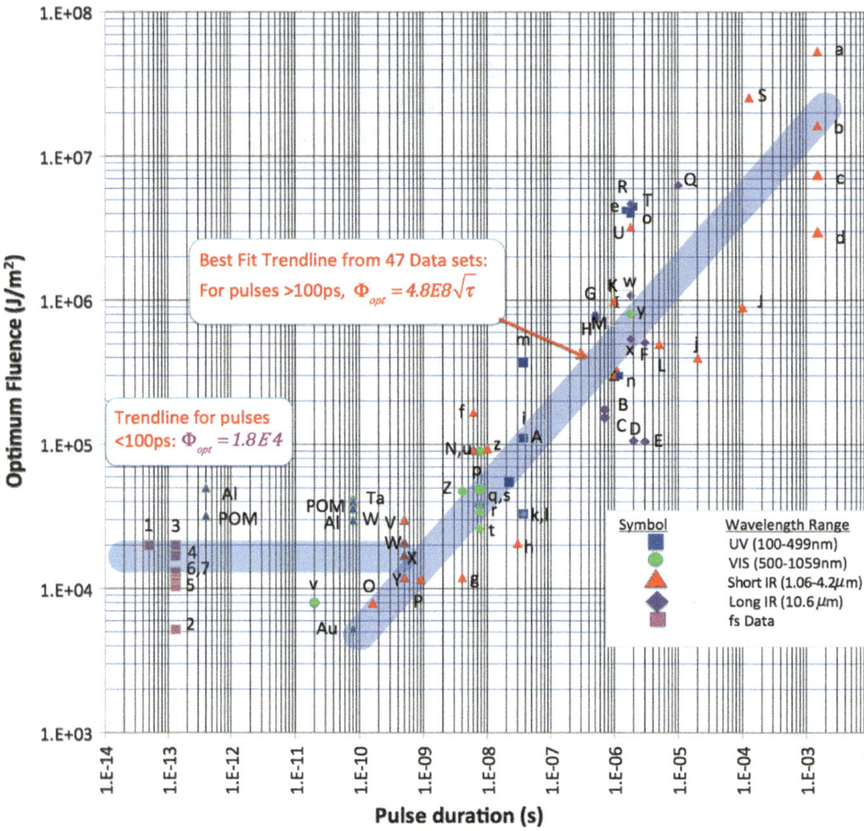

Fig. 8.11 Optimum fluence across the range 10 fs to 10 ms. New data points are blue triangles, labeled for the materials POM, Ta, W and Al. Pink squares to the left are earlier fs data. Optimum fluence for ultrashort pulses is approximately constant

36] give $C_{mopt} = 35$ N/MW at 800 nm, considerably less. Scharring's simulations for circularly-polarized 1064 nm light [37] agree with this value for C_{mopt}. Figure 8.11 shows that optimum fluence for ultrashort pulses remains approximately constant below 100 ps. A complete listing of references for the data labeled with letters [14, 38, 39] would be confusing here. However, the data ranges from ms to sub-ns duration, and from 100 nm to 10.6 μm wavelength, and still follows the $\tau^{0.5}$ trend of (8.24) within a rms deviation equivalent to a factor of 1.5. This is because Φ_{opt} is usually equivalent to surface damage fluence, which is a thermal phenomenon (Table 8.2)

New data points for Al, POM, Ta, W and Au are the blue triangles [40].

In the short pulse regime, Fig. 8.12 shows the scatter in C_m values [17, 35, 36, 40]. The upward-slant of the trendline in the figure is due to inclusion of data at 400 fs

Table 8.2 New values for C_m and Φ_{opt} for short pulses

Pulse duration (s)	400 fs	Φ (kJ/m^2)	80 ps	Φ (kJ/m^2)
POM	125±12	32±6	773±70	40±8
Al	28±3	50±20	28±3	30±7
W	31±3	260±40	36±7	34±6
Au	–	–	37±4	5.3±1
Ta	–	–	29±3	42±8

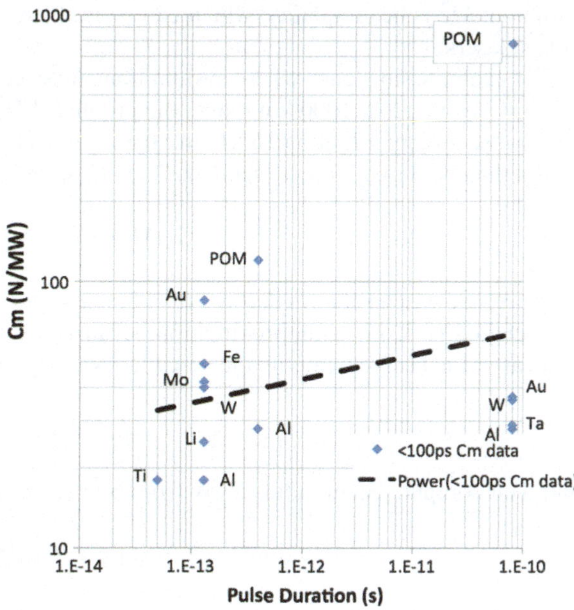

Fig. 8.12 C_m comparison. 800 nm data at 50 fs, 130 fs, 400 fs and 80 ps data. The trendline is only a guide, not based on theory

and 80 ps from the very atypical material POM, which has played a major role in 10-μm laser propulsion concepts, and is destined to do so at 1 and 0.5 μm.

The results we obtained allow making a mixture of Al powder and POM to obtain 300 N/MW, or any other value we want in the range from 30 to 770 at 80 ps. The required fluence (~30 kJ/m^2) is about the same for both materials.

8.13 Diffraction and Range as They Affect Space System Design

The goal in laser space propulsion is to get the optimum fluence—whatever it is for a particular problem—delivered to the target located at range z. The product of pulse energy and squared effective aperture diameter WD_{eff}^2 transmitting aperture required to deliver fluence Φ at wavelength λ to a target at range z is given by

$$WD_{\mathrm{eff}}^2 = \frac{\pi\, M^4 a_d^2 \lambda^2 z^2 \Phi}{4T_{\mathrm{eff}}}. \tag{8.31}$$

Diffraction causes beam spread with distance that increases linearly with wavelength and range and decreases with aperture diameter. In (8.31), a_{d} is a diffraction parameter which is 2.44 for a hard aperture and $4/\pi$ for a Gaussian radial intensity variation. M^2 is beam quality, which can be in the range 1.5–2.0 for well-designed pulsed lasers.

For example, in a case where system transmission $T_{\mathrm{eff}} = 0.8$, in order to deliver 8 kJ/m^2 to a target at 1000 km range, the product WD_{eff}^2 must be at least 140 kJ-m^2, laser pulse energy might be 15 kJ, and the necessary effective mirror diameter D_{eff} would then be 3 m, giving a "spot size" d_{s} on target

$$d_{\mathrm{s}} \approx a_{\mathrm{d}} M^2 \lambda z / D_{\mathrm{eff}}. \tag{8.32}$$

Of course we can always defocus to obtain a *larger* spot size than given by (8.32).

Equation (8.32) is an approximation because it is not accurate when $d_{\mathrm{s}} \approx D$. In that case, it is best to compute the so-called Rayleigh range z_{R}

$$z_{\mathrm{R}} = \pi\, w_0^2/(M^2\lambda) \tag{8.33}$$

and then compute beam radius away from best focus as

$$w^2(z) = w_{\mathrm{o}}^2\left[1 + (z/z_{\mathrm{R}})^2\right] \tag{8.34}$$

(see Fig. 8.13). When range z is such that $2w = D_{\mathrm{b}}$, you're done with that part of the optical system design.

8.14 Thermal Coupling with Repetitive Pulses

The *thermal* coupling coefficient $C_{\mathrm{th}} = W_{\mathrm{th}}/W$ is the fraction of incident laser energy that ends up as heat in the irradiated material, by whatever route. These include surface plasma reradiation and absorption, ordinary thermal conduction and mechanical stresses. It is perhaps the most important parameter that we do not know very well.

Thermal coupling data, at whatever wavelength, show a peak value of C_{th} at relatively low fluence, followed by a dramatic decrease as fluence rises toward large values, because of plasma shielding.

C_{th} is a critical parameter, because many of our applications require hundreds of thousands of pulses and even if $C_{\mathrm{th}} = 1\%$, temperature in a 1 mm sheet of aluminum can quickly reach the melting point.

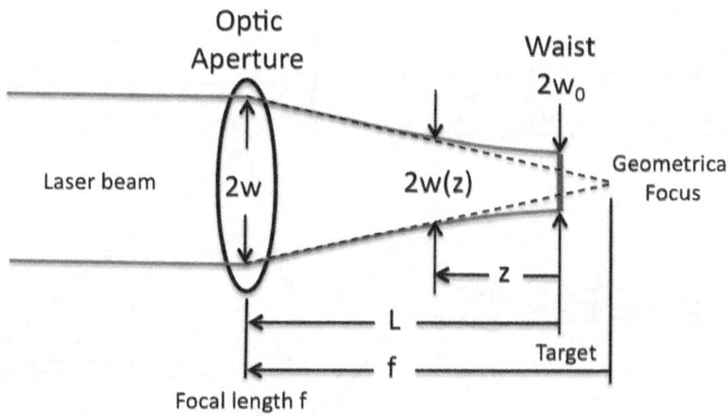

Fig. 8.13 Focusing a laser beam using an optic with aperture $D_b = 2w$

This suggests working with fs pulses, since observation of clean post-pulse features indicate low thermal coupling. Literature is full of photos of clean holes, precise ablation and no surrounding melt on targets with ultrashort pulses. However, precise C_{th} data is not available. Our instinct is to use fs pulses for minimum C_{th}, but do we really need fs pulses to be sure C_{th} is small enough?

Fortunately, Scharring has completed simulations [37] showing how C_{th} should vary with pulse duration (Fig. 8.14). The figure shows a strong advantage for ultrashort pulses as regards thermal coupling. It is clearly seen in the Figure that 10 ps single pulses at a fluence of 30 kJ/m^2 (3 J/cm^2—near optimum from our point of view) are expected to produce C_{th} of about 4% [see the "practical case" below], but longer pulses can produce C_{th} values as large as 30%.

Weber [41] shows that the maximum tolerable incident laser power $\mathrm{P_L}$ in an infinite string of pulses is given by

$$P_L = \frac{\Delta T_{\max} c_{\mathrm{FOM}}}{\sqrt{f_L} a} \tag{8.35}$$

where

$$c_{\mathrm{FOM}} = \frac{\rho\, c_p \sqrt{(4\pi\kappa)^3}}{5.2} \tag{8.36}$$

Taking aluminum as an example of target material and using $T_{\max} = T_{\mathrm{melt}} = 966$ K and other parameters for aluminum, $c_{\mathrm{FOM}} = 10.8$ and absorptivity $a = C_{\mathrm{th}} = 1.42\%$, we find P_L=73 kW for maximum power in a continuous pulse string. Because the laser we desire for propulsion is transient rather than continuous, the laser rocket example following is still realistic, even if the average power is higher than given by (8.35).

Other materials would give different results.

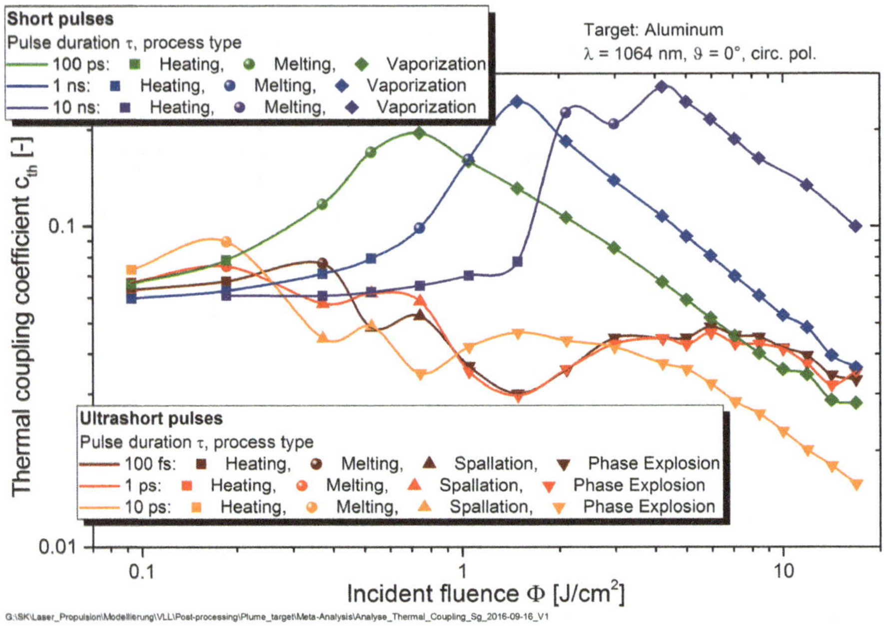

Fig. 8.14 Simulation: thermal coupling on aluminum at 1064 nm, normal incidence, circular polarization. Stefan Scharring, used by permission

8.15 Practical Case: Thermal Coupling for a Laser Rocket

In the laser rocket problem, treated in the next section, where the propelled object is a 25 kg sphere, 293 k pulses of 5 kJ are incident on the target, for a total 1.45 GJ [1.2 MW average] during an 1100 s flight. Ignoring reradiation [about 2% of incident power at T_{melt} with black body opacity], C_{th} as large as 1.4% could melt the craft. We can see that measurement of C_{th} is critical. At this time, there are no such measurements for a continuous string of ultrashort pulses.

8.16 Applications

8.16.1 Interplanetary Laser Rocket

An exciting new application is accelerating the payload in a 25 kg spherical aluminum probe from LEO to 3.6 km/s in 18 min, sufficient to achieve aphelion at Mars, with 27% mass loss (Figs. 8.15 and 8.16; Table 8.3).

Table 8.3 Parameters for Laser Rocket

Mission	Interplanetary	LEO orbit change
Parameter		
Laser wavelength	355 nm (Nd 3rd harmonic)	355 nm (Nd 3rd harmonic)
Pulse duration	100 ps	100 ps
Pulse energy	5 kJ	500 J
Probe diameter d_s	89 cm	28 cm
Probe material	Al	Al
Probe shell thickness ablated	2.7 mm	0.66 mm
Probe ablator shell mass	15 kg	15 kg
Probe mass after shell discarded	10 kg	10 kg
Pulse fluence on probe	8 kJ/m^2	8 kJ/m^2
Pulse repetition frequency	400/100 Hz	200 Hz
Pulse interval	2.5/10 ms	5 ms
Average power	2 MW/500 kW	100 kW
Optical energy in burst	1.45GJ	355 MJ
Stored in batteries	3.9GJ	1.5GJ
Battery mass	4000 kg	1520 kg
Station total mass	50,000 kg	8000 kg
Time to recharge	<1 day	<1 day
Recharge power	100 kW	17 kW
Beam quality M^2	2	2
Mirror diameter D	3 m	3 m
Rayleigh range	880 km	880 km
Used optical range	2570 km	68 km
Required aiming precision	35 nrad	4 μrad
Thickness ablated/shot	38 nm	38 nm
Assumed ablation efficiency	40%	40%
Specific impulse	816 s	816 s
Coupling coefficient C_m	70 N/MW	100 N/MW
Δv/shot	3 cm/s typ.	1.3 mm/s
Acceleration	2.6–5.3 m/s^2	0.27 m/s^2
Final probe Δv	3.6 km/s	190 m/s
Max allowable C_{th}	1.35% (incl. reradiation)	5.5% (incl. reradiation)

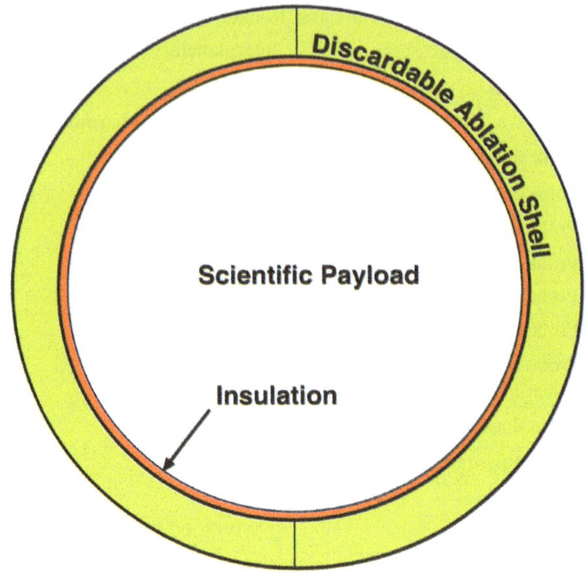

Fig. 8.15 Laser-propelled interplanetary flyer construction

In this case we create an Al/POM mixture of 5% POM and 95% Al with $C_m =$ 70 N/MW for long propellant life. Density of the combination is 2640 kg/m^3, only slightly different from that of Al.

Maximum laser range is 1650 km. We keep the laser spot size fixed at the diffraction-limit value of 9 cm at maximum range throughout the flight, by defocusing the beam as necessary at shorter range. We use a spherical probe because Δv will always be along the laser beam axis so long as the object is centered in the beam. The probe consists of a 10 kg payload surrounded by a discardable ablator shell 2 mm thick with an internal thermal insulating layer to protect instruments during acceleration. The flyer is spinning about an axis perpendicular to the laser beam and slowly precessing so as to expose the entire surface to the laser beam. Afterward, the shell is jettisoned.

Another application is direct launch of small spacecraft from Earth to LEO using lasers. This needs MW of average power to achieve useful results. But such lasers are just around the corner.

Best efficiency is obtained from "burst mode" operation of the laser in which the laser operates at 2 MW average [5 kJ, 400 Hz] for 380 s and 500 kW for 380 s, for a total of 12.7 min. Waste heat during the burst is stored onboard the mother ship and reradiated later. We averaged σT^4 over the flight assuming a linear temperature rise to determine reradiation, which was always $<2\%$ of the total during laser illumination. Table 8.3 shows the details. The laser is more powerful than the L'ADROIT laser [42] but otherwise has similar parameters. It is powered from a battery which recharges in a day from a 100 kW array [43].

Figure 8.17 shows distance s(t), velocity v(t) and acceleration a(t) during the 18-min flight.

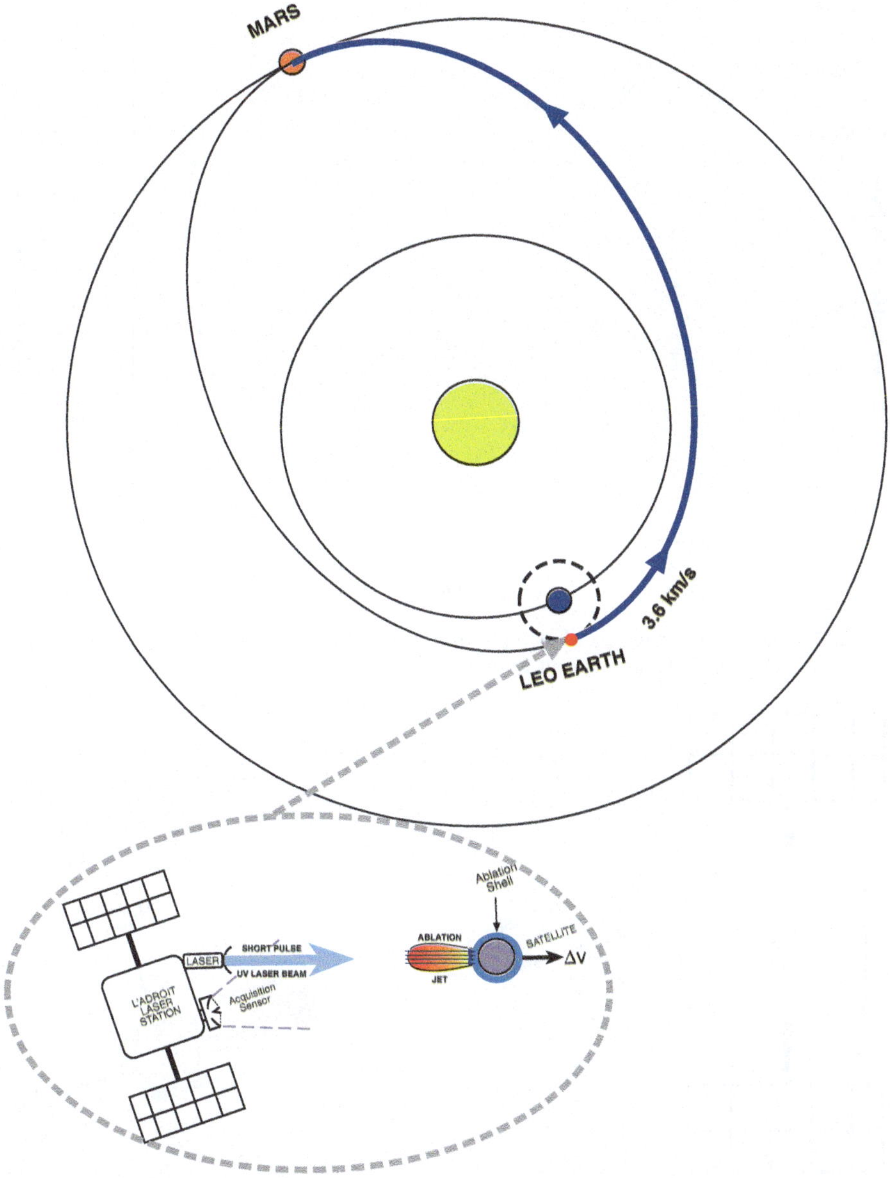

Fig. 8.16 A cis-Mars orbit starting from LEO requires $\Delta v = 3.6$ km/s

8.16.2 L'ADROIT

Remote action at the speed of light is the unique advantage of laser ablation propulsion. The L'ADROIT concept (Fig. 8.18) can generate thrust on a target 2000 km

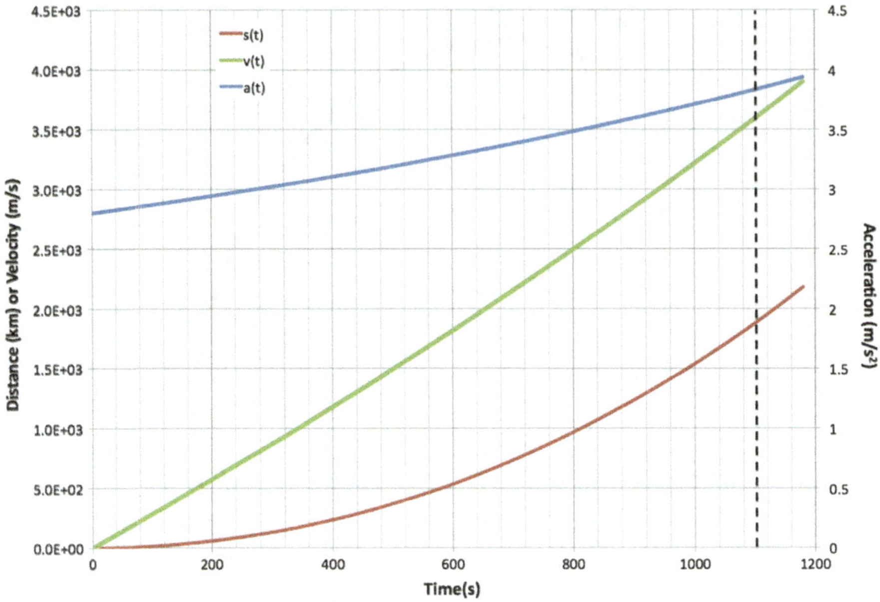

Fig. 8.17 Parameters for the cis-Mars flight from LEO. <P> = 1.25 MW [250 Hz, 5 kJ], C_m = 70 N/MW, achieving 3.6 km/s in about 18 min

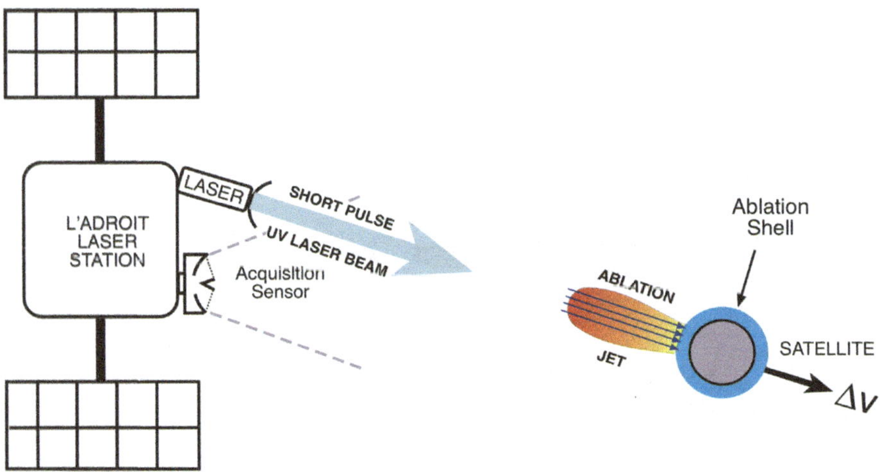

Fig. 8.18 Parameters for L'ADROIT. <P> = 21 kW [50 Hz, 400 J], causing re-entry of small LEO debris in 10 s

distant without having to be adjacent to it. This is a unique way to re-enter even tiny space debris, nudge satellites already in orbit to avoid predicted collisions, and lift derelict craft in geosynchronous orbit into higher, graveyard orbits.

L'ADROIT [42] (Laser Ablative Space Debris Removal by Orbital Impulse Transfer) was first presented at a Paris workshop in 2014 (Fig. 8.18). The laser is also operated in burst mode from batteries storing solar energy.

The version that would lift derelict geosynchronous (GEO) satellites into graveyard orbits in a day [8 kW, 3 kJ, 2.5 Hz] could raise 10 GEO satellites by 300 km in 3 months to a year, making valuable parking places in GEO available. It is a fact that 70% of objects in GEO are uncontrolled junk [32, 44].

Both laser systems operate at 355 nm wavelength (3rd Nd harmonic) with 100 ps pulses using monolithic diode-pumped Nd media. Laser electrical efficiency is about 25%.

8.16.3 Something Good for the Environment

We recently reported [45] that it is possible to directly launch small satellites to LEO from the Earth with a large ratio *m*/*M* approaching 60% of mass m delivered to orbit compared to the mass *M* on the ground. It requires very high time-average laser powers of 5–15 MW from kHz pulse trains of 100 ps duration at very high intensity on target (35 kJ/m^2), and a design like that in Fig. 8.15. Further, it requires a two-step launch process consisting of one power burst to get the satellite above the atmosphere, followed by a burst of 2–3 times normal power during about 60 s at the end of the flight in order to minimize excessive radial velocity, which can cause the satellite to crash at perigee (Fig. 8.19).

This technology will allow us to launch a fleet of small Earth-observing satellites in order to more carefully monitor global climate change and its consequences, in order to spot trends at the earliest possible time and to develop very highly detailed global models. Another application is to sending inspection craft to geosynchronous (GEO) orbit.

8.16.4 Fiber Laser Arrays Versus Monolithic Solid State Lasers

Giant arrays of pulsed fiber lasers are attractive alternative to monolithic laser media. The geometry of fibers makes diode pumping and heat removal more efficient. Fibers have about 100 times the energy output per kg of laser medium as monolithic solid state lasers. Figure 8.20 shows the "ICAN" [International Coherent Amplifier Network] idea [46].

However, for laser space propulsion applications, fibers have one major drawback. Progress has been slow on phasing large numbers of pulsed optical fibers (64 to date), and at least 25,000 phased fibers would be needed for ICAN to effectively deliver momentum from 100 ps pulses to objects at a distance of 100 km [47]. This is because

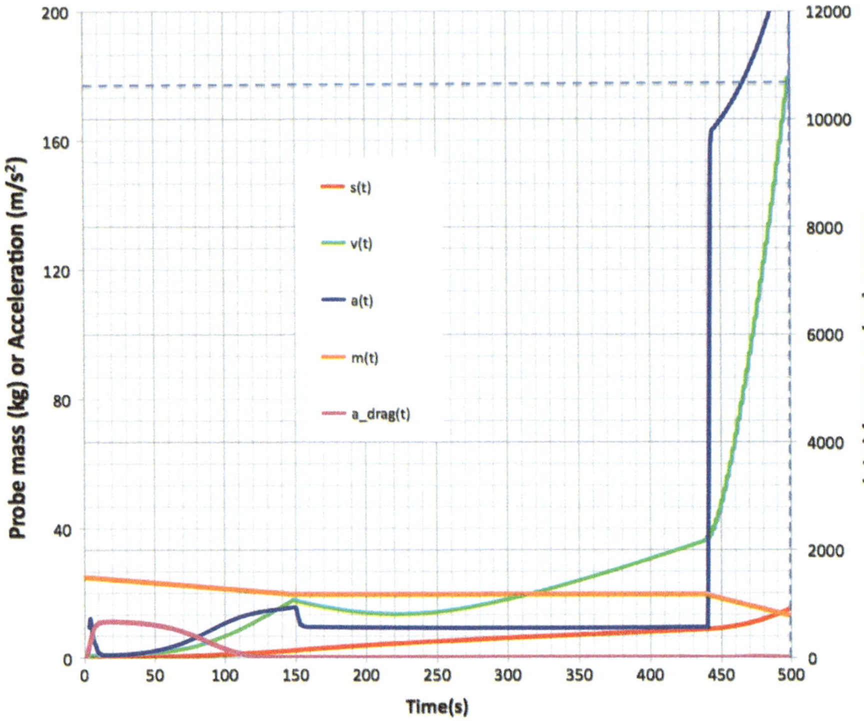

Fig. 8.19 Case 11B. Fascinating launch from ground (1 km altitude) with $C_m = 120$ N/MW. A 300 s coast followed by a 15 MW burst during 60 s at the very end gets our craft into orbit. The coast which begins at 150 s allows the craft to develop downward vertical velocity sufficient to counteract unavoidable vertical components from a groundbased laser at the end of a 1000 km flight around a curved Earth. This profile gives *m*/*M* delivered to orbit of 54% (13.5 kg). Initial zenith angle 45.5°, final zenith angle 90°. Final radial velocity is 389 m/s, final velocity 10.4 km/s. Perigee 104 km, apogee 41,700 km, 117% of geosynchronous altitude. Insertion slope is 2.12°. Minimal energy is wasted in drag, even though we are starting from the ground

we need a pulse energy of 100 J at 100 ps for this operation [to see this, apply (8.31) with $\Phi = 30$ kJ/m^2, z = 100 km, $\lambda = 355$ nm and reasonable mirror sizes]. However, a single fiber can deliver only 4 mJ without catastrophic failure due to self-focusing and other serious nonlinear optical effects under these conditions. These effects include stimulated Brillouin scattering, stimulated Raman scattering, photo-darkening and four-wave mixing.

Going from 4 phased fibers to 25,000 will be a difficult jump.

A first step to testing fibers for application to the debris problem will be forthcoming tests of debris reëntry using the EUSO telescope on board the Space Station, together with one of the ICAN fiber lasers [48].

We have learned never to say something is impossible!

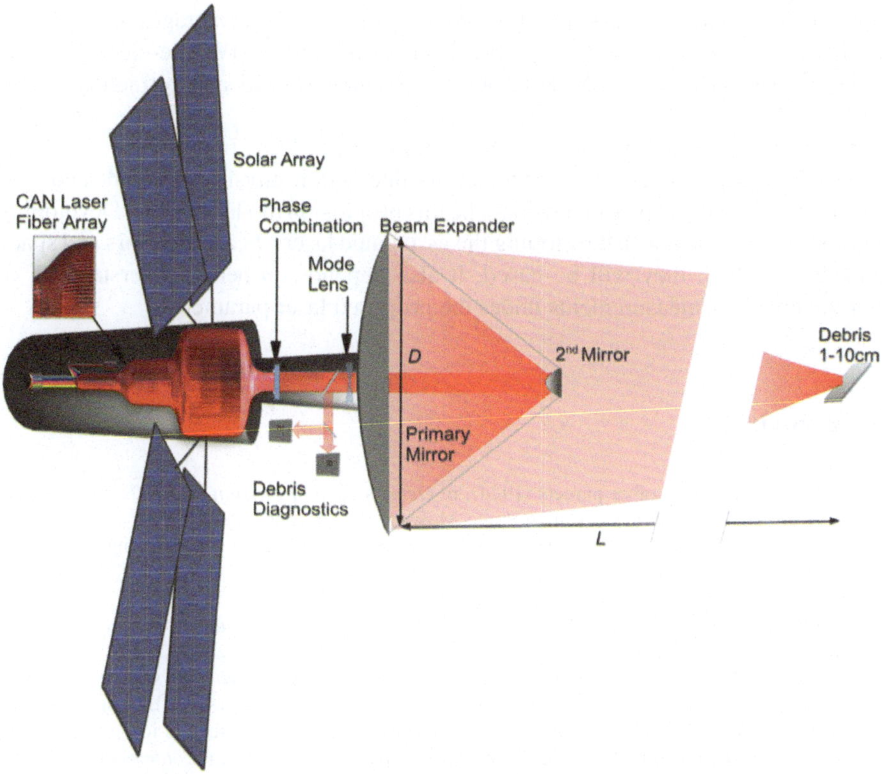

Fig. 8.20 ICAN concept

8.16.5 *Repetitive Pulse Monolithic Diode Pumped Solid State Lasers*

The state of the art in the lasers we currently need to achieve all of these applications is represented in the HiLASE program [49], where the Rutherford Appleton Laboratory's "DiPOLE 100" laser achieved its full design performance of 1 kW average power with 10 Hz, 100 J pulses at 10 ns pulse duration. Just a factor of 10 increase in output power will enable many of our applications in space.

8.16.6 *Perspective*

In this brief review, we have discussed one of the most exciting future applications of pulsed lasers. Laser ablation propulsion can re-enter space debris, nudge large satellites out of harm's way, and create a fleet of Earth-observing satellites and lift derelict GEO stations to parking orbits. We discussed the disadvantages of CW lasers for this task. However, it is also possible that new ideas for 1 MW free-electron lasers with a format consisting of 1 MHz 1 J pulses will change this idea for refractory targets [50].

It can also get a 10 kg probe launched on its way to Mars in minutes.

We discussed the related laser-materials interaction physics in detail, and new results at 400 fs and 80 ps which will make this propulsion mode possible. Developing this capability depends on developing the associated lasers, beam directors and space stations on which they will be based. It also depends on better understanding of thermal coupling measurements under the proposed laser parameters.

References

1. F. Tsander, Flight to other planets (1924), in *Development of Russian Rocket Technology*, ed. by Y. Moshkin (Mashinostroyeniye Press, Moscow, 1973) (in Russian)
2. K. Tsiolkovsky, Plan of space exploration, 1926 (in Russian), available in English in "Exploration of the Universe with Reaction Machines: Exploring the Unknown," NASA History Series. NASA SP 4407, Washington, D.C. (1995)
3. H. Oberth, *Die Rakete zu den Planetenräumen (The Rocket to the Planet Spaces)* (Oldenbourg Verlag, München, 1923)
4. E. Sänger, Zur Theorie der Photonenraketen, in *Probleme der Weltraumforschung*, (IV. Internationaler Astronautischer Kongress, Zürich 1953; S. 32), Laubscher, Biel-Bienne (1955)
5. H. Yano, Cosmic dust detection by the IKAROS large area dust detectors ion interplanetary space from the Earth to Venus, in *42nd Lunar and Planetary Science Conference* (2011) (in Japanese)
6. A. Kantrowitz, Propulsion to orbit by ground-based lasers. Astronaut. Aeronaut. **10**(5), 74–76 (1972)
7. F.V. Bunkin, A.M. Prokhorov, Use of a laser energy source in producing a reactive thrust. Sov. Phys. Usp. **19**(7), 561–573 (1976)
8. C. Phipps, J. Luke, W. Helgeson, Laser-powered, multi-newton thrust space engine with variable specific impulse, in *High-Power Laser Ablation VII, Proceedings of SPIE* **7005**, 1X1–1X-8 (2008)
9. L.N. Myrabo, D.G. Messitt, F.B. Mead Jr., Ground and flight tests of a laser propelled vehicle, in paper AIAA 98-1001, *36th AIAA Aerospace Science Meeting and Exhibit*, Reno, NV 12–15 January 1998
10. L. N. Myrabo, World Record Flights of Beam-Riding Rocket Lightcraft: Demonstration of 'Disruptive' Propulsion Technology, in paper AIAA 2001-3798, *37th AIAA/ASME/ SAE/ASEE Joint Propulsion Conference*, Salt Lake City, UT, 8–11 July 2001
11. R.A. Liukonen, Laser jet propulsion, in *XII International Symposium on Gas Flow and Chemical Lasers and High-Power Laser Conference*, Proceeding of SPIE **3574**, 470–474 (1998)
12. W.L. Bohn, Laser lightcraft performance, in *High-Power Laser Ablation II*, Proceeding of SPIE **3885**, 48–53 (1999)
13. A. Sasoh, Laser-driven in-tube accelerator. Rev. Sci. Instrum. **72**(3), 1893–1898 (2001)
14. C. Phipps, M. Birkan, W. Bohn, H.-A. Eckel, H. Horisawa, T. Lippert, M. Michaelis, Y. Rezunkov, A. Sasoh, W. Schall, S. Scharring, J. Sinko, Review: laser ablation propulsion. J. Propul. Power **26**(4), 609–637 (2010)
15. C.R. Phipps, J.R. Luke, Laser Space Propulsion, in *Laser Ablation and its Applications*, Chap. 16, *Springer Series in Optical Sciences*, vol. 129, 407–434 (2007)
16. T.K.M. Lippert, Materials for laser propulsion, *SPIE* **7005** paper 7005–38 (2008)

17. C. Phipps et al., Appl. Surf. Sci. **252**, 4838–4844 (2006)
18. B. Esmiller, C. Jacquelard, H.-A. Eckel, and E. Wnuk, Space debris removal by ground-based lasers: main conclusions of the European project CLEANSPACE. Appl. Opt. **53**(31), I45–I54 (2014)
19. C. Phipps et al., Optimum parameters for laser-launching objects into low Earth orbit. Laser Part. Beams **18**(4), 661–695 (2000)
20. C. Phipps, C. Bonnal, F. Masson, M. Boustie, L. Berthe, M. Schneider, S. Baton, E. Brambrink, J.-M. Chevalier, L. Videau and S. A. E. Boyer, Transfers from Earth to LEO and LEO to interplanetary space using lasers, Acta Astronaut. **146**, 92–102 (2018)
21. J. Mason et al., Orbital debris-debris collision avoidance. Adv. Space Res. **48**, 1643–1655 (2011)
22. D. Overbye, *Reaching for the Stars, Across 4.37 Light-years*, New York Times 12 April 2016
23. C. Phipps, Comparing Laser and Electric Propulsion, in *Proceeding 4th International Workshop on Space Debris Modeling and Remediation*, Paris, 6–8 June 2016
24. C.R. Phipps Jr, T.P. Turner, R.F. Harrison, G.W. York, W.Z. Osborne, G.K. Anderson, X.F. Corlis, L.C. Haynes, H.S. Steele, K.C. Spicochi, T.R. King, Impulse Coupling to Targets in Vacuum by KrF, HF and CO2 Lasers. J. Appl. Phys. **64**, 1083 (1988)
25. C. Phipps et al., Removing orbital debris with lasers. Adv. Space Res. **49**, 1283–1300 (2012)
26. B. Poling et al., *The Properties of Gases and Liquids*, 5th edn. (McGraw-Hill, New York, 2001), pp. 7.9–7.11
27. The SESAME equation-of-state database is maintained by group T-1 at Los Alamos National Laboratory (sesame@lanl.gov); see S.P. Lyon and J.D. Johnson, SESAME: The Los Alamos National Laboratory Equation of State Database, LANL Report No. LA-UR-92-3407, 1992 for additional information
28. J. Sinko, C. Phipps, Modeling CO_2 laser ablation impulse of polymers in vapor and plasma regimes. Appl. Phys. Lett. **95**, 131105 (2009)
29. C. Phipps, An alternate treatment of the vapor-plasma transition. Int. J. Aero. Innovations **3**, 45–50 (2011)
30. C. Phipps, High power laser systems for space debris clearing, in *Fifth International School on Lasers in Material Science,* San Servolo, 10–17 July 2016
31. M. Saha, Ionization in the solar chromosphere. Phil. Mag. **40**, 472 (1920)
32. C.R. Phipps, C. Bonnal, A spaceborne, pulsed UV laser system for re-entering or nudging LEO debris, and re-orbiting GEO debris. Acta Astronaut. **118**, 224–236 (2016)
33. C. Phipps, High Power Laser Systems for Space Debris Clearing, in *Fifth International School on Lasers in Materials Science*, San Servolo, Italy, 10–17 July 2016
34. K. Fournier, LASNEX calculations of laser-coupling coefficients for Al targets, UCRL-Pres-226849, p. 29 (2006)
35. E. Loktionov et al., Thermophysical and gas-dynamic characteristics of laser-induced gas-plasma flows under femtosecond laser ablation. Q. Electron. **44**, 225–232 (2014)
36. X. Zhu, N. Zhang, Investigation of ultrashort pulse laser propulsion using time-resolved shadowgraphy and torsion pendulum, in *Proceeding of International Symposium on Photoelectric Detection and Imaging*, SPIE 7382, 73208 (2009)
37. S. Scharring et al., Numerical simulations on laser-ablative micropropulsion with short and ultrashort laser pulses. Trans. JSASS Aerospace Tech. **14**, 69–75 (2016)
38. C. Phipps, Comparative Performance of Laser and Electric Space Propulsion, Final report, CNES document AVP-NT-3250000-ZZ-1606-CNRS (2016), available from CNES
39. C. Phipps, J. Luke, Diode laser-driven microthrusters: a new departure for micropropulsion. J. AIAA **40**(20), 310–318 (2002)
40. C. Phipps, M. Boustie, J.-M. Chevalier, S. Baton, E. Brambrink, L. Berthe, M. Schneider, L. Videau, S.A.E. Boyer and S. Scharring, Laser impulse coupling measurements at 400 fs and 80 ps using the LULI facility at 1057 nm wavelength. J. Appl. Phys. **122**, 193103. (2017) https://doi.org/10.1063/1.4997196
41. R. Weber et al., Heat accumulation during pulsed laser materials processing. Opt. Express **22**, 11312–11324 (2014)

42. C. Phipps, L'ADROIT—a spaceborne ultraviolet laser system for space debris clearing. Acta Astronaut. **104**, 243–255 (2014)
43. Bosch working on 50 kWh battery pack weighting 190 kg, http://www.electric-vehiclenews.com/2015/10/bosch-working-on-50-kwh-battery-packs.html
44. C. Phipps, Pulsed Laser ADR Strategy, in talk #5.2, *3rd European Workshop on Space Debris Modeling and Remediation*, CNES-HQ, Paris, 16–18 June (2014)
45. C. Phipps et al., Small Payload Transfers from Earth to LEO and LEO to Interplanetary Space using Lasers, in *Proceeding of 7th European Conference for Aeronautics and Space Sciences*, Politecnico di Milano, Milan, 3–6 July 2017
46. R. Soulard et al., ICAN: a novel laser architecture for space debris removal. Acta Astronaut. **105**, 192–200 (2014)
47. C. Phipps, Concerns for Phased Fiber laser arrays in space, in *4th European Workshop on Space Dberis modeling and Remediation, CNES-HQ,* Paris, 6–8 June 2016
48. T. Ebisuzaki et al., Remediation of cm-size space debris from the International Space Station, in *Proceeding Workshop on Laser Solutions for Orbitral Debrs*, Université Paris Diderot, 27–28 April 2015
49. http://www.hilase.cz/en/advanced-dpssl-laser-dipole-100-delivers-1kw-performance/
50. C. Phipps, C. Bonnal, F. Masson and P. Musumeci, Launching swarms of microsatellites using a 100 kW average power pulsed laser, JOSA B **35**(10), B20–B26 (2018) https://doi.org/10.1364/JOSAB.35.000B20

Chapter 9
Laser Structuring of Soft Materials: Laser-Induced Forward Transfer and Two-Photon Polymerization

Flavian Stokker-Cheregi, Alexandra Palla-Papavlu, Irina Alexandra Paun, Thomas Lippert and Maria Dinescu

Abstract This chapter discusses recent progress in 2D and 3D printing technologies, in particular laser-induced forward transfer and two-photon polymerization (TPP). We explore their potential for applications in micro-electronics, protein microarrays, sensors and biosensors, and tissue engineering. An overview of the factors that affect patterning, miniaturization and functionality is presented. A special focus is placed on laser direct writing via TPP for the fabrication of 3D vertical microtubes acting as microreservoirs for an osteogenic drug.

9.1 Introduction

The efforts undertaken both in research and industry to apply novel direct writing techniques to new materials with tailored properties have surged in recent years. There is a strong interest towards integrating such materials in two and three-dimensional patterns for applications in various fields, e.g. pharmacy, bioengineering, or micro-electronics, to name a few.

F. Stokker-Cheregi · A. Palla-Papavlu · M. Dinescu (✉)
Lasers Department, National Institute for Lasers, Plasma, and Radiation Physics, Atomistilor 409, MG 36, 077125 Magurele, Romania
e-mail: dinescum@nipne.ro

I. A. Paun
CETAL Department, National Institute for Lasers, Plasma, and Radiation Physics, Atomistilor 409, MG 36, 077125 Magurele, Romania

I. A. Paun
Faculty of Applied Sciences, University Politechnica of Bucharest,, RO 060042 Bucharest, Romania

T. Lippert
Research with Neutrons and Muons Division, Paul Scherrer Institut, 5232 Villigen PSI, Switzerland

T. Lippert
Laboratory of Inorganic Chemistry, ETH Zürich, Zürich, Switzerland

P. M. Ossi (ed.), *Advances in the Application of Lasers in Materials Science*, Springer Series in Materials Science 274,
https://doi.org/10.1007/978-3-319-96845-2_9

From a research perspective, the appeal of using lasers for small-scale material applications has been fueled by their favorable properties which make significant achievements in areas such as thin films and optoelectronics, possible. In industry, however, lasers have been predominantly used for welding, drilling, cladding and micromachining applications, due to their high precision and cost-efficiency.

The increasing interest towards miniaturization of devices has resulted in scientists and engineers working on solutions for more compact and more efficient designs through micro-manufacturing techniques. Examples of micro-fabrication processes include lithography, chemical vapor deposition, dry and wet etching, and others. Although each technique has its own advantages, most of them are multi-step processes, which inherently gives rise to complications in the production chain. Due to these limitations and because of the small range of materials that can be processed, there has been an extensive search for alternative methods for manufacturing microstructures and micro-devices.

Processing of "soft" materials, e.g. polymers and biological compounds, with high accuracy and low cost is of particular interest, due to their potential applications in various fields, e.g. from the fabrication of micro-optical components, to sensors or novel bio-devices. However, such materials are highly sensitive, and therefore their successful printing requires a special approach relying on key technological developments. In this respect, the use of laser based techniques is particularly appealing, as they provide both flexibility and complexity in the design of soft-material structures, which ultimately allows the creation of reproducible patterns with micro and nano precision.

One of the direct write techniques which has shown great potential for the spatial transfer of various materials, such as polymers, microbeads, liposomes, etc., and also for the fabrication of devices, for example sensors and organic light emitting diodes (OLEDs), is laser-induced forward transfer (LIFT). This technique, which was initially named Laser Writing or Material Laser Recording, was first developed for applications in the graphic industry, i.e. printing of different dried inks onto glass substrates [1–3]. A few years later, LIFT (as named by the authors of the respective study) was used to deposit copper metal pixels onto fused silica substrates under vacuum, and, a couple of years later, the same procedure was demonstrated under atmospheric conditions [4, 5].

The principle of LIFT consists in focusing or imaging a laser beam through a transparent support onto a thin film material to be transferred (also named "donor"). Every laser pulse initiates the transfer of the donor material onto a substrate (known as "receiver") that is usually placed parallel and facing the thin film at very short distances, i.e. less than 100 μm. A pattern of the transferred material can be "written" on the substrate by applying successive laser shots while adjusting the position of the donor-receiver system with respect to the laser beam. Conventional LIFT has been applied to transfer a wide range of materials, from simple ones, like metals [6–8] to more complex materials, such as oxides [9, 10] and high-temperature superconductors [11]. However, conventional LIFT has some limitations. Biological compounds for example may be damaged by direct exposure to the energetic laser pulses. Therefore, different modifications to LIFT have been proposed, most of them focusing on the introduction of an additional layer between the donor substrate and the thin film to be transferred, which has the role of protecting the sensitive transfer layer. To this end, different intermediate layers have been used, from metals to polymers and nanoparticles, which gave rise to variations of the LIFT technique, named, e.g. dynamic release layer (DRL)—LIFT, blister actuated (BA)—LIFT, etc., but all of them still relying on the same basic principle.

In the next section we will describe most of the intermediate type of layers used for the transfer of solid and liquid phase materials and also for the fabrication of different devices. Although LIFT is a well-established method for the realization of 2D patterns, its ability to develop 3D structures has only recently been investigated [12–14].

Computer-controlled techniques are gaining popularity due to their ability to produce 3D structures replicating a computer-aided design [15]. For example, cell matrices have been manufactured from Ti–6Al–4V powders by selective electron beam melting [16]. Ceramic or metallic structures were constructed by depositing molten materials and multiphase solidification jets [17]. Robust ceramics based on tricalcium phosphate were manufactured by micro-stereolithography [18]. Porous scaffold networks were produced by lithography using bio-glasses, based on additive manufacturing techniques [19].

Another promising three-dimensional micro-fabrication method that has recently attracted attention is based on two-photon polymerization by ultra-short laser pulses. The method, called laser direct writing via two-photon polymerization (LDW via TPP) , allows to produce well-defined 3D matrices with micro and nano-sized features that mimic in vivo environments [20–25]. The cascade polymerization is initiated in the volume of a photosensitive material (also known as "photoresist") by the ultrashort laser pulses following two-photon absorption processes. After the 3D pattern is executed, the unexposed photoresist is washed off to reveal the desired structure. A major advantage of LDW via TPP is the ability to achieve complex 3D architectures, with spatial resolutions ranging from sub-micrometers to tens of nanometers [26]. The first 3D structures for tissue engineering produced by LDW via TPP were manufactured from a commercially available hybrid polymer, Ormocomp® (technology GmbH). Due to their growing impact in the field of biomedical applications, materials suitable for LDW via TPP are now expanding to biocompatible materials (e.g. polyethylene (glycol) diacrylate), different triacrylate mixtures (e.g. tris(2-hydroxy ethyl) isocyanurate triacrylate), sol-gel hybrids, hydrolytically degradable polycaprolactone polymers, methacrylamide modified gelatin, and water-soluble photoinitiators, such as 2-hydroxy-2-methylpropiophenone (commercial name Irgacure) [21–25].

This chapter discusses the complexities arising from working with these classes of materials by giving an overview on the current progress and future prospects of 2D and 3D micro/nano-fabrication methods, with an emphasis on laser induced forward transfer and two-photon polymerization.

An analysis of the characteristics and process parameters of LIFT for transferring a wide range of materials in solid and liquid phase is presented, along with possible applications of LIFT in biology (e.g. fabrication of microarrays) and microelectronics (e.g. fabrication of micro devices, such as sensors and organic light emitting diodes).

In addition, a brief history and the basic physics behind the direct laser direct writing via two photon polymerization techniques are also presented. We give a detailed description of the 3D fabrication of vertical polymer microtubes by LDW via TPP and we discuss their viability as micro-reservoirs for an osteogenic drug. Furthermore, we show how the microstructures fabricated by LDW via TPP are able to promote osteogenesis through electrical stimulation. The results demonstrate the immense potential of LDW via the TPP method towards bio-applications through the use of environment-friendly, precise and versatile laser-assisted processes, which have the ability to produce complex 3D architectures that mimic in vivo media, with a spatial resolution in the range of tens of nanometers.

9.2 Laser-Induced Forward Transfer (LIFT)

9.2.1 LIFT in Solid Versus Liquid Phase

As outlined in the *Introduction*, conventional LIFT relies on the laser beam to push a small volume of material from a donor film onto a receiver substrate. The dimension and shape of the transferred pattern is given by the size and shape of the beam spot and/or the laser projection mask. In the case of LIFT of solids, the transfer mechanism involves either melting or ablation of the solid donor film. Although this is acceptable for materials which can be exposed to a high heat load, such as metals, degradation and loss of functionality may occur in the case of polymers or biological compounds upon their direct laser irradiation.

Therefore, numerous variations to original LIFT process have been proposed. Most of the variations to LIFT include the addition of a layer, "sandwiched" between the donor substrate and the donor thin film material, which is often named sacrificial layer or dynamic release layer (DRL). Other approaches focus on embedding the material to be transferred into an absorbing matrix; this process is known as matrix-assisted pulsed laser evaporation-direct write (MAPLE-DW) [27–30]. MAPLE-DW has been demonstrated for printing different materials and devices, i.e. biomolecules, sensors and different electronic components [27–30]. However, the main issue related to this technique is the use of specially designed donor films, for example in the case of printed inks or pastes from a specially designed suspension containing the ink/paste to be deposited.

The first studies available in literature, in which sacrificial layers were used, are focused on metallic layers as DRLs [31–33]. Generally, the use of a metallic layer, usually Au or Pt, as they are biocompatible, is particularly useful for the transfer of biomolecules, because the metal nanoparticles generated as a result of the ablation process do not influence the functionality of the transferred material. For example, in [31–36] the authors use thin (50–100 nm) metal films to transfer different

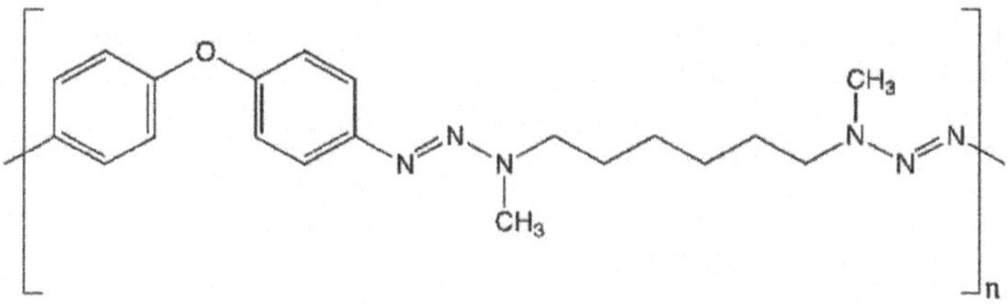

Fig. 9.1 Chemical structure of the triazene polymer which was used for most LIFT experiments

types of bio-compounds and cells, while in [37] the authors name the process AFA-LIFT (absorbing film assisted-LIFT). In the works presented in [38, 39], the authors use thick (several μm) absorbing polyimide layers and name the process BA-LIFT (blister-actuated-LIFT) to transfer different materials, mostly in liquid phase (e.g. embryonic stem cells, 9-anthracenemethanol and tris8-hydroxyquinoline aluminum-Alq_3-organic luminophores).

Furthermore, another approach involves the use of a photo-decomposing polymer, i.e. triazene polymer (TP) (see Fig. 9.1 for the chemical structure), which upon laser irradiation decompose completely into small and gaseous fragments, and thus only a minimum contamination of the transferred material may be expected. These polymers, in most cases, were demonstrated to be particularly useful both for the transfer of materials in solid, as well as in liquid phase [40–44]. A list of the most representative materials transferred by LIFT, together with the intermediate layers applied for the transfer are presented in Table 9.1.

LIFT (and its variations) has been proven as an additive manufacturing method that exceeds the possibilities of conventional printing methods through its ability to miniaturize and embed a wide range of materials (see Table 9.1), from solids to liquids, to devices into different substrates. The rest of this section of the chapter discusses the dynamic mechanism of LIFT, with particular applications for solids (polymers), and liquids. In order to advance the technique to 3D printing and even further to industrial applications, there is great interest to get an insight into the underlying mechanisms of LIFT. Therefore, different time-resolved pump-probe imaging techniques have been used to visualize material ejection from both solid and liquid donors. The experimental setup for such investigations is shown in Fig. 9.2.

In the case of LIFT of solids, in particular polymers aiming at sensor applications [57–60], the laser transfer is dominated by a "microexplosion" of the flyer, and thus a shockwave is generated, as for example shown in Fig. 9.3, where the ejection of a polymer layer (60 nm PECH on top of 150 nm TP layer) in ambient atmosphere is visualized together with its shockwave. The sequence of images is taken at a laser fluence of 300 mJ/cm^2 and increasing time intervals (step of 100 ns) after the pump laser pulse.

After 100 ns the ejected flyer appears, as a dark, flat shape, which at 200 ns is clearly visible together with the shockwave, which propagates with a higher speed than the flyer. The shape of the flyer doesn't change for 300 ns, as depicted in Fig. 9.3. However, after 400 ns the flyer starts to lose cohesion, bends, and finally

Table 9.1 List of the most representative materials transferred by LIFT, together with the intermediate layers applied for the transfer

Material	Intermediate layer	Laser	Substrate	References
Au and Al	Poly (methacrylonitrile) and Teflon AF (1600) (Du Pont)	ArF 193 nm	Si wafer	[45]
Salmon sperm DNA	Ti	Nd:YAG 355 nm	poly-L-lysine coated glass	[34]
Conidia of the Trichoderma strain	Ag	KrF 248 nm	Glass	[37]
HRP enzyme solution	Au	KrF 248 nm	Glass	[46]
Embryonic stem cells	polyimide	$Nd:YVO_4$ 355 nm	Petri dish coated with a thick layer of Matrigel®	[38]
Self-assembled octapeptides	Au	KrF 248 nm	Au coated glass	[47]
Photobiotin	Au	KrF 500 fs	Glass slides, nitrocellulose coated glass slides, agarose and ORMOCER treated glass slides	[48]
OLED (Al/MEH-PPV)	TP	XeCl 308 nm	ITO covered glass	[49, 50]
Mammalian cells	TP	ArF 193 nm	Glass	[42]
Quantum dots	TP	ArF 193 nm	Glass wafers coated with ITO	[51]
Methylcellulose	TP	XeCl 308 nm	Glass	[52]
Poly(2-hydroxyethyl methacrylate); polyacrylic acid; poly (4-vinylpyridine)	Cr	Nd:YAG 266 nm	Low temperature oxide (LTO)/Si, and capacitive sensors	[33]
Mesotetraphenylporphyrin	Ti	Nd:YAP (500 ps) 1078 nm	Glass N/A	[53]
PEDOT:PSS	Au and Ag	KrF 248 nm (35 ns) and Nd:YAG, 355 nm (50 ps)	Si or quartz suprasil	[54]
$Gd_3Ga_5O_{12}$	TP	Ti:sapphire 800 nm	Si wafer	[41]
Polymer light emitting diodes (Al/PFN)	TP	XeCl 308 nm	Glass/ITO/PEDOT:PSS/PVK	[55, 56]

(continued)

Table 9.1 (continued)

Material	Intermediate layer	Laser	Substrate	References
Polyisobutylene (PIB), Polyethylenimine (PEI), Polyepichlorhydrin (PECH)	TP	XeCl 308 nm Nd:YAG 355 nm	Glass, Al interdigital transducers	[57–60]
Polystyrene microbeads	TP	XeCl 308 nm	Thermanox coverslips	[61, 62]
Liposomes	TP	ArF 193 nm	Glass	[40]
Al	TP	XeCl 308 nm	Glass	[43]
Semiconducting oligomer	TP	KrF 248 nm, (ns pulses) Nd:YAG 355 nm (pulses)	Si/SiO_2	[63]
Ag paste	TP	355 nm (ps pulses)	Glass	[64]
Single walled carbon nanotubes	TP	XeCl 308 nm	Pt interdigitated transducers	[65]
$Ca_3Co_4O_9$	TP	XeCl 308 nm	PDMS	[66]
$SnCl_2(acac)_2 \rightarrow SnO_2$	TP	XeCl 308 nm	Pt electrodes	[67, 68]
Ferrocene	TP	XeCl 308 nm	PDMS	[69]

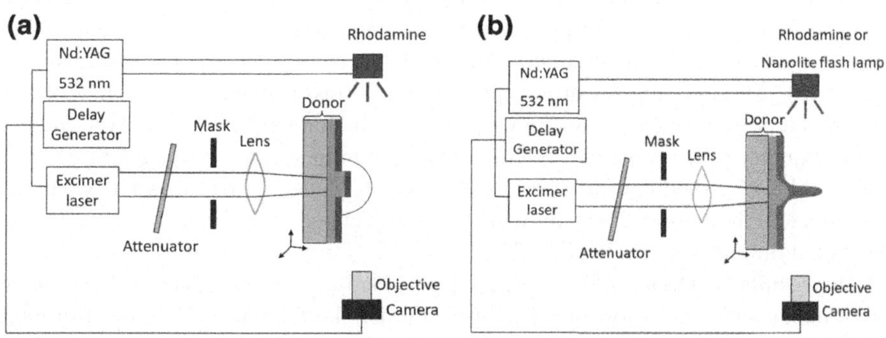

Fig. 9.2 Setup for time resolved imaging experiments for LIFT of **a** solid donors and **b** liquid donors

disintegrates. This type of investigation is very important as it allows obtaining information regarding the individual propagation speed of both shockwave and flyer, thus shedding light onto the energy conversion. This information ultimately allows tun-

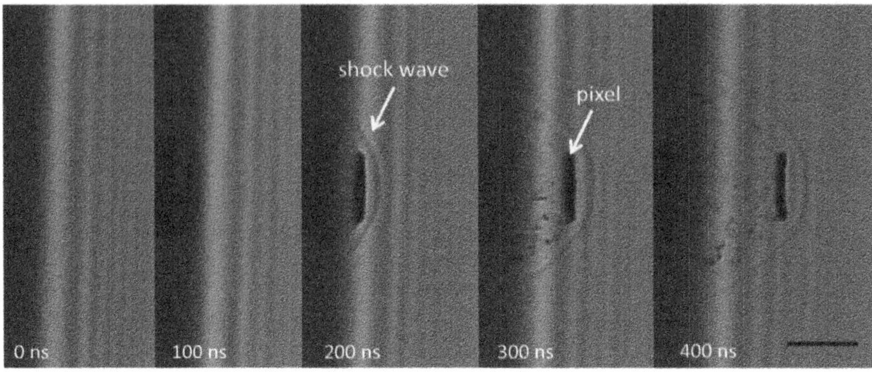

Fig. 9.3 Sequence of photos taken for **a** back side ablation of a PECH thin film on top of a 150 nm triazene polymer film at 300 mJ/cm^2 laser fluence. The scale bar is 500 μm

ing the process parameters, e.g. donor-receiver distance, pressure during processing, etc., for successful LIFT printing of materials, e.g. polymers for functional device applications.

Unlike solids, in the case of LIFT of (Newtonian) liquids, the transfer mechanism is dominated by the formation of long liquid needle-like jets. The first papers on time resolved imaging study of liquid LIFT were published by Young and coworkers [70]. The authors of the study investigated a viscous ink (composed of barium titanate powders solved in α-terpineol) which they irradiated with a 355 nm laser and acquired time-resolved images of the transfer process. They found three different regimes depending on the laser fluences, i.e. plume (for high laser fluence, where only splashed material is observed), jetting, and sub-threshold (low fluences, no material removal). The jetting regime was characterized by the appearance of an irregular jet which fragmented into multiple droplets after traveling a short distance (~50 μm). Another step towards explaining the transfer dynamics in LIFT of liquids was reported by Barron and coworkers [71]. In that work, time-resolved images of the transfer process using a bovine serum albumin solution were acquired. The authors also observed the formation of a jet which broke into free droplets when the jet had a length of about 200 μm, which they correlated to the well-defined droplets deposited onto the substrate [71, 72].

The results of Duocastella et al. [73] provided a more detailed explanation of the transfer and deposition mechanism in the LIFT of liquids. They used in their experiments a solution of water and glycerol (50:50 v/v) which was blade-coated on a microscope slide previously coated with a 50 nm thick Ti layer (which acts as DRL). As irradiation source they used a Nd:YAG laser operating at 355 nm. They suggested that the ejection mechanism in the LIFT of aqueous solutions is driven by the formation of a high-pressure vapor bubble which expands. The bubble expansion process can result into three different situations, depending on the applied laser fluence. At low fluences, a needle-like jet is developed. However, this jet is not energetic enough to overcome the surface tension forces, and thus the jet does not

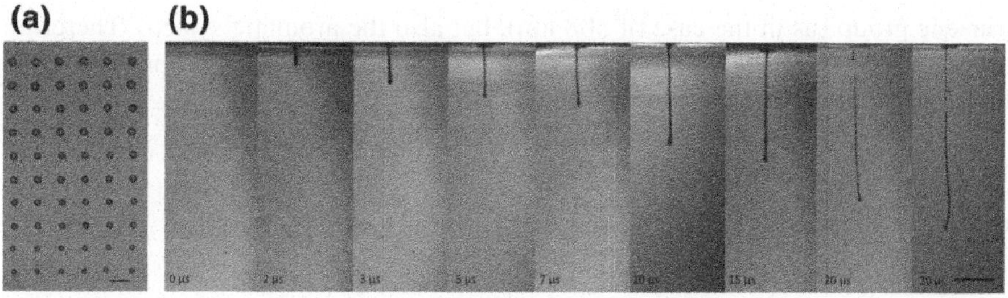

Fig. 9.4 **a** Optical microscopy image of a microarray printed by TP-LIFT (TP thickness is 60 nm) on glass. The laser used for the transfer is XeCl and the laser fluence increases from bottom to top of the image (from 20 to 38 mJ/cm^2). Scale bar is 100 µm. **b** Time resolved images of TP-LIFT. The laser used for irradiation is the XeCl (308 nm), the TP layer is 60 nm, and the laser fluence is 25 mJ/cm^2. Scale bar is 500 µm. Reproduced with permission [74]

contact the receptor and no material is transferred. At intermediate fluences, the same jet is formed, but in this case it advances at constant velocity while progressively thinning. At high fluences, the bubble expansion is so violent that it overcomes the surface tension, resulting in bubble bursting, a situation which generates splashing on the receiver substrate [73]. The results obtained by Duocastella et al. [73] and Barron et al. [71] provide an explanation of the transfer mechanism in the liquid phase LIFT when using a metallic DRL layer.

Other experiments using a polymer (TP) as sacrificial DRL layer are complementary to the existing results on the dynamics of liquid phase LIFT. In [74] the material used in the study was also a model liquid solution consisting of a mixture of water, glycerol and SDS blade coated onto TP layers with different thicknesses. The first step in the time-resolved study of LIFT is the imaging of the transfer process at the conditions in which well-defined droplets are obtained. This study was carried out using two lasers, i.e. an ArF and a XeCl excimer laser, and employing TP layers having different thicknesses. An optical microscopy image of the transferred microarrays (with the XeCl laser and 60 nm TP layer thickness) is presented in Fig. 9.4a: circular droplets are obtained, with droplet diameter increasing with increasing laser fluence. Immediately after deposition, the receiver substrate was removed, and time-resolved images of the LIFT process were acquired using the same liquid film as in the preparation of the previous microarray, and under identical irradiation conditions. A series of time resolved images of liquid phase LIFT are presented in Fig. 9.4b). In these images, the surface of the donor is located at the top of the image. The flyer expands towards the bottom part of the images. These shadowgraphy images correspond quite well to the fluence onset and fluence range for the array fabrication for metallic DRLs [72].

The absorption of the laser beam takes place in the triazene layer, which is vaporized, and a large amount of gas is released during the laser pulse [75]. The differences between the processes, as revealed by the shadow graphic images for 193 and 308 nm irradiations, are most probably due to the different photon energies of the two lasers wavelengths. The higher energetic photons, at 193 nm, are not only decomposing the triazene group (as in the case of 308 nm), but also the aromatic system. Therefore, more gaseous ablation products are created, which result in a higher pressure and a faster, more violent ejection of the liquids.

9.2.2 LIFT for Device Fabrication: Towards Industrial Applications

During the last years, there is growing interest in LIFT for complex material printing and patterning, and furthermore there are many requirements for fabrication of micro-devices at a reasonable speed and without post-processing methods. A wide range of devices have been described in the literature, based on inorganic, organic, and macromolecular materials, and even biological systems. A compilation of examples of LIFT variations for printing different device systems, such as sensors, biosensors, or organic light emitting diodes, is presented in the following section.

The first example is the use of LIFT with metal intermediate layers in bio-sensing applications. In recent studies [77, 78] the capability to print odorant binding proteins as sensitive membrane for the detection of food contaminants was demonstrated. LIFT was used towards the development of a surface-acoustic wave (SAW) based bio-electronic nose, by transferring odorant binding proteins (OBPs) on the active area. This approach made it possible to take advantage of both the tunable selectivity of the OBPs, as well as the fast response times and high selectivity characteristic to most SAW-based sensors. OBPs are a type of small extracellular proteins, which play an important role in odor detection.

The realization of the "bio-electronic nose" reported in [58] involved the use of three SAW resonators, which were coated using LIFT, as well as an uncoated SAW device which was used as reference. Three different OBPs were used as respective coatings: the wild-type OBP from pig, a double mutant of the OBP from bovine, and the wild-type OBP from bovine. Each of these proteins is characterized by different binding specificities. The functionality of the obtained bio-sensing arrays was demonstrated through exposure of compounds commonly used in food processing, i.e. to various concentrations of octanol and carvone.

As shown in Fig. 9.5a, the use of LIFT allowed the transfer of odorant binding proteins with a very high lateral resolution and using very low amounts of detection agent, whereas the obtained SAW-based biosensor arrays exhibited excellent functional properties. The sensitivity and detection limits achieved by the obtained microstructures were found to be comparable to those of conventional detectors. These results show the feasibility of LIFT towards the realization of biosensors that can be used in applications ranging from the control of food quality (through detection of contaminants, such as molds), to evaluation of indoor air quality.

Due to the advancements in material fabrication and material processing techniques, the field of sensors has seen increased interest in recent years, both in terms of industrial development and scientific research. In this context, LIFT is proving

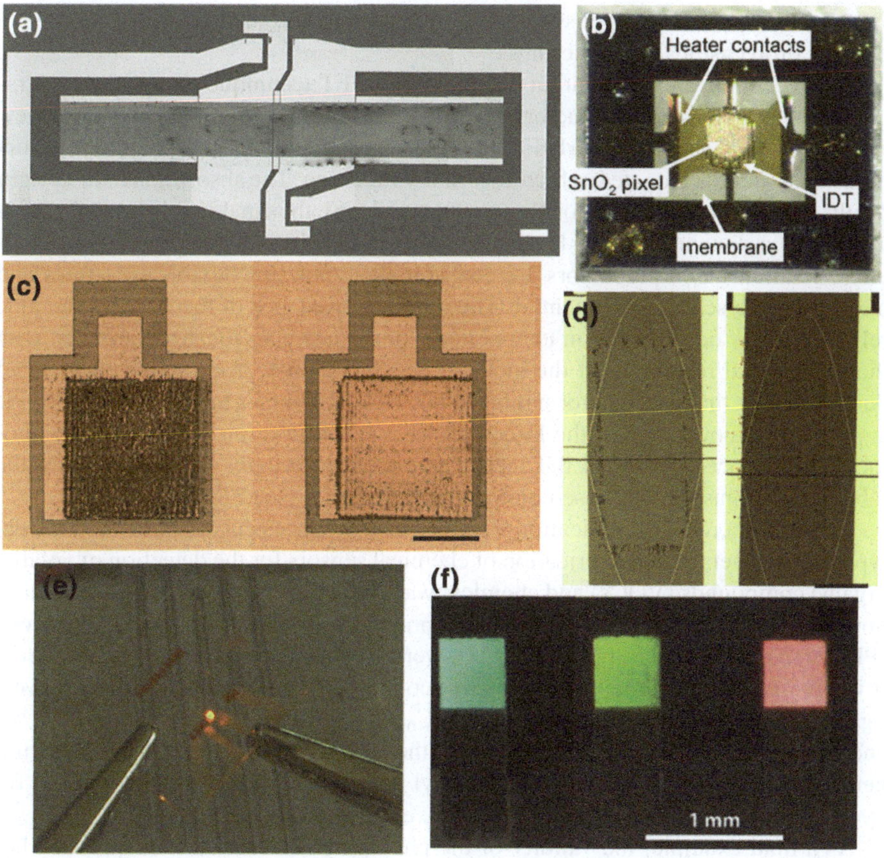

Fig. 9.5 **a** Optical microscopy images of a SAW sensors printed by LIFT, by 50% overlap (ratio of overlapped length between two neighboring droplets to the individual droplet diameter. Scale bar is 100 μm; **b** SnO_2 rLIFTed pixel onto a commercial sensor structure; **c** Optical microscopy images of solidly mounted resonators (SMR) devices coated with PEI polymer deposited at 400 (left) and 300 (right) mJ/cm^2 laser fluence. Scale bar is 150 μm; **d** PECH pixels transferred by TP-LIFT at different laser fluences (left 500 and right 400 mJ/cm^2) onto the resonating cavity of the SAW resonators. The scale bar is 200 μm; **e** Image of the successful operation of an Al/MEH-PPV (poly[2-methoxy-5-(2′-ethylhexyloxy)-p-phenylene vinylene) transferred system onto an ITO substrate; **f** Examples of polymer light emitting diodes transferred by LIFT, i.e. blue, red, and green pixels with an Al/TBA cathode transferred at approximately 70 mJ/cm^2. Reproduced with permission **a** [77]; **b** [68]; **c** [59]; **d** [44]; **e** [50]; **f** [76]

to be a promising method for the coating of commercial sensors. It can be applied to transfer a large variety of materials, without the use of solvents, onto an equally large number of substrate types, with high lateral resolution. LIFT is also being considered as a replacement or alternative to ink-jet printing and photolithography methods. Another of its notable advantages is the possibility to use large area donor substrates prepared by various well-established methods, such as evaporation, spin coating, drop casting, or sputtering.

One recent advanced variation based on the LIFT technique allows the integration of metal oxide semiconductor materials (e.g. SnO_2) in micro-sensors. This new method, called reactive LIFT (rLIFT), relies on using metal complex precursor donors (e.g. metal acetylacetonates) which absorb UV radiation. The absorption of laser light results in their partial decomposition to SnO_2 [67, 68], thus making the transfer layer also act as a dynamic release layer. Due to their low deposition (growth) temperatures, solution based precursors are considered suitable for flexible substrates and have favorable scalability potential. The unique advantages of the rLIFT technique rely on the low decomposition temperatures of the acetylacetonates combined with the high spatial resolution of the method. The fabrication processed is accelerated due to the fact that the transformation of the metal precursor to its respective oxide is partially achieved during the laser transfer phase as a result of thermal and photochemical processes. The rLIFT technique has been recently demonstrated for the fabrication of gas sensors based on SnO_2 and Pd:SnO_2 (see Fig. 9.5b) [68].

Another interesting application of LIFT is the successful transfer of different types of polymers for the fabrication of chemical sensors for the detection of volatile organic compounds (VOCs) and chemical warfare agents (see Fig. 9.6c, d). Up until now, different types of polymers, such as polyethylenimine (PEI), polyisobutylene (PIB), and polyepichlorohydrin (PECH), were applied for the realization of sensors using the LIFT technique, and their physico-chemical properties have been investigated. PEI, with the molecular formula $(C_6H_{21}N_5)_n$, PECH $[CH(CH_2Cl)CH_2O]_n$, and PIB $[CH_2(CH_3)_2]_n$, can be utilized for the detection of nerve agents (e.g. sarin), nerve agent simulants (e.g. dimethyl methyl phosphonate—DMMP), as well as for hydrocarbons (e.g. acetone) and esters (e.g. ethyl acetate).

As a final example, the validity of the photopolymer-based LIFT approach has been shown, by fabricating working miniaturized model OLED devices in 2007 [50]. An example of the first reported device fabricated by LIFT is shown in Fig. 9.5e). The OLED devices were fabricated by printing either the cathode/electroluminescent material (EL) system as a bilayer, i.e. Aluminum/MEH-PPV (poly[2-methoxy-5-(2-ethylhexyloxy)-1,4-phenylenevinylene)]), onto the anode [transparent tin-doped indium oxide (ITO)], or by printing the cathode/EL material in sequential steps. Good functionality and pixel morphology were observed both for the single and bilayer printed pixels. The first works on LIFT printing of sensitive materials, such as MEH-PPV, for OLED fabrication without damaging their functional properties opened a new field for device manufacturing by laser assisted methods.

Later works have shown that by optimizing the laser transfer process, i.e. by adding a thin conductive film (of PEDOT:PSS) onto the ITO receiver substrate it was possible to reduce the laser fluence required for the transfer, i.e. from 250 to 80 mJ/cm^2, and optimized devices could be fabricated. Furthermore, the overall performance of the LIFT-ed red and green OLED materials was slightly lower than that of conventional devices, and for the blue OLEDs it was better than conventional devices; however, these works represent a proof of concept that it is possible to apply

LIFT to fabricate micro-devices for the field of electronics. Detailed information of the fabrication of OLEDs by TP-LIFT is given in [79].

The viability of a processing technique in terms of its perspective applications in the field of micro-electronics relies heavily on its ability to achieve the patterning and printing of materials with high reproducibility and spatial resolution. In recent years, laser-induced forward transfer has been seen more and more as an alternative to established industrial techniques, such as inkjet or screen printing, and therefore is classified as an emerging additive printing technique. The LIFT technique was first reported 30 years ago, and has since branched out into several variations that were adapted to suit a range of material phases, from liquid to solid, and integration approaches towards the achievement of functional devices. One of the most significant breakthroughs in the development of this technique was the addition of the dynamic release layer, which opened the way for a new range of applications by facilitating the intact transfer of a wide range of materials that were previously inaccessible.

The results presented in this section focus on the advantages provided by the inclusion of an intermediate layer as a DRL in LIFT experiments, and it also discusses some of the factors that limit the use of this approach towards the patterning and printing of solid and liquid materials and devices (sensors and organic light emitting diodes). We show how the TP-LIFT technique is an ideal candidate for the development of novel manufacturing processes in which high resolution and printing speed can be integrated alongside existing additive manufacturing techniques in order to improve patterning and printing performance.

The benefits stemming from such integration would render additive printing processes more cost-effective and environmentally friendly than current methods. However, there still is a general need to address the challenges posed by patterning and printing with TP-LIFT and LIFT in delivering materials with the desired structure and functional performance.

9.2.3 Conclusions and Future Prospects

Current needs for advanced electronics, miniaturized devices on a chip, such as sensor systems, or biomolecule microarrays require new strategies for rapid and cost-effective material processing and integration. Therefore, new direct write techniques are needed, which on one hand would compete with photolithography for size and scale, and on the other hand reduce the number of fabrication steps.

Successful "soft" material printing requires several technological developments, i.e. a combination of laser system benefits with the flexibility of soft sensitive material design which ultimately leads to the creation of reproducible patterns with micro and nano precision. Laser induced forward transfer for material printing and patterning is an old technology with new approaches, which show great potential for applications ranging from micro-electronics, i.e. devices, interconnects, etc., to microarrays and biosensors in biology.

In this chapter we have tried to show the potential of LIFT for different applications, including its potential in 3D printing and industrial applications. Special focus is set on the dynamic mechanism of LIFT both for solid and liquid phase materials, together with examples of materials transferred by TP-LIFT, illustrating the potential of this technique as a contactless laser-assisted printing method.

9.3 Laser Direct Writing Via Two Photon Polymerization (LDW Via TPP)

9.3.1 3D Biomimetic Structures for Tissue Engineering

Tissue engineering is an ambitious and relatively new field of science, in which scientists try to develop biomimetic structures that support cell growth until the state of functional tissues is reached. At present, a growing interest is dedicated to the fabrication of porous structures that mimic the complexity of the natural cell environment [80].

Porous architectures, including holes with controllable shapes, positions, and dimensions, are key to achieving human-like bone structures by facilitating cell growth in a desired manner that defines the final shape of the new bone. The newest approaches are based on 3D implants that trigger the osteogenic differentiation of the cells and provide therapeutic regimes that reduce inflammation and post-implantation infections. To facilitate the osteointegration of implants, one must effectively administrate drugs that fight against post-implant complications. For this purpose, versatile drug delivery systems that deliver therapeutic agents in a controlled manner are developed [80, 81].

For example, a major issue in clinical applications of orthopedic implants is the decrease in chronic inflammation. The most common pharmaceutical agent that fights inflammatory processes is dexamethasone (Dex), a synthetic glucocorticoid used for the treatment of rheumatoid arthritis, inflammatory bowel disease, post-transplant immunotherapy and other inflammatory diseases [82]. However, its oral administration often leads to severe side effects such as glucocorticoid-induced osteoporosis [15]. Therefore, alternative strategies rely on the localized delivery of drugs from carrier materials designed in the form of 3D biomimetic architectures. Their attachment to the implants makes higher local drug concentrations possible and reduces the risk of systemic toxicity, as compared to traditional routes of administration [16]. Implantable drug delivery systems are even more effective when the drug is loaded onto the surface of the implant and released on demand. The release of the drug from such intelligent systems can be triggered by UV, visible and near infrared radiation, magnetic fields, ultrasounds, and electrical stimuli (ES) [18, 19], to which special attention is paid [83]. ES emerged as a powerful tool for tissue engineering, with positive effects on cell adhesion, proliferation and differentiation [84–87]. The positive roles of electrical stimuli applied through conductive substrates, such as metals, graphene, and conductive polymers, on the bone regeneration process has been extensively investigated [85–87]. In this regard, we have recently developed titanium-polypyrrole/poly (lactic-co-glycolic) structures and we have shown that their 200 mA electrical stimulation applied for 4 h gave rise to osteogenesis in osteoblast cell types [88, 89]. Moreover, intrinsic conductive polymers such as polypyrrole (PPy) have a positive impact on the differentiation and mineralization of cells, even in the absence of external electrical stimuli [16, 90]. The most popular conductive materials for building delivery systems for electrically controlled drug release are conductive polymers, due to favorable properties such as high conductivity/weight ratio and good biocompatibility at room temperature [91, 92]. In orthopedics, the most popular conductive polymer is polypyrrole (PPy) , whose intrinsic conductivity stimulates bone growth [93–95]. For example, polypyrrole membranes were used for the controlled release of glutamate, dopamine, and adenosine triphosphoric acid [96–98]. Recently, a polypyrrole film was used for potentially controlled release of chlorpromazine [99]. Polypyrrole-based systems loaded with drugs have also been used in actuators, such as microfluidic pumps [100, 101]. The major limitations of polypyrrole drug delivery systems are: (i) the small amount of drug that can be loaded and released from the polymer; and (ii) the drug delivery interval, which is limited by the size and polarity of the drug molecule. To overcome these problems, drug molecules have been attached to the surface of polypyrrole by binding to biotin-streptavidin [102]. Alternative strategies used polypyrrole as an active element that triggers the release of drugs from degradable hydrogels [22]. In addition, PPy is of interest due to its capabilities to control the release of Dex in cells grown onto 3D scaffolds.

In this section we will review the direct laser writing via two photon polymerization of polymeric 3D structures, with emphasis on tissue engineering applications. After a brief overview of the potential and the basic principle of the LDW via the TPP technique, we present a particular application for bone tissue engineering, where LDW via TPP was used for the fabrication of vertical microtubes acting as microreservoirs for osteogenic drugs. We demonstrate that these microstructures can be efficiently used for electrically stimulated osteogeneisis. The results demonstrate the versatility of the LDW via the TPP method, which uses environment-friendly, precise, and versatile laser-assisted processes, and has the ability to produce 3D complex architectures that mimic in vivo media with spatial resolution of tens of nanometers.

9.3.2 Basics of LDW via TPP

The process of two-photon absorption is related to changing the electronic state of an atom or molecule that is excited from a lower energy level to a higher electronic state through the simultaneous absorption of either two identical photons or two photons having different energies.

The probability of the two-photon absorption process has a quadratic dependence on the number of photons, and therefore its success relies on using high-repetition rate, ultra-short laser sources, such as femtosecond (fs) lasers. Fs laser sources provide a large number of photons per unit volume, and therefore it is important that the irradiated material does not show linear absorption at the working wavelength.

The experimental assembly used in the two-photon polymerization typically uses femtosecond lasers with nano-Joule pulse energies. Although the average energy is low, the peak intensity of the pulse is high enough to promote the absorption of two photons. The average power depends on the nature of the used materials (such as the type of polymer, photoinitiator, dopants, etc.). The intrinsic properties of the employed materials have to be accounted for when choosing an appropriate laser power. The laser power has to be large enough to exceed the polymerization threshold, but must be small enough to avoid material degradation. Sample preparation usually consists of placing a drop of photopolymer on a pre-cleaned substrate, followed by irradiation with the focused laser beam. The photopolymerization occurs only in the area of the spot in which the laser intensity exceeds the polymerization threshold. The triggered chemical reaction transforms low molecular weight molecules (monomers) into macromolecules of repeating units. The minimum volume of solidified material created through this process around the focal point is called a 'voxel'. Controlled three-dimensional microstructures can be created by scanning the laser beam or by moving the sample in X, Y and Z directions. The precision with which these structures can be processed depends predominantly on the voxel's geometry, which can be controlled by adjusting the average laser power and the scanning speed.

Microfabrication by two-photon polymerization has unique advantages over conventional microfabrication processes. The TPP reaction is usually achieved by pulsed laser systems operating in the near infrared range, where most curable monomers are transparent. Because of this, the depth of light penetration is high, thus providing the possibility to polymerize only a small region of the liquid monomer, without inducing any modifications on the surface or in the surrounding regions.

The non-linear character of the two-photon polymerization allows a localized excitation which gives the LDW via the TPP method a high resolution. This resolution depends mainly on the shape and size of the beam at the focal site. For plane waves, the minimal on-site dimension is expressed using Abbe's equation. However, in the case of pulsed lasers, the shape of the beam can generally be approximated to a Gaussian profile. The waist of the beam at the focal plane can be expressed as:

$$w_0 = \frac{\lambda}{\pi \mathrm{NA}}$$

where λ is the wavelength and NA is the numerical aperture of the system. The beam waist is defined as the distance from the Gaussian beam center to the point where the amplitude of the electric field decreases to 1/e of its maximum value. The equation above is an approximation that works well for small NAs (NA << 1).

While the resolution of optical systems is limited by the diffraction of their components, the process of two photons absorption pushes the manufacturing resolution below the diffraction limit. The two-photon polymerization configuration uses large numerical aperture optical systems, such as microscope lenses. For such systems, the beam waist can be written as:

$$w_0 = \frac{\lambda}{\pi \mathrm{NA}} \sqrt{n^2 - \mathrm{NA}^2}$$

The equation above shows that the smallest feature is limited by the numerical aperture of the microscope lens and the wavelength. However, since the rate of absorption of two photons is proportional to the square of incident beam intensity, and therefore depends on the Gaussian intensity profile of the square, the beam has a narrower waist. Therefore, due to the nonlinear nature of this process, it is possible to achieve a resolution below the diffraction limit. This phenomenon reduces the waist of the beam by a factor of $\sqrt{2}$.

Another aspect that contributes to improving the resolution of the two-photon polymerization process is the polymerization threshold. There is a minimum power below which there is no polymerization. This usually occurs due to the presence of oxygen in the resin, which inhibits the action of the photoinitiators and therefore prevents the polymerization reaction. Hence, by adjusting the laser power close to the polymerization threshold, it is possible to create structures with a resolution well below the diffraction limit.

9.3.3 LDW Via TPP of 3D Structures

The LDW via the TPP technique usually combines computer-controlled positioning systems with a near-infrared femtosecond laser source, e.g. a Ti: Sapphire laser working at 800 nm (Fig. 9.6). Such a setup can be an effective approach to achieve three-dimensional microstructuring of photosensitive materials by two-photon polymerization, with a resolution in the range of a few hundreds of nm. In order to better control the resolution, high precision positioning systems are used (e.g. piezoelectric stage/scanner). However, piezoelectric systems have a limited range, i.e. of few hundred microns in each direction. Alternatively, optical scanning systems may be used to move the laser beam over larger areas, but with inferior precision and higher aberrations.

In the following section, we discuss experiments in which the structures were fabricated by LDW via TPP of the photoresist IP-L 780 (Nanoscribe GmbH). The structures were polymerized on the surface of a glass slide using laser powers between 20 and 44 mW at a scanning velocity of 50 μm/s. Complex geometries were designed, with vertical microtubes arranged in triangular lattices having different constants. The structures were manufactured using the lithography system Photonic Professional 3D. The polymer was placed on a glass cover and irradiated by focused laser pulses with a wavelength of 780 nm, a pulse duration of 120 fs, and a repetition rate of 80 MHz. The precise 3D scanning of the samples was achieved using piezo stages.

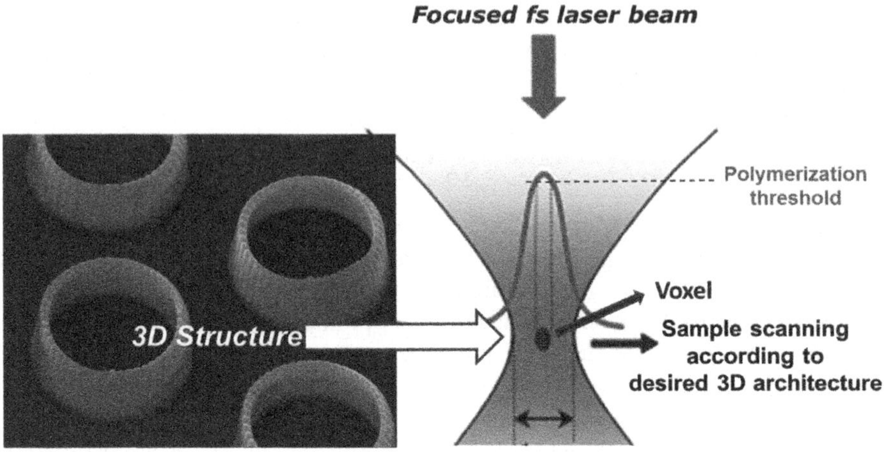

Fig. 9.6 Principle of direct laser writing by two photon polymerization. Scale bar in the left hand side image is 5 μm

Coarse positioning was achieved through XY motorized steps. Each vertical micro tube was produced by a rapid prototyping algorithm [88, 89] implemented through the General Writing Script (GWL) language developed by Nanoscribe.

A CCD camera was coupled to the experimental assembly to monitor the procedure in real time. After the microstructure is fabricated, the sample is immersed in isopropyl alcohol to wash and remove the non-consolidated monomer.

9.3.3.1 Fabrication and Functionality of 3D Microtube Arrays

The motivation behind fabricating 3D microtube arrays by LDW via TPP stemmed from the idea that the mechanical stimulation of bone-forming cells (osteoblasts) promotes the growth, differentiation, and mineralization of cells until new and functional bone forms. To verify the role of substrate stiffness on cellular behavior, micropillars having different heights were initially manufactured by LDW via TPP of ORMOCORE. It was found that the taller (i.e. more flexible) micropillars generated mechanical forces on the seeded osteoblasts, promoting their growth.

Therefore, the mechanical actuation of such microstructures via electrical stimulation could be a solution towards stimulating bone regeneration. Moving further, if the microstructures could incorporate osteogenic drugs and act as electrically activated microreservoirs, we could positively influence the process of osteogenesis by controlled drug release in conjunction with electrical stimulation of cells. For this purpose, we made vertical microtubes and checked whether they are favorable for the growth of bone-forming cells (osteoblasts) . The microtubes were designed to function as biomimetic structures that support cell growth and mineralization, as well as microreservoirs for loading osteogenic drugs to be released on demand by elec-

Fig. 9.7 SEM images of microtubes arrays. Note the microtubes arranged in lattices with different constants: (left panel) 8 μm (tightly packed); (middle panel) 12 μm (medium packed); (right panel) 24 μm (loosely packed). Reproduced with permission [20]

trical stimulation. To this end, vertical microtubes arranged in lattices of triangular units having 8 (tightly packed), 12 (medium packed) and 24 μm (loosely packed) constants, respectively, were produced (Fig. 9.7).

To assess the potential of the microstructures for the growth of bone tissue, MG63 cell cultures were evaluated by the cell viability trypan blue assay. The total dead and living cells per sample were counted and the viability was calculated. For all packing densities, the cell viability was higher than 90%, indicating that the arrays support cell viability.

The cells were afterwards stained with acridine orange (green)/Hoechst (blue) and visualized under a fluorescence microscope (Fig. 9.8). The staining of the cells grown on glass slides is shown for comparison (control). The insets show only the HO stained cell nuclei. Fluorescent staining showed that the arrays promoted the osteoblasts adhesion and that the cellular morphology was affected by the underlying microstructures. The cells seeded on the tightly packed microtube arrays retained their typical polygonal to spindle-like morphology (Fig. 9.8b) and looked similar to the cells grown on glass (Fig. 9.8a). The cells from the medium packed arrays showed elongated shapes, while some of the cells fell between the microtubes and squeezed through the narrow spaces between them (Fig. 9.8c). The cells from the loosely packed microtubes arrays can be seen on the bottom glass substrate; therefore, they underwent minimum exposure to the micro-relief structures (Fig. 9.8d).

The images also reveal that the cell nuclei were deformed in response to the respective micro-architectures, as revealed by the insets in Fig. 9.8. For the medium packed arrays, the nuclei underwent dramatic deformations, by following the walls of the microtubes (inset in Fig. 9.8c). On the tightly packed arrays, the shape changes

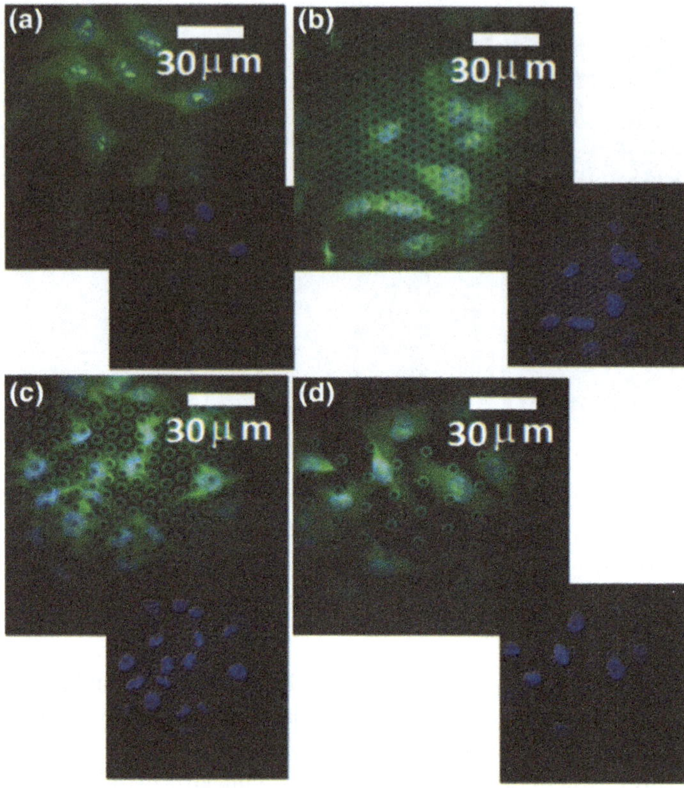

Fig. 9.8 MG63 osteoblast-like cells after 1 day of cell culture on **a** glass substrates (controls) and on arrays of different microtubes packing densities; **b** tightly packed; **c** medium. packed; **d** loosely packed. Reproduced with permission [20]

of the cells nuclei were subtler (inset in Fig. 9.8b). The loosely packed architecture had no effects on the cell nuclei.

The degree of induced cell-based mineralization was also evaluated. Cell mineralization was confirmed by EDX monitoring of the calcium deposited in the seeded osteoblasts. Calcium was detectable in all samples starting with Day 7. The cells from the tightly packed arrays showed the highest calcium deposition levels, while those from the loosely packed ones had the lowest level of mineralization, i.e. close to that of the glass seeded cells, about 3.4 times less than for the tightly packed microtube arrays. The medium packed arrays induced an intermediate level of cell mineralization. The improvement of osteogenesis on the tightly packed microtubes was most likely due to the changes in cell cytoskeletal arrangement and nucleus-induced microtubule architecture.

9.3.3.2 Electrically Controlled Release of Dex from the 3D Biomimetic Structures

The following step was to transform the microtubes in microreservoirs for the incorporation of dexamethasone (Dex), an anti-inflammatory and osteogenic drug. To control the release of Dex in the seeded cells, the microstructures were made electrically conductive by adding polypyrrole (PPy) to the microtubes. We refer to these microsystems as Electrically Responsive Microreservoirs (ERRs).

The PPy/Dex mixture (1/1 weight ratio) was loaded within the microtubes via a simple immersion process. To prevent the passive drug diffusion, the PPy/Dex loaded microtubes were sealed with a thin PLGA layer.

The release of Dex from the ERRs was monitored for 350 h. This time interval was selected based on previous studies indicating that osteoblastic differentiation reaches a maximum after 7–14 days of treatment [103–105].

The experimental results for Dex release from the ERRs are presented in Fig. 9.9. Without applied voltage, for all ERRs, Dex release began after approximately 50 h of immersion in the release medium. This time scale is consistent with previously published studies on PLGA layers used as coatings for Indometacin [106]. The passive release mechanism of the drug through PLGA coatings is mainly determined by bulk erosion of the polymer by randomly splitting the polymeric chains. The subsequent drug release dynamic was almost the same for all ERRs, regardless of the volume of the microtubes. The linear growth in the first 120 h was followed by stabilization, with cumulative release of Dex reaching only about 30%. On the other hand, electrical stimuli caused visible changes in the drug release profile. At the end of the ES protocol, approximately 98% of Dex was released from each ERR. With the increase in microtube volume, the drug release was completed faster and required less applied voltage cycles. The concentrations of released Dex increased slightly with the increase in microtubule volume.

Subsequently, the 3D electrically-stimulated biomimetic structures obtained by LDW via TPP were evaluated for tissue engineering applications. Thus, the first step was to evaluate the cell viability of the MG-63 cells by growing them on the ERRs. Cell viability was evaluated by trypan blue exclusion. Based on the observed cell counts, the calculated viability of the MG-63 cells grown at 24 h of cells culture on all ERRs was higher that 90%, proving that the ERRs allowed cell growth. At later stages of cell culture, corresponding to advanced differentiation stages, the cells proliferation process ceases [107, 108].

Further biological tests included the investigation of cell morphology and adhesion onto the ERR structures. The cell morphology attached to the ERRs was visualized with an F-actin, integrin, and nucleus immuno-fluorescence staining (Fig. 9.10). The fluorescence microscopy examination demonstrates that all ERRs facilitate the adhesion of the MG-63 osteoblasts at the same extent as the control surfaces (glass). Cell populations are characterized by randomized polygonal cells and short cytoplasmic processes [109], which were not significantly altered compared to cells grown on control surfaces (glass). It can be observed that the cells appear well-spread, with normal morphology, both on the ERRs as well as on the glass substrate. Integrins

Fig. 9.9 Cumulative release of Dex from ERRs with different microtubes volumes: (grey data points) without electrical stimulation; (colored data points) electrical stimulation with voltage cycles between −1 V and +1 V applied at time intervals indicated by the arrows. Reproduced with permission [89]

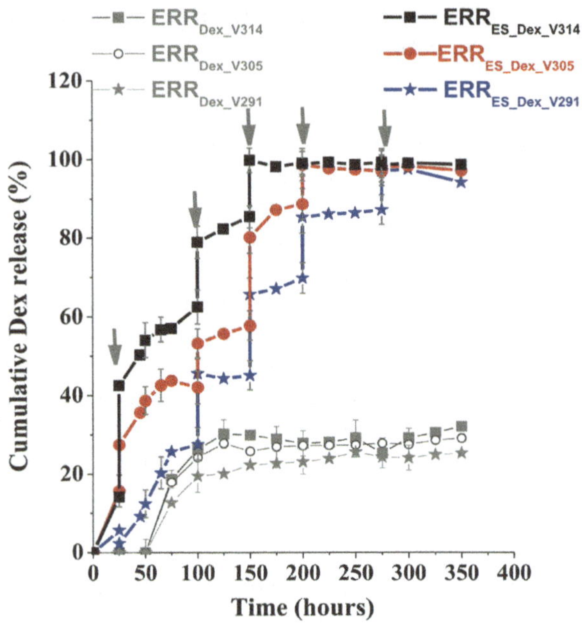

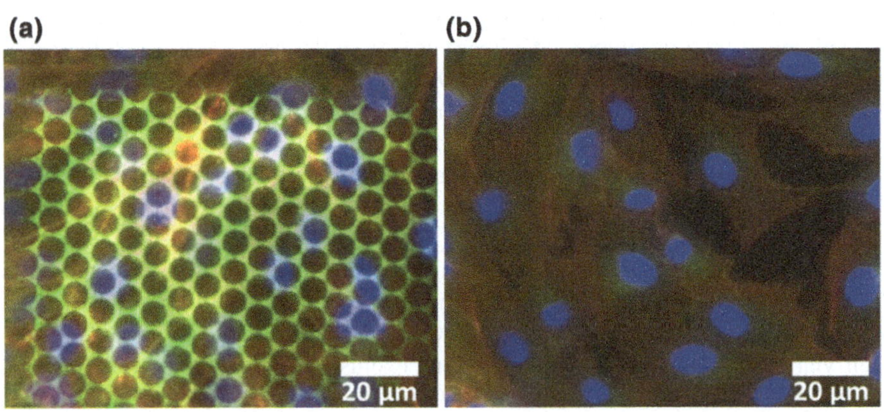

Fig. 9.10 Fluorescence microscope images of MG-63 cells grown on: **a** ERRs and **b** glass, as visualized by immunofluorescence staining of integrin (green: FTICa; red: F-actin cytoskeleton; blue: cells nuclei). Reproduced with permission [89]

(visible as green background) were present in the MG-63 cell cytoplasm, suggesting that cells adhered on the ERRs.

The osteogenic cell differentiation process was further monitored at time scales relevant for the osteogenic role of Dex release [103], which is 6 and 12 days, respectively, after electrical stimulation. The investigations were based on the measurement of ALP activity, Alizarin Red S staining, and rhRT-PCR reverse transcription polymerase chain reaction (RT-PCR).

It has been found that proliferative cells, after reaching confluence, enter the differentiation stage by producing ALP, one of the biochemical markers of long-term osteoblastic differentiation involved in regulating the availability of phosphate for the production of hydroxyapatite [110–113]. After 6 days of culture, there was no significant difference in ALP activity between stimulated and unstimulated samples. However, during 12 days of culture, ALP activity from the samples where Dex was released by electrical stimulation was almost twice as high as in non-stimulated samples. This indicates the beneficial role of the electrically controlled Dex release for promoting the cells osteoblastic differentiation.

One of the early indicators of the osteogenic differentiation phase is osteocalcin. Following electrical stimulation of the Dex-loaded ERRs, the qRT-PCR analysis showed that the level of osteocalcin gene expression in MG-63 cells increased by 2.2 times as compared to the cells grown on unstimulated control samples. This indicates that the controlled release of Dex from the electrically stimulated ERRs caused a stronger response of cells osteogenic differentiation. The mineralization was further confirmed by cell staining with Alizarin Red S, which is retained by the mineralized extracellular matrix. It was found that on both stimulated and unstimulated ERRs, the cells produced a mineralized extracellular matrix and revealed mature osteoblastic features. The red staining was significantly more pronounced for the electrically stimulated samples than for the unstimulated ones, proving the positive influence of the electrically-controlled Dex release on cell mineralization. The quantification of the mineralization level detected by Alizarin red S staining shows that the level of mineralization was 1.6 times higher for the stimulated samples than for the unstimulated ones.

9.3.4 Conclusions and Future Prospects

This section of the chapter addresses the direct laser writing via two photon polymerization (LDW via TPP) fabrication of polymeric 3D structures, with an emphasis on tissue engineering applications. LDW by TPP is a promising rapid prototyping fabrication method based on two-photon polymerization with ultrashort laser pulses that allows the fabrication of custom 3D architectures, with a spatial resolution down to 100 nm, by direct laser 'recording' of the desired structure into the volume of a photopolymer. After a brief overview of the potential and the basic principle of the LDW via TPP technique, we presented a specific application for bone tissue engineering, where LDW via TPP was used for the fabrication of vertical microtubes acting as microreservoirs for osteogenic drugs. We demonstrated that these microstructures can be efficiently used for electrically stimulates osteogenesis. The results show the versatility of the LDW via the TPP method, which uses environment-friendly, precise and versatile laser-assisted processes, and has the ability to produce 3D complex architectures that mimic in vivo media with spatial resolution in the range of tens of nanometers.

Acknowledgements This work was supported by the Romanian National Program 4N/2016, the Paul Scherrer Institut (Switzerland), and grants of the Romanian National Authority for Scientific Research and Innovation, CNCS—UEFISCDI, project number PN-II-RU-TE-2014-4-2311 (Flex-SENS), PNII-PT-PCCA-2013-4-1992 (SOLE), PN-II-RU-TE-2014-4-2534 and PN-III-P2-2.1-PED-2016-1715 (HERMESH). A part of this work (two photon polymerization experiments) was carried out in the CETAL laser facility, supported by the National Program PN 1647—LAPLAS IV.

The authors would like to thank Dr. Catalin Luculescu for the SEM analysis, Catalin Stelian Tuta for the drug release measurements by HPLC, and Dr. Marian Zamfirescu from CETAL-INFLPR for assistance with the software design of the microstructures.

Conflict of Interest The authors declare no competing interests.

References

1. R.S. Braudy, Proc. IEEE **57**, 1771–1772 (1969)
2. A.D. Brisbane, U.S. Patent 3,560,258, filled and issued May 31, 1967 issued 2 Feb 1971 (1971)
3. M.L. Levene, R.D. Scott, B.W. Siryj, Appl. Opt. **9**(10), 2260–2265 (1970)
4. J. Bohandy, B.F. Kim, F.J. Adrian, J. Appl. Phys. **60**(4), 1538–1539 (1986)
5. J. Bohandy, B.F. Kim, F.J. Adrian, A.N. Jette, J. Appl. Phys. **63**(4), 1158–1162 (1988)
6. V. Sametoglu, V.T.K. Sauer, Y.Y. Tsui, Opt. Express **21**(15), 18525–18531 (2013)
7. V. Schultze, M. Wagner, Appl. Surf. Sci. **52**, 303 (1991)
8. Z. Kantor, Z.S. Toth, T. Szorenyi, A.L. Toth, Appl. Phys. Lett. **64**, 3506 (1994)
9. J.A. Greer, T.E. Parker, SPIE Proc. **998**, 113 (1988)
10. I. Zergioti, S. Mailis, N.A. Vainos, P. Papakonstantinou, C. Kalpouzos, C.P. Grigoropoulous, C. Fotakis, Appl. Phys. A **66**, 579–582 (1998)
11. E. Fogarassy, C. Fuchs, F. Kerherve, G. Hauchecorne, J. Perrière, J. Mater. Res. **4**, 1082 (1989)
12. C.W. Visser, R. Pohl, C. Sun, G.-W. Römer, B. Huis in't Veld, D. Lohse, Adv. Mat. **27**, 4087–4092 (2015)
13. M. Zenou, Z. Kotler, Opt. Express **24**(2), 1431–1446 (2016)
14. J. Luo, R. Pohl, L. Qi, G.-W. Römer, C. Sun, D. Lohse, C.W. Visser, Small 1602553 (2017)
15. S. Walsh, G.R. Jordan, C. Jefferiss, K. Stewart, J.N. Beresford, Rheumatology **40**, 74–83 (2001)
16. Y. Peck, D.A. Wang, Expert Opin. Drug Deliv. **10**, 369–383 (2013)
17. D.A. LaVan, T. McGuire, R. Langer, Nat. Biotech. **21**, 1184–1191 (2003)
18. S. Mura, J. Nicolas, P. Couvreur, Nat. Mater. **12**, 991–1003 (2013)
19. B.P. Timko, T. Dvir, D.S. Kohane, Adv. Mater. **22**, 4925–4943 (2010)
20. M. Mihailescu, I.A. Paun, M. Zamfirescu, C.R. Luculescu, A.M. Acasandrei, M. Dinescu, J. Mat. Sci. **61**, 4262–4273 (2016)
21. H.B. Sun, T. Tanaka, S. Kawata, Appl. Phys. Lett. **80**, 3673–3680 (2002)
22. H.B. Sun, S. Kawata, Adv. Polym. Sci. **170**, 169–273 (2004)
23. D.S. Correa, L. De Boni, A.J.G. Otuka, V. Tribuzi, C.R. Mendonça, Two-Photon Polymerization Fabrication of Doped Microstructures, in *Polymerization*, Chap. 15, 12 Sept 2012
24. O. Kufelt, A. El-Tamer, C. Sehring, M. Meißner, S. Schlie-Wolter, B.N. Chichkov, Acta Biomater. **18**, 186–195 (2015)
25. A. Marino, C. Filippeschi, G.G. Genchi, V. Mattoli, B. Mazzolai, G. Ciofani, Acta Biomater. **10**, 4304–4313 (2014)

26. M. Staples, Microchips and controlled-release drug reservoirs, wiley interdiscip. Rev. Nanomed. Nanobiotechnol **2**, 400–417 (2010)
27. A. Piqué, D.B. Chrisey, R.C.Y. Auyeung, J. Fitz-Gerald, H.D. Wu, R.A. McGill, S. Lakeou, P.K. Wu, V. Nguyen, M. Duignan, Appl. Phys. A **69**(7), S279–S284 (1999)
28. A. Pique, D.B. Chrisey, R.C.Y. Auyeung, S. Lakeou, R. Chung, R.A. McGill, P.K. Wu, M. Duignan, J. FitzGerald, H.D. Wu, SPIE **3618**, 330–339 (1999)
29. R.C.Y. Auyeung, H. Kim, A.J. Birnbaum, M. Zalalutdinov, S.A. Mathews, A. Pique, Appl. Phys. A Mater. Sci. Process. **97**(3), 513–519 (2009)
30. A.J. Birnbaum, H. Kim, N.A. Charipar, A. Pique, Appl. Phys. A Mater. Sci. Process. **99**(4), 711–716 (2010)
31. W.A. Tolbert, I.-Y.S. Lee, M.M. Doxtader, E.W. Ellis, D.D. Dlott, J. Imaging Sci. Technol. **37**(4), 411–421 (1993)
32. V. Dinca, R. Fardel, F. Di Pietrantonio, D. Cannatà, M. Benetti, E. Verona, A. Palla-Papavlu, M. Dinescu, T. Lippert, Sens. Lett. **8**(3), 436–440 (2010)
33. C. Boutopoulos, V. Tsouti, D. Goustouridis, S. Chatzandroulis, I. Zergiotti, Appl. Phys. Lett. **93**, 191109 (2008)
34. J.M. Fernandez-Pradas, M. Colina, P. Serra, J. Dominguez, J.L. Morenza, Thin Solid Films **453–454**, 27–30 (2004)
35. B. Hopp, T. Smausz, N. Kresz, N. Barna, Z. Bor, L. Kolozsvári, D.B. Chrisey, A. Szabó, A. Nógrádi, Tissue Eng. **11**, 1817–1823 (2005)
36. J.A. Barron, P. Wu, H.D. Ladouceur, B.R. Ringeisen, Biomed. Microdevices **6**(2), 139–147 (2004)
37. B. Hopp, T. Smausz, Zs Antal, N. Kresz, Zs Bor, D. Chrisey, J. Appl. Phys. **96**(6), 3478–3481 (2004)
38. N.T. Kattamis, P.E. Purnick, R. Weiss, C.B. Arnold, Appl. Phys. Lett. **91**, 171120 (2007)
39. N.T. Kattamis, N.D. McDaniel, S. Bernhard, C.B. Arnold, Appl. Phys. Lett. **94**, 103306 (2009)
40. A. Palla-Papavlu, I. Paraico, J. Shaw-Stewart, V. Dinca, T. Savopol, E. Kovacs, T. Lippert, A. Wokaun, M. Dinescu, Appl. Phys. A Mater. Sci. Process. **102**(3), 651–659 (2011)
41. D. Banks, K. Kaur, R. Gazia, R. Fardel, M. Nagel, T. Lippert, R. Eason, EPL (Europhys. Lett.) **83**, 38003 (2008)
42. A. Doraiswamy, R. Narayan, T. Lippert, L. Urech, A. Wokaun, M. Nagel, B. Hopp, M. Dinescu, R. Modi, R. Auyeung, D. Chrisey, Appl. Surf. Sci. **252**, 4743–4747 (2006)
43. T. Mattle, J. Shaw-Stewart, C.W. Schneider, T. Lippert, A. Wokaun, Appl. Surf. Sci. **258**, 9352–9354 (2012)
44. F. Di Pietrantonio, M. Benetti, D. Cannata, E. Verona, A. Palla-Papavlu, V. Dinca, M. Dinescu, T. Mattle, T. Lippert, Sens. Actuators, B **174**, 158–167 (2012)
45. S. Latsch, H. Hiraoka, Mat. Res. Soc. Symp. Proc. **304**, 103–110 (1993)
46. M. Farsari, V. Dinca, M. Dinescu, T.S. Drakakis, C. Fotakis, Proc. of SPIE **6606**, 66061F (2007)
47. V. Dinca, E. Kasotakis, J. Catherine, A. Mourka, A. Mitraki, A. Popescu, M. Dinescu, M. Farsari, C. Fotakis, Appl. Surf. Sci. **254**, 1160–1163 (2007)
48. V. Dinca, A. Ranella, A. Popescu, M. Dinescu, M. Farsari, C. Fotakis, Appl. Surf. Sci. **254**, 1164–1168 (2007)
49. R. Fardel, P. Feurer, T. Lippert, M. Nagel, F. Nuesch, A. Wokaun, Appl. Surf. Sci. **254**, 1332–1337 (2007)
50. R. Fardel, M. Nagel, F. Nüesch, T. Lippert, A. Wokaun, Appl. Phys. Lett. **91**, 061103-(1–3) (2007)
51. J. Xu, J. Liu, D. Cui, M. Gerhold, A. Wang, M. Nagel, T. Lippert, Nanotechnology **18**(2), 025403 (2007)
52. R. Fardel, M. Nagel, F. Nuesch, T. Lippert, A. Wokaun, Appl. Surf. Sci. **254**, 1322–1326 (2007)
53. T.V. Kononenko, I.A. Nagovitsyn, G.K. Chudinova, Bull. Lebedev. Phys. Inst. **36**(2), 29–33 (2009)
54. L. Rapp, C. Cibert, A.P. Alloncle, P. Delaporte, Appl. Surf. Sci. **255**, 5439–5443 (2009)

55. J. Shaw-Stewart, T. Lippert, M. Nagel, F. Nuesch, A. Wokaun, ACS Appl. Mater. Interfaces **3**, 309–316 (2011)
56. J. Shaw-Stewart, T. Lippert, M. Nagel, F. Nuesch, A. Wokaun, ACS Appl. Mater. Interfaces **4**, 3535–3541 (2012)
57. V. Dinca, A. Palla Papavlu, M. Dinescu, J. Shaw-Stewart, T. Lippert, F. Di Pietrantonio, D. Cannata, M. Benetti, E. Verona, Appl. Phys. A **101**, 559–565 (2010)
58. V. Dinca, A. Palla Papavlu, A. Matei, C. Luculescu, M. Dinescu, T. Lippert, F. Di Pietrantonio, D. Cannata, M. Benetti, E. Verona, Appl. Phys. A **101**, 429–434 (2010)
59. D. Cannata, M. Benetti, F. Di Pietrantonio, E. Verona, A. Palla Papavlu, V. Dinca, M. Dinescu, T. Lippert, Sens. Actuators, B **173**, 32–39 (2012)
60. V. Dinca, T. Mattle, A. Palla Papavlu, L. Rusen, C. Luculescu, T. Lippert, M. Dinescu, Appl. Surf. Sci. **278**, 190–197 (2013)
61. A. Palla-Papavlu, V. Dinca, I. Paraico, A. Moldovan, J. Shaw-Stewart, C.W. Schneider, E. Kovacs, T. Lippert, M. Dinescu, J. Appl. Phys. **108**, 033111–033116 (2010)
62. A. Palla Papavlu, V. Dinca, C. Luculescu, J. Shaw-Stewart, M. Nagel, M. Dinescu, J. Opt. **12**, 124114–124119 (2010)
63. L. Rapp, A.K. Diallo, S. Nénon, A.P. Alloncle, C. Videlot-Ackermann, F. Fages, M. Nagel, T. Lippert, P. Delaporte, Thin Solid Films **520**, 3043–3047 (2012)
64. T. Inui, R. Mandamparambil, T. Araki, R. Abbel, H. Koga, M. Nogia, K. Suganuma, RSC Advances **5**, 77942–77947 (2015)
65. A. Palla-Papavlu, M. Dinescu, A. Wokaun, T. Lippert, Appl. Phys. A **117**, 371–376 (2014)
66. J. Chen, A. Palla Papavlu, Y. Li, L. Chen, M. Dobeli, D. Stender, S. Populoh, A. Weidenkaff, C.W. Schenider, A. Wokaun, T. Lippert, Appl. Phys. Lett. **104**, 231907-(1–3) (2014)
67. T. Mattle, A. Hintennach, T. Lippert, A. Wokaun, Appl. Phys. A **110**, 309–316 (2013)
68. A. Palla Papavlu, T. Mattle, S. Temmel, U. Lehmann, A. Hintennach, A. Grisel, A. Wokaun, T. Lippert, Sci. Rep. **6**, 25144 (1–9) (2016)
69. B. Mitu, A. Matei, M. Filipescu, A. Palla Papavlu, A. Bercea, T. Lippert, M. Dinescu, J. Phys. D Appl. Phys. **50**, 115601 (2017)
70. D. Young, R.C.Y. Auyeung, A. Pique, D.B. Chrisey, D. Dlott, Appl. Phys. Lett. **78**, 3169 (2001)
71. J.A. Barron, H.D. Young, D.D. Dlott, M.M. Darfler, D.B. Krizman, B.R. Ringeisen, Proteomics **5**, 4138 (2005)
72. M. Colina, M. Duocastella, J.M. Fernandez-Pradas, P. Serra, J.L. Morenza, J. Appl. Phys. **99**, 084909 (2006)
73. M. Duocastella, J.M. Fernandez-Pradas, J.L. Morenza, P. Serra, J. Appl. Phys. **106**, 084907 (2009)
74. A. Palla-Papavlu, J. Shaw-Stewart, T. Mattle, V. Dinca, T. Lippert, A. Wokaun, M. Dinescu, Appl. Surf. Sci. **278**, 180–184 (2013)
75. M. Nagel, R. Fardel, P. Feurer, M. Häberli, F. Nüesch, T.K. Lippert, A. Wokaun, Appl. Phys. A **92**, 781–789 (2008)
76. J. Shaw-Stewart, T. Mattle, T. Lippert, M. Nagel, F. Nuescu, A. Wokaun, Appl. Surf. Sci. **278**, 341–346 (2013)
77. F. Di Pietrantonio, M. Benetti, D. Cannatà, E. Verona, A. Palla-Papavlu, J.M. Fernández-Pradas, P. Serra, M. Staiano, A. Varriale, S. D'Auria, Biosens. Bioelectron. **67**, 516–523 (2015)
78. A. Palla Papavlu, V. Dinca, M. Dinescu, Chapter 14 Laser printing of proteins and biomaterials, in *Laser Printing of Functional Materials: Electronics, 3D Microfabrication and Biomedicine*, 1st edn, ed. by A. Piqué, S. Pere (©2017 Wiley-VCH Verlag GmbH & Co. KGaA), pp. 329–348. Published 2017 by Wiley-VCH Verlag GmbH & Co. KGaA
79. A. Palla Papavlu, T. Lippert, Chapter 3 LIFT using a dynamic release layer, in *Laser Printing of Functional Materials: Electronics, 3D Microfabrication and Biomedicine*, 1st edn, ed. by A. Piqué, S. Pere (©2017 Wiley-VCH Verlag GmbH & Co. KGaA). Published 2017 by Wiley-VCH Verlag GmbH & Co. KGaA
80. R.B. Greenwald, J. Contr, Release **74**, 159–171 (2001)

81. J. Meng, B. Xiao, Y. Zhang, J. Liu, H. Xue, J. Lei, H. Kong, Y. Huang, Z. Jin, N. Gu, H. Xu, Sci Rep. **3**, 2655 (2013)
82. F. McLaughlin, J. Mackintosh, B.P. Hayes, A. McLaren, I.J. Uings, P. Salmon, J. Humphreys, E. Meldrum, S.N. Farrow, Bone **30**, 924–930 (2002)
83. S.C. Balmert, S.R. Little, Adv. Mater. **24**, 3757–3778 (2012)
84. L. Ghasemi-Mobarakeh, M.P. Prabhakaran, M. Morshed, M.H. Nasr-Esfahani, H. Baharvand, S. Kiani, S.S. Al-Deyab, S. Ramakrishna, J. Tissue Eng. Regen. Med. **5**, 17–35 (2011)
85. J. Zhang, K.G. Neoh, X. Hu, E.T. Kang, W. Wang, Biotechnol. Bioeng. **110**, 1466–1475 (2013)
86. W.W. Hu, Y.T. Hsu, Y.C. Cheng, C. Li, R.C. Ruoh, C.C. Chien, C.A. Chung, C.W. Tsao, Mater. Sci. Eng., C **37**, 28–36 (2014)
87. S. Meng, Z. Zhang, M. Rouabhia, J. Bone, Miner. Metab. **29**, 535–544 (2011)
88. I.A. Paun, F. Stokker-Cheregi, C.R. Luculescu, A.M. Acasandrei, V. Ion, M. Zamfirescu, C.C. Mustaciosu, M. Mihailescu, M. Dinescu, Mater. Sci. Eng. C Mater. Biol. Appl. **55**, 61–69 (2015)
89. I.A. Paun, M. Zamfirescu, C.R. Luculescu et al., Appl. Surf. Sci. **392**, 321–331 (2017)
90. I.A. Paun, A.M. Acasandrei, C.R. Luculescu, C.C. Mustaciosu, V. Ion, M. Mihailescu, E. Vasile, M. Dinescu, Appl. Surf. Sci. **357**, 975–984 (2015)
91. X. Jiang, Y. Marois, A. Traoré, D. Tessier, L.H. Dao, R. Guidoin, Z. Zhang, Tissue Eng. **8**, 635–647 (2002)
92. G. Shi, M. Rouabhia, Z. Wang, L.H. Dao, Z. Zhang, Biomaterials **25**, 2477–2488 (2004)
93. S.S. Mihardja, R.E. Sievers, R.J. Lee, Biomaterials **29**, 4205–4210 (2008)
94. E. De Giglio, L. Sabbatini, S. Colucci, G. Zambonin, J. Biomater. Sci. Polym. Edn **11**, 1073–1083 (2000)
95. E. De Giglio, S. Cometa, C.D. Calvano, L. Sabbatini, P.G. Zambonin, S. Colucci, A.D. Benedetto, G. Colaianni, J. Mater. Sci. Mater. Med. **18**, 1781–1789 (2007)
96. B. Zinger, L.L. Miller, J. Am. Chem. Soc. **106**, 6861–6863 (1984)
97. L.L. Miller, G.A. Smith, A.C. Chang, Q.X. Zhou, J. Contr. Rel. **6**, 293–296 (1987)
98. M. Pyo, G. Maeder, R.T. Kennedy, J.R. Reynolds, J. Electroanal. Chem. **368**, 329–332 (1994)
99. S.B. Adeloju, S.J. Shaw, G.G. Wallace, Anal. Chim. Acta **341**, 155–160 (1997)
100. M.R. Abidian, D.H. Kim, D.C. Martin, Adv. Mater. **8**, 405–409 (2006)
101. P.M. George, D.A. LaVan, J.A. Burdick, C.Y. Chen, E. Liang, R. Langer, Adv. Mater. **18**, 577–581 (2006)
102. Y. Li, K.G. Neoh, E.T. Kang, J. Biomed, Mater. Res. Part A **73**, 171–181 (2005)
103. B. Massoumi, A.A. Entezami, Eur. Polym. J. **37**, 1015–1020 (2001)
104. F. Langenbach, J. Handsche, Stem Cell Res. Ther. **4**, 117–123 (2013)
105. S. Walsh, G.R. Jordan, C. Jefferiss, K. Stewart, J. Beresford, Rheumatology **40**, 74–83 (2003)
106. I.A. Paun, A. Moldovan, C.R. Luculescu, M. Dinescu, Appl. Surf. Sci. **257**, 10780–10788 (2011)
107. M.R. Abidian, D.H. Kim, D.C. Martin, Adv. Mater. **18**, 405–409 (2006)
108. I. Kalajzic et al., J. Biol. Chem. **280**, 24618–24626 (2005)
109. G. Teti, A. Bigi, M. Mattioli-Belmonte, R. Giardino, M. Fini, A. Mazzotti, M. Falconi, J. Life Sci. **7**, 965–970 (2013)
110. H.C. Anderson, Lab. Invest. **60**, 320–330 (1989)
111. A. Nakamura, Y. Dohi, M. Akahane, H. Ohgushi, H. Nakajima, H. Funaoka, Y. Takakura, Tiss. Eng. Part C: Meth. **15**, 169–180 (2009)
112. S.L. Cheng, J.W. Yang, L. Rifas, S.F. Zhang, L.V. Avioli, Endocrinology **134**, 277–286 (1994)
113. J.J. Alm, T.J. Heino, T.A. Hentunen, H.K. Väänänen, H.T. Aro, Tissue Eng. Part C: Meth. **18**, 658–666 (2012)

Chapter 10
UV- and RIR-MAPLE: Fundamentals and Applications

Anna Paola Caricato, Wangyao Ge and Adrienne D. Stiff-Roberts

Abstract Laser beams are powerful tools for diagnostic purposes and growth of thin films. However, the interaction of lasers with organic or biological materials may result in laser-induced photo-chemical and photo-thermal damage to the materials of interest. In order to reduce these effects, the matrix-assisted pulsed laser evaporation (MAPLE) technique was introduced for thin film deposition with the possibility to use two different wavelength ranges to induce target ablation: laser absorption by the matrix (UV-MAPLE) or direct laser excitation of vibrational frequency of the matrix constituent bonds (RIR-MAPLE). In this chapter, the UV-MAPLE physical working principles will be outlined together with its applications and latest results. The RIR-MAPLE working principle and applications to materials of interest for optoelectronics applications will be described and discussed also, with particular attention to the emulsion-based RIR-MAPLE approach relating the film properties with the physical-chemical properties of the emulsion components (primary and secondary solvent, surfactant and matrix).

10.1 Introduction

In the last decades, organic materials have attracted increasing attention as interesting alternatives to inorganic materials in many device applications like field effect transistors [1], light-emitting diodes (LEDs) [2, 3], photovoltaic devices [4], sensors [5] and white light sources for indoor and outdoor lighting [6].

A. P. Caricato (✉)
Department of Mathematics and Physics "E. De Giorgi", University of Salento, 73100 Lecce, Italy
e-mail: Annapaola.Caricato@le.infn.it

W. Ge · A. D. Stiff-Roberts
Department of Electrical and Computer Engineering, Duke University, Durham, NC 27708-0291, USA

P. M. Ossi (ed.), *Advances in the Application of Lasers in Materials Science*, Springer Series in Materials Science 274,
https://doi.org/10.1007/978-3-319-96845-2_10

The interest in these materials relies on their low cost and the possibility of a wide modulation of their chemical, optical and electronic properties by tailoring the organic molecular structure [7–9].

To intelligently design new and high-performance systems and devices, the deposition of organic materials in thin film form is necessary and requires accurate control over their morphological, chemical and structural properties. Many applications, like those for the electronics industry, require the deposition of films with good uniformity containing single or multilayer structures of organic or polymeric materials, homogeneous composite materials, or materials with graded compositions. More stringent requirements are necessary in other applications, like polymer-LED applications [10], such as film surface coverage uniformity, thickness control and good interfacial properties. Moreover, conformality[1] has become an increasingly important characteristic in the fabrication of optoelectronic and medical devices having high aspect ratio features, 3D geometries, and textured/nanostructured surfaces. Conformal coatings are necessary also for modifying the internal surfaces of porous materials (including membranes, foams, and textiles) or irregular surface geometries, as well as for encapsulating fibers, nanowires, or particles [11].

Conventional techniques for depositing organic materials are generally non-vacuum, solution-based methods and include drop casting [12], spin coating [13], ink-jet printing [14], spraying [15] and Langmuir Blodgett depositions [16].

Although many successful devices have been demonstrated using these techniques, they suffer from poor thickness control and are unable to provide controlled deposition on patterned substrates, inhibiting conformal coverage. The interaction energies between solution components coupled with the overall interfacial energy with the substrate can result in either meniscus formation inside a feature or capillary bridge formation over a feature [11].

Moreover, these solution-based techniques are unable to control inherent, solvent-induced conformational defects [17] and are not compatible with the realization of heterostructured devices [18] because an underlying layer can be damaged by the solvent during the deposition of the overlying layer. In fact, with these techniques, only bilayer structures can be deposited using orthogonal solvents [19].

Also, vacuum methods have been applied to thin polymer film preparation: plasma polymerization [20], electron beam irradiation [21, 22], ultraviolet (UV) light vapor irradiation [23, 24], thermally assisted vacuum evaporation [25] and polymer sputtering [26]. With respect to traditional solution-based methods (e.g., spin casting, dip coating), vapor phase methods of polymer thin films offer the advantage of forming conformal films on high aspect ratio structures; however, these techniques can be applied only for the deposition of polymers with a low molecular weight such that the polymers evaporate as polymer units and condense on the substrate without a reaction.

In this scenario, a new non-destructive laser-based approach was developed at the Naval Research Laboratory in the late 90s [27] for the deposition of organic

[1] Conformal coverage is achieved when a film of uniform thickness precisely follows the geometry of the underlying substrate.

materials, even with complex chemical structures, including bio-organic molecules, coordination compounds and polymers [28, 29]: the matrix-assisted pulsed laser evaporation (MAPLE) technique. This technique can operate with lasers emitting in the UV (UV-MAPLE) or IR (RIR-MAPLE) range of the electromagnetic spectrum. In the following, a brief description of the main characteristics of the UV-MAPLE technique is provided together with a review of its recent applications, referring to previous publications for more details [30, 31]. More attention will be devoted to the description of the RIR-MAPLE technique and, in particular, the emulsion-based RIR-MAPLE approach. The working principle of the technique will be illustrated and discussed, relating the film properties with the physical-chemical properties of the emulsion components.

10.2 Conventional UV-MAPLE

The MAPLE technique is a derivative of the pulsed laser deposition technique (PLD) [32] in which some stratagems are used in order to avoid/minimize the photo-thermal and photo-chemical damage induced by the high energies and temperatures characteristic of a laser-material interaction process under ablation conditions.

Operatively, in MAPLE experiments, the material of interest (solute) is dissolved/suspended in a volatile and absorbing (at the used laser wavelength) solvent with concentration of a low weight percentage in order to produce a homogeneous solution [28]. The solution (MAPLE-matrix) is frozen at the liquid nitrogen temperature forming the target to be irradiated [28] by ultra-violet (UV) or infra-red (IR) laser sources [33].

The use of few wt% of the solute and the absorbing and volatile solvent are such to reduce/minimize the photochemical and photothermal damage respectively. The ablation onset is defined by the thermodynamic parameters of the solvent and for this reason, depositions can proceed at much lower fluencies (laser energy divide by laser spot area at the target surface J/cm^2), as compared to conventional PLD (0.05–0.5 J/cm^2).

A detailed description of the MAPLE deposition set up is reported in [30].

Like all deposition techniques, the MAPLE technique presents advantages and disadvantages.

The main UV-MAPLE advantages are quite simple procedures; in principle, good control of film thickness; different independent deposition parameters; deposition on non-planar substrates with good uniformity; and the possibility to deposit multilayer and composite films.

Meanwhile, drawbacks of the UV-MAPLE technique include: high roughness values (very often the surface of MAPLE-deposited films presents a high density of micron sized droplets); low deposition rates with UV laser (one order of magnitude lower than those for PLD); and possible formation of reactive radicals due to photochemical reactions of the solvent molecules which can react with the solute material.

Table 10.1 Main deposition parameters in molecular dynamic simulation and in a typical MAPLE experiment [40]

	Simulation	Experiments
τ_p	50 ps	20-30 ns
L_p	50 nm	4 μm
τ_{th}	7 ns	181 μs
τ_s	20 ps	3 ns

Initially, an evaporative process was assumed as the working principle of the MAPLE process: the laser energy is mainly absorbed by the solvent and converted into thermal energy, which causes the vaporization of the solvent layer and only a moderate heating of the solute. The collective motion of the many solvent molecules, evaporated by a single laser pulse, carries the few solute molecules present in the evaporated layer to the substrate. The solvent molecules (volatile) are pumped away from the deposition chamber during the time of flight (low sticking coefficient).

However, this mechanism was in contradiction with the high surface roughness values and the presence of aggregates or clusters with sizes ranging from tens of nanometers to tens of microns which characterized the surface of MAPLE deposited films as evidenced by AFM and SEM images [34–39]. In fact, the formation of large polymer features was rather unexpected since the polymer concentration in the target is low and the polymer molecules are dissolved in the matrix down to the molecular level and the expanding plasma plume should not provide a suitable environment for condensation of polymer clusters.

Molecular dynamic simulation studies shed light on the process [35, 40]. Leveugle and Zhigilei considered that the concentrations of polymer molecules (typically 0.1–5 wt%) and the collective behaviour of polymer molecules may play an important role in defining the mechanisms of molecular ejection and the morphological characteristics of the deposited films.

Although molecular dynamic simulations and conventional UV-MAPLE experiments are characterized by different values of the experimental conditions (typical values of deposition parameters are reported in Table 10.1), they are characterized by similar physical conditions and for this reason the ejection mechanisms revealed in the simulations also work in experiments (albeit at much larger time and length scales[2]).

Both simulations and experiments are generally characterized by the following conditions:

$$\tau_s < \tau_p < \tau_{th}, \tag{10.1}$$

which means that the MAPLE process proceeds under the regimes of

[2]The natural limit of the applicability of the model are defined by the conditions for the onset of multiphoton absorption, photochemical fragmentation, ionization and plasma formation.

(1) Thermal confinement:

$$\tau_p < \tau_{th} = \frac{L_p^2}{A\,D_T} \tag{10.2}$$

and

(2) No stress confinement:

$$\tau_p > \tau_s \approx \frac{L_p}{C_S} \tag{10.3}$$

In the equations above, τ_p is the laser pulse duration, D_T the thermal diffusivity, L_p the laser penetration depth, A is a constant defined by the geometry of the absorption region and C_S is the speed of sound in target material.

In the thermal confinement regime, the heat conduction does not contribute to the energy redistribution during the laser pulse and the thermal energy is largely confined within the absorbing region. While, in the absence of stress confinement, the rate of the laser energy deposition is sufficiently low to allow for the mechanical relaxation (expansion) of the absorption region of the target during the laser pulse.

From molecular dynamic simulations, at the early stage of the laser-target interaction, it results that two regimes characterize the MAPLE process: below and above a threshold fluence value. The existence of a well-defined threshold fluence is related to the transition from a metastable, superheated liquid to a two-phase mixture of liquid and vapor at a temperature of approximately 90% of the critical temperature.

Below a threshold fluence value an evaporative process takes place where only thermal evaporation of individual matrix molecules and no ejection of polymer molecules is observed.

Above a threshold fluence value an explosive process occurs in which a prompt ejection of small cluster and liquid droplets of matrix-polymer structures as well as of matrix molecules take place. Polymer molecules are only ejected as part of matrix-polymer clusters [40]. A complete and detailed description of the process is reported in [40].

Clearly, the high surface roughness values observed in MAPLE-deposited films is an intrinsic aspect of the MAPLE process, and many papers consider the influence of various MAPLE deposition parameters on roughness, such as matrix solvent [41–43], substrate temperature [37], laser fluence [38, 41–47], number of laser pulses [48], polymer concentration [49, 50], laser-target interaction volume and laser wavelength [46, 51]. However, with good control of the deposition parameters, it is possible to greatly reduce the surface roughness values, as demonstrated for the green emitting polymer poly[9,9-dioctylfluorenylene-2,7-diyl)-co-(1,4-diphenylene-vinylene-2-methoxy-5-2{ethylhexyloxy}benzene)] (commercially-named ADS125GE). Experimental results showed that a good interplay among the deposition parameters, such as polymer concentration, target—substrate distance and substrate rotation frequency, makes it possible to deposit a very flat (9 nm-rough) polymeric layer with a thickness of about 40 nm without modifying the emission properties of the polymer [52].

As described in the above, the solvent plays an important role in the MAPLE mechanism because it: (i) must dissolve the solute without interacting with it (also under laser irradiation); (ii) must be volatile; (iii) must absorb the laser radiation and (iv) must "transport" the solute from the target to the substrate. The solvent provides the conditions for "soft" ejection and deposition of fragile macromolecules, but it can also have an effect on the film surface and deleterious effects on the quality of the deposited films, which presents a number of problems for applications sensitive to solvent contamination and/or carried out in UHV conditions.

So, the choice of the solvent is critical, and in addition to the aforementioned aspects, it must be considered that some solvent-related effects may impact deposition, like the co-deposition of organic material [37, 53], the formation of solvent by-product, which may interact with the solute [37], and the presence of solvent in the deposited films. In fact, molecular dynamic simulations have demonstrated that part of the solvent reaches the substrate. This aspect has been demonstrated experimentally using a substrate on which a polymer (polydimethylsiloxane-PDMS) "sensitive" to the solvent has been used as test and placed in front of the substrate and irradiating with an ArF laser at the conventional fluences used in MAPLE depositions (60–250 mJ/cm^2) a frozen toluene target placed in front of it [54].

A study of the minimum amount of solvent necessary for ablation without destroying the solute material was reported in [55] in which molecular dynamic simulations and experiments were carried on a lysozyme–water system. The simulations were performed below and above the threshold fluence value, considering only the solute material (lysozyme) as target and the lysozyme–water system (water content from 0 to 20 wt%). From that study, considering only the lysozyme target, it resulted that (i) below the threshold fluence value only the desorption of several small molecular fragments without ablation was observed, while (ii) above the threshold fluence value ablation was observed but without intact lysozime globules. The pyrolytic decomposition of polymer molecules drives the ejection of the larger polymer fragments and molecular droplets. Adding small wt% of water in the target resulted in a decrease of the ablation threshold and the vaporization of water pockets resulting in the target expansion and generation of large bubbles filled with superheated water vapor without lysome ejection. Just above the ablation threshold expanding water vapor breaks the resistance of the interconnected viscous network of lysozyme molecules, entrains droplets of polymer material, and results in the ejection of almost the entire target. For higher values of the laser fluence (3 times higher than the ablation threshold) the ejection of large polymer droplets is driven by the expansion of a mixture of water vapor and small volatile products of polymer and thermal decomposition was observed. The final result of the study was a phase diagram for obtaining ablation of intact lysozime molecules considering the smallest amount of water at different values of laser fluence.

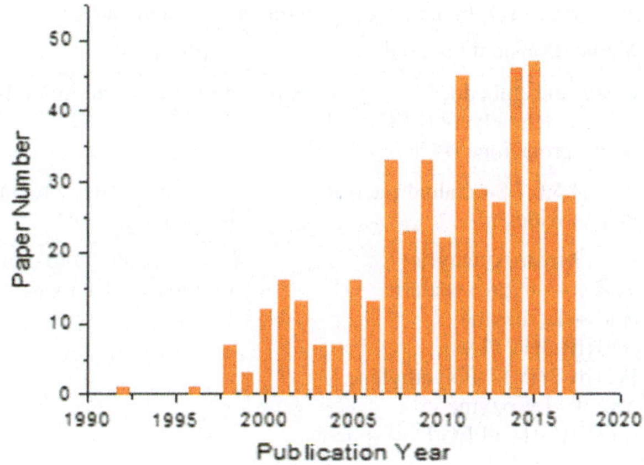

Fig. 10.1 Time evolution of MAPLE publications since its development

10.3 UV-MAPLE: Applications

The successful deposition of many polymers by MAPLE [56] and the very interesting properties reported for MAPLE deposited films have recently prompted a lot of interest towards this technique.

Further, the working principle of the technique has been successfully applied to the deposition of colloidal nanoparticles [30, 57].

It is precisely the characteristic MAPLE process that gives the films special properties. A nanostructured glassy polymer film (poly(methyl methacrylate)–PMMA) has been deposited with exceptional thermal and kinetic stability. The films were 40% less dense and presented a 40 K higher glass transition temperature respect to standard films. Moreover, the kinetic stability in glassy state was enhanced by 2 order of magnitude [58]. These properties are strictly connected to the temperature gradient experienced by the ejected polymer and the film structures and morphology were explained in term of the Zhigilei's model [40].

Particularly interesting, also, is the possibility offered by the MAPLE process in controlling molecular ordering and delicately manipulating crystalline morphology during film growth thanks to the prolonged time of deposition, with respect to other solvent-based methods, which allows for the manipulation of temperature during deposition [59, 60]. Playing with two processing parameters, deposition time and temperature, a model polymer, poly(ethylene oxide) (PEO), was deposited atop a temperature-controlled substrate with an average growth rate of less than 10 nm/h obtaining the formation of two-dimensional (2D) PEO crystals, composed of monolamellar crystals laterally grown from larger nucleus droplets [59, 60].

In Fig. 10.1 the evolution of the MAPLE publications since its invention up to date is reported. While, in Table 10.2 a detailed list of the latest deposited materials is listed.

Most of the recent applications consider MAPLE deposition of proteins for *biomedical applications* [61–107]. In particular, the *antimicrobial activity*

Table 10.2 Maple-deposited material, application and reference

Maple deposited material	Application	References
Lysozyme embedded into a matrix of polyethylene glycol (PEG) and polycaprolactone (PCL)	Biomedical (antimicrobial)	[61]
1,2,3,4,5,6,7,8-octahydroacridine-N(10)-oxides	Biomedical (microbicidal and anti-pathogenic)	[63]
Poly(3-hydroxybutyric acid-co-3-hydroxyvaleric acid)—Lysozyme (P(3HB-3HV)/Lys); P(3HB-3HV)—Polyethylene glycol—Lysozyme (P(3HB-3HV)/PEG/Lys) spheres	Biomedical (cytotoxicity and antimicrobial activity)	[64]
4-[(4-chlorobenzyl)oxy]-3,4′-dichloroazobenzene azoic dye (CODA)	Optoelectronic and organic electroluminescent devices	[108]
Yb^{3+}/Er^{3+} co-doped LiYF4	Optoelectronic	[109]
Ferrocene	Flexible electronics, lab-on-a-chip devices, MEMS, and medical implants	[130]
Tungsten trioxide (WO_3)	Electrochromic applications	[131]
Polymer nano-composite: – Poly(methyl methacrylate) PMMA doped with $NaYF_4$:Yb^{3+} , Er^{3+}; – PMMA with aluminum doped ZnO (AZO)		[134]*, [135]
Arrays of vertical microtubes of IP-L780 filled with polypyrrole (conductive)/dexamethasone (PPy/Dex) in poly(lactic-co-glycolic acid)(PLGA)	Biomedical (drug delivery)	[65]
Composite coatings based on poly(3-hydroxybutyrate-co-hydroxyvalerate) (PHBV) and commercial calcium phosphates (CaPs)	Biomedical (tissue engineering and regenerative medicine)	[69]
Poly(3-hexylthiophene) (P3HT)	Devises based on carrier transport	[132, 133]
Poly(lactic-co-glycolic acid) copolymer and quercetin (PLGA/Q) microspheres	Biomedical (biocompatibility)	[67]
Dispersed diamonds (UDDs) with doxorubicin (DOX) onion-like carbon (OLCs) with doxorubicin (DOX)	Biomedical (drug delivery)	[68]

(continued)

Table 10.2 (continued)

Maple deposited material	Application	References
Heterostructures or blend of [6]-phenyl-C61 butyric acid butyl ester ([C60]PCB-C4) and 2,5-diamino-3,4-dicyanothiophene as central unit and triphenylamine (LV5) or carbazol (LV4)	Photovoltaic	[119]
Heterostructures or blend poly(3-hexylthiophene) (P3HT) and fullerene (C60)	Photovoltaic	[120]
Chitosan/biomimetic apatite	Biomedical (antimicrobial)	[62]
Ruthenium complex N_3, melanin nanoparticle (MNP), and porphyrin-based donor-π-acceptor dye YD2-o-C8	Dye-sensitized solar cell (DSSC)	[121]
4-(ferrocenylmethylimino)-2-hydroxy-benzoic acid	Optoelectronic	[128]
Biopolymeric thin films containing flavonoid natural compounds and silver nanoparticles	Biomedical (Antimicrobial activity)	[71, 88]
– Blend of arylenevinylene oligomers, 1,4-bis [4-(N,N′-diphenylamino)phenylvinyl] benzene (L78) and 3,3′-bis(N-hexylcarbazole)vinylbenzene (L13) and buckminsterfullerene (C60) – Blend and multilayer of arylene based polymer AMC16 and AMC22 and (AMC16:C60) and (AMC22:C60)	Photovoltaic	[122, 124]
Heterostructures based on zinc phthalocyanine (ZnPc), magnesium phthalocyanine (MgPc) and 5,10,15,20-tetra(4-pyrydil)21H,23H-porphine (TPyP)	Devices based on Photogeneration process	[129]
ZnO nanoparticles functionalized with sodium stearate (C18) and usnic acid (UA)	Biomedical, Food science	[72]
Poly(ethylene glycol)-block-poly(ϵ-caprolactone) methyl ether (PEG-PCL Me copolymer)	Biomedical	[73]
Zinc oxide, cyclodextrin and cefepime (ZnO/CD/Cfp)	Biomedical (antimicrobial surfaces with biomedical applications)	[75]

(continued)

Table 10.2 (continued)

Maple deposited material	Application	References
Zinc oxide (ZnO) nanoparticles incorporating with polyethylene glycol (PEG)	Biomedical (antimicrobial)—contact lens	[76, 93]
(Ag) nanoparticles and polyvinylpyrrolidone (PVP)	Biomedical (antimicrobial)—contact lens	[77]
Lipase and Lipase doped with m-DOPA (3-(3,4-dihydroxyphenyl)-2-methyl-l-alanine)	Biosensors	[78, 84, 92]
– Calcium alendronate monohydrate; – Calcium alendronate monohydrate/octacalcium phosphate composite	Biomedical (tissue regeneration)	[79]
– Polypyrrole (PPy) embedded in poly(lactic-co-glycolic)acid (PLGA); – Polypyrrole (PPy) embedded in polyurethane (PU)	Biomedical (bone regeneration)	[103]
Bioactive glasses, composite polymer-bioactive glass nanostructures	Biomedical (protective coating against corrosion for stainless steel implants)	[74]
Magnetite nanocomposite	Biomedical (antimicrobial)	[80]
Sr-substituted hydroxyapatite (SrHA) and ZOL modified hydroxyapatite (ZOLHA)	Biomedical (bone prostheses)	[81]
Titanium-polypyrrole/poly(lactic-co-glycolic)acid (Ti-PPy/PLGA)	Biomedical (osteogenesis)	[82]
Polyethylene glycol (PEG)	Biomedical (antimicrobial)	[83]
ADS125GE	Organic photonics and opto-electronics	[52]
Poly(3,3$_V$didodecylquater thiophene) (PQT-12)	Organic thin-film transistors (OTFT) devices	[111]
– Polyvinyl chloride (PVC) – Polyacrylic acid (PAA) – Polyaniline (PAni)	Optoeletronics	[112]
Poly (9,9-dioctylfluorene)—PFO	Optoeletronics	[113]
γ-cyclodextrin/usnic acid (γCD/UA)	Biomedical (antimicrobial)	[85]
Laccase	Biomedical (enzyme)	[86]
Kanamycin functionalized 5 nm-magnetite (Fe_3O_4@KAN)	Biomedical (antimicrobial)	[87]

(continued)

Table 10.2 (continued)

Maple deposited material	Application	References
Silk fibroin-poly(3-hydroxybutyric-acid-co-3-hydroxyvaleric-acid) (SF-PHBV) biodegradable coatings	Biomedical (tissue engineering and drug delivery)	[104]
Fibronectin (FN)	Biomedical (osseointegration)	[89]
Polymer blends comprised polyurethane (PU), poly(lactic-co-glycolic acid) (PLGA), and polylactide-polyethylene glycol-polylactide (PPP)	Biomedical (tissue engineering)	[99]
– Magnetite (Fe_3O_4) and eugenol (E) nanoparticles embedded in poly(3-hidroxybutyric acid-co-3-hidroxyvaleric acid)-polyvinyl alcohol (P(3HB-3HV)-PVA) microsphere – P(3HB-3HV)-PVA-Fe_3O_4@E	Biomedical (antimicrobial)	[90]
– Ferrocene – Ferrocene carboxaldehyde	Optics (Second harmonic generation)	[114, 117]
Hydroxypropyl methylcellulose (HPMC) and ethylcellulose (EC)	Biomedical (drug delivery)	[102]
Bovine odorant-binding proteins (bOBPs)	Biosensors	[107]
AZO-derivative pyrazolone derivatives, namely 1-phenyl-3-methyl-4-(1′-azo-2′-sodium carboxylate)-pyrazole-5-one	Optics (Second harmonic generation)	[115]
Zinc phthalocyanine (ZnPc) and 1,4,5,8-naphthalene-tetracarboxylic dianhydride (NTCDA)	Photovoltaic	[125]
Sensitive poly(N-isopropylacrylamide), (pNIPAAm)	Biomedical (tissue engineering)	[105]
Arylenevinylene oligomers based on triphenylamine (P78)/carbazole (P13) group and tris(8-hydroxyquinolinato) aluminum salt (Alq3)		[116]
– PLA-PVA microspheres loaded with usnic acid (UA) – Polylactic-co-glycolic acid-polyvinyl alcohol (PLGA-PVA) microspheres loaded with usnic acid – UA	Biomedical (antimicrobial)	[91, 97]

(continued)

Table 10.2 (continued)

Maple deposited material	Application	References
Nanocrystalline apatite	Biomedical (tissue engineering)	[100]
Iron-free (Apo-rLf)], cisplatin (Cis) or their combination embedded within a biodegradable polycaprolactone (PCL)	Biomedical (antitumoral)	[101]
Eu-doped LiYF4	Optoelectronics	[118]
Ribonuclease A (RNase A)	Biomedical	[94]
Polylactic acid-chitosan-magnetite-eugenol (PLA-CS-Fe_3O_4@EUG)	Biomedical (antimicrobial)	[95]
Arylenevinylene compound, 4,4′,4″-tris[(4′-diphenylamino)styryl] triphenylamine and fullerene derivative, [6, 6]-phenyl C61 butyric acid butyl ester	Photovoltaic	[126]
Cinnamomumverum-functionalized Fe_3O_4 nanoparticles	Biomedical (antimicrobial drug delivery)	[96]
Levan and oxidized levan	Tissue regeneration for the engineering of cell instructive scaffolds	[106]
PVP/Fe_3O_4 nanostructures and 2-((4-Ethylphenoxy)methyl)-N-(arylcarbamothioyl)benzamides	Biomedical (antimicrobial)	[98]

of many proteins has been studied, for example, lysozyme embedded in a matrix of polyethylene glycol (PEG) and polycaprolactone (PCL) [61], poly(3-hydroxybutyric acid-co-3-hydroxyvaleric acid)—lysozyme (P(3HB-3HV)/Lys) and p(3HB-3HV)—PEG—lysozyme (P(3HB-3HV)/PEG/Lys) spheres [64], zinc oxide (ZnO) nanoparticles incorporated with polyethylene glycol (PEG) [76], and magnetite nanocomposite [80].

Together with the antimicrobial activity, many materials were deposited by MAPLE for *controlled drug delivery* [65, 68, 102], *tissue engineering, regenerative medicine* [66, 69, 73, 79, 99, 100, 103–106] and *biosensors* [78, 107].

Another class of materials deposited by MAPLE includes materials for *optoelectronic, photovoltaics and organic devices* [52, 108–133].

As stressed above, the MAPLE technique offers the possibility to deposit easily heterostructures and multilayer films, including on flexible substrates [123, 129].

Thanks to the MAPLE technique, it was possible to deposit, for the first time, a white light OLED [134]. Three layers of a blue-light-emitting polymer (BP) poly[9,9-dioctylfluorenyl-2,7-diyl], a green-light-emitting polymer (GP)poly[(9,9-dioctylfluorenyl-2,7-diyl)-co-(1,4-diphenylene-vinyl-ene-2-methoxy-5-{2-ethylhexyloxy}benzene)] and a red-light-emittingpolymer (RP) poly[2-methoxy-5-(2-ethylhexyloxy)-1,4-phenylenevinylene] were deposited, in

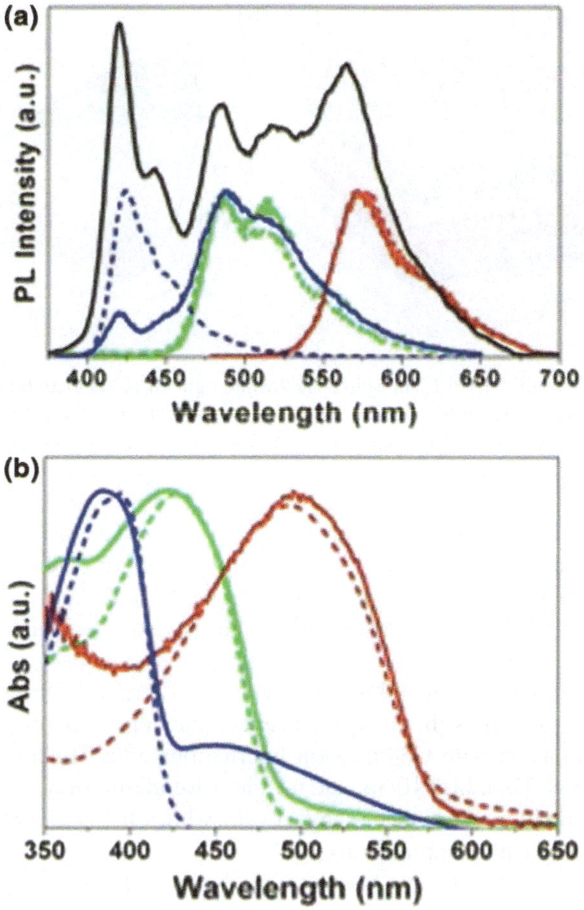

Fig. 10.2 **a** PL spectra of BP (blue line), GP (green line) and RP (red line) deposited by spin-coating (dashed lines) and MAPLE (solid lines). PL whiteemission (black line) from the three-layers stack (BP + RP + RP). **b** Normalized absorption spectra of BP (blue line), GP (green line) and RP(red line) deposited by spin-coating (dashed lines) and MAPLE (solidlines) [134]

combination with the spin coating technique (for the deposition of the blue emitting polymer) on a hole-transporting layer (≈80 nm thick of PEDOT:PSS) which was spin-coated onto an indium-tin-oxide (ITO)-coated glass substrate.

The emitting polymers presented good surface roughness values (after an optimization study) and emission properties comparable with those of spin-coated films, as shown in Fig. 10.2.

A root mean square surface roughness of about 1 nm was achieved for the BP. After the MAPLE deposition of the other layers, the mean roughness value increased to 6.9 ± 0.9 nm for the bilayer device (BP + RP) and to 10.3 ± 1.6 nm for the complete three-layers (BP + RP + GP) polymeric stack. However, these roughness values were

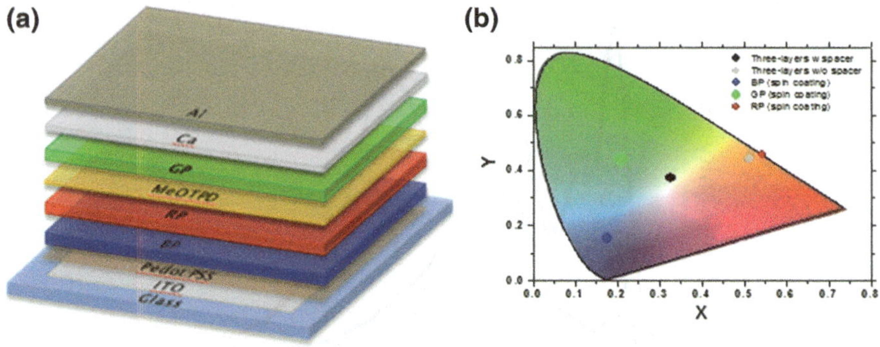

Fig. 10.3 **a** sketch of the MAPLE and spin coated white light OLED device; **b** Commission Internationale de l'Éclairage coordinates with indication of the colour coordinates for the MAPLE deposited device

still compatible with OLED fabrication which was realized depositing atop electrode made by Calcium (10 nm) covered by Aluminum (100 nm), as shown in Fig. 10.3a.

The choice of the blue/red/green polymer sequence in the active layer stack was due to the deep energy levels of red-light-emitting polymer, which confined the excitons.

By controlling the carriers and the energy transfer across the three light-emitting layer interfaces, as well as the interplay between the deposition parameters, a pure white colour emission with Commission Internationale de l'Éclairage coordinates ($X = 0.327$, $Y = 0.374$), Fig. 10.3b, and a Colour Rendering Index of 70 have been achieved. The maximum luminance shown by the white OLED was about 200 cd/m^2, which is good for display applications [134].

Finally, many strategies have been used in order to deposit, using the MAPLE technique, films with *complex stoichiometry* and graded properties, like the use of a *double and triple-beam triple-target* [77, 135–139] and *combinatorial* MAPLE [81, 106]. The use of a multi-target system was stimulated by the fact that mixing components of different nature, such as organic polymers and inorganic dopants, in the same target at a certain proportion and exposing them to the same laser beam does not necessarily result in good quality composite films. The laser pulse energy and wavelength is not optimized for each component individually and the mixing proportion in the composite film is dictated by the initial proportion of the target and thus cannot be changed in the process. These limitations were removed in the recently proposed method of double or triple-beam deposition using MAPLE polymer and inorganic targets, each being simultaneously exposed to a laser beam of a different wavelength.

10.4 RIR-MAPLE: Motivation for Emulsion Targets

The key innovative idea of MAPLE is the incorporation of solvents as a target matrix to absorb incident laser energy (from UV or IR light) during physical vapor deposition of organic thin films, especially for polymers and other macromolecules. In MAPLE-based deposition, the target polymer is dissolved in the solvent matrix, and the entire solution is subsequently frozen by liquid nitrogen to make a solid target for laser irradiation. The use of a solvent matrix to absorb laser energy leads to three possible deposition scenarios (Fig. 10.4). In the first scenario (Fig. 10.4a), the laser energy is not able to completely vaporize the frozen solvent matrix such that a chunk of iced target polymer is ejected by the laser to the substrate, which then melts and dries to form the thin film. In the second scenario (Fig. 10.4b), the laser energy is completely and solely absorbed by the solvent matrix such that it vaporizes, leaving isolated polymer chains to deposit on the substrate intact. In the third scenario (Fig. 10.4c), the laser energy is absorbed by both the solvent matrix and target polymer such that the polymer undergoes photochemical reactions during deposition, leading to fragmented species co-existing in a degraded film. Thus, the choice of solvent matrix, the composition of the target solution, and the laser properties (energy, fluence, and pulse width) can play important roles in determining which deposition scenario occurs. In fact, O'Malley, et al. demonstrated that the laser pulse width, for example, has very little impact on the final thin film properties for hydrophobic polymers [140]. Ideally, MAPLE-deposited organic thin films should benefit from the second scenario because the polymer film is not degraded and does not have solvent contamination.

For the case of resonant infrared MAPLE, or RIR-MAPLE, the use of low energy IR laser irradiation can help avoid the degradation that occurs in the third scenario and can provide gentle deposition of polymeric materials. In order to achieve the second, ideal deposition scenario using RIR-MAPLE, the wavelength of the IR laser must be tuned to be resonant only with molecular vibrational modes in the solvent matrix that are not present in the polymer to be deposited. However, for a given organic target material, the chosen solvent matrix must also account for the solubility of the organic material. If the polymer solubility is poor in a given solvent, the deposition will not be uniform because the material is not evenly dispersed within the target at a molecular level. This dual need for the solvent matrix to provide good solubility for the target polymer and to provide unique resonant absorption of the IR laser energy can limit the ability of RIR-MAPLE to be applied generally to different organic material systems.

To help make RIR-MAPLE a more universal technique that can deposit most organic material systems, Pate, et al. developed a novel approach to prepare the deposition target. Instead of using a frozen solution, a frozen emulsion, or mixture of two or more liquids that are normally immiscible, is used as the deposition target. This deposition approach is referred to as emulsion-based RIR-MAPLE, and it establishes a new research direction for the MAPLE technique by exploring the impact of more complex target chemistry as an important deposition parameter [46, 141, 142]. For example, a specific organic target material may have many good solvents from which

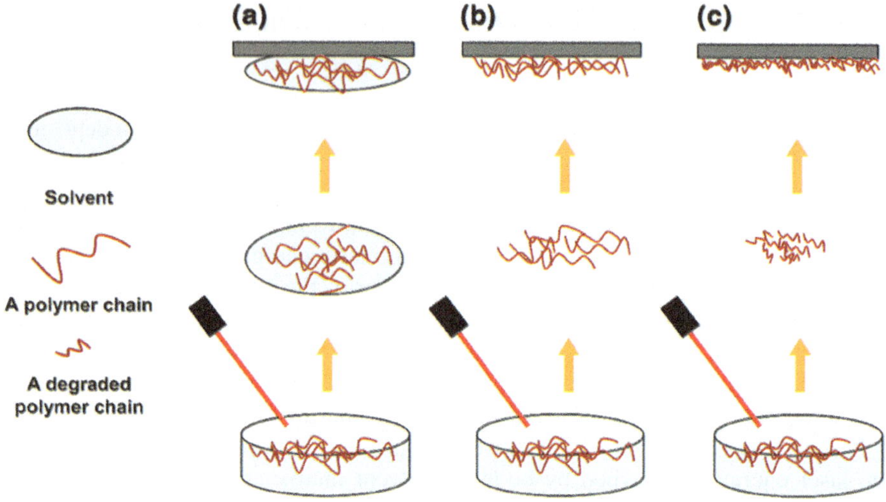

Fig. 10.4 A schematic diagram of three possible deposition scenarios that can occur during MAPLE deposition from a frozen solution. **a** The laser energy is not absorbed efficiently by the solvent matrix, ejecting iced polymer target to the substrate. **b** The laser energy is absorbed completely by the solvent matrix, leaving dry polymer chains to deposit onto the substrate (ideal). **c** The laser energy is not only absorbed by the solvent matrix, but also by the polymer, which breaks polymer chains and leaves fragmented polymer to deposit on the substrate

to choose, and the selection of a particular primary solvent is somewhat arbitrary for a frozen target solution in which phase explosion is the primary mechanism for film growth [143].Yet, in the case of frozen emulsion targets, a different film formation mechanism is proposed, and each component of the frozen emulsion target can have a profound effect on the resulting thin film properties, as described herein. Root-mean-square (RMS) surface roughness, as determined by atomic force microscopy (AFM) measurements, is presented as one example of thin film properties because it has a direct impact on the applicability of organic thin films in electronic and photonic devices.

10.5 RIR-MAPLE: Frozen Emulsion Targets

For a traditional two-phase emulsion (without any polymer), two liquid phases are not miscible (water is a polar phase and oil is a non-polar phase). From the perspective of thermodynamics, the classic Gibbs–Deuhem equation [144] describing the Gibbs free energy for liquid-liquid interfaces can be used to predict emulsion behavior, and is shown as (10.4):

$$dG^{\sigma} = -S^{\sigma}dT + Ad\gamma + \sum n_i d\mu_i \tag{10.4}$$

where $-S^{\sigma}\mathrm{d}T$ is the entropy term, $A\mathrm{d}\gamma$ is the interfacial energy term (A is the interface area, γ is the interfacial tension), and $\sum n_i \mathrm{d}\mu_i$ is component term (n_i is the number of moles of component i in the interface layer, μ_i is the chemical potential of component i). At constant temperature and composition, the entropy term and component term can be ignored (10.5):

$$\gamma = \left(\frac{\partial G^{\sigma}}{\partial A}\right)_{T,n_i} \tag{10.5}$$

For the interface between two dissimilar phases, such as water and oil, the interfacial tension, γ, is positive. As a result, the interfacial area between the water and oil tends to be reduced naturally in order to keep the Gibbs free energy as low as possible. As a result, the oil phase tends to separate from the water phase by forming a distinct layer above or below the water (depending on the relative density of the oil to water). This planar interface yields the lowest interfacial area and represents the extreme case of reducing interfacial free energy.

By shaking or sonication (energy input to the system), the separate layer of oil can be broken into small droplets and dispersed in the water temporarily, leading to the formation of an oil-in-water (o/w) emulsion. In the case of emulsion-based RIR-MAPLE, the oil phase is anon-polar primary solvent used to dissolve an organic target material. If a surfactant is present in the emulsion, the interfacial tension of primary solvent and water can be reduced by the adsorption of surfactant at the interface. Due to steric hindrance, the surfactant also prevents primary solvent droplets from coalescing and leading to emulsion flocculation. While the primary solvent droplet size in a two-phase o/w emulsion is not monodisperse, the statistical average of the droplet size can be determined, and it is partially affected by the miscibility of the primary solvent in water (solubility-in-water). In general, for a two-phase emulsion, if the solubility-in-water of the primary solvent is lower, the solvent droplet size is larger because the interfacial tension, γ, is larger and the two phases tend towards a planar interface. As a result, smaller solvent droplets are driven to coalesce quickly into larger droplets in order to reduce the interfacial area with water.

If a third phase (e.g., hydrophobic polymer) is added to a traditional two-phase, o/w emulsion, the interplay among the primary solvent, water and polymer can change the emulsion properties. Figure 10.5 shows the concept of competitive bonding in a three-phase, o/w emulsion containing polymer, primary solvent and water. In this case, the larger the bonding strength between primary solvent and polymer molecules, compared to that between primary solvent and water molecules, the smaller the oil phase droplet size (comprising primary solvent and polymer) in the corresponding emulsion. Alternatively, the presence of the hydrophobic polymer within the primary solvent can reduce the interfacial tension of the oil phase with water. Therefore, the average droplet size of a three-phase, o/w emulsion for a primary solvent with low solubility-in-water can be smaller than the droplet size expected for a traditional two-phase, o/w emulsion. For emulsion-based RIR-MAPLE deposition, the three-phase o/w emulsion is flash frozen in a target cup that has been pre-cooled to −196 °C (77 K) by liquid nitrogen. This flash freezing enables the emulsion to solidify before

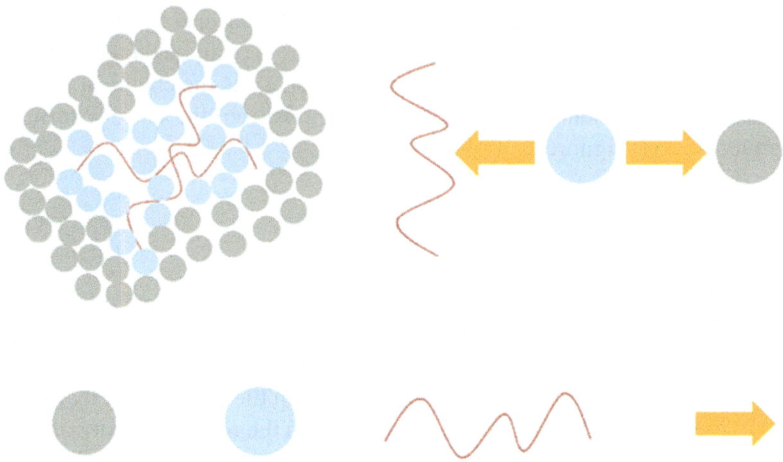

Fig. 10.5 Graphic representation of three-phase, o/w emulsion. The top left shows solvent molecules that bond with polymer chains and envelop them, forming a droplet inside the water molecules. The top right shows the competition between the bonding of a solvent molecule to a polymer chain and the bonding of a solvent molecule to a water molecule. The bonding of water molecules to hydrophobic polymer molecules is ignored because hydrophobic polymers are insoluble in water

it separates (~10–30 s) so that the emulsified droplets are maintained in the final target.

10.6 RIR-MAPLE: Film Formation from Emulsion Targets

The droplet size of the hydrophobic polymer/primary solvent in a three-phase, o/w emulsion is important to emulsion-based RIR-MAPLE because of the nature of film formation for this process (Fig. 10.6). Similar to RIR-MAPLE of frozen solutions, the target matrix resonantly absorbs the laser energy, but contrary to frozen solution targets, the absorption of the laser energy is completely decoupled from the organic target material and its primary solvent. Instead, resonant absorption of the laser energy occurs only in the continuous water-ice phase of the emulsion, which serves as the target matrix. This resonant absorption by water-ice is achieved by using an Er:YAG laser with a peak wavelength at 2.94 μm, which is resonant with the vibrational modes of hydroxyl bonds [141, 145]. While water comprises two hydroxyl bonds, many polymers and corresponding primary solvents, especially those germane to photonic and electronic applications, do not contain hydroxyl bonds. Thus, when phase explosion occurs in the water-ice matrix for a frozen emulsion target, instead of transferring a wide variety of uncontrolled polymer chain configurations as in frozen

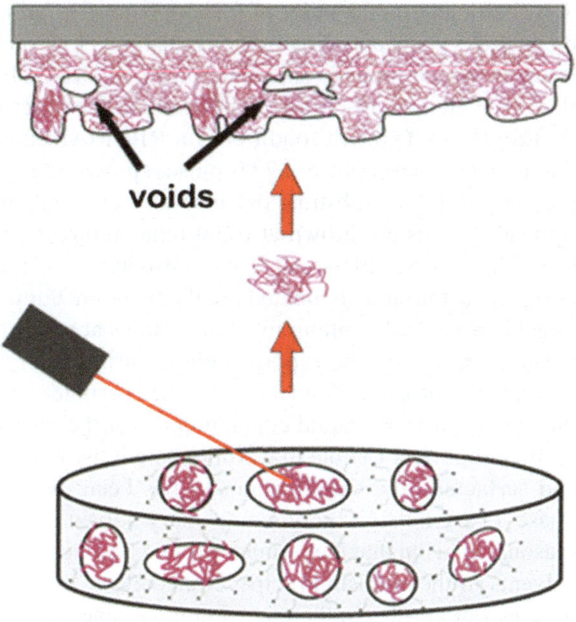

Fig. 10.6 Schematic illustration of the film formation mechanism in emulsion-based RIR-MAPLE process. The surface roughness and internal voids of the film are highlighted by blackline

solutions [35, 40, 146–148], the emulsified polymer/primary solvent droplets are transferred, and the morphology of these droplets can be controlled by the emulsion target chemistry [149]. The film deposited from the frozen emulsion target results from the cluster-by-cluster stacking of the polymer, as shown in Fig. 10.6. These polymer clusters do not necessarily exhibit close packing, which leads to uneven surfaces and voids inside the film. Therefore, the surface and internal morphologies of films deposited by emulsion-based RIR-MAPLE depend on the emulsified droplet size and the polymer cluster morphology within the droplet. As a result, by creating frozen three-phase, o/w emulsions, the gentle deposition of polymer thin films can be achieved for almost any hydrophobic polymer, as well as other hydrophobic soluble or dispersed materials, such as small molecules, oligomers and nanoparticles.

Pate empirically determined a standard emulsion-based deposition recipe for most hydrophobic organic materials [150]. This emulsion recipe contains a primary solvent to dissolve the target organic material, a secondary solvent (typically phenol or benzyl alcohol) to stabilize the frozen target in vacuum and to enrich the hydroxyl bond concentration, and de-ionized (DI) water to provide the majority of hydroxyl bonds for resonant laser absorption. In addition, 0.001wt% surfactant, typically sodium dodecyl sulfate (SDS), is prepared in the DI water to help stabilize the emulsion. The typical volume ratio of the primary solvent, secondary solvent and DI water is 1:1:4, which determines the concentration of hydroxyl bonds in the emulsion and the depth of the laser-target interaction volume. For the standard emulsion recipe, the hydroxyl bond concentration leads to "evaporative" deposition in which the laser energy density is large and the water-ice matrix is vaporized (laser wavelength absorption coefficient on the order of $\alpha = 1000\ \text{cm}^{-1}$). "Ablative" deposition occurs when the hydroxyl concentration in the emulsion is low such that the laser energy density is small, which leads to splashing of un-vaporized liquid ejecta onto the substrate (laser wavelength absorption coefficient on the order of $\alpha = 10\ \text{cm}^{-1}$).

The standard growth conditions for RIR-MAPLE deposition are: 1.46 J/cm^2 laser fluence (90 μs laser pulse at 2 Hz pulse repetition rate), 1E−5 to 1E−4 Torr chamber pressure, target-to-substrate distance of 7 cm, and substrate temperatures <200 °C (typically, films are grown at room temperature). The pulsed laser rasters radially across the target, which rotates at a constant rate (usually 3–4 RPM), in a pattern that ensures that the target is ablated evenly. In the vacuum chamber, the substrate faces the target in a vertical configuration and rotates at a constant rate (usually 10–15 RPM) during deposition. The vaporous host matrix (water), and some primary/secondary solvent and surfactant from the emulsified particles, are pumped away by the vacuum, thereby minimizing liquid contamination of the deposited film.

It is important to note that while the emulsion target contains secondary solvent and surfactant, the target chemistry has been explained in the context of a three-phase, o/w emulsion (polymer, primary solvent, and water). This simplification is reasonable given that the primary solvent accounts for 80% by volume of the organic solvents in the emulsion recipe. Nonetheless, the secondary solvent and surfactant have essential roles in the emulsion target chemistry, and their impact must also be considered in the overall emulsion morphology.

10.7 RIR-MAPLE: Impact of Primary Solvent, Secondary Solvent, Surfactant and Matrix in Frozen Emulsion Targets

As mentioned earlier, the unique approach of emulsion-based RIR-MAPLE is that the polymer morphology within the target is controlled by the emulsion chemistry. The emulsified polymer droplets are directly transferred to the substrate by the incident laser irradiation, thereby determining the thin film properties. While this process is straightforward, the emulsions are complex due to the presence of at least five different components (hydrophobic polymer, primary solvent, secondary solvent, surfactant, and water).Thus, the ability to control the resulting thin film properties depends on understanding this target emulsion chemistry.

Regarding the primary solvent, chemicals with lower solubility-in-water and lower vapor pressure yield smoother films deposited by emulsion-based RIR-MAPLE. As described in the discussion of three-phase, o/w emulsions, primary solvents with lower solubility-in-water yield smaller solvent droplets containing smaller polymer clusters within the water matrix. The films deposited from smaller polymer clusters are smoother due to smaller surface features and higher packing density of the polymer clusters. Primary solvents with lower vapor pressure lead to smoother films because the sublimation rate of the frozen target in vacuum is reduced and the deposition is more consistent over time. Due to its vapor pressure, the primary solvent can sublimate in vacuum over time, even without laser irradiation. As more frozen solvent molecules sublimate, the polymer concentration on the surface of the frozen target emulsion increases, and it has been demonstrated that films deposited from targets containing highly concentrated polymers are rougher than films deposited from targets with low concentrations of polymers [40]. A secondary effect of a low vapor pressure primary solvent is that the reduced sublimation rate can prevent the target composition from changing over time.

The desired properties of the primary solvent were determined by investigating two families of solvents, alkyl aromatic solvents (toluene, o-xylene, pseudocumene) and chlorinated aromatic solvents (chlorobenzene (CB), 1,2-dichlororbenzene (o-DCB), 1,2,4-trichlorobenzene (TCB)) [151]. Poly(3-hexylthiophene-2,5-diyl) (P3HT) was deposited using these chemicals as the primary solvent in emulsion-based RIR-MAPLE because this hydrophobic conjugated polymer has been widely studied for organic solar cells (OSCs) and has many well-known properties [152, 153]. For each solvent family, as the number of side groups increases (CH_3- or Cl- for alkyl aromatic solvents and chlorinated aromatic solvents, respectively), the vapor pressure and solubility-in-water decrease. The RMS surface roughness of the P3HT films decreased with increasing number of side groups (for each solvent family), with TCB yielding the smoothest films, and correspondingly, the highest OSC power conversion efficiency.

Regarding the secondary solvent (typically phenol), it was initially included in the emulsion target to increase the hydroxyl bond concentration for "evaporative" deposition and to reduce the overall vapor pressure of the emulsion so that it remains stable in vacuum over time. However, a deposition study of poly-[2,6-(4,4-bis(2-ethylhexyl)-4H-cyclopenta [2,1-b;3,4-b′]-dithiophene)-alt-4,7(2,1,3-benzothiadiazole)] (PCPDTBT), another common polymer used in OSCs, demonstrates that the secondary solvent could play another role. Figure 10.7 shows two emulsions of PCPDTBT, with and without the secondary solvent phenol, using TCB as the primary solvent. Figure 10.7 demonstrates that a good emulsion cannot be formed without the secondary solvent phenol, indicating that phenol also acts as a "surfactant" to help stabilize the emulsion. To better understand the impact of the secondary solvent, PCPDTBT films deposited from TCB emulsions with different amounts of phenol were characterized by AFM. The RMS surface roughness values are shown in Table 10.3, demonstrating comparable film roughness for a phenol volume ratio in the range of 0.1–0.5. However, if phenol is not added at all, a stable emulsion cannot be formed, and if too much phenol is added, the film roughness increases significantly due to its high solubility-in-water. Thus, phenol may behave as a surfactant in small amounts, remaining at the interface between the primary solvent and water, and as a primary solvent in large amounts. In fact, the phenol molecule contains a benzene ring (hydrophobic part) and a hydroxyl group (hydrophilic part) (Fig. 10.7c). The exact role of the secondary solvent, the ideal solvent choice, and the best concentration within the emulsion remain as open research questions.

Regarding the surfactant (SDS), an important issue has been to determine the optimum surfactant concentration that produces a meta-stable emulsion (~10 to 30 s

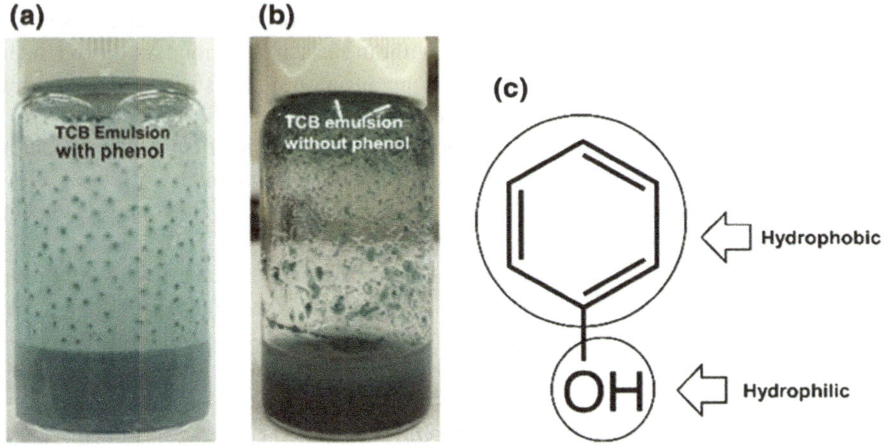

Fig. 10.7 PCPDTBT emulsion prepared from TCB as the primary solvent with **a** phenol and **b** without phenol as the secondary solvent. **c** Chemical structure of the phenol molecule

Table 10.3 RMS roughness of PCPDTBT films deposited from TCB as primary solvent with different amount of phenol as the secondary solvent

TCB:Phenol:DI water (volume ratio)	RMS roughness (nm)
1:0:3	NA (not an emulsion)
1:0.1:3	11.87 (2.43)
1:0.25:3	11.10 (0.89)
1:0.5:3	15.23 (0.33)
1:1:3	22.23 (1.79)

The roughness of the films was obtained by averaging three AFM images from different areas of each film. The values in parentheses are the standard deviation of the RMS surface roughness for each film from three AFM images

for flash-freezing), yet is not transferred to the substrate in any significant amount. Emulsion-based RIR-MAPLE deposition of P3HT was conducted using different amounts of SDS surfactant: no surfactant, 1E−3 wt%, 1E−2 wt%, and 1E−1 wt%. Specular x-ray diffraction (XRD) measurements were conducted to identify characteristic peaks for SDS, and these peaks were observed for the 1E−1 wt% and 1E−2 wt% films. However, the spectrum for the 1E−3 wt% surfactant film was featureless, similar to the film with no surfactant, indicating that there was not enough surfactant material in the film to form regularly spaced structures. As a result of this study, 1E−3 wt% surfactant in DI water has been identified as sufficient concentration to maintain the emulsion stability until it is frozen without being deposited on the substrate in a significant quantity.

Regarding the water-ice matrix, for some polymers of interest, exposure to water can cause degradation. While the water-ice matrix is vaporized upon laser irradiation and is not incorporated into the deposited film, it is not clear how much the

Table 10.4 Surface morphology of P3HT thin films deposited from alcohol/water host matrix containing 25 vol% of different alcohols: pure water, methanol, ethanol, 1-propanol, and 1-butanol

Host matrix	Roughness (nm)	Thickness (nm)	Roughness/thickness ratio
Pure water	35.9	74.9	0.48
Methanol	20.5	63.8	0.32
Ethanol	23.7	68.8	0.34
1-propanol	24.0	107.0	0.22
1-butanol	42.8	105.0	0.40

water matrix may impact the polymer within the emulsified droplets. Therefore, a co-matrix of alcohol and water has been investigated to maintain the hydroxyl bond concentration while reducing the total amount of water within the emulsion. P3HT was deposited using emulsion ratios (by volume) of 1:0.2:1:3 (primary solvent (TCB): secondary solvent (phenol): alcohol: DI water). Four alcohols were chosen for the co-matrix with water: methanol, ethanol, 1-propanol, and 1-butanol. The surface morphology was determined by AFM (25 μm × 25 μm scan size), as shown in Fig. 10.8. Table 10.4 summarizes the surface morphology details for the P3HT films deposited from alcohol/water co-matrices with different alcohols. Compared to the P3HT film deposited from pure water, adding alcohol to the matrix reduces the RMS surface roughness/thickness ratio significantly, which gives an overall idea of the film smoothness. Among the four alcohols under consideration, 1-propanol yields the smoothest film with RMS surface roughness/thickness ratio of 0.22. While the possible degradation of polymers due to water exposure in the emulsion remains an open research question, this alcohol/water co-matrix study does indicate that inclusion of an alcohol in the continuous water phase can improve film deposition, and it may be possible to completely replace water with an appropriate alcohol.

10.8 RIR-MAPLE: Emulsion Targets for Hydrophilic Polymers

So far, emulsion-based RIR-MAPLE has been described in the context of hydrophobic polymer thin film deposition, yet the frozen emulsion target can be modified for hydrophilic polymer deposition. One obvious question is if the same o/w emulsion target is appropriate for hydrophilic polymers that are water soluble. As a first step, thin film deposition of polyethylene glycol (PEG) using an o/w emulsion recipe is compared to deposition from a solution of PEG in water. Figure 10.9 shows the corresponding AFM images. Counterintuitively, the emulsion target leads to smoother films than the water solution target. Regarding the water solution, the polymer molecules are no longer deposited as polymer clusters defined by the emulsion droplet size. On the contrary, the polymers are deposited directly from the contin-

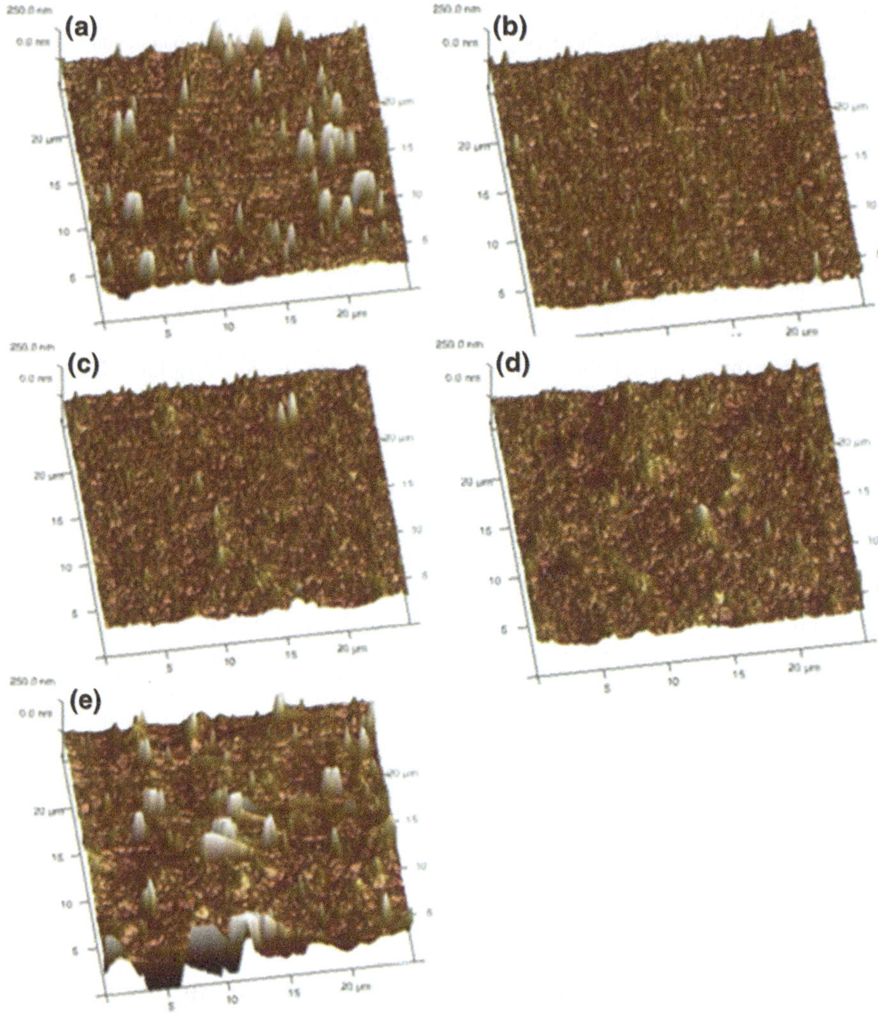

Fig. 10.8 AFM images of P3HT films deposited from different host matrices containing 25 vol% matrix: 75 vol% DI water, where the matrix is: **a** DI water, **b** methanol, **c** ethanol, **d** 1-propanol, and **e** 1-butanol

uous water phase, which can result in polymer chain collisions when frozen water in the target is evaporated by absorbing incident laser energy. The polymer chains form large polymer aggregates that contribute to the large surface roughness of the deposited films. In fact, this deposition description can be applied to any MAPLE system in which a good solvent, alone, is used as the matrix for resonant laser absorption such that the solvent is removed during deposition. Correspondingly, rough films of

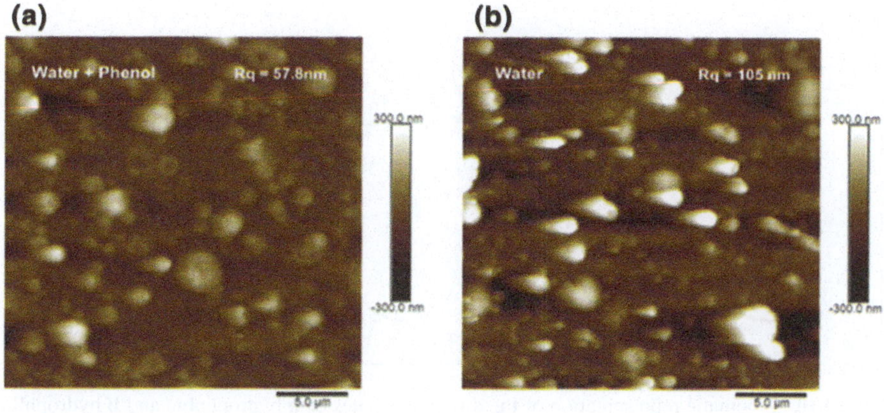

Fig. 10.9 AFM images of PEG polymer films deposited from **a** an emulsion target prepared by the mixed solution of water and phenol and from **b** a water solution. The RMS surface roughness, Rq, is shown in each figure

water solutions have been observed previously for different polymers and different MAPLE variations [45, 154].

Yet, the concept of an o/w emulsion target benefits hydrophilic polymer deposition, as indicated by the film surface roughness, and the appropriate emulsion recipe for hydrophilic polymers must be determined. First, it is important to note that, unlike hydrophobic polymers, hydrophilic polymer molecules in an o/w emulsion are dispersed in the continuous water phase. Figure 10.10 describes the different dispersion states of hydrophobic and hydrophilic polymers in an o/w emulsion. Even though the solvent droplets do not directly confine the polymer into smaller clusters, they can divide the continuous water phase into different regions. Thus, the solvent droplets can act as barriers to reduce the aggregation of polymers during the water vaporization. Nonetheless, the surface roughness of hydrophilic polymer films deposited from o/w emulsions is higher than the roughness of hydrophobic polymer films in which the polymers are completely confined within organic solvent droplets.

To use the concept of frozen emulsion targets effectively for hydrophilic polymers, the emulsion type should be inverted to water-in-oil (w/o) emulsions, in which water droplets are dispersed within the organic solvent continuous phase. A w/o emulsion should be ideal for deposition of smooth hydrophilic polymer films because the size and morphology of the water droplets containing polymer clusters are determined by the emulsion chemistry. One challenge to using w/o emulsions is that the continuous oil phase, i.e. organic solvents, must be the dominant phase (in volume). However, RIR-MAPLE requires water to be the dominant phase in order to efficiently absorb the laser energy.

To investigate emulsion-based RIR-MAPLE deposition of hydrophilic polymers, poly(3,4-ethylenedioxythiophene) polystyrene sulfonate (PEDOT:PSS), which is a common hole transport layer in organic solar cells, is used as a model material for

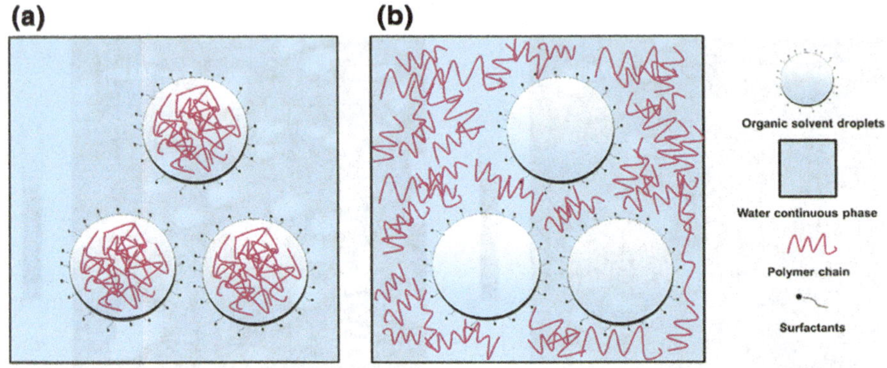

Fig. 10.10 Schematic representation of the dispersion states of **a** hydrophobic and **b** hydrophilic polymers in the emulsion targets

deposition from a w/o emulsion. In order to balance the trade-off between more water in the emulsion to absorb laser energy and more organic solvent in the emulsion to get a w/o type, the standard emulsion recipe was modified to a volume ratio of 2:0.25:2 (primary solvent: secondary solvent: DI water (with polymer)).It should be noted that the PEDOT:PSS water solution contains 1 volume starting PEDOT:PSS water solution and 1 volume DI water is added. In addition, in order to achieve a w/o emulsion, the SDS surfactant is no longer ideal because the hydrophilic-lipophilic balance (HLB) number of SDS is too high. The HLB number of a surfactant indicates whether the surfactant is more hydrophilic or lipophilic. In general, a surfactant with a high HLB value (such as SDS) is more hydrophilic and is used to make an o/w emulsion, while a low HLB number indicates the surfactant is more lipophilic and used to make a w/o emulsion. Thus, three surfactants with different HLB numbers were chosen for comparison: SDS (HLB = 40), poly(ethylene glycol)-poly(propylene glycol)-poly(ethylene glycol) (PEG-PPG-PEG) (HLB = 24), and polyethylene glycol hexadecyl ether (BriJ@52) (HLB = 5). SDS and PEG-PPG-PEG are water soluble and were prepared in PEDOT:PSS water solutions at a concentration of 0.001 wt%. In contrast, BriJ@52 is soluble in organic solvents and was prepared in the primary solvent (TCB) at a concentration of 0.001 wt%. A reference sample was also included in which no surfactant was used.

Figure 10.11 shows the AFM images of PEDOT:PSS films deposited from (a) no surfactant, (b) SDS, (c) PEG-PPG-PEG, and (d) BriJ@52 as the surfactants. Table 10.5 shows the details of the recipes and the corresponding RMS surface roughness obtained from the AFM images. The surface roughness for the BriJ@52 film is much lower compared to the other films, demonstrated by the smaller polymer clusters observed on the surface, providing indirect evidence that a w/o emulsion was achieved. A direct proof of the emulsion type achieved using different types of surfactants is desirable, which should be studied in the future using cryo-transmission electron microscopy to directly image the emulsion or by measuring the conductivity of the emulsion indirectly.

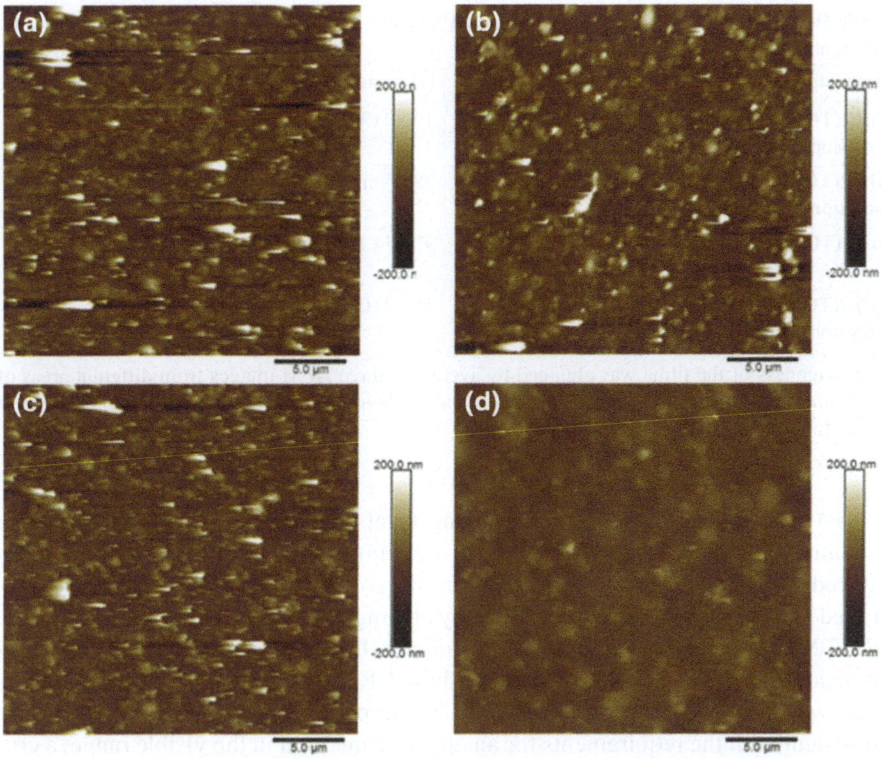

Fig. 10.11 AFM images of PEDOT:PSS films deposited from **a** no surfactant, **b** SDS as the surfactant, **c** PEG-PPG-PEG as the surfactant, and **d** BriJ@52 as the surfactant

10.9 RIR-MAPLE: Applications Using Emulsion Targets

The impacts of emulsion target chemistry on the properties of deposited polymer films have been identified, which is crucial to further development of emulsion-based RIR-MAPLE. With the role of primary solvent on hydrophobic polymer deposition being articulated, fundamental deposition mechanisms using emulsions as the target are proposed to help identify the roles of secondary solvent, surfactant, and alcohol/water co-matrices. These mechanisms also are explored by using the proposed hypotheses to improve the deposition of hydrophilic polymers. These studies provide insight and direction to determine the best emulsion target chemistry to control film properties for different applications. Further studies should focus on the in situ characterization of the emulsion morphology and theoretical simulation to validate the proposed hypotheses.

While understanding of emulsion-based RIR-MAPLE is far from complete, several applications have already been demonstrated showing the unique capabilities of the technique. Ryan Pate demonstrated an all-polymer distributed Bragg reflector

Table 10.5 RMS surface roughness of PEDOT:PSS films deposited from the emulsion target using different types of surfactants

Deposition recipe	RMS surface roughness (nm)
(a) 2(TCB):0.25(Phenol):2(1 PEDOT solution + 1 water)	46.80 (3.77)
(b) 2(TCB):0.25(Phenol):2(1 PEDOT solution + 1 SDS water)	38.75 (1.77)
(c) 2(TCB):0.25(Phenol):2(1PEDOT solution + 1 PEG-PPG-PEG water)	37.65 (2.47)
(d) 2(TCB):0.25(Phenol):2(1PEDOT solution + 1 BriJ water)	16.00 (0.28)

The roughness of the films was obtained by averaging three AFM images from different areas of each film. The values in parentheses are the standard deviation of the RMS surface roughness for each film from three AFM images

(DBR) based on eight pairs of alternating layers of P3HT and PMMA deposited by emulsion-based RIR-MAPLE [142], showing the capability to deposit multi-layered structures with polymers having similar solubility. Ryan McCormick demonstrated anti-reflection polymer coatings by etching away PMMA from a polystyrene (PS)/PMMA blended film to create a gradient refractive index structure [155]. This application demonstrates that emulsion-based RIR-MAPLE can blend two polymers with nanoscale domain sizes because the pore sizes in the etched film were consistent with the requirements for an effective medium in the visible range, a crucial requirement to achieve the gradient refractive index structure for application to anti-reflection coatings. Ryan McCormick also initiated bulk-heterojunction organic solar cell fabrication by emulsion-based RIR-MAPLE. Although the efficiencies of the initial organic solar cells fabricated using emulsion-based RIR-MAPLE were below 1%, the device performance already demonstrated significant improvement over organic solar cells fabricated by UV-MAPLE [127]. This improvement shows the potential of emulsion-based RIR-MAPLE for application to organic optoelectronic device fabrication, which extends beyond solar cells to include organic light emitting diodes (OLEDs) and organic photodetectors.

10.10 Conclusions

The MAPLE technique has demonstrated its ability to deposit organic and biological materials with minimum modification of the chemical structure and functionality of the material of interest, above all for medical applications. For this reason, there has been increasing interest in the technique for the last several years, also motivated by the interesting properties presented by some polymers, like PMMA, together with the possibility to have better control of molecular ordering and crystalline morphology during film growth due to the prolonged time of deposition compared to other solvent-based methods. Moreover, of relevant importance, is the possibility offered by the MAPLE technique to deposit polymer multilayers with respect to conventional deposition techniques used for polymer deposition.

Most of the advantages and peculiarity of this laser-based technique come from the possibility to change many independent deposition parameters. This aspect is particularly true for the emulsion-based RIR-MAPLE technique which offers an especially gentle mechanism for the deposition of organic and biological materials. Although all aspects of the process are not completely understood, it provides high versatility in deposition due to the additional degrees of freedom associated with the emulsion components, as demonstrated by the encouraging results for the deposition of materials for optoelectronic applications that open the way to realize new, high-performance devices.

References

1. H. Sirringhaus, P.J. Brown, R.H. Friend, M.M. Nielsen, K. Bechgaard, B.M.W. Langeveld-Voss, A.J.H. Spiering, R.A.J. Janssen, E.W. Meijer, P. Herwig, D.M. de Leeuw, Nature **401**, 685 (1999)
2. C.W. Tang, S.A. VanSlyke, Appl. Phys. Lett. **51**, 913 (1987)
3. H.E. Katz, J. Huang, J.Annu. Rev. Mater. Res. **39**, 71 (2009)
4. C.J. Brabec, N.S. Sariciftci, J.C. Hummelen, Adv. Funct. Mater. **11**, 15 (2001)
5. C. Bartic, G. Borghs, Anal. Bioanal. Chem. **384**, 354 (2006)
6. S. Reineke, F. Lindner, G. Schwartz, N. Seidler, K. Walzer, B. Lüssem, K. Leo, Nature **459**, 234 (2009)
7. A.G. MacDiarmid, J.C. Chiang, M. Halpern, W.S. Huang, S.L. Mu, N.L. Somasiri, W. Wu, S.I. Yaniger, Mol.Cryst Liq. Cryst. **121**, 173 (1985)
8. T.A. Sktheim, R.L. Elsenbaumer, J.R. Reynolds, *Handbook of Conducting Polymers* (Wiley, New York, 1997)
9. K.G. Neoh, E.T. Kang, K.L. Tan, Polymer **33**, 2292 (1992)
10. G. Witte, C.J. Wöll, Mater. Res. **19**, 1889 (2004)
11. S. Baxamusa, Conformal polymer CVD, in *CVD Polymers: Fabrication of Organic Surfaces and Devices*, ed. by K.K. Gleason (Wiley-VCH Verlag GmbH & Co. KGaA, Weinheim, 2015)
12. Y. Shi, J. Liu, Y. Yang, J. Appl. Phys. **87**, 4254 (2000)
13. A.R. Inigo, C.C. Chang, W. Fann, J.D. White, Y.-S. Huang, U.-S. Jeng, H.-S. Sheu, K.-Y. Peng, S.-A. Chen, Adv. Mater. **17**, 1835 (2005)
14. H. Sirringhaus, T. Kawase, R.H. Friend, T. Shimoda, Science **290**, 2123 (2000)
15. R. Saf, M. Goriup, T. Steindl, T.E. Hamedinger, D. Sandholzer, G. Hayn, Nat. Mater. **3**, 323 (2004)
16. M.I. Sluch, C. Pearson, M.C. Petty, M. Halim, I.D.W. Samuel, Synth. Met. **94**, 285 (1998)
17. B.J. Schwartz, Ann. Rev. Phys. Chem. **54**, 141 (2003)
18. H. Usui, Preparation of polymer thin films by physical vapor deposition, in *Functional Polymer Films: 2 Volume Set*, ed. by W. Knoll, R.C. Advincula (Wiley-VCH Verlag GmbH & Co. KGaA, Weinheim, 2011)
19. J. Park, C. Lee, J. Jung, H. Kang, K.H. Kim, B. Ma, B.J. Kim, Adv. Funct. Mater. **24**, 7588 (2014)
20. F.F. Shi, Surf. Coat. Technol. **82**, 1 (1996)
21. P. Favia, R. d'Agostino, Surf. Coat. Technol. **98**, 1102 (1998)
22. N. Inagaki, S. Tasaka, M. Makino, J. Appl. Polym. Sci. **64**, 1031 (1997)
23. M.J. Sowa, M.E. Littau, V. Pohray, J.L. Cecchi, J. Vac. Sci. Technol. **A18**, 2122 (2000)

24. G.H. Hishmeh, T.I. Barr, A. Skylarov, S. Hardcastle, J. Vac. Sci. Tech-nol. **A14**, 1330 (1996)
25. K. D'Almeida, J.C. Bernede, F. Ragot, A. Godoy, F.R. Diaz, S. Lefrant, J. Appl. Poly. Sci. **82**, 2042 (2001)
26. H. Biederman, J. Vac. Sci. Technol. **A18**, 1642 (2000)
27. R.A. McGill, D.B. Chrisey, Method of producing a film coating by matrix assisted pulsed laser deposition *Patent* No. 6.025.036 (2000)
28. D.B. Chrisey, A. Pique, R.A. McGill, J.S. Horwitz, B.R. Ringeisen, D.M. Bubb, P.K. Wu, Chem. Rev. **103**, 553 (2003)
29. A. Piqué, App. Phys. A: Mater. Sci. Process. **105**, 517 (2011)
30. A. Luches, A.P. Caricato, Fundamentals and Applications of MAPLE in *Laser-Surface Interactions for New Materials Production* (Springer Series in Materials Science 130, Springer, Berlin, 2010)
31. A.P. Caricato, MAPLE and MALDI: theory and experiments in *Lasers in Materials Science* (Springer Series in Materials Science 191, Springer International Publishing, Basel, 2014)
32. R. Eason, *Pulsed Laser Deposition of Thin Films: Applications-Led Growth of Functional Materials* (Wiley, New York, ISBN: 978–0-471-44709-2, 2006)
33. J.A. Greer, Appl. Phys. A Mater. Sci. Process. **105**, 661 (2011)
34. K.B. Shepard, Appl. Phys. A Mater. Sci. Process. **110**, 771 (2013)
35. E. Leveugle, L.V. Zhigilei, A. Sellinger, J.M. Fitz-Gerald, Appl. Surf. Sci. **253**, 6456 (2007)
36. A.P. Caricato, M. Lomascolo, A. Luches, F. Mandoj, M.G. Manera, M. Mastroianni, M. Martino, R. Paolesse, R. Rella, F. Romano, T. Tunno, D. Valerini, Appl. Phys. A Mater. Sci. Process. **93**, 651 (2008)
37. A.P. Caricato, M. Epifani, M. Martino, F. Romano, R. Rella, A. Taurino, T. Tunno, D. Valerini, J. Phys. D Appl. Phys. **42**, 095105 (2009)
38. K. Rodrigo, P. Czuba, B. Toftmann, J. Schou, R. Pedrys, Appl. Surf. Sci. **252**, 4824 (2006)
39. G. Socol, I.N. Mihailescu, A. Albu, S. Antohe, F. Stanculescu, A. Stanculescu, L. Mihut, N. Preda, M. Socol, O. Rasoga, Appl. Surf. Sci. **255**, 5611 (2009)
40. E. Leveugle, L.V. Zhigilei, J. Appl. Phys. **102**, 074914 (2007)
41. D.M. Bubb, P.K. Wu, J.S. Horwitz, J.H. Callahan, M. Galicia, A. Vertes, R.A. McGill, E.J. Houser, B.R. Ringeisen, D.B. Chrisey, J. Appl. Phys. **91**, 2055 (2002)
42. D.M. Bubb, M. Papantonakis, B. Collins, E. Brookes, J. Wood, U. Gurudas, Chem. Phys. Lett. **448**, 194 (2007)
43. A.P. Caricato, G. Leggieri, M. Martino, A. Vantaggiato, D. Valerini, A. Cretì, M. Lomascolo, M.G. Manera, R. Rella, M. Anni, Appl. Phys. A Mater. Sci. Process. **101**, 759 (2010)
44. A.P. Caricato, M. Anni, M.G. Manera, M. Martino, R. Rella, F. Romano, T. Tunno, D. Valerini, Appl. Surf. Sci. **255**, 9659 (2009)
45. D.M. Bubb, A.O. Sezer, J. Gripenburg, B. Collins, E. Brookes, Appl. Surf. Sci. **253**, 6465 (2007)
46. R. Pate, A.D. Stiff-Roberts, Chem. Phys. Lett. **477**, 406 (2009)
47. A. Sellinger, E. Leveugle, M. Tabetah, Appl. Phys. A Mater. Sci. Process. **105**, 529 (2011)
48. A. Luches, A.P. Caricato, Appl. Phys. B **105**, 503 (2011)
49. D.B. Bubb, J. Corgan, S.Y. Yi, M. Khan, L. Hughes, U. Gurudas, M. Papantonakis, R.A. McGill, Appl. Phys. A Mater. Sci. Process. **100**, 523 (2010)
50. R. Pate, K.R. Lantz, A.D. Stiff-Roberts, Thin Solid Films **517**, 6798 (2009)
51. A.L. Mercado, C.E. Almond, J.G. Hoekstra, J.M. Fitz-Gerald, Appl. Phys. A Mater. Sci. Process. **81**, 591 (2005)
52. A.P. Caricato, M. Cesaria, C. Leo, M. Mazzeo, A. Genco, S. Carallo, T. Tunno, A. Massafra, G. Gigli, M. Martino, J. Phys. D Appl. Phys. **48**, 135501 (2015)
53. E. György, A.P. del Pino, J. Roqueta, B. Ballesteros, L. Cabana, G. Tobias, J. Nanopart. Res. **15**, 1852 (2013)
54. A.P. Caricato, V. Arima, M. Cesaria, M. Martino, T. Tunno, R. Rinaldi, A. Zacheo, Appl. Phys. B **113**, 463 (2013)
55. M. Tabetah, A. Matei, C. Constantinescu, N.P. Mortensen, M. Dinescu, J. Schou, L.V. Zhigilei, J. Phys. Chem. B **118**, 13290 (2014)

56. A.P. Caricato, A. Luches, Appl. Phys. A Mater. Sci. Process. **105**, 565 (2011)
57. E. György, A.P. Caricato, MAPLE deposition of nanomaterials, in *Pulsed Laser Ablation: Advances and Applications in Nanoparticles and Nanostructuring Thin Films* (Pan Stanford Publishing Pte. Ltd., Singapore, 2017)
58. Y. Guo, A. Morozov, D. Schneider, J.W. Chung, C. Zhang, M. Waldmann, N. Yao, G. Fytas, C.B. Arnold, R.D. Priestley, Nat. Mater. **11**, 337 (2012)
59. H. Jeong, K.B. Shepard, G.E. Purdum, Y. Guo, Y.-L. Loo, C.B. Arnold, R.D. Priestley, Macromolecules **49**, 2860 (2016)
60. H. Jeong, S. Napolitano, C.B. Arnold, R.D. Priestley, J. Phys. Chem. Lett. **8**, 229 (2017)
61. A. Visan, R. Cristescu, N. Stefan, M. Miroiu, C. Nita, M. Socol, C. Florica, O. Rasoga, I. Zgura, L.E. Sima, M. Chiritoiu, M.C. Chifiriuc, A.M. Holban, I.N. Mihailescu, G. Socol, Appl. Surf. Sci. **417**, 234 (2017)
62. A. Visan, G.E. Stan, C. Ristoscu, G. Popescu-Pelin, M. Sopronyi, C. Besleaga, C. Luculescu, M.C. Chifiriuc, M.D. Hussien, O. Marsan, E. Kergourlay, D. Grossin, F. Brouillet, I.N. Mihailescu, Int. J. Pharm. **511**, 505 (2016)
63. M. Marinescu, L.O. Cinteza, G.I. Marton, L.G. Marutescu, M.-C. Chifiriuc, C. Constantinescu, J. Mol. Struct. **1144**, 14 (2017)
64. V. Grumezescu, A.M. Holban, L.E. Sima, M.B. Chiritoiu, G.N. Chiritoiu, A.M. Grumezescu, L. Ivan, F. Safciuc, F. Antohe, C. Florica, C.R. Luculescu, M.C. Chifiriuc, G. Socol, Int. J. Pharm. **521**, 184 (2017)
65. I.A. Paun, M. Zamfirescu, C.R. Luculescu, A.M. Acasandrei, C.C. Mustaciosu, M. Mihailescu, M. Dinescu, Appl. Surf. Sci. **392**, 321 (2017)
66. S. Koo, S.M. Santoni, B.Z. Gao, C.P. Grigoropoulos, Z. Ma, J. Mater. Res. **32**, 128 (2017)
67. O. Fufă, M. Socol, N. Preda, S. Grigorescu, S. Croitoru, G. Socol, Digest J. Nanomater. Biostruct. **12**, 73 (2017)
68. O.M. Perepelytsina, O.M. Yakymchuk, M.V. Sydorenko, O.N. Bakalinska, F. Bloisi, L.R.M. Vicari, Nanoscale Res. Lett. **11**, 314 (2016)
69. O. Rasoga, L. Sima, M. Chiritoiu, G. Popescu-Pelin, O. Fufă, V. Grumezescu, M. Socol, A. Stănculescu, I. Zgură, G. Socol, Appl. Surf. Sci. **417**, 204 (2017)
70. F. Sima, E. Axente, C. Ristoscu, O. Gallet, K. Anselme, I.N. Mihailescu, Adv. Mater. Interface **427** (2016)
71. R. Cristescu, A. Visan, G. Socol, A.V. Surdu, A.E. Oprea, A.M. Grumezescu, M.C. Chifiriuc, R.D. Boehm, D. Yamaleyeva, M. Taylor, R.J. Narayan, D.B. Chrisey, Appl. Surf. Sci. **374**, 290 (2016)
72. M.S. Stan, S. Constanda, V. Grumezescu, E. Andronescu, A.M. Ene, A.M. Holban, B.S. Vasile, L. Mogoantă, T.-A. Bălşeanu, G.D. Mogoşanu, G. Socol, A.M. Grumezescu, A. Dinischiotu, V. Lazar, M.C. Chifiriuc, Appl. Surf. Sci. **374**, 318 (2016)
73. L. Rusen, P. Neacsu, A. Cimpean, I. Valentin, S. Brajnicov, L.N. Dumitrescu, J. Banita, V. Dinca, M. Dinescu, Appl. Surf. Sci. **374**, 23 (2016)
74. L. Floroian, C. Ristoscu, N. Mihailescu, I. Negut, M. Badea, D. Ursutiu, M.C. Chifiriuc, I. Urzica, H.M. Dyia, C. Bleotu, I.N. Mihailescu, Molecules **21**, 740 (2016)
75. A.E. Oprea, L.M. Pandel, A.M. Dumitrescu, E. Andronescu, V. Grumezescu, M.C. Chifiriuc, L. Mogoantă, T.-A. Bălşeanu, G.D. Mogoşanu, G. Socol, A.M. Grumezescu, F. Iordache, H. Maniu, M. Chirea, A.M. Holban, Molecules **21**, 220 (2016)
76. G. Huang, Y. Chen, J. Zhang, Appl. Surf. Sci. **360**, 383 (2016)
77. G. Huang, W.H. Tse, J. Zhang, RSC Adv. **6**, 67166 (2016)
78. V. Califano, G. Ausanio, F. Bloisi, A. Aronne, L.R.M. Vicari, L. Nasti, Sens. Bio-Sens. Res. **6**, 103 (2015)
79. E. Boanini, P. Torricelli, L. Forte, S. Pagani, N. Mihailescu, C. Ristoscu, I.N. Mihailescu, A. Bigi, Colloids Surf., B **136**, 449 (2015)
80. A.M. Holban, A.M. Grumezescu, C.M. Saviuc, Adv. Struct. Mater. **74**, 311 (2015)
81. E. Boanini, P. Torricelli, F. Sima, E. Axente, M. Fini, I.N. Mihailescu, A. Bigi, J. Colloid Interface Sci. **448**, 1 (2015)

82. I.A. Paun, F. Stokker-Cheregi, C.R. Luculescu, A.M. Acasandrei, V. Ion, M. Zamfirescu, C.C. Mustaciosu, M. Mihailescu, M. Dinescu, Mater. Sci. Eng., C **55**(5494), 61 (2015)
83. P. Yin, G.B. Huang, W.H. Tse, Y.G. Bao, J. Denstedt, J. Zhang, J. Mater. Chem. B **3**, 3234 (2015)
84. A. Aronne, F. Bloisi, R. Calabria, V. Califano, L.E. Depero, E. Fanelli, S. Federici, P. Massoli, L.R.M. Vicari, Appl. Surf. Sci. **336**, 196 (2015)
85. F. Iordache, V. Grumezescu, A.M. Grumezescu, C. Curuţiu, L.M. Diţu, G. Socol, A. Ficai, R. Truşcə, A.M. Holban, Appl. Surf. Sci. **336**, 407 (2015)
86. N. Cicco, A. Morone, M. Verrastro, M. Dinescu, A. Matei, B. Mitu, D. Centonze, Lecture Notes Electr. Eng. **319**, 47 (2015)
87. V. Grumezescu, E. Andronescu, A.M. Holban, L. Mogoantə, G.D. Mogoşanu, A.M. Grumezescu, A. Stənculescu, G. Socol, F. Iordache, H. Maniu, M.C. Chifiriuc, Appl. Surf. Sci. **336**, 188 (2015)
88. R. Cristescu, A.V. Surdu, A.M. Grumezescu, A.E. Oprea, R. Trusca, O. Vasile, G. Dorcioman, A. Visan, G. Socol, I.N. Mihailescu, D. Mihaiescu, M. Enculescu, M.C. Chifiriuc, R.D. Boehm, R.J. Narayan, D.B. Chrisey, Appl. Surf. Sci. **336**, 234 (2015)
89. F. Sima, P.M. Davidson, J. Dentzer, R. Gadiou, E. Pauthe, O. Gallet, I.N. Mihailescu, K. Anselme, ACS Appl. Mater. Interfaces **7**, 911 (2015)
90. V. Grumezescu, A.M. Holban, F. Iordache, G. Socol, G.D. Mogoşanu, A.M. Grumezescu, A. Ficai, B.Ş. Vasile, R. Truşcă, M.C. Chifiriuc, H. Maniu, Appl. Surf. Sci. **306**, 16 (2014)
91. V. Grumezescu, G. Socol, A.M. Grumezescu, A.M. Holban, A. Ficai, R. Truşcă, C. Bleotu, P.C. Balaure, R. Cristescu, M.C. Chifiriuc, Appl. Surf. Sci. **302**, 262 (2014)
92. V. Califano, F. Bloisi, A. Aronne, S. Federici, L. Nasti, L.E. Depero, L.R.M. Vicari, Biosensors **4**, 329 (2014)
93. G. Huang, Y. Chen, J. Zhang, *14th IEEE International Conference on Nanotechnology, IEEE-NANO*, vol **6967984**, 815 (2014)
94. C. Popescu, A.C. Popescu, I. Iordache, M. Motoc, D. Pojoga, A. Simon-Gruita, N. Constantin, G. Duta Cornescu, E. Gyorgy, J. Mater. Sci. **49**, 4371 (2014)
95. A.M. Holban, V. Grumezescu, A.M. Grumezescu, B.S. Vasile, R. Truşcă, R. Cristescu, G. Socol, F. Iordache, J. Nanotechnol. **5**, 872 (2014)
96. A.G. Anghel, A.M. Grumezescu, M. Chirea, V. Grumezescu, G. Socol, F. Iordache, A.E. Oprea, I. Anghel, A.M. Holban, Molecules **19**, 8981 (2014)
97. V. Grumezescu, A.M. Holban, A.M. Grumezescu, G. Socol, A. Ficai, B.S. Vasile, R. Truscă, C. Bleotu, V. Lazar, C.M. Chifiriuc, G.D. Mogosanu, Biofabrication **6**, 035002 (2014)
98. C. Limban, A.V. Missir, A.M. Grumezescu, A.E. Oprea, V. Grumezescu, B.Ş. Vasile, G. Socol, R. Truşcă, M.T. Caproiu, M.C. Chifiriuc, B. Gălăţeanu, M. Costache, L. Moruşciag, G. Pîrcălăbioru, D.C. Nuţă, Molecules **19**, 12011 (2014)
99. I.A. Paun, M. Zamfirescu, M. Mihailescu, C.R. Luculescu, C.C. Mustaciosu, I. Dorobantu, B. Calenic, M. Dinescu, J. Mater. Sci. **50**, 923 (2014)
100. A. Visan, D. Grossin, N. Stefan, L. Duta, F.M. Miroiu, G.E. Stan, M. Sopronyi, C. Luculescu, M. Freche, O. Marsan, C. Charvilat, S. Ciuca, I.N. Mihailescu, Mater. Sci. Eng. B Solid-State Mater. Adv. Technol. **181**, 56 (2014)
101. V. Dinca, P.E. Florian, L.E. Sima, L. Rusen, C. Constantinescu, R.W. Evans, M. Dinescu, A. Roseanu, Biomed. Microdevice **16**, 11 (2014)
102. A. Palla-Papavlu, L. Rusen, V. Dinca, M. Filipescu, T. Lippert, M. Dinescu, Appl. Surf. Sci. **302**, 87 (2014)
103. I.A. Paun, A.M. Acasandrei, C.R. Luculescu, C.C. Mustaciosu, V. Ion, M. Mihailescu, E. Vasile, M. Dinescu, Appl. Surf. Sci. **357**, 975 (2015)
104. F.M. Miroiu, N. Stefan, A.I. Visan, C. Nita, C.R. Luculescu, O. Rasoga, M. Socol, I. Zgura, R. Cristescu, D. Craciun, G. Socol, Appl. Surf. Sci. **355**, 1123 (2015)
105. L. Rusen, V. Dinca, B. Mitu, C. Mustaciosu, M. Dinescu, Appl. Surf. Sci. **302**, 134 (2014)
106. E. Axente, F. Sima, L.E. Sima, M. Erginer, M.S. Eroglu, N. Serban, C. Ristoscu, S.M. Petrescu, E.T. Oner, I.N. Mihailescu, Biofabrication **6**, 35010 (2014)

107. F. Di Pietrantonio, M. Benetti, V. Dinca, D. Cannatà, E. Verona, S. D'Auria, M. Dinescu, Appl. Surf. Sci. **302**, 250 (2014)
108. A. Rotaru, A. Moanţă, C. Constantinescu, M. Dumitru, H.O. Manolea, A. Andrei, M. Dinescu, J. Therm. Anal. Calorim. **128**, 89 (2017)
109. C.E. Secu, M. Secu, F. Stokker-Cheregi, V. Ion, S. Brajnicov, M. Dinescu, Thin Solid Films **625**, 6 (2017)
110. A. Matei, M. Marinescu, C. Constantinescu, V. Ion, B. Mitu, I. Ionita, M. Dinescu, A. Emandi, Appl. Surf. Sci. **374**, 206 (2016)
111. C. Constantinescu, L. Rapp, P. Rotaru, P. Delaporte, A.P. Alloncle, Chem. Phys. **450–452**, 32 (2015)
112. C. Constantinescu, A. Rotaru, A. Nedelcea, M. Dinescu, Mater. Sci. Semicond. Process. **30**, 242 (2015)
113. A.P. Caricato, M. Anni, M. Cesaria, S. Lattante, G. Leggieri, C. Leo, M. Martino, A. Perulli, V. Resta, Appl. Phys. B: Lasers Opt. **119**, 453 (2015)
114. C. Constantinescu, A. Matei, V. Ion, B. Mitu, I. Ionita, M. Dinescu, C.R. Luculescu, C. Vasiliu, A. Emandi, Appl. Surf. Sci. **302**, 83 (2014)
115. C. Constantinescu, A. Matei, I. Ionita, V. Ion, V. Marascu, M. Dinescu, C. Vasiliu, A. Emandi, Appl. Surf. Sci. **302**, 69 (2014)
116. M. Socol, N. Preda, L. Vacareanu, M. Grigoras, G. Socol, I.N. Mihailescu, M. Jelinek, A. Stanculescu, M. Stoicanescu, Appl. Surf. Sci. **302**, 216 (2014)
117. A. Matei, C. Constantinescu, V. Ion, B. Mitu, I. Ionita, M. Dinescu, C. Vasiliu, A. Emandi, J. Organomet. Chem. **751**, 638 (2014)
118. F. Stokker-Cheregi, A. Matei, M. Dinescu, C.E. Secu, M. Secu, J. Phys. D Appl. Phys. **47**, 045304 (2014)
119. A. Stanculescu, O. Rasoga, M. Socol, L. Vacareanu, M. Grigoras, G. Socol, F. Stanculescu, C. Breazu, E. Matei, N. Preda, M. Girtan, Appl. Surf. Sci. **417**, 183 (2017)
120. M. Socol, N. Preda, A. Stanculescu, C. Breazu, C. Florica, F. Stanculescu, S. Iftimie, M. Girtan, G. Popescu-Pelin, G. Socol, Appl. Surf. Sci. **417**, 196 (2017)
121. C.-P. Yen, P.-F. Yu, J. Wang, J.-Y. Lin, Y.-M. Chen, S.-Y. Chen, AIP Adv. **6**, 085011 (2016)
122. A. Stanculescu, G. Socol, L. Vacareanu, M. Socol, O. Rasoga, C. Breazu, M. Girtan, F. Stanculescu, Appl. Surf. Sci. **374**, 278 (2016)
123. M. Socol, N. Preda, O. Rasoga, C. Breazu, I. Stavarache, F. Stanculescu, G. Socol, F. Gherendi, V. Grumezescu, G. Popescu-Pelin, M. Girtan, N. Stefan, Appl. Surf. Sci. **374**, 403 (2016)
124. F. Stanculescu, O. Rasoga, A.M. Catargiu, L. Vacareanu, M. Socol, C. Breazu, N. Preda, G. Socol, A. Stanculescu, Appl. Surf. Sci. **336**, 240 (2015)
125. A. Stanculescu, M. Socol, O. Rasoga, I.N. Mihailescu, G. Socol, N. Preda, C. Breazu, F. Stanculescu, Appl. Surf. Sci. **302**, 169 (2014)
126. A. Stanculescu, G. Socol, M. Grigoras, T. Ivan, L. Vacareanu, M. Socol, O. Rasoga, C. Breazu, I.N. Mihailescu, I. Iordache, N. Preda, F. Stanculescu, Appl. Phys. A Mater. Sci. Process. **117**, 261 (2014)
127. A.P. Caricato, M. Cesaria, G. Gigli, A. Loiudice, A. Luches, M. Martino, V. Resta, A. Rizzo, A. Taurino, Appl. Phys. Lett. **100**, 073306 (2012)
128. A. Matei, M. Marinescu, C. Constantinescu, V. Ion, B. Mitu, I. Ionita, M. Dinescu, A. Emandi, Appl. Surf. Sci. **374**, 206 (2016)
129. M. Socol, N. Preda, O. Rasoga, C. Breazu, I. Stavarache, F. Stanculescu, G. Socol, F. Gherendi, V. Grumezescu, G. Popescu-Pelin, M. Girtan, N. Stefan, Appl. Surf. Sci. **374**, 403 (2016)
130. B. Mitu, A. Matei, M. Filipescu, A. PallaPapavlu, A. Bercea, T. Lippert, M. Dinescu, J. Phys. D Appl. Phys. **50**, 115601 (2017)
131. Boyadjiev S.I. Boyadjiev, N. Stefan, I.M. Szilágyi, N. Mihailescu, A. Visan, I.N. Mihailescu, G.E. Stan, C. Besleaga, M.T. Iliev, K.A. Gesheva, J. Phys Conf. Ser. **780**, 012013 (2017)
132. B.X. Dong, A. Li, J. Strzalka, G.E. Stein, P.F. Green, J. Polym. Sci. Part B: Polym. Phys. **55**, 39 (2017)
133. A. Li, B.X. Dong, P.F. Green, MRS Commun. **5**, 593 (2015)

134. F. Mariano, A.P. Caricato, G. Accorsi, C. Leo, M. Cesaria, S. Carallo, A. Genco, D. Simeone, T. Tunno, M. Martino, G. Gigli, M. Mazzeo, J. Mater. Chem. **32**, 7667 (2016)
135. A.M. Darwish, S. Wilson, A. Blackwell, K. Taylor, S. Sarkisov, D. Patel, P. Mele, M.W. Johnson, X. Zhang, B. Koplitz, *Proceedings of SPIE—The International Society for Optical Engineering*, vol **9586**, 958602 (2015)
136. A.M. Darwish, A. Burkett, A. Blackwell, K. Taylor, S. Sarkisov, D. Patel, B. Koplitz, D. Hui, Compos. B Eng. **68**, 355 (2015)
137. A.M. Darwish, S. Wilson, A. Blackwell, K. Taylor, S. Sarkisov, D. Patel, P. Mele, B. Koplitz, *Proceedings of SPIE—The International Society for Optical Engineering*, vol **9586**, 958605 (2015)
138. A.M. Darwish, S. Moore, A. Mohammad, D. Alexander, T. Bastian, W. Dorlus, S. Sarkisov, D. Patel, P. Mele, B. Koplitz, D. Hui, Compos. B Eng. **109**, 82 (2017)
139. A.M. Darwish, A. Burkett, A. Blackwell, K. Taylor, V. Walker, S. Sarkisov, B. Koplitz, *Proceedings of SPIE—The International Society for Optical Engineering*, *vol* **9200**, 92000C (2014)
140. S.M. O'Malley, J. Schoeffling, R. Jimenez, B. Zinderman, S. Yi, D.M. Bubb, Appl. Phys. A **117**, 1343–1351 (2014)
141. R. Pate, K.R. Lantz, A.D. Stiff-Roberts, IEEE J. Sel. Top. Quantum Electron. **14**, 1022 (2008)
142. R. Pate, R. McCormick, L. Chen, W. Zhou, A.D. Stiff-Roberts, Appl. Phys. A **105**, 555 (2011)
143. R.D. Torres, S.L. Johnson, R.F. Haglund, J. Hwang, P.L. Burn, P.H. Holloway, Crit. Rev. Solid State Mater. Sci. **36**, 16 (2011)
144. T.F. Tadros, *Emulsion Science and Technology: A General Introduction* pp. 1–56 (2009)
145. G.M. Hale, M.R. Querry, Appl. Opt. **12**, 555 (1973)
146. D.M. Bubb, S.M. O'Malley, C. Antonacci, D. Simonson, R.A. McGill, J. Appl. Phys. **95**, 2175 (2004)
147. A. Piqué, R.A. McGill, D.B. Chrisey, D. Leonhardt, T.E. Msina, B.J. Spargo, J.H. Callahan, R.W. Vachet, R. Chung, M.A. Bucaro, Thin Solid Films **355**, 536 (1999)
148. E. Leveugle, A. Sellinger, J.M. Fitz-Gerald, L.V. Zhigilei, Phys. Rev. Lett. **98**, 216101 (2007)
149. W. Ge, N.K. Li, R.D. McCormick, E. Lichtenberg, Y.G. Yingling, A.D. Stiff-Roberts, ACS Appl. Mater. Interfaces **8**, 19494 (2016)
150. R. Pate, *Matrix-Assisted Pulsed Laser Evaporation of Conjugated Polymer and Hybrid Nanocomposite Thin Films: A Novel Deposition Technique for Organic Optoelectronic Devices* (Department of Electrical and Computer Engineering, Duke University, Durham, NC, 2011)
151. S.L. Murov, *Properties of Organic Solvents*. Available from: http://murov.info/orgsolvents.htm (2014)
152. C.S. Hendrik, M.N. Odireleng, S.M. Pontsho, S.D. Mokhotjwa, B.M. Bakang, J. Mater. Sci. Eng. *B***5** (2015)
153. J. Peet, J.Y. Kim, N.E. Coates, W.L. Ma, D. Moses, A.J. Heeger, G.C. Bazan, Nat. Mater. **6**, 497 (2007)
154. B. Toftmann, K. Rodrigo, J. Schou, R. Pedrys, Appl. Surf. Sci. **247**, 211 (2005)
155. R.D. McCormick, E.D. Cline, A.S. Chadha, W. Zhou, A.D. Stiff-Roberts, Macromol. Chem. Phys. **214**, 2643 (2013)

Chapter 11
Combinatorial Laser Synthesis of Biomaterial Thin Films: Selection and Processing for Medical Applications

Emanuel Axente, Carmen Ristoscu, Adriana Bigi, Felix Sima and Ion N. Mihailescu

Abstract A new generation of implantable biomaterials should possess smart surfaces and interfaces able to modulate cellular behaviour and directly address specific clinical issues. This chapter provides an overview of recent advances in the field of laser-based combinatorial synthesis of thin biomaterial films with gradient of composition on solid substrates for medical applications. Laser processing methods and selected applications for tissue engineering and regenerative medicine are reviewed in sequel.

11.1 Introduction

In materials science, the substance properties can be optimized by tuning the composition. The idea of combinatorial materials science is to extend and utilize advanced fabrication technologies of materials, which are able to produce a large number of different combinations on a single substrate in one-step experiment under identical conditions [1]. After the combinatorial deposition, the material libraries are examined for desired physical properties by adequate characterization methods. By using this approach, a fast development of new materials can be anticipated.

As known, Combinatorial chemistry refers to a new methodology by which one can simultaneously synthesize tens, hundreds or even thousands of different micro-samples, often called (molecular) libraries [2]. These compound libraries can be mixtures, sets of individual compounds or chemical structures. In fact, combinatorial

---

E. Axente · C. Ristoscu · I. N. Mihailescu (✉)
Lasers Department, National Institute for Lasers, Plasma and Radiation Physics (INFLPR), Magurele, Ilfov, Romania
e-mail: ion.mihailescu@inflpr.ro

A. Bigi
Department of Chemistry “G. Ciamician”, University of Bologna, Bologna, Italy

F. Sima
Center for Advanced Laser Technologies (CETAL), National Institute for Lasers, Plasma and Radiation Physics (INFLPR), Magurele, Ilfov, Romania

P. M. Ossi (ed.), *Advances in the Application of Lasers in Materials Science*, Springer Series in Materials Science 274,
https://doi.org/10.1007/978-3-319-96845-2_11

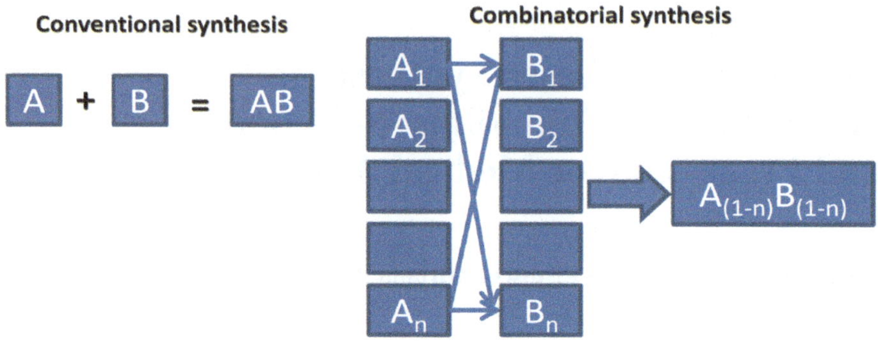

Fig. 11.1 Schematic representation of conventional and combinatorial syntheses

chemistry is a combination of chemistry and biology. It can generate array of different, but structurally similar molecules, in order to investigate a specific molecule along with its derivatives. The basic principle of combinatorial chemistry is to concurrently prepare and monitor a large number of different molecules.

In conventional synthesis, a compound is obtained directly, $A + B \rightarrow AB$, while in combinatorial approach we have $A_{(1-n)} + B_{(1-n)} \rightarrow A_{(1-n)}B_{(1-n)}$ (Fig. 11.1).

A comprehensive review of high throughput combinatorial synthesis available in [3] discusses the application of combinatorial approaches to the discovery of new electronic, ferroelectric, piezoelectric, multiferroic and magnetic materials. The combinatorial method based on the use of sputter deposition to develop new libraries includes materials which exhibit Colossal Magnetoresistance (CMR) [4]. These authors obtained a library of 128 distinct compositions of Y, La, Ba, Sr, Ca, and Co containing films such as $La_{0.60}YCa_{0.33}MnO_x$ and $Nd_{0.7}Sr_{0.33}MnO_{3-x}$. A library of 25,000 oxide compositions based on Y, Al, La, Eu, and V, e.g. $Y_{0.845}Al_{0.070}La_{0.060}Eu_{0.025}VO_4$ with very high quantum efficiency similar to commercial red phosphors was demonstrated in [5]. Other luminescent inorganic oxides containing elements from the Sr–Ce–O system were obtained using thin film physical vapour deposition methods followed by prolonged heat treatment to form single-phase compositions [6].

A special UHV deposition system based on the M600 MBE was developed for the synthesis of solid state materials combinatorial libraries [7]. These include the synthesis of wide compositional ranges of mixed oxides, hydrides and nitrides.

Combinatorial Magnetron Sputtering (CMS) demonstrated to act as a viable tool for the rapid development of enormous libraries of complex materials [8].

Binary Ag–Te thin films to be used as a cation supply layer in conductive bridge random access memory (CBRAM) devices were synthesized by combinatorial sputter deposition technique [9]. Switching properties investigated and compared to cells with only Ag cation supply layer, demonstrated an improved cycling behaviour when Te is added to pure Ag.

Ti–Ta alloys, recognized as appropriate materials for applications in actuators as well as biomedical implants, were manufactured by glancing angle deposition (GLAD) at room temperature in form of a nanocolumnar (of ~150 to 160 nm in width) Ti–Ta library [10].

Combinatorial Pulsed Laser Deposition (C-PLD) have been used to grow discontinuous Fe/MgO multilayers with increased tunnelling magnetoresistance (TMR) and field sensitivity at room temperature [11].

C-PLD was applied for embedding Ag into Hydroxyapatite (HA) with the aim to enhance the antimicrobial performances of coatings for load bearing implants [12]. The Ag content along the length of the combinatorial structure increased up to a maximum of ~1 at.%. Ag content values up to 0.6 at. % into HA coatings sustained antimicrobial activity and were found nontoxic for mesenchymal stem cells (MSC) growth. Ag, recognised as antibacterial material, was used to obtain Zr–Cu–Al–Ag thin film metallic glasses by confocal magnetron co-sputtering with DC power supplies [13]. The antibacterial activity proved to be strongly dependent on alloy compositions: sensitive to Cu and Ag, but insensitive to Zr and Al. More precisely, the alloys with higher antibacterial activity were in the compositional range of 30–50% Zr, 33–36% Cu, 10–25% Al, and 5–10% Ag.

Antimicrobial Ag-doped Carbon structures were fabricated by C-PLD [14]. The optimal combination between convenient physical-chemical properties, efficient shielding against microbial colonization and positive effects on MG63 cells was found for Ag-doped Carbon films containing 2–7 at.% Ag. These mixtures can be used to fabricate efficient coatings of metallic implants, with the goal to decrease the risk of implant associated biofilm infections which are difficult to treat and habitually connected with implant failure.

A review on high throughput combinatorial thin film material growth and characterization is given in [15]. The selection of relevant combinatorial libraries and extrapolation from small-scale deposition techniques to industrially relevant processes must be implemented with care.

In 2012, we proposed a single-step combinatorial approach based on Matrix-Assisted Pulsed Laser Evaporation (MAPLE) for both the immobilization and blending of organic compounds [16]. Combinatorial MAPLE (C-MAPLE) was applied to obtain a compositional library of levan and oxidized levan in form of thin film. We proved that by C-MAPLE one can rapidly generate discrete areas of organic film compositions with improved properties than starting materials. Moreover, compositional gradients could be used to identify specific nanostructured surfaces in order to control cell proliferation or modulate intracellular signalling pathways [17].

The proposed C-MAPLE generated combinatorial gradients proved remarkable potential in the screening of new bioactive interfaces for tissue regeneration. Compositional gradients of Sr-substituted HA (SrHA) and Zolendronate (ZOL) modified HA (ZOLHA) were synthesized on Titanium substrates with the aim to modulate the composition of the thin films and hence the promotion of bone growth and the inhibition of bone resorption [18]. Antimicrobial orthopedic maps from chitosan and biomimetic apatite powders have been obtained by C-MAPLE [19]. These last results are presented and discussed in the next sections of this chapter, along with in situ, single-step, embedding of active substances (Fibronectin–FN) in biodegradable polymeric (PDLLA–poly-dl-lactide) matrices for local release [20].

11.2 Combinatorial Laser Synthesis Approaches

The research progress at interface among material science, physics, chemistry and biology demonstrated the permanent need to create smart, bioresponsive surfaces in order to control cell behavior, and consequently to respond to particular clinical problems [21]. During last decades, combinatorial science along its whole applications spectrum emerged as a forefront domain for sustained progress achievements in modern drug discovery and delivery [22–24] and new materials synthesis for various fields of nanoscience and nanotechnology [25, 26], such as semiconductors, magnetic materials, shape memory alloys, and energy materials. Indeed, in the simplest description, the method involves a direct bifunctional mechanism where at least two materials having different properties play distinct synergetic roles for a specific application. The key concept of combinatorial materials science relies on the fabrication of compositional material libraries consisting in well-defined gradients that can be used in high-throughput screening of structure-activity relationships (SAR) for biomedical applications.

Different methodologies are constantly explored for the identification or discovery of novel materials with promising properties for nano-biotechnology. Buenconsejo et al. [27] report on the development of a method for multilayer sputter deposition which allows obtaining multiple ternary shape memory alloys libraries on a single 4 inch SiO_2/Si wafer substrate. The parallel synthesis and biological high-throughput evaluation of more than 800 analogues of Procaspase-Activating Compound 1 (PAC-1) was performed by Hsu et al. [28], using combinatorial libraries. The protocol for library construction involved the condensation of 31 hydrazides in parallel with 27 aldehydes to generate 837 PAC-1 analogues. The authors evidenced that six analogue compounds were found to be substantially more potent than basic PAC-1 to induce apoptosis of cancer cells.

Indeed, combinatorial methods and high-throughput screening are extremely promising approaches for optimizing the SAR of both organic and inorganic compositional library of materials [29]. We have recently demonstrated that micro- and nano-fabrication of coatings with compositional gradient by laser-based techniques could be a promising alternative to the existing procedures [30, 31]. Laser processing allows control of the morphology and/or chemistry features of biomaterials and the fabrication of hybrid compositional libraries.

The first laser attempt was the introduction, ten years ago, of "*an unconventional concept*" of combinatorial chemistry, namely the combinatorial-pulsed laser deposition [32]. This technology, developed as an efficient but cost-effective and single-step processing technique for the synthesis of coatings with gradient of composition and morphology, was applied to a broad variety of inorganic materials [33, 34]. Here, the compositional gradient is naturally obtained by the simultaneous ablation of

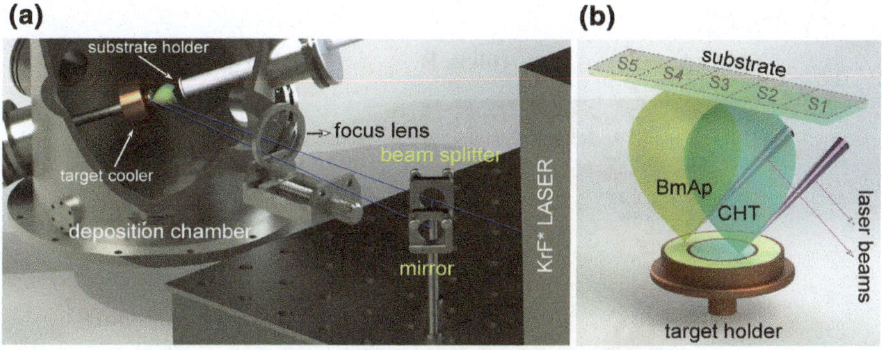

Fig. 11.2 **a** C-MAPLE experimental setup **b** Schematic representation of thin films compositional library obtained by C-MAPLE. Reproduced from [19]

two materials with two laser beams followed by their growth as a library thin film on a solid substrate. C-PLD was also applied to biomaterials, namely doped CaP gradient coatings and reported by Socol et al. [12] for discovering optimal composition/morphology characteristics for tissue engineering.

A development of C-PLD was the introduction of Combinatorial-MAPLE technique by Sima and Axente et al. [16, 17] for the fabrication of organic and composite compositional libraries. Indeed, C-PLD is restricted to the synthesis of inorganic thin film libraries only (high-power laser irradiation induces irreversible damage to organic materials), while C-MAPLE approach allows preserving organic material properties after transfer. Moreover, we recently demonstrated the possibility of tailoring materials properties (morphology and/or chemistry features) to generate bioresponsive surfaces, able to modulate and control cell behaviour [17–19, 30, 31].

Both experimental and theoretical aspects of MAPLE technique were extensively addressed in the literature [30, 31, 35–40], therefore in the following we will refer to C-MAPLE only. Similar to C-PLD, in the basic irradiation geometry, two cryogenic targets are simultaneously evaporated by two pulsed laser beams (Fig. 11.2a). In our experimental design, the beam of one laser is optically split into two beams, then focused onto the surface of the targets (Fig. 11.2b), each one containing different frozen solutions. Alternatively, different lasers could be considered, having different characteristics such pulse duration, wavelength, repetition rate in order to optimize the absorption of the frozen solvent of each target.

The evaporated materials are collected and assembled on solid substrates, a natural compositional gradient being thus obtained in-between, due to substance fluxes intermixing (Fig. 11.2b). Several experimental parameters could be independently adjusted for assembling uniform and functional hybrid thin films. The high versatility of the method allows controlling the key deposition parameters, as laser fluence and repetition rate, dynamical pressure inside the vacuum reaction chamber, number of laser pulses to be applied for each compound, targets to substrate separation distance, distance between the two laser spots focused on targets to control materials

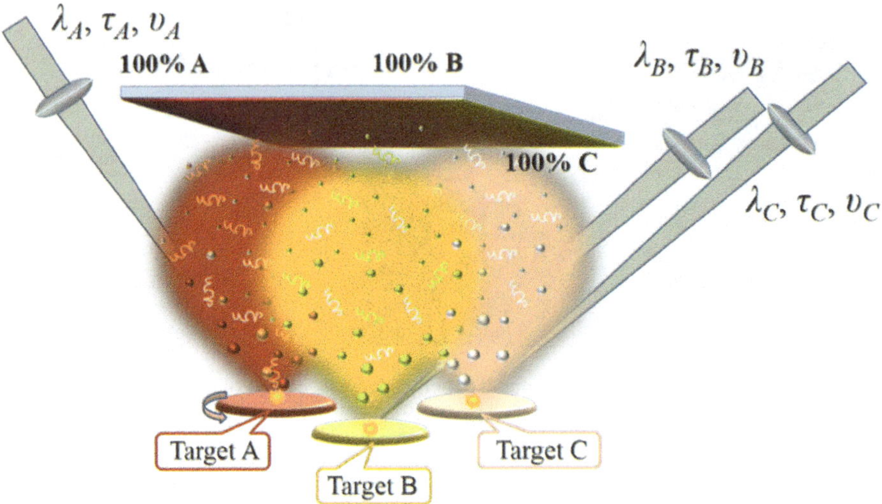

Fig. 11.3 Schematic representation of the C-MAPLE experimental setup using three targets

spreading. Moreover, various compositional distributions inside libraries could be easily achieved by simply modifying the initial concentration of the solute in the frozen targets.

A special arrangement could be also designed in which three or more targets could be irradiated in the same time (Fig. 11.3). Here, the interplay between the experimental parameters depicted above could result in the fabrication of compositional diagrams, in which the composition-structure-properties relationship could be investigated in each consecutive point. Accordingly, it is obvious that laser-combinatorial technologies could bring unique advantages for further developments in the synthesis of new materials, surface functionalization or design of innovative properties in the biomedical field. C-MAPLE is an innovative concept for the in situ biofabrication of compositional libraries thin films of organic, inorganic and composite materials in a single-step process. Several prospective research directions are the fabrication of *cell-instructive surfaces* for tissue repair, supports for cell signalling response and the possibility of drugs and proteins release from biodegradable polymeric coatings which are extensively addressed in the following sections.

11.3 Biomaterials Selection for Biomedical Applications

11.3.1 Compositional Gradient Thin Films of Sr-Substituted and ZOL Modified HA

Due to high similarity with the inorganic phases composition of biomineralized tissues of vertebrates, Calcium Phosphates (CaPs) exhibit excellent biocompatibility and bioactivity [41–44]. Accordingly, CaPs coatings on metal implants are nowadays wide scale used in the fields of tissue engineering and regenerative medicine, as well as biomimetic biomaterials interacting with the "stem cell niche" [45]. The complete list of CaPs, including their detailed synthesis approaches and advanced properties for biomedical applications, has been comprehensively reviewed in several studies [41, 42] and references therein. It is generally accepted that the main mineral phase of bone and teeth is a CaP, which could be assimilated to synthetic hydroxyapatite (HA, $Ca_{10}(PO_4)_6(OH)_2$), having the Ca/P ratio ~1.67. However, biological apatites differ from stoichiometric HA, e.g. containing significant amounts of different ions, either incorporated in the apatite crystal structure or just adsorbed on the crystal surface, as reviewed by Boanini et al. [41]. Correlating the composition with other structural properties such as degree of crystallinity, size and shape of crystallites grains, and solubility allows for controlling the osteoinductive properties of HA-coated titanium implants [41, 43].

Continuous developments in surface coatings biotechnologies are explored to improve osseointegration, mitigate possible adverse effects and implant infection. Indeed, during the whole lifetime, bone is continuously subjected to remodeling mainly by pro-formative osteoblasts and pro-resorptive osteoclasts. It was recently shown that bisphosphonates (BPs) are among the most active inhibitors for bone degradation [46–49]. Biofunctional alendronate–HA thin films deposited by MAPLE on titanium substrates were reported by Bigi et al. [46]. It is shown that such hybrid coatings inhibit osteoclast proliferation and differentiation, and promote their apoptosis. In addition, alendronate has a valuable influence on osteoblast growth, viability and earlier differentiation. More recently, the same group investigated the influence of the simultaneous presence of two inhibitors of bone degradation, strontium and zoledronate, on the direct synthesis of HA structure and bone cell responses [49]. The results of in vitro co-culture tests put into evidence that Sr plays a major beneficial role on osteoblast proliferation and differentiation, whereas the inhibition of osteoclast formation and differentiation is greater for ZOL than for Sr.

In view of discovering an optimum between dopants, we applied C-MAPLE, to deposit thin films with variable compositions in gradient of Sr-substituted HA (SrHA) and ZOL modified HA (ZOLHA) on Titanium substrates [18].

SrHA nanocrystals were chemically synthesized using 50 ml solution obtained by dissolving appropriate amounts of $Ca(NO_3)_2{\cdot}4H_2O$ and $Sr(NO_3)_2$ in CO_2-free deionized water in order to get a Sr/(Ca + Sr) ratio of 0.1. The reaction was carried out in N_2 atmosphere, at pH adjusted to 10 with NH_4OH. The total concentration of $[Ca^{2+}] + [Sr^{2+}]$ was 1.08 M. In the following, the solution was heated at 90 °C and

Table 11.1 Experimental conditions for C-MAPLE samples fabrication

Target	Substrate	Substrate temperature	Laser fluence (J cm^{-2})	Dynamic pressure (Pa)	Target-substrate distance (cm)	No. of laser pulses
ZOLHA (0.2 g in 20 ml H_2O)	5 × Ti	RT	1.6	1	5	20,000
SrHA (0.2 g in 20 ml H_2O)			0.7			

50 ml of 0.65 M $(NH_4)_2HPO_4$ solution (pH 10 adjusted with NH_4OH), was added drop-wise under stirring. The resulting precipitate was maintained in contact with the reaction solution for 5 h at 90 °C under stirring, then centrifuged and repeatedly washed with distilled water. The final product was dried at 37 °C overnight. The same procedure (without $Sr(NO_3)_2$) was applied to synthesize ZOLHA. In this case, disodium zoledronate tetrahydrate was added to the phosphate solution. The ZOL concentration was 14 mM, as calculated on final volume.

Compositional gradient thin films were obtained by simultaneous laser vaporization of the two distinct material targets using the protocol described in Sect. 11.2 and reviewed in [30, 31]. The experiments of thin coatings deposition were carried out in a vacuum reaction chamber [16, 17]. Briefly, 0.2 g of each nanocrystal powder, SrHA and ZOLHA, were both homogenously suspended in 20 ml distilled water by ultrasonically stirring. Then, 3 ml of each solution were poured into a two concentric ring holder. This way, any possible mixing was avoided. The holder was immersed in liquid nitrogen for 15 min and the solutions were frozen. They were further used as solid targets in the reaction chamber where a cooler supplied with liquid nitrogen flow kept them frozen during multi-pulse laser irradiation and vaporization. The experimental setup used in this study is schematically depicted in Fig. 11.2. In a single-step processing protocol, the synchronized evaporated materials were collected onto 5 distinct Ti substrates of 12 mm diameter. We applied 20,000 laser pulses to grow~400 nm thin films, which is sufficient for the in vitro tests under consideration. The samples were labelled C-1 to C-5, where the composition varies from SrHA (C-1) to ZOLHA (C-5). The experimental conditions for MAPLE transfer and immobilization of SrHA and ZOLHA were carefully adjusted to prevent any thermal damage and irreversible decomposition of materials and are collected in Table 11.1.

The X-ray diffraction (XRD) patterns of the starting powders were recorded using a PANalytical X'Pert PRO powder diffractometer equipped with a fast X'Celerator detector using Ni-filtered Cu Kα radiation (λ =0.154 nm, 40 mA, 40 kV). Figure 11.4a presents the XRD patterns of SrHA and ZOLHA showing that both samples consist of HA as unique crystalline phase (ICDD PDF n.9–432). The Sr concentration in SrHA inferred by ICP spectrometry was about 8.4 at.%. The incorporation of Sr into HA structure is demonstrated by the values of the lat-

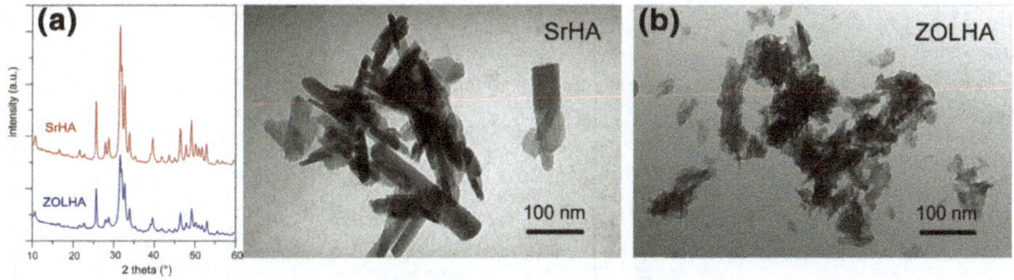

Fig. 11.4 Powder X-ray diffraction patterns of the as-prepared SrHA and ZOLHA materials (**a**). TEM images of SrHA and ZOLHA nanocrystals (**b**). Adapted from [18]

tice parameters of SrHA, $a = 9.443(1)$ Å, $c = 6.905(1)$ Å, which are enlarged when compared to undoped, stoichiometric HA ($a = 9.4269$ Å, $c = 6.8840$ Å) [50], in agreement with the bigger ionic radius of Sr than Ca. However, zoledronate incorporation (7.6 wt% as measured by spectrophotometric analysis) has no significant influence on the values of the lattice constants, in agreement with our previous study [49].

Samples for Transmission Electron Microscopy (TEM) analyses of the powders were prepared by dispersing a small amount of both materials in ethanol, followed by ultrasonication. Further, a microdrop of the suspension was pipetted onto holey carbon foils placed on standard copper microgrids. A Philips CM 100 transmission electron microscope operating at 80 kV was used. Our investigations revealed that the morphology of HA nanocrystals is considerably influenced by the presence of the bisphosphonate. Indeed, ZOLHA composite nanocrystals display significantly reduced dimensions, ill-defined shapes and edges, in comparison to SrHA nanocrystals (Fig. 11.4b), which morphology does not differ significantly from the characteristic "plate-like" morphology of HA nanocrystals [51].

Figure 11.5 shows the XRD patterns of the combinatorial thin films deposited by C-MAPLE. In agreement with powder structures, the patterns evidenced the presence of HA as the unique crystalline phase. A qualitative evaluation of the crystallite sizes in the c-axis direction was calculated from the width at half maximum intensity ($\beta_{1/2}$) of the (002) reflection plane using the classical Scherrer equation. The results indicate that the coherent length of the perfect crystalline domains exhibits a mean value of about 37.4 nm, with no significant variation as a function of composition (Table 11.2).

Morphological investigations of the combinatorial coatings were performed using a Philips XL-20 scanning electron microscope operating at 15 kV, coupled with Energy dispersive X-ray spectrometry (EDS). SEM images in Fig. 11.6 show an image in center of the combinatorial library (sample C-3) where it is possible to appreciate that the coatings surface exhibits a granular morphology, with closely packed grains of tens of nanometers size. A previous investigation [52] on the biomimetic apatite layers and Ti substrates interface, evidenced the uniformity, homogeneity and compactness of MAPLE coatings. EDS analyses clearly evidenced the successful fabrication of the compositional gradient, the decrease of Sr concentration being

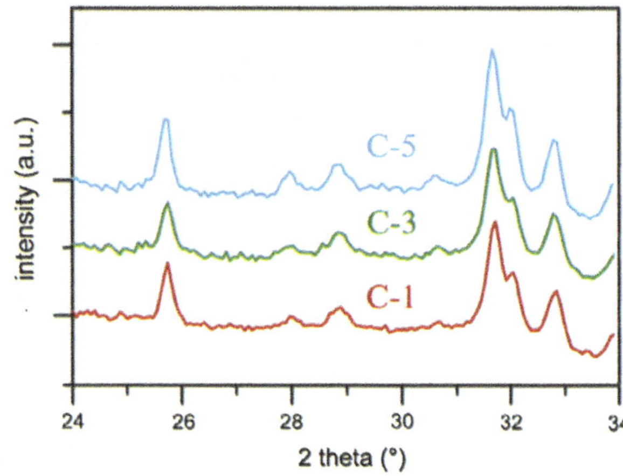

Fig. 11.5 Powder X-ray diffraction patterns of the C-1, C-3 and C-5 thin films. Reproduced from [18]

Table 11.2 Coherent lengths (τ_{hkl}) of the perfect crystalline domains in the direction normal to 002 plane calculated using the Scherrer method, and Sr content (%) evaluated by EDS measurements of coatings

Sample code	τ_{002} (nm)	Sr content (atom%)
C-1	38.8 (6)	8.4 (3)
C-2	36.7 (2)	5.8 (3)
C-3	38.1 (11)	4.0 (1)
C-4	37.0 (7)	1.4 (1)
C-5	36.4 (11)	–

evidenced in the coatings (samples C-1 to C-5) and summarized in Table 11.2. The stoichiometric transfer by C-MAPLE is validated by comparing the initial Sr content in the SrHA target with that measured in sample C-1. Moreover, one can notice the homogeneous Sr and Ca distribution, as shown by EDS elemental maps (see inset in Fig. 11.6 for C-3 sample).

Osteoblast (OB) and osteoclast (OC) were co-cultured on SrHA and ZOLHA combinatorial coatings, and their proliferation and viability (10 and 21 days) were evaluated by WST1 colorimetric reagent test (Fig. 11.7a). The detailed protocols used for the evaluation of cells activity, differentiation and morphology are reported in [18]. Co-cultures were used since allowed evaluating the whole interaction system e.g. cells, CaPs and drugs, and for simulating the in vivo microenvironment and bone remodelling processes where cells influence each other. After 10 days, CTR OB/OC shows a significant enhancement of cell proliferation when compared to CTR OB and bare Ti. However, C-5 sample (ZOLHA) shows a significant reduction of OB viability at 10 days with respect to experimental groups and controls, while at 21 days the difference is maintained only with respect to CTR OB/OC.

The evaluation of the most common markers for OB differentiation is presented in Fig. 11.7b, c. Here, the activity of Alkaline Phosphatase (ALP) was found signifi-

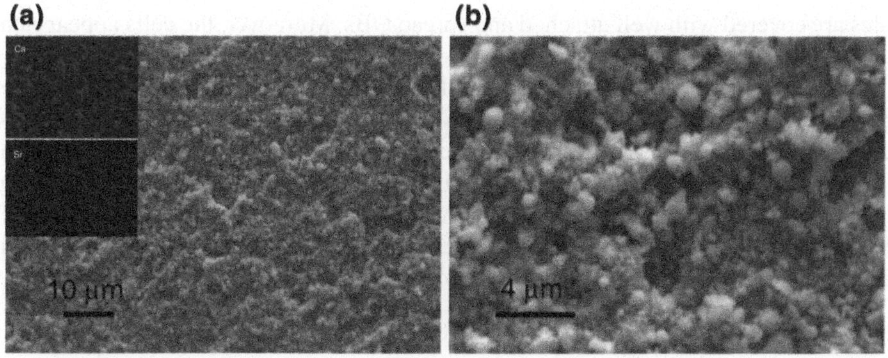

Fig. 11.6 SEM micrographs of C-3 sample at two different magnifications. Inset—the EDS maps show the homogeneous distribution of Ca and Sr throughout the surface of MAPLE deposited coating. Adapted from [18]

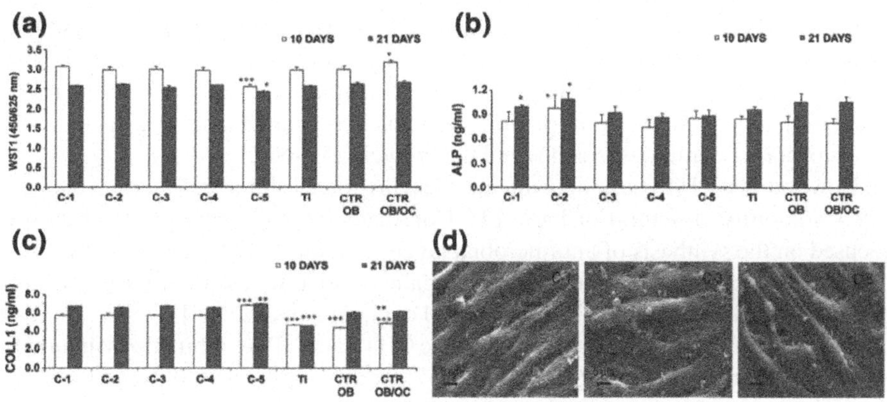

Fig. 11.7 Osteoblast proliferation (WST1) and activity (ALP, COLL1, OPG/RANKL ratio) in co-cultures with osteoclasts after 10 and 21 days of culture on Ti with different combination of Sr and Zol (*$p < 0.05$; **$p < 0.005$; ***$p < 0.0001$). **a** WST1. 10 days: ***C-5 versus C-1, C-2, C-3, C-4, Ti, CTR OB, CTR OB/OC; *CTR OB/OC versus Ti, CTR OB; 21 days: *C-5 versus CTR OB/OC. **b** ALP. 10 days: *C-2 versus C-4; 21 days: *C-1 versus C-4, C-5, /C-2 versus C-3, C-4, C-5. **c** COLL1: 10 days: ***C-5 versus C-1, C-2, C-3, C-4, Ti, CTR OB, CTR OB/OC; ***Ti, CTR OB versus C-1, C-2, C-3, C-4; **CTR OB/OC versus C-2; ***CTR OB/OC versus C-1, C-3, C-4; 21 days: *C-5 versus CTR OB/OC; **C-5 versus CTR OB; ***Ti versus C-1, C-2, C-3, C-4, CTR OB, CTR OB/OC. **d** SEM micrographs of osteoblasts grown on C-1, C-3 and C-5. Adapted from [18]

cantly higher on C-1 (SrHA) group than on C-4 and C-5 (21 days), and on C-2 group when compared to C-4 (10 and 21 days), C-3, and C-5 (21 days). The production of Type I Pro-Collagen (COLL1) appears enhanced on C-5 group in comparison with all other groups (10 days) and CTRs (10 and 21 days). Ti (10 and 21 days), CTR OB and CTR OB/OC (21 days) are significantly lower than the experimental groups. SEM investigations at the end of the experimental times evidenced that all the samples are covered with well attached and spread OBs. Moreover, the cells appear rich of filopodia, as could be seen in Fig. 11.7d for samples C-1, C-3 and C-5.

These results demonstrate that C-MAPLE technique is a powerful tool for the generation of coatings able to modulate osteoblast and osteoclast behavior when co-cultured. In fact, the possibility to fabricate combinatorial libraries of crystalline thin films with a gradient composition of Sr-substituted HA and zoledronate containing HA provides materials with cellular activity depending on the relative content of Sr and ZOL, and as a consequence, able to tailor promotion of bone growth and prevention of bone resorption according to specific requirements.

11.3.2 Combinatorial Maps Fabricated from Chitosan and Biomimetic Apatite for Orthopaedic Applications

Chitosan (CHT) is known to be an exceptionally biodegradable, non-toxic and biocompatible cationic polysaccharide which can be synthesized from chitin by alkaline deacetylation [53–56]. It has a well recognized potential to be used as an antimicrobial agent [57], along with delivery of antibiotics, such as beta-lactams (e.g., penicillins, cephalosporins), aminoglycosides and daptomycin [58–60].

However, the CHT use as anti-infective driving force is limited because it presents a low solubility at neutral or basic pH. Consequently, the centre of attention was focused on the synthesis of antimicrobial layers containing CHT and its derivatives. They proved excellent inhibitory activity against a wide spectrum of *Gram-positive* and *Gram-negative* bacteria, including food contaminants [61–64].

There are in literature many reports on the CHT and CaPs combinations in form of powders [65], membranes [66], scaffolds [67], or microspheres [68]. Nevertheless, a few only were devoted to procedures developed for the simultaneous preparation of a composite biomaterial with two components, predicted to guarantee a closer contact between them [69–71]. Our approach proceeds by fabrication of compositional maps of CHT and biomimetic apatite (BmAp) by C-MAPLE technique. An excimer laser source (KrF*, $\lambda = 248$ nm, $\tau_{FWHM} = 25$ ns) operated at 10 Hz frequency repetition rate was used for the cryogenic targets evaporation. 12 mm diameter Ti (grade 4) disks, Si wafers or glass slides were used as deposition substrates. The coated samples area formed by five consecutive 12 mm Ti disks were further denoted S1 to S5; where S1 stands for the coating area having CHT as major component, S5 represents the coating area with richer content of BmAp. S2–S3–S4 series indicates blended coating areas with decreasing CHT/BmAp ratio (see Fig. 11.2b). For comparison reasons, simple CHT and BmAp films have been also deposited on the same type of substrates [19]. Deionized water was selected as solvent to obtain solutions of 1 and 2 wt% for CHT and BmAp, respectively. The optimum laser energy was of 100 mJ in the case of CHT target and of 70 mJ for the BmAp one.

C-MAPLE composite layers exhibit a homogeneous spongy aspect all over the substrate as revealed by SEM micrographs [19]. This feature is known to promote cell adhesion. On S1 one can observe CHT particulates having a spherical morphology, while in the blended regions, S2–S4, elongated filiform structures can be noticed. One can suppose that these structures may sustain toughness improvement similar to that provided by the biocomposite materials reinforced with fibers, ensuring the desired mechanical properties when implanted [72].

The CHT-BmAp synthesized biocomposite thin films are amorphous and rough as demonstrated by high resolution AFM investigations [73]. One notices that a progressive increase of roughness (R_{RMS}) occurs with the CHT concentration in the C-MAPLE composite films [19].

The Ca/P molar ratio for each combinatorial surface varied in the range ~1.3 to 1.5, lower than the stoichiometric value of HA (i.e., 1.67), pointing to the synthesis of a calcium-deficient biomimetic apatite.

FTIR spectra of BmAp, HA (pure and highly crystalline), and CHT powders are comparatively shown in Figs. 11.8a, b. The first important difference is the broader IR bands of BmAp powder with respect to the HA material. In addition to the distinctive [74, 75] ν_4 bending, ν_1 symmetric stretching and ν_3 asymmetric stretching vibration bands of orthophosphate units of HA, the BmAp powder elicited additional IR bands at ~872, 1418, 1480 and 1635 cm^{-1}. The well-defined band at ~872 cm^{-1} could be attributed to the superimposed contributions of the ν_2 bending modes of carbonate groups and the vibrations exhibited by hydrogen phosphate ions $(HPO_4)^{2-}$ (characteristic to non-apatitic domains). The weak band at ~1635 cm^{-1} is assigned to the bending mode of water molecules. It signalises the existence of hydration, and correlates well with the very broad band ranging from 3600 to 2500 cm^{-1}, caused by the stretching vibrations of adsorbed water molecules. The broad IR absorbance maxima of the BmAp powder spectrum, and the absence of the libration (~628 cm^{-1}) and stretching (~3571 cm^{-1}) modes of structural $(OH)^-$ groups, point towards an apatite-like compound with a short-range order.

In case of CHT powder, the IR spectrum (Figs. 11.8a, b) exhibits two prominent bands, placed in the wave numbers regions (i) 3600–2500 cm^{-1} and (ii) 1100–950 cm^{-1}. They are assigned to the (i) overlapped stretching vibrations of adsorbed water and N-H bonds, and to the (ii) symmetric stretching vibrations of C–O bonds in groups like COH, COC and CH_2OH, respectively [76–79]. The sharp bands but with reduced intensity, centred at (iii) ~1150 cm^{-1} and (iv) ~890 cm^{-1}, are ascribed to the (iii) asymmetric stretching vibrations of C–O, and (iv) wagging vibrations of C–H bonds of the saccharide groups of CHT [78, 79]. The peaks at 1650, 1591 and 1316 cm^{-1} are characteristic to amides I, II and III groups vibrations, whilst the bands at 1420 and 1376 cm^{-1} are attributed to CH_3 symmetric deformation modes [76–79].

Figure 11.8c, d present the comparative FTIR spectra acquired for the pure (CHT and BmAp) layers and blended (CHT-BmAp) thin films on titanium substrates. One general observation is that the spectra of pure CHT and BmAp layers present less defined IR envelopes with broader bands with respect to the corresponding powders. This is indicative for a decreased structural order, a rather expected behaviour for structures fabricated at room temperature in non-equilibrium conditions by MAPLE. Surprisingly, all major vibration bands of CHT and BmAp were recorded in the

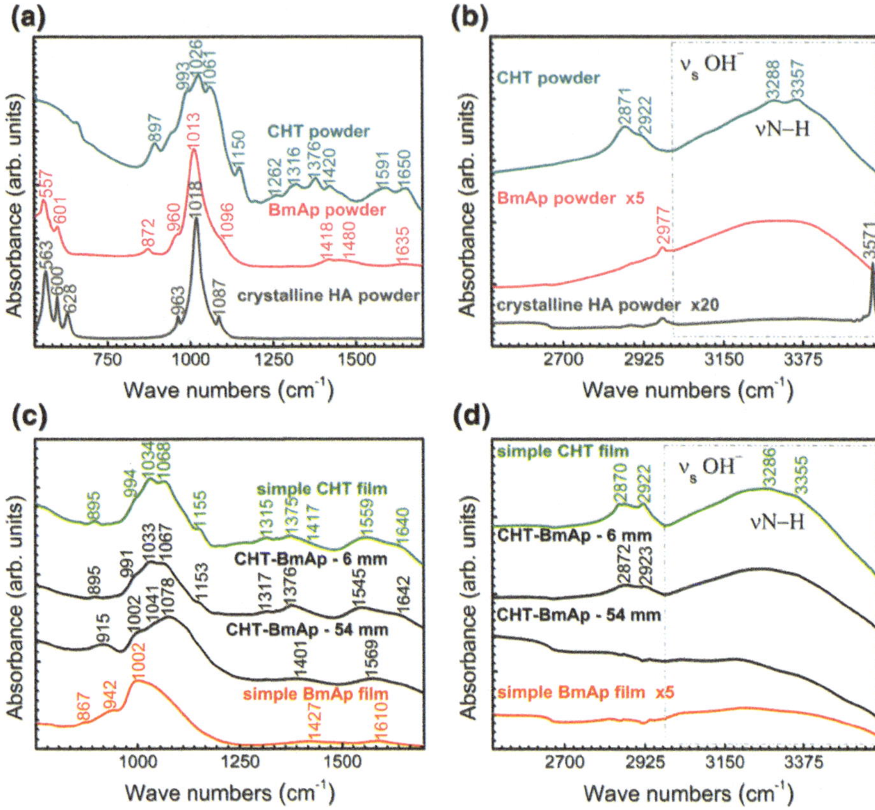

Fig. 11.8 Comparative FTIR spectra of CHT, BmAp, and crystalline HA powders (**a**, **b**), and simple CHT, BmAp, and CHT-BmAp films (**c**, **d**) in the fingerprint (**a**, **c**) and functional groups (**b**, **d**) regions. For a better visual evaluation, the spectra were normalized to the intensity of the most prominent band centred at ~1040 to 1010 cm^{-1}. Reproduced with permission from [19]

case of coatings, besides similar intensity ratio and insignificant wavelength shifts. Another aspect which has to be emphasized is the preservation of similar levels of hydration (for both types of films) as the ones found in the parent materials.

XPS analysis validated the chemical composition of the C-MAPLE thin films. Accordingly, the survey spectra acquired on the surface of CHT-BmAp samples exhibit C 1s, O 1s, N 1s, Ca 2s, Ca 2p, P 2s and P 2p photoelectron peaks [19]. The atomic ratio (N 1s/N 1s + Ca2p) is decreasing from CHT to BmAp compounds, indicating the composition gradient of combinatorial layers. On the other hand, the Ca 2p/P 2p ratio for CHT-BmAp compounds varied between 0.9 and 1.03 along the sample length, slightly lower than the values obtained by EDS (~1.3 to 1.5).

The composition gradient of the CHT-BmAp has been confirmed all along the combinatorial films by FTIR and XPS analyses, validating the deposition technique.

The development of biofilms consisting of the Gram-positive (*S. aureus*) and Gram-negative (*E. coli*) presented a different dynamic on C-MAPLE bio-composite coatings. Both *S. aureus* and *E. coli* biofilms on the bare Ti substrate revealed an analogous dynamic, a decreasing trend of biofilm embedded viable cells, quantified at the three consecutive time intervals [19]. The high surface roughness of bare Ti substrate is a prerequisite for an improved osseointegration but also favours the rapid and firm adhesion of *S. aureus* and *E. coli*. This behavior sustains the leading position held by these strains in the etiology of implant-associated infections [80].

In case of microbial biofilms development on the CHT-BmAp surfaces, a more distinct inhibitory effect was observed against the *S. aureus* biofilm, as compared to *E. coli*, for all tested time intervals (Fig. 11.9). S2, S3 and S4 samples inhibited *S. aureus* microbial adherence in the first time interval (6 h), while none of the tested samples inhibited the adherence of the Gram-negative *E. coli* strain (Fig. 11.9a). At 24 h, all tested CHT-BmAp samples decreased the number of *S. aureus* biofilm (Fig. 11.9b); the most intensive inhibitory effect was observed for the S4 sample in the case of *S. aureus* biofilm. The *E. coli* biofilm development was inhibited by the S1–S5 samples, but the most intensive antibiofilm effect was noticed for S4 and S5 (Fig. 11.9b).

An intensive anti-biofilm effect was exhibited after 48 h against *S. aureus* in the case of all CHT-BmAp tested samples (Fig. 11.9c) but the most significant activity was observed for S1 and S2 samples. The *E. coli* biofilm was inhibited only by S3–S5 samples, the most intensive effect being noticed for S5 (Fig. 11.9c).

These results recommend S3 and S4 as the most promising compositions, assuring an anti-biofilm protection on long time against both bacterial species (Fig. 11.10).

Although the mechanisms of antimicrobial action of CHT are not entirely elucidated, the literature states that its main action takes place at the microbial cell surface, as suggested by electron microscopy findings. Cellular wall disorganization both in Gram-positive and Gram-negative bacteria, as well as in fungal strains was shown, via the electrostatic interaction of positively-charged CHT with the negatively charged bacterial components (proteins, phospholipids) [81–83]. This action mechanism is responsible for a wide spectrum of the antimicrobial activity of CHT and its derivatives including filamentous fungi, yeasts and bacteria [84]. Nevertheless, it seems that CHT is more active against Gram-positive than Gram-negative bacteria, as revealed also by our study. In this regard, *S. aureus* biofilm proved to be more susceptible than *E. coli*, probably due to the stronger interaction of CHT with the amino acids rich fraction of the Gram-positive bacterial wall [85]. In exchange, the Gram-negative strains have a more complex cellular wall due to the negatively-charged lipopolysaccharides which can act as a barrier in the interaction of CHT with other structures of the cellular wall and membrane [86].

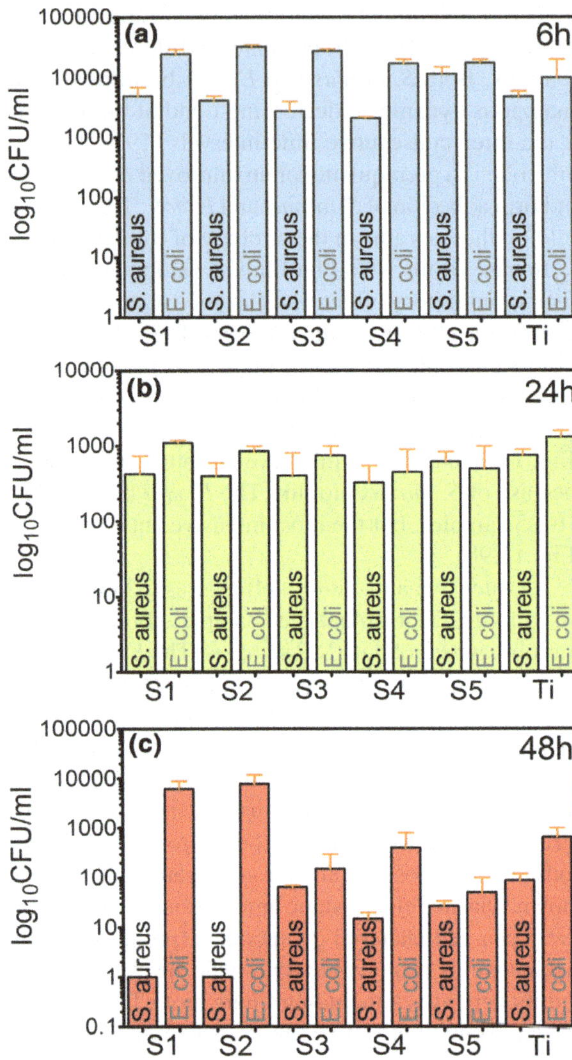

Fig. 11.9 Graphic representation of the number of viable microbial cells adhered to the surface of CHT-BmAp composite film after **a** 6 h, **b** 24 h and **c** 48 h of incubation. Reproduced with permission from [19]

11.3.3 Combinatorial Fibronectin Embedded in a Biodegradable Matrix by C-MAPLE

The last developments in tissue engineering field, revealed that artificial implant materials to be considered for in vivo tissue regeneration should mimic to a large extent the natural extracellular matrix (ECM). Indeed, cells interact with the ECM which is an interlocking mesh of fibrous proteins, glycoproteins and glycosamynoglycans [87]. Among various ECM molecules, fibronectin (FN) plays out a key role for tissue regeneration [88], being directly involved in the regulation of cell behav-

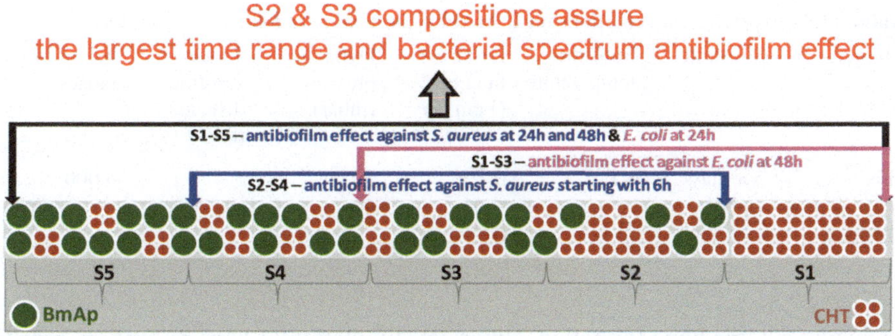

Fig. 11.10 CHT-BmAp gradient coating on Ti surface

iors (e.g. attachment, spreading, migration, differentiation and proliferation) [89, 90]. FN-cell interactions are mainly mediated by integrins, which are transmembrane receptors responsible for cell signaling [91].

Grigorescu et al. [92] reported on the fabrication of functional fibronectin patterns by nanosecond excimer laser direct writing onto Titanium substrates for tissue engineering applications. The authors evidenced the successful surface biofunctionalization with FN spots that preserve the global conformation of the protein and consequently the bio-chemical properties. The in vitro assays showed that the laser-transferred FN patterns were non-cytotoxic and exhibited a high affinity towards two representative cell lines (i.e. MC3T3-E1 osteoblasts and Swiss 3T3 fibroblasts).

Sima et al. [93, 94] were first that reported on the MAPLE synthesis and advanced characterization of FN and vitronectin functional coatings on various substrates. Recently, the same group [95], demonstrated the possibility to fabricate hybrid inorganic–organic thin implant coatings by laser-based techniques. PLD and MAPLE were subsequently applied for assembling HA and FN and synthesizing a synergistic interface for biomimetic implant applications. The authors proved that less than 7 μg FN per cm^2 onto HA surface is appropriate for improving adhesion, spreading, and differentiation of osteoprogenitor cells.

Very recently, Brigaud et al. [96] reported on the first direct comparison of the in vitro synergistic osteogenic potential of three ECM-associated members, the osteogenic BMP-2, -6 and -7 when combined with native human plasma FN. Two strategies were considered in this study: (i) direct delivering in solution or (ii) use of laser-generated coatings of BMPs and/or FN onto titanium-hydroxyapatite surfaces. It was evidenced that BMPs-FN association improves the osteogenic activity of the growth factors, but with essential inconsistencies between groups, depending on each BMP member, in close relation with the affinity of BMPs for FN.

Obviously, the cell-surface interactions are, at least in the first stage, strongly influenced by both the surface properties of the material and the cells type. Accordingly, surface functionalization with specific bioactive molecules plays a crucial role in current research. Nevertheless, if the active compounds release is too fast, in a single high-dose burst mode, [97, 98] this could induce critical clinical complications

Table 11.3 Experimental conditions for C-MAPLE fabrication of FN-PDLLA libraries

Target	Substrate	Substrate temperature	Laser fluence ($J cm^{-2}$)	Dynamic pressure (mbar)	Target-substrate distance (cm)	No. of laser pulses
FN	5 × glass and Si	RT	0.7	10^{-2}	5	5.000
PDLLA			0.55			

including ectopic bone formation, abnormal stimulation of bone resorption, cancer induction, inflammation or life-threatening swelling (as reviewed in [99, 100] for BMPs).

Consequently, the fabrication of surface coatings with advanced properties that stimulate controlled and gradual release of active biomolecules could definitely maximize their efficiency while minimizing or eventually eliminating possible undesired side effects. Laser-based techniques could be a viable solution to achieve this goal. C-MAPLE was recently introduced for in situ fabrication of a binary protein-polymer combinatorial thin film system [20]. Thus, laser-assembling of a composite library consisting of a biodegradable polymer (PDLLA) and a protein (FN) in a single-step procedure was demonstrated. This protein-polymer system was selected due to well-known potential of FN to mediate the initial cell attachment, while PDLLA is a biodegradable matrix with well-studied behavior in biological environments. A broad variety of materials having distinct biodegradation kinetic properties is currently explored, including but not limited to both natural and synthetic biopolymers that mimic the micro-bio-environment [20, 38]. For the moment, the most studied ones as poly(lactic acid-co-glycolic acid) (PLGA), poly(ε-caprolactone) (PCL), poly-DL-lactide (PDLLA) and polyethylene glycol (PEG) are usually grown as thin film for drug delivery due to their recognized biodegradability and biocompatibility [101–103]. Human FN was purified from cryoprecipitated blood plasma using the protocol reported in [104] while fluorescent-FN was prepared as described in [105].

This study is particularly difficult, since the two compounds are dissolved in distinct solvents, the challenge being related to finding the laser irradiation parameters in respect to the thermo-physical properties of the two targets. The first target was prepared following the protocol described in [93], the solutions consisting of FN dissolved in distilled water and saline buffer (50 mM Tris, pH 7.4, 150 mM NaCl). Commercial PDLLA powder (0.04 g) was dissolved in 10 ml chloroform to obtain the second homogenous solution. These were accommodated in a two concentrically copper holders and simultaneously frozen in liquid nitrogen to obtain two cryogenic targets. The laser fluence was adjusted for each target material after a systematic parametric study and fixed at 0.7 J cm^{-2} for FN and 0.55 J cm^{-2} for PDLLA, respectively. A KrF* excimer laser beam (λ = 248 nm, τ = 25 ns) was used for C-MAPLE, and the other experimental parameters are collected in Table 11.3.

Surface morphology, along the median direction of the combinatorial library, was investigated by optical microscopy (Zeiss Axio Imager Z1M with an Axio Cam

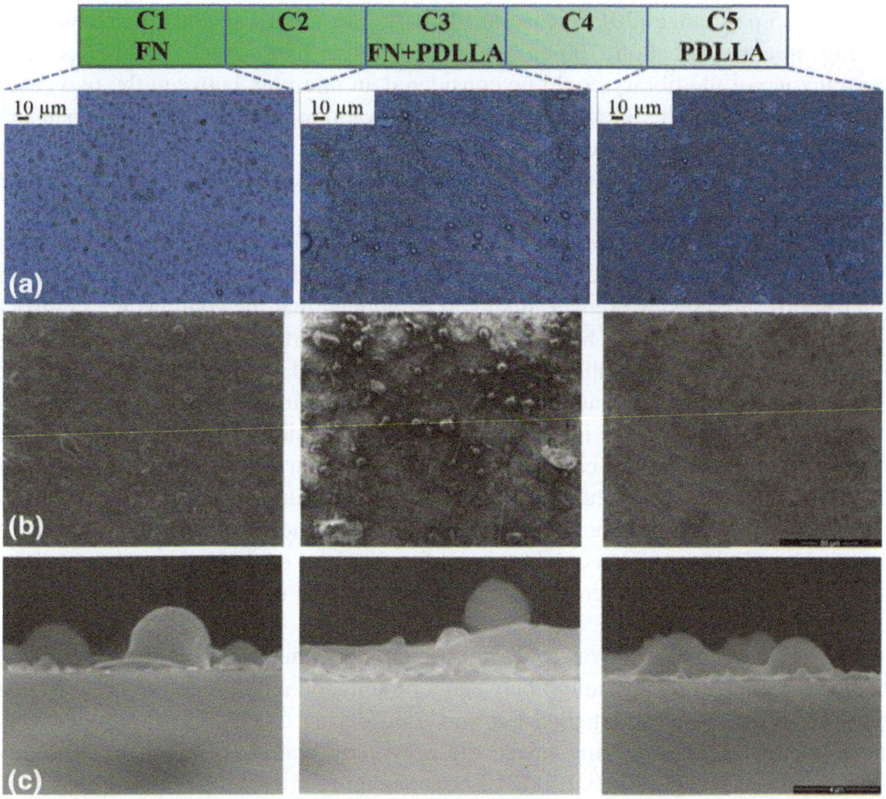

Fig. 11.11 Optical images of FN (left), FN + PDLLA (centre) and PDLLA (right) thin films (scale bar = 10 μm) (**a**). SEM images of the same regions of the combinatorial library (scale bar = 50 μm) (**b**) and the corresponding cross-section SEM recordings (**c**). Adapted from [20]

MRc 5-HR) and Scanning Electron Microscopy (SEM) (FEI Co., model Inspect S, 0–30 kV accelerating voltage, working distance 0–30 mm, with ECON 4/6 Energy Dispersive Detecting (EDAX) Unit). Both, planar and cross-section images have been recorded on distinct locations of the library corresponding to protein, polymer and their mixtures (Fig. 11.11).

Indeed, for tissue engineering applications, the physical-chemical properties of the coatings (e.g. topography, composition and structure) have a strong influence on the proteins adsorption which in turn influences the cellular responses. Accordingly, a 3D control of surface features is a must for the synthesis of bioactive coatings. We present in Fig. 11.11a specific optical images of the gradient coatings revealing areas covered by either FN (left), PDLLA (right) and a clear mixture of the two compounds (center). Micrometer-sized, quasi-uniformly distributed spherical particles are characteristic for surface profile after FN transfer, in agreement with our previous MAPLE studies [93]. Optical profilometry measurements revealed that particulates sizes are within the 1 ÷ 1.5 μm range [20]. In case of PDLLA side, a homogenous matrix with an average thickness of about 700 nm, embedding larger particulates, is observed. The central zone of the library could be considered an interplay between the two cases depicted above, by the intermixing of the two evaporation fluxes during expansion and consequent assembling on the substrate. The average thickness of this region was evaluated at about 650 nm, with larger particulates formed by a possible coalescence effect.

Complementary SEM investigations of samples C1 (FN side), C3 (central library) and C5 (PDLLA side) given in Fig. 11.11b, are in good agreement with the optical microscopy analyses. Moreover, it should be noticed that the coatings are compact and free of micro-cracks. The abundant presence of particulates embedded in thinner uniform matrix was confirmed by cross-section SEM analyses (Fig. 11.11c). The sharp interface between coatings and substrates, without any visible signs of peeling and/or delamination, is indicative for a good adhesion, a critical parameter for implant coatings. One can correlate by tailoring the surface features the improved coating adherence with the higher specific surface area (due to the presence of microparticles), which is also beneficial for enhancing the adhesion capacity of binding to tissue cells. Highly uniform combinatorial film libraries could be thus fabricated, by simply adjusting the number of the laser pulses applied to each target, in respect to their specific ablation rate.

The goal of this study was to demonstrate the synthesis of composite coatings, having a compositional gradient between a protein and a polymeric matrix. To this purpose, a fluorescently labelled FN was used during C-MAPLE experiments, in order to facilitate monitoring inside the polymeric matrix. Confocal microscopy was employed using a Zeiss LSM710 laser scanning microscope (Carl Zeiss, Jena Germany) with 100× and 1.4 numerical aperture objective in a multitrack mode by dual excitation for the fluorescent FN and visualization of PDLLA, respectively. Tens of Z-stacks images were collected in depth, probing the location on samples. Zen 2010 software was used for image overlapping and 3D reconstruction. A representative example is presented in Fig. 11.12a, where distinct micrometric-sized fluorescent FN packages are evidenced inside the polymeric matrix. This was achieved by exciting either simultaneously or individually the FN and PDLLA in a 3D topography. One could notice the preservation of the FN fluorescence inside the polymeric matrix, which stands for a clear confirmation that the protein is conserving the biological conformation after the laser transfer and immobilization.

The compositional investigation of combinatorial library along the longitudinal direction was performed by FTIR and μFTIR. Spectra were recorded in absorbance mode with a Shimadzu 8400S instrument equipped with an automated IR microscope (AIM) 8800 apparatus. The spectral range investigated in this study was set to $4000 \div 500\,\text{cm}^{-1}$ with a resolution of $4\,\text{cm}^{-1}$ wavenumber. Both FN and PDLLA (C1 and C5, corresponding to library borders) were found unaffected after C-MAPLE experiments since the main absorption peaks present in the spectra confirmed the safe laser transfer [20]. Briefly, the intact transfer of the polymer structure is confirmed by the identification of the absorption peaks centred at 1090, 2800–3000, 1754, 1450, 1381, 1269, and 1186 cm^{-1} wavenumbers, in agreement with other studies reported

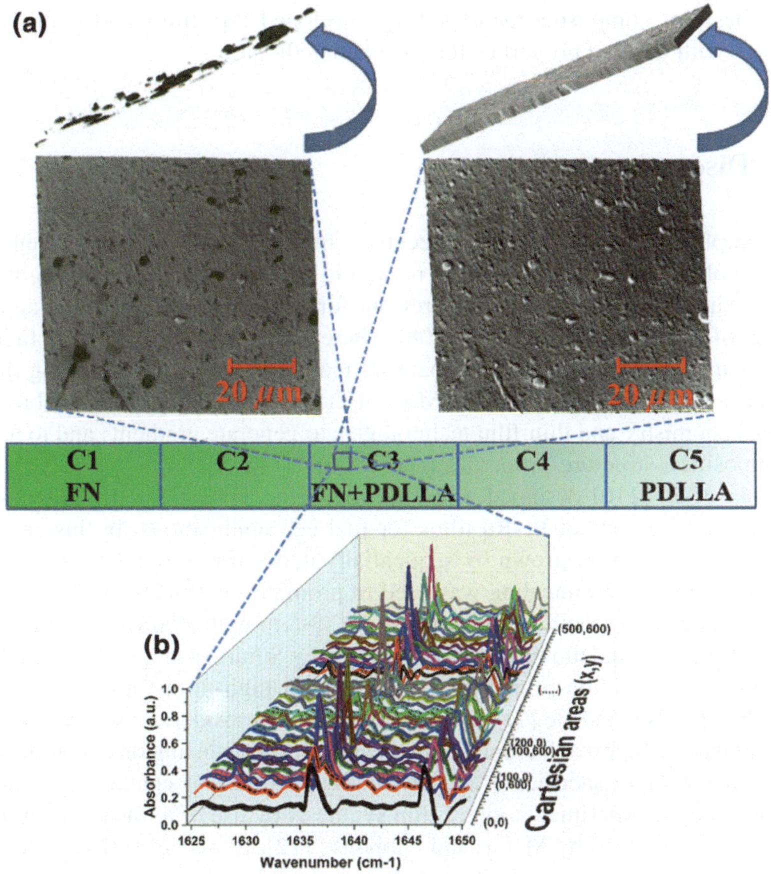

Fig. 11.12 **a** Confocal microscopy images of sample C3 (FN + PDLLA) and the corresponding cross-section image recordings of the respective films. **b** μFTIR studies of sample C3 (FN+ PDLLA). Adapted from [20]

in literature [106]. The main result was revealed by μFTIR analysis, when it is shown that the protein preserved the structural integrity inside the polymeric matrix. Indeed, the two peaks at 1645 and 1635 cm^{-1} (Fig. 11.12b) are indicative for the congruent MAPLE transfer of FN structures with higher level of organization, required for biological activity [93].

The innovative design of this combinatorial laser approach for the synthesis of biomaterial thin films has a double advantage. On one hand, frozen solutions consisting of different material-solvent mixtures having distinct absorption characteristics at specific laser wavelengths could be easily assembled in form of functional combinatorial libraries in a simultaneous manner. On the other hand, in a single-step process, easy control is possible of the thickness profile along library by monitoring the number of laser pulses and laser fluence to be applied to each target. This allows the synthesis of composite coatings, gradient-doped thin films and multi-layered structures with tailored properties for bio-nanomedicine.

11.4 Discussion

Firstly implemented for the pharmaceutical industry, combinatorial science has rapidly evolved for various purposes related to materials science and engineering being nowadays considered as the forefront for discovery, development, and optimization of new materials [15, 107, 108]. There are several approaches in literature dealing with the synthesis of combinatorial materials libraries, displaying distinct advantages and drawbacks [1, 109]. Most of them resort to coupling combinatorial materials chemistry and thin-film technologies to generate gradients and to address the composition-structure-properties relationship.

Cooper et al. [110] designed a plasma sputtering system for the deposition of combinatorial libraries of Pt–Ru films for fuel cell applications. In this approach, multilayered films were grown by sequentially depositing through various masks, while post-deposition annealing was used to promote interdiffusion of the layered structures. A custom combinatorial sputtering system with a built-in in situ metrology for deposition rate monitoring was reported by Suram et al. [8]. It enabled measurement of composition-property relationships for high-throughput combinatorial investigations. Tsui and He [111] discussed the basic considerations for implementing combinatorial approach to molecular beam epitaxy, focusing on extending applications area without sacrificing any existing capabilities of conventional method. High-throughput experimentation of film synthesis by use of a linear tape transport system was introduced by Matias and Gibbons [112]. Ion-beam assisted deposition technique was used to growth epitaxial films on flexible, polycrystalline metal tapes.

Laser-based synthesis techniques are a powerful tool for the generation of combinatorial coatings due to several specific advantages depicted in Sect. 11.2 and reviewed by Koinuma and Takeuchi [107]. Christen et al. [113] introduced a method for continuous compositional-spread thin-film fabrication [$(Sr_{1-x}Ca_x)RuO_3$] based on pulsed-laser deposition on large substrate areas. Bendersky et al. [114] fabricated $Zn_{1-x}Mg_xO$ epitaxial thin-film composition spreads using layer-by-layer PLD method where the composition across the chip is linearly varied from ZnO to MgO. Craciun et al. [33] reported on the structural investigations of ITO-ZnO films grown by the combinatorial pulsed laser deposition technique. Socol et al. [115] demonstrated the possibility of tailoring of optical, compositional and electrical properties of the $In_xZn_{1-x}O$ thin films obtained by C-PLD An extension of C-PLD was explored by Bassim et al. [116] that deposited ternary continuous spread thin film libraries from Al_2O_3, HfO_2, and Y_2O_3 targets, at two different background pressures. The crystalline growth of yttrium iron garnet (YIG) films doped with bismuth and cerium by combinatorial pulsed laser deposition, co-ablating a YIG target and either a Bi_2O_3 or a CeO_2 target, for application in microwave and optical communications, was reported by Sposito et al. [117]. Very recently, Holder et al. [118] evidenced the novel phase diagram behavior and materials design in heterostructural semiconductor alloys, using an innovative approach combining RF co-sputtering (SnS and CaS targets) and C-PLD ($Mn_{1-x}Zn_xO$ libraries).

Although proving the successful growth and spreading of materials with complex stoichiometry, in all the examples depicted above only inorganic materials were considered. Indeed, none of the methods mentioned are able to fabricate organic thin films while preserving their structure-properties unaffected. Consequently, C-MAPLE could be a viable approach for fabricating combinatorial polymer libraries and cell-instructive platforms for biomedical applications.

We have recently shown that laser micro/nanofabrication of gradient coatings by C-MAPLE is a promising alternative approach to the classical procedures that generates controlled morphology and chemistry surface features for tissue engineering. We have explored Levan (L) properties because it is a promising biopolymer with huge potential for biomedical applications [119, 120]. Levan is produced by microbial fermentation of *Halomonas smyrnensis* AAD6^T batch cultures grown on pretreated sugar beet molasses [121], and can be considered a renewable resource. Recent studies suggest challenging applications for Levan and its derivates due to anticancer activity, anti-inflammatory, anticytotoxic, and antitumoral properties [122–125] as recently reviewed by Toksoy Öner et al. in [120] and references therein.

C-MAPLE technique was used for the fabrication of gradient coatings of L and Oxidized Levan (OL) in a single-step process on glass slide substrates [16, 17] using the protocol described in Sect. 11.2. It was shown that laser-generated compositional gradients could be used to screen for specific nanostructured surfaces cues to control cell proliferation or to modulate intracellular signalling pathways. The cell signalling response to the surface composition and roughness gradients evidenced that mixtures of OL and L increase ERK activation of osteoblasts as compared to simple polymers. Our studies revealed a structure-function relationship and beneficial design guidelines (e.g. the existence of an optimum mixture). Further developments for the screening of optimum dosages of combined signalling molecule mixtures that could guide cell fate towards the phenotypes required for tissue regeneration stands for next challenges.

Last achievements in the field were reported by Darwish et al. [126] that employed concurrent triple-beam matrix-assisted and direct pulsed laser deposition techniques for the synthesis of polymer nano-composite films with inorganic up-conversion phosphor and electro-optic additives. The proposed triple-beam triple-target MAPLE/PLD method can be potentially extended for the fabrication of a broad variety of nano-composite coatings for a wide spectrum of applications in nanomedicine and nanosciences in general.

11.5 Conclusions and Perspectives

In this chapter we described laser-based combinatorial synthesis of coatings with variable composition as a reliable, fast and single-step process. Moreover, it is a green approach which can be extensively employed for surface and interface functionalization at micro- and nano-scale. Laser-assisted fabrication of hybrid organic/inorganic gradient coatings consisting of various CaPs, drugs, proteins and polymers can be obtained for biomedical applications.

C-MAPLE demonstrated the capability to control the composition of bioactive coatings and consequently to stimulate bone growth and, in turn, inhibit bone resorption (Sect. 3.1). Thin films with compositional gradient of SrHA and ZOLHA were deposited on titanium substrates. Osteoblast-like MG63 cells and human osteoclasts co-cultured up to 21 days showed that Sr counterbalances the negative effect of relatively high ZOL content on osteoblasts viability. Both active compounds in the library influence ECM deposition: ZOL supports type I collagen production, whereas Sr increases alkaline phosphatase activity. Our study evidenced that osteoprotegerin/RANKL ratio is influenced to a higher extent by ZOL than Sr, and subsequently, induced a decreased proliferation and activity of osteoclasts.

The synthesis of composite chitosan-biomimetic hydroxyapatite coatings with compositional gradient has been successfully achieved by C-MAPLE (Sect. 3.2). The structures were found amorphous, with specific roughness, a morphology characteristic to MAPLE thin films assembling. The gradient concentration of chitosan into the inorganic matrix influenced the antimicrobial activity of the coatings. Samples containing about 20–40% chitosan (S3 and S4 regions in the library) exhibit the best activity when considered as long-term protection against initial adhesion and biofilms developed by Gram-positive and Gram-negative strains.

Single-step synthesis of a combinatorial library consisting in a biodegradable polymer and a protein was demonstrated (Sect. 3.3). Such composite coatings were obtained by simultaneous laser-irradiation of two cryogenic targets, with distinct absorption properties at 248 nm laser wavelength. Surface morphology along the longitudinal direction of the coating was strongly influenced by each material evaporated, resulting in a rough but compact layer embedding micrometer-sized particulates. The initial features of the compounds were preserved after laser transfer, as revealed by confocal microscopy of stained FN and FTIR micro-analyses.

This combinatorial laser approach could open new perspectives for the development of innovative biomaterials with tailored pre-designed properties after fast, clean and single-step generation of multicomponent compositional diagrams of CaPs, polymers, proteins, drugs, growth factors and nanoparticles.

Acknowledgements The authors acknowledge with thanks the support of NATO and UEFISCDI under the contracts NATO SPS G-4890 and 43 NATO/2017. F.S. and E.A. acknowledge the support of UEFISCDI under the contracts PNIIRU-TE-2014-4-1790 and -1273, 131PED/2017 and Nucleus programme-4 N/2016.

References

1. X.-D. Xiang, I. Takeuchi, Introduction, in *Combinatorial Materials Synthesis* (CRC Press, Boca Raton, FL, 2003)

2. P.C. Seoane, Combinatorial synthesis: an important methodology in medicinal chemistry, in *Anales de la Real Academia Nacional de Medicina* (2005)
3. M.L. Green, I. Takeuchi, J.R. Hattrick-Simpers, Applications of high throughput (combinatorial) methodologies to electronic, magnetic, optical, and energy-related materials. J. Appl. Phys. **113**(23), 9_1 (2013)
4. G. Briceno et al., A class of cobalt oxide magnetoresistance materials discovered with combinatorial synthesis. Science pp. 273–275 (1995)
5. E. Danielson et al., A combinatorial approach to the discovery and optimization of luminescent materials. Nature **389**(6654), 944 (1997)
6. E. Danielson et al., X-ray powder structure of Sr_2 CeO_4: a new luminescent material discovered by combinatorial chemistry. J. Mol. Struct. **470**(1), 229–235 (1998)
7. https://www.southampton.ac.uk/chemistry/research/projects/physical_vapour_deposition.page
8. S.K. Suram et al., Combinatorial thin film composition mapping using three dimensional deposition profiles. Rev. Sci. Instrum. **86**(3), 033904 (2015)
9. W. Devulder et al., Combinatorial study of Ag–Te thin films and their application as cation supply layer in CBRAM cells. ACS Comb. Sci. **17**(5), 334–340 (2015)
10. Y. Motemani et al., Nanostructured Ti–Ta thin films synthesized by combinatorial glancing angle sputter deposition. Nanotechnology **27**(49), 495604 (2016)
11. A. García-García et al., Combinatorial pulsed laser deposition of Fe/MgO granular multilayers. Appl. Phys. A **107**(4), 871–876 (2012)
12. G. Socol et al., Combinatorial pulsed laser deposition of Ag-containing calcium phosphate coatings. Digest J. Nanomater. Biostruct. **7**(2), 563–576 (2012)
13. Y. Liu et al., Combinatorial development of antibacterial Zr–Cu–Al–Ag thin film metallic glasses. Sci. Rep.s **6**, 26950 (2016)
14. I.N. Mihailescu et al., Fabrication of antimicrobial silver-doped carbon structures by combinatorial pulsed laser deposition. Int. J. Pharm. **515**(1), 592–606 (2016)
15. S.S. Mao, P.E. Burrows, Combinatorial screening of thin film materials: an overview. J. Materiomics **1**(2), 85–91 (2015)
16. F. Sima et al., Combinatorial matrix-assisted pulsed laser evaporation: single-step synthesis of biopolymer compositional gradient thin film assemblies. Appl. Phys. Lett. **101**(23), 233705 (2012)
17. E. Axente et al., Combinatorial MAPLE gradient thin film assemblies signalling to human osteoblasts. Biofabrication **6**(3), 035010 (2014)
18. E. Boanini et al., Strontium and zoledronate hydroxyapatites graded composite coatings for bone prostheses. J. Colloid Interface Sci. **448**, 1–7 (2015)
19. A. Visan et al., Combinatorial MAPLE deposition of antimicrobial orthopedic maps fabricated from chitosan and biomimetic apatite powders. Int. J. Pharm. **511**(1), 505–515 (2016)
20. F. Sima et al., Combinatorial matrix assisted pulsed laser evaporation of a biodegradable polymer and fibronectin for protein immobilization and controlled release. Appl. Surf. Sci. **306**, 75–79 (2014)
21. T. Hanawa, Biofunctionalization of metallic materials: creation of biosis–abiosis intelligent interface, in *Interface Oral Health Science 2014* (Springer, Berlin, 2015), pp. 53–64
22. M. Håkanson, E. Cukierman, M. Charnley, Miniaturized pre-clinical cancer models as research and diagnostic tools. Adv. Drug Deliv. Rev. **69**, 52–66 (2014)
23. R. Liu et al., Tumor-targeting peptides from combinatorial libraries. Adv. Drug Deliv. Rev. **110**, 13–37 (2017)
24. A.J. Moy, J.W. Tunnell, Combinatorial immunotherapy and nanoparticle mediated hyperthermia. Adv. Drug Deliv. Rev. (2017)
25. X. Wang, X. Sun, G. Briceno, Y. Lou, K.-A. Wang, H. Chang, W.G. Wallace-Freedman, S.-W. Chen, P.G. Schultz, Science **268**, 1738–1740 (1995)
26. P.J.S. Buenconsejo et al., A new prototype two-phase (TiNi)–(β-W) SMA system with tailorable thermal hysteresis. Adv. Func. Mater. **21**(1), 113–118 (2011)

27. P.J.S. Buenconsejo et al., Preparation of 24 ternary thin film materials libraries on a single substrate in one experiment for irreversible high-throughput studies. ACS Comb. Sci. **14**(1), 25–30 (2011)
28. D.C. Hsu et al., Parallel synthesis and biological evaluation of 837 analogues of procaspase-activating compound 1 (PAC-1). ACS Comb. Sci. **14**(1), 44–50 (2011)
29. O.Z. Fisher et al., Bioinspired materials for controlling stem cell fate. Acc. Chem. Res. **43**(3), 419–428 (2009)
30. E. Axente et al., Biopolymer thin films synthesized by advanced pulsed laser techniques, in *Recent Advances in Biopolymers* (InTech, 2016)
31. F. Sima et al., Bioresponsive surfaces and interfaces fabricated by innovative laser approaches. Adv. Mater. Interfaces pp. 427–462 (2016)
32. I. Takeuchi, Combinatorial pulsed laser deposition. Pulsed Laser Deposition Thin Films: Appl.-Led Growth Funct. Mater. p. 161 (2007)
33. D. Craciun et al., Structural investigations of ITO-ZnO films grown by the combinatorial pulsed laser deposition technique. Appl. Surf. Sci. **255**(10), 5288–5291 (2009)
34. K.A. Sloyan et al., Growth of crystalline garnet mixed films, superlattices and multilayers for optical applications via shuttered combinatorial pulsed laser deposition. Opt. Express **18**(24), 24679–24687 (2010)
35. A. Luches, A.P. Caricato, Fundamentals and applications of MAPLE, in *Laser-Surface Interactions for New Materials Production: Tailoring Structure and Properties*, ed. by A. Miotello, P.M. Ossi (Springer, Berlin, 2010), pp. 203–233
36. I.N. Mihailescu et al., Advanced biomimetic implants based on nanostructured coatings synthesized by pulsed laser technologies, in *Laser-Surface Interactions for New Materials Production: Tailoring Structure and Properties*, ed. by A. Miotello, P.M. Ossi (Springer, Berlin, 2010), pp. 235–260
37. B. Riggs et al., Matrix-assisted pulsed laser methods for biofabrication. MRS Bull. **36**(12), 1043–1050 (2011)
38. F. Sima, I.N. Mihailescu, Biomimetic assemblies by matrix-assisted pulsed laser evaporation, in *Laser Technology in Biomimetics* (Springer, Berlin, 2013), pp. 111–141
39. I.N. Mihailescu et al., Biomaterial thin films by soft pulsed laser technologies for biomedical applications, in *Lasers in Materials Science* (Springer, Berlin, 2014), pp. 271–294
40. V. Dinca et al., Bio-interfaces engineering using laser-based methods for controlled regulation of mesenchymal stem cell response in vitro, in *Recent Advances in Biopolymers* (InTech, 2016)
41. E. Boanini, M. Gazzano, A. Bigi, Ionic substitutions in calcium phosphates synthesized at low temperature. Acta Biomater. **6**(6), 1882–1894 (2010)
42. S.V. Dorozhkin, Bioceramics of calcium orthophosphates. Biomaterials **31**(7), 1465–1485 (2010)
43. U. Ripamonti, L.C. Roden, L.F. Renton, Osteoinductive hydroxyapatite-coated titanium implants. Biomaterials **33**(15), 3813–3823 (2012)
44. S.B. Goodman et al., The future of biologic coatings for orthopaedic implants. Biomaterials **34**(13), 3174–3183 (2013)
45. D. Williams, *The Continuing Evolution of Biomaterials* (Elsevier, Amsterdam, 2011)
46. A. Bigi et al., Biofunctional alendronate–hydroxyapatite thin films deposited by matrix assisted pulsed laser evaporation. Biomaterials **30**(31), 6168–6177 (2009)
47. R.G.G. Russell, Bisphosphonates: the first 40years. Bone **49**(1), 2–19 (2011)
48. E.V. Giger, B. Castagner, J.-C. Leroux, Biomedical applications of bisphosphonates. J. Controlled Release **167**(2), 175–188 (2013)
49. E. Boanini et al., Combined effect of strontium and zoledronate on hydroxyapatite structure and bone cell responses. Biomaterials **35**(21), 5619–5626 (2014)
50. C. Capuccini et al., Interaction of Sr-doped hydroxyapatite nanocrystals with osteoclast and osteoblast-like cells. J. Biomed. Mater. Res., Part A **89**(3), 594–600 (2009)
51. C. Capuccini et al., Strontium-substituted hydroxyapatite coatings synthesized by pulsed-laser deposition: in vitro osteoblast and osteoclast response. Acta Biomater. **4**(6), 1885–1893 (2008)

52. A. Visan et al., Biomimetic nanocrystalline apatite coatings synthesized by matrix assisted pulsed laser evaporation for medical applications. Mater. Sci. Eng., B **181**, 56–63 (2014)
53. D. Vllasaliu et al., Absorption-promoting effects of chitosan in airway and intestinal cell lines: a comparative study. Int. J. Pharm. **430**(1), 151–160 (2012)
54. I.A. Sogias, A.C. Williams, V.V. Khutoryanskiy, Chitosan-based mucoadhesive tablets for oral delivery of ibuprofen. Int. J. Pharm. **436**(1), 602–610 (2012)
55. K. Derakhshandeh, S. Fathi, Role of chitosan nanoparticles in the oral absorption of Gemcitabine. Int. J. Pharm. **437**(1), 172–177 (2012)
56. Q. Gan, T. Wang, Chitosan nanoparticle as protein delivery carrier—systematic examination of fabrication conditions for efficient loading and release. Colloids Surf., B **59**(1), 24–34 (2007)
57. A.F. Martins et al., Antimicrobial activity of chitosan derivatives containing N-quaternized moieties in its backbone: a review. Int. J. Mol. Sci. **15**(11), 20800–20832 (2014)
58. S.P. Noel et al., Chitosan films: a potential local drug delivery system for antibiotics. Clin. Orthop. Relat. Res. **466**(6), 1377–1382 (2008)
59. A.M. Grumezescu et al., Synthesis, characterization and in vitro assessment of the magnetic chitosan–carboxymethylcellulose biocomposite interactions with the prokaryotic and eukaryotic cells. Int. J. Pharm. **436**(1), 771–777 (2012)
60. A.M. Grumezescu et al., Magnetic chitosan for drug targeting and in vitro drug delivery response. Biointerface Res. Appl. Chem. **1**, 160 (2011)
61. M. Kong et al., Antimicrobial properties of chitosan and mode of action: a state of the art review. Int. J. Food Microbiol. **144**(1), 51–63 (2010)
62. P. Dutta et al., Perspectives for chitosan based antimicrobial films in food applications. Food Chem. **114**(4), 1173–1182 (2009)
63. I. Leceta et al., Characterization and antimicrobial analysis of chitosan-based films. J. Food Eng. **116**(4), 889–899 (2013)
64. T. Dai et al., Chitosan preparations for wounds and burns: antimicrobial and wound-healing effects. Expert Rev. Anti-infect. Ther. **9**(7), 857–879 (2011)
65. A. Yoshida et al., Preparation of bioactive chitosan-hydroxyapatite nanocomposites for bone repair through mechanochemical reaction. Mater. Trans. **45**(4), 994–998 (2004)
66. M. Ito et al., Effect of hydroxyapatite content on physical properties and connective tissue reactions to a chitosan–hydroxyapatite composite membrane. J. Biomed. Mater. Res., Part A **45**(3), 204–208 (1999)
67. Y. Zhang, M. Zhang, Calcium phosphate/chitosan composite scaffolds for controlled in vitro antibiotic drug release. J. Biomed. Mater. Res., Part A **62**(3), 378–386 (2002)
68. M. Sivakumar, I. Manjubala, K.P. Rao, Preparation, characterization and in-vitro release of gentamicin from coralline hydroxyapatite–chitosan composite microspheres. Carbohyd. Polym. **49**(3), 281–288 (2002)
69. Q. Hu et al., Preparation and characterization of biodegradable chitosan/hydroxyapatite nanocomposite rods via in situ hybridization: a potential material as internal fixation of bone fracture. Biomaterials **25**(5), 779–785 (2004)
70. N. Davidenko et al., Chitosan/apatite composite beads prepared by in situ generation of apatite or Si-apatite nanocrystals. Acta Biomater. **6**(2), 466–476 (2010)
71. W. Thein-Han, R. Misra, Biomimetic chitosan–nanohydroxyapatite composite scaffolds for bone tissue engineering. Acta Biomater. **5**(4), 1182–1197 (2009)
72. B. Boyan et al., Effect of surface roughness and composition on costochondral chondrocytes is dependent on cell maturation state. J. Orthop. Res. **17**(3), 446–457 (1999)
73. A.I. Visan, C.G. Ristoscu, I.N. Mihailescu, Composite coatings based on renewable resources synthesized by advanced laser techniques, in *Composites from Renewable and Sustainable Materials* (InTech, 2016)
74. M. Markovic, B.O. Fowler, M.S. Tung, Preparation and comprehensive characterization of a calcium hydroxyapatite reference material. J. Res. Nat. Inst. Stand. Technol. **109**(6), 553 (2004)

75. L.E. Sima et al., Differentiation of mesenchymal stem cells onto highly adherent radio frequency-sputtered carbonated hydroxylapatite thin films. J. Biomed. Mater. Res., Part A **95**(4), 1203–1214 (2010)
76. K.S. Katti, D.R. Katti, R. Dash, Synthesis and characterization of a novel chitosan/montmorillonite/hydroxyapatite nanocomposite for bone tissue engineering. Biomed. Mater. **3**(3), 034122 (2008)
77. Q. Yuan et al., Controlled and extended drug release behavior of chitosan-based nanoparticle carrier. Acta Biomater. **6**(3), 1140–1148 (2010)
78. C. Paluszkiewicz et al., FT-IR study of montmorillonite–chitosan nanocomposite materials. Spectrochim. Acta Part A Mol. Biomol. Spectrosc. **79**(4), 784–788 (2011)
79. S.M. Silva et al., Application of infrared spectroscopy to analysis of chitosan/clay nanocomposites, in *Infrared Spectroscopy-Materials Science, Engineering and Technology* (InTech, 2012)
80. C.R. Arciola et al., Biofilm-based implant infections in orthopaedics, in *Biofilm-Based Healthcare-Associated Infections* (Springer, Berlin, 2015), pp. 29–46
81. R.G. Cuero, Antimicrobial action of exogenous chitosan. Exs **87**, 315 (1999)
82. R. Muzzarelli et al., Antimicrobial properties of N-carboxybutyl chitosan. Antimicrob. Agents Chemother. **34**(10), 2019–2023 (1990)
83. H.K. No et al., Antibacterial activity of chitosans and chitosan oligomers with different molecular weights. Int. J. Food Microbiol. **74**(1), 65–72 (2002)
84. D. Raafat, H.G. Sahl, Chitosan and its antimicrobial potential–a critical literature survey. Microb. Biotechnol. **2**(2), 186–201 (2009)
85. A.B.V. Kumar et al., Characterization of chito-oligosaccharides prepared by chitosanolysis with the aid of papain and Pronase, and their bactericidal action against Bacillus cereus and Escherichia coli. Biochem. J. **391**(2), 167–175 (2005)
86. I. Helander et al., Chitosan disrupts the barrier properties of the outer membrane of gram-negative bacteria. Int. J. Food Microbiol. **71**(2), 235–244 (2001)
87. M.L. Tanzer, Current concepts of extracellular matrix. J. Orthop. Sci. **11**(3), 326 (2006)
88. H. Bachman et al., Utilizing fibronectin integrin-binding specificity to control cellular responses. Adv. Wound Care **4**(8), 501–511 (2015)
89. K. Webb, V. Hlady, P.A. Tresco, Relationships among cell attachment, spreading, cytoskeletal organization, and migration rate for anchorage-dependent cells on model surfaces. J. Biomed. Mater. Res. **49**(3), 362 (2000)
90. B.G. Keselowsky, D.M. Collard, A.J. García, Surface chemistry modulates fibronectin conformation and directs integrin binding and specificity to control cell adhesion. J. Biomed. Mater. Res., Part A **66**(2), 247–259 (2003)
91. R.O. Hynes, Integrins: bidirectional, allosteric signaling machines. Cell **110**(6), 673–687 (2002)
92. S. Grigorescu et al., Fabrication of functional fibronectin patterns by nanosecond excimer laser direct write for tissue engineering applications. J. Mater. Sci.: Mater. Med. **24**(7), 1809–1821 (2013)
93. F. Sima et al., Fibronectin layers by matrix-assisted pulsed laser evaporation from saline buffer-based cryogenic targets. Acta Biomater. **7**(10), 3780–3788 (2011)
94. F. Sima et al., Thin films of vitronectin transferred by MAPLE. Appl. Phys. A Mater. Sci. Process. **105**(3), 611–617 (2011)
95. F. Sima et al., Inorganic–organic thin implant coatings deposited by lasers. ACS Appl. Mater. Interfaces. **7**(1), 911–920 (2014)
96. I. Brigaud et al., Synergistic effects of BMP-2, BMP-6 or BMP-7 with human plasma fibronectin onto hydroxyapatite coatings: a comparative study. Acta Biomater. **55**, 481–492 (2017)
97. Z.S. Haidar, R.C. Hamdy, M. Tabrizian, Delivery of recombinant bone morphogenetic proteins for bone regeneration and repair. Part A: current challenges in BMP delivery. Biotech. Lett. **31**(12), 1817 (2009)

98. Z.S. Haidar, R.C. Hamdy, M. Tabrizian, Delivery of recombinant bone morphogenetic proteins for bone regeneration and repair. Part B: delivery systems for BMPs in orthopaedic and craniofacial tissue engineering. Biotech. Lett. **31**(12), 1825–1835 (2009)
99. E.J. Carragee, E.L. Hurwitz, B.K. Weiner, A critical review of recombinant human bone morphogenetic protein-2 trials in spinal surgery: emerging safety concerns and lessons learned. Spine J. **11**(6), 471–491 (2011)
100. A.W. James et al., A review of the clinical side effects of bone morphogenetic protein-2. Tissue Eng. Part B: Rev. **22**(4), 284–297 (2016)
101. I.A. Paun et al., Thin films of polymer blends for controlled drug delivery deposited by matrix-assisted pulsed laser evaporation. Appl. Phys. Lett. **96**(24), 243702 (2010)
102. R.K. Averineni et al., PLGA 50:50 nanoparticles of paclitaxel: development, in vitro antitumor activity in BT-549 cells and in vivo evaluation. Bull. Mater. Sci. pp. 1–8 (2012)
103. R. Cristescu et al., Antimicrobial activity of biopolymer–antibiotic thin films fabricated by advanced pulsed laser methods. Appl. Surf. Sci. **278**, 211–213 (2013)
104. L. Poulouin et al., Plasma fibronectin: three steps to purification and stability. Protein Expr. Purif. **17**(1), 146–152 (1999)
105. C. Hoffmann et al., Fluorescein isothiocyanate-labeled human plasma fibronectin in extracellular matrix remodeling. Anal. Biochem. **372**(1), 62–71 (2008)
106. R. Cristescu et al., Matrix assisted pulsed laser evaporation of poly (d, l-lactide) thin films for controlled-release drug systems. Appl. Surf. Sci. **253**(19), 7702–7706 (2007)
107. H. Koinuma, I. Takeuchi, Combinatorial solid-state chemistry of inorganic materials. Nat. Mater. **3**(7), 429 (2004)
108. I. Takeuchi, J. Lauterbach, M.J. Fasolka, Combinatorial materials synthesis. Mater. Today **8**(10), 18–26 (2005)
109. J.C. Grunlan, A.R. Mehrabi, R.A. Potyrailo, Introduction: combinatorial instruments and techniques. Rev. Sci. Instrum. **76**(6), 62101–62200 (2005)
110. J.S. Cooper, G. Zhang, P.J. McGinn, Plasma sputtering system for deposition of thin film combinatorial libraries. Rev. Sci. Instrum. **76**(6), 062221 (2005)
111. F. Tsui, L. He, Techniques for combinatorial molecular beam epitaxy. Rev. Sci. Instrum. **76**(6), 062206 (2005)
112. V. Matias, B.J. Gibbons, Linear combinatorial approach to thin film research. Rev. Sci. Instrum. **78**(7), 072206 (2007)
113. H. Christen et al., An improved continuous compositional-spread technique based on pulsed-laser deposition and applicable to large substrate areas. Rev. Sci. Instrum. **74**(9), 4058–4062 (2003)
114. L. Bendersky et al., Microstructural study of epitaxial $Zn_{1-x}Mg_xO$ composition spreads. J. Appl. Phys. **98**(8), 083526 (2005)
115. G. Socol et al., Tailoring of optical, compositional and electrical properties of the $In_xZn_{1-x}O$ thin films obtained by combinatorial pulsed laser deposition. Digest J. Nanomater. Biostruct. **6**(1), 107–115 (2011)
116. P.K. Schenck et al., Design and spectroscopic reflectometry characterization of pulsed laser deposition combinatorial libraries. Appl. Surf. Sci. **254**(3), 781–784 (2007)
117. A. Sposito et al., Combinatorial pulsed laser deposition of doped yttrium iron garnet films on yttrium aluminium garnet. J. Appl. Phys. **115**(5), 053102 (2014)
118. A.M. Holder et al., Novel phase diagram behavior and materials design in heterostructural semiconductor alloys. Sci. Adv. **3**(6), e1700270 (2017)
119. M.S. Bostan et al., Comprehensive characterization of chitosan/PEO/levan ternary blend films. Carbohyd. Polym. **102**, 993–1000 (2014)
120. E.T. Öner, L. Hernández, J. Combie, Review of levan polysaccharide: from a century of past experiences to future prospects. Biotechnol. Adv. **34**(5), 827–844 (2016)
121. A. Poli et al., High level synthesis of levan by a novel Halomonas species growing on defined media. Carbohyd. Polym. **78**(4), 651–657 (2009)
122. H.K. Sarilmiser, E.T. Oner, Investigation of anti-cancer activity of linear and aldehyde-activated levan from Halomonas smyrnensis AAD6 T. Biochem. Eng. J. **92**, 28–34 (2014)

123. A.D. Sezer et al., Development and characterization of vancomycin-loaded levan-based microparticular system for drug delivery. Pharm. Dev. Technol. **22**(5), 627–634 (2017)
124. M. Erginer et al., Sulfated levan from Halomonas smyrnensis as a bioactive, heparin-mimetic glycan for cardiac tissue engineering applications. Carbohyd. Polym. **149**, 289–296 (2016)
125. E.A. Queiroz et al., Levan promotes antiproliferative and pro-apoptotic effects in MCF-7 breast cancer cells mediated by oxidative stress. Int. J. Biol. Macromol. **102**, 565–570 (2017)
126. A.M. Darwish et al., Polymer nano-composite films with inorganic upconversion phosphor and electro-optic additives made by concurrent triple-beam matrix assisted and direct pulsed laser deposition. Compos. B Eng. **109**, 82–90 (2017)

Chapter 12
Laser Synthesized Nanoparticles for Therapeutic Drug Monitoring

Matteo Tommasini, Chiara Zanchi, Andrea Lucotti, Enza Fazio, Marco Santoro, Salvatore Spadaro, Fortunato Neri, Sebastiano Trusso, Emilio Ciusani, Ugo de Grazia, Marina Casazza and Paolo M. Ossi

Abstract In this chapter are illustrated the context and activities that have been pursued to develop a surface enhanced Raman spectroscopy technique for clinical applications in therapeutic drug monitoring with focus on anti-epileptic and anti-Parkinson drugs. We discuss how the surface nanostructure of sensors made of arrays of noble metal nanoparticle obtained by pulsed laser ablation in inert gas and in liquid affects their optical properties, in particular the wavelength, intensity and width of the surface plasmon. These in turn are shown to strongly influence the intensity and the spatial reproducibility of the enhanced Raman signals as functions of drug concentration.

M. Tommasini · C. Zanchi · A. Lucotti
Dipartimento di Chimica, Materiali e Ing. Chimica "G. Natta", Politecnico di Milano, Piazza Leonardo da Vinci 32, 20133 Milan, Italy

C. Zanchi · P. M. Ossi
Dipartimento di Energia, Politecnico di Milano,
Via Ponzio 34-3, 20133 Milan, Italy

E. Fazio · M. Santoro · S. Spadaro · F. Neri
Dipartimento di Scienze Matematiche e Informatiche,
Scienze Fisiche e Scienze della Terra, Viale Ferdinando
Stagno d'Alcontres 31, 98166 Messina, Italy

S. Trusso (✉)
CNR-Istituto per i Processi Chimico-Fisici, Viale Ferdinando Stagno d'Alcontres 37, 98158 Messina, Italy
e-mail: trusso@ipcf.cnr.it

E. Ciusani · U. de Grazia · M. Casazza
Fondazione I.R.C.C.S. Istituto Neurologico Carlo Besta,
Via Celoria 11, 20133 Milan, Italy

P. M. Ossi (ed.), *Advances in the Application of Lasers in Materials Science*,
Springer Series in Materials Science 274,
https://doi.org/10.1007/978-3-319-96845-2_12

12.1 Historical Background

Noble metal nanoparticles, in particular gold nanoparticles, are nowadays relevant for many technological applications, due to their unique optical and chemical properties [1–4]. It is intriguing that gold was used already in the antiquity in water suspensions that were considered for health applications with variable outcomes. Indeed, since gold was assumed to be incorruptible and indestructible, it was believed to have strong therapeutic efficacy. Although the metal was known to be insoluble in water as well as in other solvents available to medical doctors, it was possible to ground gold to powders fine enough to allow for preparing metal suspensions in aqueous solvents. At that time, there was no idea about the very nature of gold in such solutions, nor there was any control on them. "Soluble (or drinkable, or *potable*) gold" (PG) was independently introduced in China and in Egypt presumably in the same period, between the fifth and the fourth Century BC. The supposed curative properties of such gold colloids intruded into the fantastic and ranged from the solution of heart problems to diarrhoea, epilepsy, headache, tumors, venereal diseases. However, there are no reports on relevant therapeutic success of such oldest colloidal PG.

It seems appropriate to mention that the word colloid, that originates from the French *colle*, was invented and introduced in the literature quite recently [5] even if we use extensively throughout this retrospective.

Throughout the Middle Ages, across Europe the dream of alchemists was to obtain gold in such a form that it was soluble in water, thus leading to truly PG. An Arabian alchemist, known to Europeans as Geber, was the first who succeeded to dissolve gold in *aqua regia*, a mixture of nitric and hydrochloric acids, obtaining terchloride of gold (gold(III) chloride, IUPAC) [6]. Around the end of the XV Century European alchemists were able to manage the production of *aqua fortis*, i.e. nitric acid; among them most notably the Swiss Philippus Aureolus Theophrastus Bombastus von Hohenheim, called Paracelsus, or Paracelso (Einsiedeln, CH, 1493, Salzburg, A, 1541), medical doctor, alchemist and astrologist handled *aqua fortis* and *aqua regia* by slightly different paths [7]. In particular, after dissolving gold in *aqua regia* the solution was distilled with water and the salt was precipitated with a solution of potassium hydroxide [6]: PG was ready to be used.

Throughout the Renaissance and in the following Centuries PG at low doses was considered beneficial in many diseases. In particular it was thought to act on the central nervous system. Paracelsus quotes PG usefulness against epilepsy, melancholy, hysteria, mania, S. Vitus dance [8]. His contemporary Giovanni Andrea left a compilation of remedies against the sick and noticed that *aurum potabile* had therapeutic effect against leprosy (psoriasis), plague, ulcer, diarrhea, fistula and epilepsy; it could also prevent apoplectic stroke [9]. Important doses of PG caused gastrointestinal irritation and prolonged use led to kidney damage and other toxic effects known as *auric fever*. The healing properties of PG were exposed with considerable detail in the initial chapter of the first book on colloidal gold, published in 1618 by the British medical doctor and philosopher Francisco Antonii [10] who reported also about the curative success of PG in a few case studies. About half a century later, in 1676, the

German chemist J. Knuckels in turn quoted the curative properties of PG for several diseases [11]. Moreover, he specified in chapter seven of his book on potable noble metals that an aqueous gold solution contains metallic gold, it appears neutral and looks slightly pink-colored. Based on visual observations Knuckels proposed that the metal should be reduced to particles of average size such that they are not visible to the human eye. Also, their concentration, that he called *contamination*, had to be diluted enough to make them not discernible. One century later, in 1769, a French dictionary [12] reported the popular concepts that PG contains the metal in elemental state, under extreme subdivision, as a suspension in a liquid. The first correlation between different colors of differently prepared PG solutions and the average size of the floating metal particles was provided around 1800 by J. B. Richter who, according to Ostwald [13], clearly pointed out that pink, or purple suspensions contain metal particles of smallest size, while yellow solutions correspond to a degree of particle agglomeration. He recognized also that water purity affects the stability of the suspensions and the coagulative effects of electrolytes. Noticeably, present studies indicate that isolated, spherical gold particles sized around 20 nm appear red, while increasing their size to 100 nm they look yellow. Thus, a correlation between particle size and optical properties of the solution was already empirically established before M. Faraday performed the first systematic investigation on the synthesis and the optical properties of gold colloids. He called them *a divided state of gold* and studied also the effect of mechanical pressure on thin films he obtained from dried solutions [14]. Faraday obtained his samples by a two-phase method, reducing an aqueous solution of gold chloride anions ($AuCl_4^-$) using red phosphorus in CS_2. While gold chloride continued to be used, most of the 19th Century European pharmacopoeias recommended mixtures of gold and sodium chlorides as the standard form of PG. In an American textbook of the same period such a PG was indicated for cirrhosis of the kidney, progressive general paralysis, tuberculosis, dyspepsia, advanced syphilis, depression, hypochondria, hysteria, impotence [15]. Injection of gold salts, including gold cyanide and gold bromide, became common since it was recognized that gold salts are scarcely absorbed from the intestinal tract. The US Pharmacopoeia included Aurii and Sodii Chloridum as an official drug in its tenth edition (1926). However, in general around the beginning of the 20th Century therapies with PG in any form were no more considered in pharmacology texts, with the exception of controversial success of gold-potassium cyanid (potassium gold cyanide, IUPAC) [16], and later of gold sodium thyomalate (gold sodium thiomalate, IUPAC) [17] in the therapy of *lupus erythematosus*. In 1949 about 67 percent of patients of *lupus* treated with intramuscular injection of the latter gold salt were observed to have beneficial effects although the mechanism of action was not established [18]. While past use of gold particles had mainly a therapeutic motivation, our present ability to exploit plasmonics allows for directing to diagnostics and drug delivery the application of such particles and the nanostructures resulting from their assembling.

12.1.1 Therapeutic Drug Monitoring (TDM)

Therapeutic Drug Monitoring (TDM) is a clinical practice to determine the drug concentration in the blood of patients. TDM is based on the idea that the concentration of a drug in the plasma, or in the blood, reflects the concentration of the drug at the target site better than the given dose [19]. Thus, monitoring the drug concentration in patients allows for adjusting the dose in prolonged therapies and helps understanding the patient dependent pharmaco-kinetics and pharmaco-dynamics. TDM is relevant for drugs characterized by a narrow therapeutic index (NTI), that is defined as the drug concentration range for which, for a given patient, the desired drug effects are obtained. Below this range the drug is not effective and above this range it has adverse or toxic effects. In the case of NTI drugs, *fluctuations* in the drug concentration could lead the patient out of the therapeutic range concentration, thus causing negative health effects. Therefore, it is relevant for medical doctors to monitor the concentration level of NTI drugs. This allows to associate the clinical conditions of the patient to the drug posology, thus ensuring the efficacy of the treatment for the minimal drug concentration, which also avoids adverse, or toxic effects. Henceforth, TDM is used since long in the clinical practice for NTI drugs, which include anti-cancer drugs [20] and anti-epileptic drugs (AEDs), for which even a small dose variation may become toxic or ineffective to control seizures [21].

In recent years nanostructured noble metal (Au, Ag) architectures have been designed with properties useful for both diagnostic and therapeutic purposes. In addition, such nanostructures have been directed also at multiplexing, namely at extracting from the same diagnostic signal information related to different analytes at the same time [22]. Such nanostructures can be synthesized to obtain plasmonic structures able to produce Surface Enhanced Raman Scattering (SERS) from target drug molecules [23].

12.1.2 Epilepsy

Epilepsy is the most common among important neurological pathologies. It is a chronical disease that affects about 65 million people throughout the world [24]. In the majority of cases epilepsy can be successfully treated, however the therapeutic gap (i.e., the difference between patients and successfully treated patients) can be fairly wide, especially in developing Countries [25], where many AEDs are not available or too expensive [26]. Several epileptic patients under pharmacological treatment may suffer from collateral effects that negatively affect the quality of their life, such as hyperactivity, confusional state, insomnia, and depression [27]. To limit these effects, each patient should be treated with a personalized dose of the drug, after the neurologist critically assessed TDM results [28]. TDM is required at the beginning of the treatment by AEDs, or when the pharmacological treatment is modified. Since seizures usually occur irregularly, it is difficult to find the optimal pharmacological dosage based just on clinical observation. Furthermore, in the absence of TDM, the symptoms of drug toxicity may be confused with other symptoms, especially in patients suffering from neurological pathologies other than epilepsy [29]. Another important issue is that the therapeutic index depends on the patient: for a given AED, there is not one single dose that can be safely adopted for all patients.

12.1.3 Parkinson's disease (PD)

Parkinson's disease (PD) is a neuro-degenerative pathology characterized by slow but progressive evolution. PD mainly affects the control of movement and equilibrium, manifesting with rigidity, tremor, bradykinesia, and postural instability. A number of non-motor symptoms may also be present as cognitive impairment, psychiatric symptoms, sleep disorders, pain and fatigue. It is the most frequent pathology among movement disorders. PD symptoms are likely known since thousands of years (there is such a mention in an old Indian writing (5000 B.C.) and in a Chinese document dated 2500 B.C.). The name of the disease is due to James Parkinson, a British pharmacist and surgeon who described most of the symptoms in his *Essay on the shaking palsy* [30]. The disease is diffused all over the world and among all ethnicities. On the average, the age of the first symptoms is 58–60 years, however about 5% of the patients may show symptoms at an earlier age, between 21 and 40 years. The most commonly used anti-Parkinson drug, namely Levodopa, tends to lose efficacy over time, with greater than 80% of patients on therapy for longer than ten years experiencing dyskinesia and "on-off" periods, that means alternations between periods of good symptom control (on-time) and periods of reduced motor symptom control (off-time). One of the major goals in PD treatment is controlling disabling symptoms. This goal is often not reached with orally delivered drugs at tolerable dosage.

In this case continuous subcutaneous injection of apomorphine (APO) is an effective treatment which reduces the off-periods, and allows to improve the movement ability of the patient [31]. In such a scenery, it is fundamental to adjust APO administration in patients, also considering pharmacokinetics and variations of drug concentration in blood. In clinical practice this aim usually results from clinical standardized observations and motor tests (when possible) [32]. Alternatively, chemical analyses on blood samples are a reliable method to determine the plasmatic drug concentration, even though such an approach is not currently used in treating PD at an advanced stage.

12.1.4 Analytical techniques

The standard techniques currently used in clinical laboratories for TDM are the immunoassay and chromatographic techniques (High Performance Liquid Chromatography, HPLC), often associated with mass spectrometry (i.e., HPLC-MS). Both

Table 12.1 Clinical analytical methods, time requirements and costs for selected AEDs (see text for details)

Active Principal Ingredient (API)	Analytical method	Technical required time (days)	Total required time (days)	Unitary cost (Euro)
Oxcarbazepine	2; 3	2	3–4	29
Perampanel	2; 3	2	3–4	29
Lacosamide	2; 3	2	3–4	29
Lamotrigine	1; 2; 3	2	3–4	29
Valproic acid	1; 2	1	1	15
Levetiracetam	1; 2; 3	2	3–4	29
Phenytoin (Dintoin)	1; 2; 3	1	1	15

1: immunoassay; 2: Liquid phase tandem Mass Spectrometry; 3: HPLC-MS . *Data source* Laboratorio di Patologia Clinica e Genetica Medica, Fondazione IRCCS Istituto Neurologico C. Besta, Milano

techniques are highly sensitive and reliable, but unfortunately the associated costs are high, the waiting time is long and sizeable blood withdrawal (a few milliliters) is required.

Table 12.1 illustrates for selected AEDs the technique used for TDM, the technical time required, the time required to obtain the result since the shipping of the blood sample to the laboratory, and the cost of the single analysis (waiting time). It is worth commenting that the difference between the technical time and the waiting time is essentially due to the need of collecting, from several patients, a number of samples which is enough to be analyzed as a single batch, which allows to keep the costs within the limits reported in Table 12.1.

The characteristics of currently available TDM techniques, summarized in Table 12.1, limit the widespread use of TDM and the population of patients which could benefit from such exams. In recent time, in TDM literature a complementary approach to the measurement of drug concentration has emerged, which considers Surface Enhanced Raman Spectroscopy (SERS) as the analytical technique [23, 33–35].

In particular, the detection of carbamazepine (CBZ), a widely used drug to treat epilepsy, is highly significant to improve the quality of life of patients, mainly in developing Countries.

The same considerations on life quality improvement hold concerning apomorphine (APO) for PD at an advanced stage. In this scenery, the straightforward use of the Surface Enhanced Raman Spectroscopy (SERS) with gold and silver colloids, even if very sensitive to qualitatively detect analytes, is not acceptable because of its strong limitations in quantitative detection, mainly arising from uncontrolled colloid aggregation. This is reflected by the lack of scientific literature on this specific subject [35].

12.2 Surface Enhanced Raman Spectroscopy (SERS)

The intrinsic limitation to Raman spectroscopy is the signal weakness, i.e. the small number of inelastically scattered photons with respect to the overwhelming majority of elastically scattered photons that originate the Rayleigh peak. Although in resonant conditions the efficiency of the Raman process increases significantly, not always laser sources suitable to exploit resonance conditions with the system of interest are available. Thus, since the discovery of SERS effect at the beginning of the 70s of XXth Century many efforts were devoted to use either spontaneously, or artificially corrugated surfaces of selected metals to significantly extend the range of application of the technique. In particular the study of an analyte dissolved in a highly diluted solution is a typical scenery suitable to highlight the characteristics of the SERS technique. To get the Raman signal intensification it is required that the analyte molecule is in close proximity to a roughened metallic surface where a surface plasmon (SP) can resonantly be excited at the energy of the probe photon beam. We recall that if a semi-infinite electron gas is confined by a plane surface, solving the Laplace equation for the admissible fluctuations of the charge density, periodic plasma oscillations are predicted in the surface plane, whose amplitude decays exponentially towards the bulk of the electron gas. The quantum of oscillation is called surface plasmon. Once we use EM periodic boundary conditions, so that the tangential component of the EM field and the normal component of the displacement are both constant across the surface—vacuum interface and we take into account the dispersion relation for the dielectric constant of the electron gas, we obtain the relation between the frequency of a SP, ω_s and that of a volume plasmon ω_p

$$\omega_s^2 = 1/2\omega_p^2 \tag{12.1}$$

The relative intensities of volume and surface plasmon excitations depend on the incidence angle, the energy and the polarization of the exciting EM radiation.

When the typical size of the metallic structure downscales to the nanometer range (5–50 nm) the relative weight of surface atoms becomes relevant with respect to that of volume atoms and we observe a displacement of the experimentally measured plasmon wavelength to higher values. Nanoparticles (NPs) of silver and gold show SP absorption peaks around 410 and 520 nm, respectively, thus in the visible range. Such SPs originate from the plasma oscillations of conduction electrons at the NP surface. It is thus possible to excite electron oscillations with some of the most commonly used laser sources operating in the visible range. For this reason, both nanostructured noble metals are very popular as SERS substrates. Compared to other metals nanostructured Ag and Au show the most relevant intensification effects of Raman signals. In many cases for both metals also the chemical intensification mechanism has been observed. The latter is ascribed to resonance effects within the target molecule and charge-transfer effects between molecular orbitals and the conduction band of the noble metal surface [36].

12.2.1 SERS Sensors Obtained by Pulsed Laser Deposition

The SP wavelength and width are the optical features of specific interest of a corrugated metallic surface whose behavior as a SERS sensor is to be tested. Since Ag and Au produce the most relevant increases of Raman intensities the problem consists of engineering the surface nanostructure controlling how it grows, shifting the wavelength and width of the SP as functions of the features of the Raman spectrum of the analyte of interest. The technique we adopted to realize a surface artificially corrugated at the nanometer scale is the pulsed laser ablation (PLA) in an ambient inert gas at high pressure. Schematically, a laser pulse in the UV, from an excimer laser (KrF; λ: 248 nm) with typical duration of some ns (25 ns) is focused to a spot on the target surface where it deposits an energy of some mJ (10 mJ). If the energy density deposited by the laser pulse at the target is above the ablation threshold the dense, highly anisotropic vapor made of atoms and small clusters ejected from the irradiated surface strongly absorbs the laser radiation, provided the number density of ablated atoms exceeds a critical value. The energized vapor is quickly converted into a fully ionized plasma that undergoes supersonic expansion normal to the target surface. When such an ablation plume expands through an inert gas at progressively higher pressure, fast photography pictures of the plume show the formation of a shock wave at the plume propagation front, whose shape becomes sphere-like. At the same time the plume is smaller than in a free expansion, it is braked, a diffusive behavior sets on at long times (microseconds) while at the same time the plume tail moves backwards to the target. This behavior leads to a meaningful interpenetration between plasma species and ambient gas atoms that results in the splitting of the plume into two particle populations: a fast particle group propagating through the gas similarly to a collisionless vacuum expansion and a slow group, delayed by the collisional interactions with gas atoms. Moving from a critical review of the models that have been proposed to describe the complex phenomenology we just summarized [37] it was possible to model the expansion of an ablation plasma through an inert gas at high pressure, predicting NP formation during the collisional stage of plume expansion and the average number N of constituent atoms of a NP that reached its stationary size

$$N = (<n_a> \sigma_{a-a} <v> t_f) \times (<n_g> \sigma_{a-g} <v> t_f) \tag{12.2}$$

where n is the number density, σ the elastic collision cross section, $<v>$ the average particle velocity and t_f is the NP formation time in the expanding plume. The indexes a and g refer to ablated species and ambient gas atoms respectively. A thorough investigation indicates the process parameters that govern the nanostructure evolution as a function of the number of laser pulses N_{LP}, i.e. of deposited ablation plumes on a rigid inert support. At low N_{LP} we observe isolated, near spherical NPs, whose degree of mutual interaction progressively increases, leading to coalescence with increasing N_{LP}. As a result, islands whose shape becomes increasingly irregular with increasing size form. Such islands are separated from each other by inter-island channels of width in the nanometer range and length up to tens of nanometers. If the deposition proceeds further a semi-continuous film is obtained. At fixed experimental conditions such as laser parameters (wavelength; pulse duration and energy), incidence angle of the laser pulse at the target surface, target-support distance, ambient gas kind, process temperature, the relevant parameter to modify the surface nanostructure of the sensor to be deposited, besides N_{LP} is the ambient gas pressure p_g. At fixed N_{LP} the SP wavelength shows a blue-shift, with increasing p_g. Conversely, at fixed p_g, an analogous SP blue-shift is observed on lowering N_{LP}.

In Fig. 12.1 the trend is shown in the case of self-assembled arrangements of Ag NPs on (100) Si supports. For Au NPs, we observe the same parametric trends just discussed. Transmission electron microscopy observations of NP assemblies of the two noble metals, deposited in conditions such as to avoid coalescence, indicate size range between 1 and 4 nm for Ag NPs and between 2 and 6 nm for Au NPs. Such data are in agreement with the model predictions (12.2). Thus, we are able to control the formation of such NPs and their deposition in conditions of integrity on the support where they form a sensor with definite optical properties. Yet a partly hindered parameter of PLA process, namely the ablated mass per laser pulse M can significantly affect the film nanostructure. At fixed energy density deposited by each laser pulse M depends on the area of the spot to which the laser beam is focused. We specifically addressed the problem taking into account both plume propagation [38] and surface nanostructure and optical properties of the growing film [39]. Using pulses that ablate target masses in a ratio about 1:2 (7.0 ng, M_s; 16.4 ng, M_l) we obtain Ag films with similar morphology, made of islands interconnected through channels of variable width and length, as shown in Fig. 12.2. Yet the films have very different optical properties: when $M = M_s$ the SP shows an evident maximum at about 590 nm, while for $M = M_l$ a broad maximum is found around 820 nm. Checking the SERS performance of the two sensors against a reference analyte such as Rhodamine 6G (R6G) in aqueous solutions at moderate molar concentrations (10^{-6} M), using an exciting radiation at 633 nm wavelength, the sensor deposited at M_l does not produce appreciable analyte signals, differently to the well-defined spectrum obtained with the sensor deposited at M_s.

Presently we are thus able to synthesize via PLA in inert gas at high pressure *elemental* noble metal (Ag; Au) SERS-active substrates to be used as sensors, tailoring their surface nanostructure and consequently their optical properties (SP wavelength and width) as functions of the Raman features of the analyte of interest.

A perspective direction towards the synthesis of noble metal based sensors with SP wavelength adjusted ad hoc between the two extremes belonging to isolated NPs either pure Ag (around 400 nm) or pure Au (around 520 nm) is provided by the use of Ag and Au colloidal solutions produced by pulsed laser ablation in a stationary liquid (PLAL). The process was introduced about twenty years ago [40, 41], and it was soon recognized as a chemically clean, simple, efficient to synthesize NPs [42, 43]. We discussed the basics of PLAL and the dependence of the concentration of Au NPs synthesized in water on the liquid layer thickness and the ablation spot diameter as well as the trend of SP position and colloid morphology versus laser pulse energy [37].

Summarizing, PLAL in a stationary liquid involves the simultaneous production of NPs by ablation of the target material and the fragmentation/assembling of dispersed

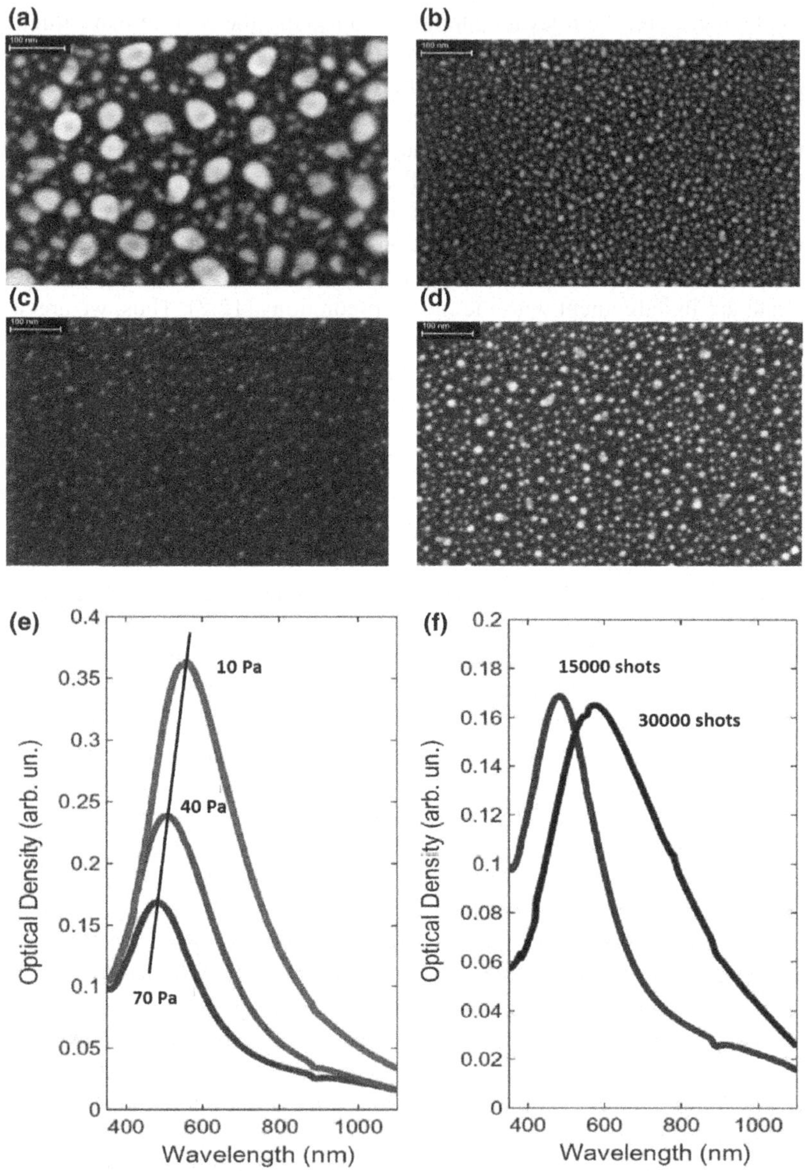

Fig. 12.1 SEM micrographs of Ag NPs arrays synthesized at fixed LP = 15,000 in Ar atmosphere at **a**, 10 Pa; **b**, 40 Pa; **c**, 70 Pa; **d**, 70 Pa and LP = 30,000. In **e**, Uv-vis spectra of films (**a**), (**b**), (**c**); in **f** UV-vis spectra of films (**c**) and (**d**)

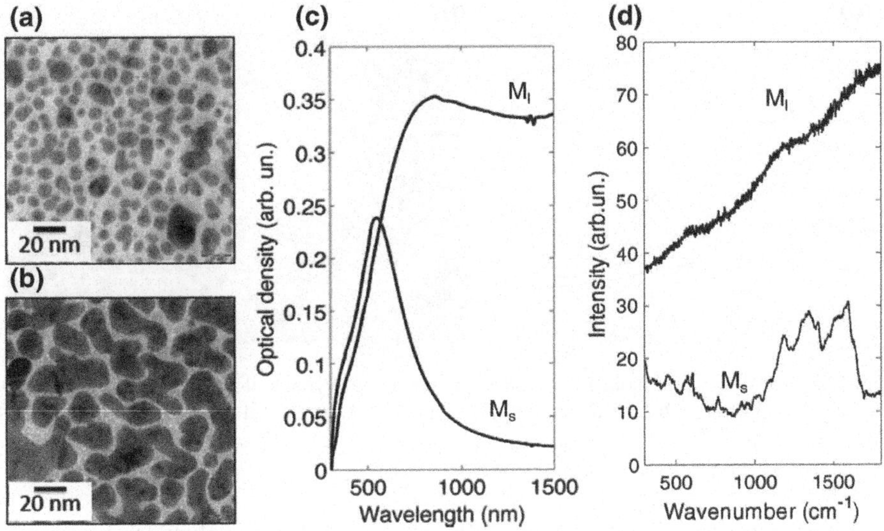

Fig. 12.2 Critical SERS dependence on the surface nanostructure of Ag films synthesized with small ($M_s = 7.0$ ng) and large ($M_l = 16.4$ ng) amounts of ablated matter per laser pulse. **a–c**, SEM micrographs and UV-vis spectra of samples M_s and M_l, respectively; **d**, SERS spectra recorded from the two samples after soaking in Rh6G solution 10^{-6}M for 60 min

NPs by continuous irradiation of the already synthesized particles along the ablation time. In this sense pulse duration τ is an important parameter for NP generation. By changing τ from nanoseconds to picoseconds the ablation mechanism changes from melting and thermal evaporation to phase explosion. The shorter τ, the more efficient the ablation process that includes a nearly instantaneous evaporation with a minimized heat affected zone [44]. A lower τ limit of about 10 ps was estimated for phase explosion. Furthermore, primary plasma shielding is negligible for ultrashort laser pulses, as compared to ns pulses [45], resulting in enhanced ablation efficiency. A further advantage is the short time required to produce the colloidal solutions. We performed PLAL in pure water, irradiating pure Au and Ag targets [99.99%, purchased from Mateck] using radiation of $\lambda = 532$ nm. This was provided by the second harmonic of a Nd:YAG source with repetition rate of 10 Hz and $\tau = 6$ ns as well as by an ultra-fast Nd:YVO_4 source with repetition rate of 10^2 kHz and $\tau =$ 8 ps. The energy density deposited at the target was fixed at $F = 1.5$ J cm^{-2} and the PLAL process duration was $t = 20$ min and $t = 5$ min for ns-PLAL and ps-PLAL, respectively. Aiming at obtaining systems with variable Ag–Au ratios, Au and Ag colloidal solutions (produced ablating high purity Au and Ag target) were mixed in different volumetric ratios. Then a drop is extracted from a given solution and dried in air before performing optical measurements and taking micrographs.

Looking at the absorbances in Fig. 12.3a we observe that the intensity ratios between Ag and Au plasmon peaks correlate with the volumetric ratios used to prepare the colloidal mixtures. Thus, neglecting to a first approximation absorptiv-

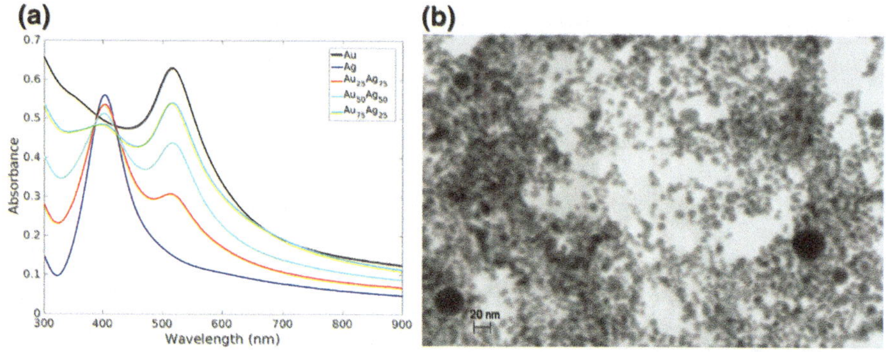

Fig. 12.3 **a** UV-vis spectra of films obtained spraying and drying colloidal suspensions on transparent inert substrates; **b** STEM micrograph of an $Au_{75}Ag_{25}$ colloid droplet dried on a STEM grid

ity differences between Ag and Au NPs, we can transpose volumetric ratios into atomic fractions. $Au_{1-x}Ag_x$ (x = 25; 50; 75) alloyed NPs were synthesized by a post-synthesis ns irradiation treatment of the starting colloidal mixtures. The optical properties of such NP mixtures were investigated as a function of stoichiometry and synthesis history. In Fig. 12.3 the position of the plasmon maximum (observed for the Ag and Au NPs at 410 and 515 nm, respectively) remains almost unchanged in the different mixtures, while their relative intensity changes according to the stoichiometric ratios of the mixtures. Such mixed colloids are mainly characterized by NPs with size less than 10 nm as shown in the micrograph from $Au_{75}Ag_{25}$ (Fig. 12.3b). In Fig. 12.4 we report, for the exemplary case of $Au_{75}Ag_{25}$, the absorbance change undergone by the solution upon a post-synthesis irradiation stage using again radiation at $\lambda = 532$ nm, with $\tau = 6$ ns pulses, depositing an energy density value $F = 1.2$ J cm^{-2}. The post irradiation lasted for a time $t = 120$ min. A single absorption peak emerges at about 505 nm, associated to the disappearance of the Ag spectral contribution at about 410 nm. Such a feature indicates that Ag-Au *alloy* NPs were formed after the post-synthesis irradiation. Support to this hypothesis is provided by the same qualitative features, with a single plasmon peak, of the profile displayed by ps-PLAL prepared colloids synthesized from a home-made $Au_{50}Ag_{50}$ alloy target as shown in the inset of Fig. 12.4a. In Fig. 12.4b is shown also a STEM image of the ns-irradiated $Au_{75}Ag_{25}$ colloids. Besides coalescence, the lack of an apparent core–shell structure in the picture strongly suggests that the NPs are alloys rather than core–shell composites.

In conclusion, sensors with further tunable SPR position can be explored by PLAL. By alloying Au with Ag the NP surface chemistry can be modified depending on the specific characteristics of the analyte.

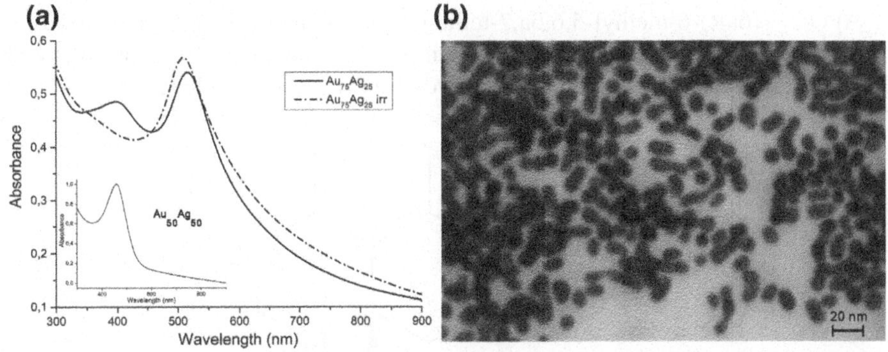

Fig. 12.4 **a** UV-vis spectra of films obtained spraying on transparent inert substrates and drying colloidal suspensions of $Au_{75}Ag_{25}$ as prepared and after post-synthesis irradiation; in the inset the spectrum of the colloid obtained ablating an alloy target. **b** STEM micrograph of the $Au_{75}Ag_{25}$ post-synthesis irradiated colloid droplet dried on a STEM grid: evidence for coalescence and lack of core-shell structures

12.3 Application of PLA-Synthesized Nanostructured Gold Sensors to Detect Apomorphine and Carbamazepine

The realistic scenery for application of PLD sensors in TDM is their use to determine drug concentration from the intensity of the SERS signal of drugs dissolved in liquids obtained from the patient blood. The typical steps may require blood centrifugation to obtain plasma and extraction of the drug from plasma using a suitable solvent (e.g., methanol, chloroform). The obtained solution constitutes the sample to be measured by SERS. We remind that a SERS measurement is based on the interaction between the drug molecule and a nanostructured, roughened metallic surface. The SERS intensity is related to the amount of adsorbed drug molecules at the surface and it can be related to the drug concentration in the plasma extract. Along this path, the analytical capability of the SERS technique has been shown to be comparable to that of HPLC [33].

12.3.1 Apomorphine (APO)

The first example deals with the development of a method for the quantification of APO in water solution by means of SERS spectroscopy, and the subsequent evaluation of the applicability of the method to the detection on samples of clinical origin. The main result has been to obtain a SERS signal of APO to a final, lower limit, concentration of 3.3×10^{-7} M in aqueous solution, and of 3.3×10^{-6} M in blood serum.

APO, (6aR)-6-methyl-5,6,6a,7-tetrahydro-4H-dibenzo[de,g]quinoline-10,11-diol (molecular weight 267.328 g/mol) is employed in the treatment of patients with Parkinson's disease. The chemical structure of APO is shown below.

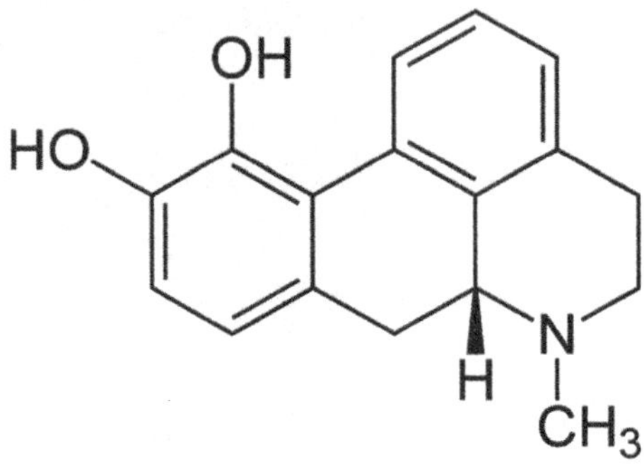

In the case of patients suffering from severe, long-term motor complications, APO is delivered via continuous subcutaneous infusion in order to reach adequately control at tolerable doses. Adjusting the APO concentration as it fluctuates with time is critical, and is monitored via standardized clinical observations and tests of motor condition. Standard chemical analysis (HPLC) on blood samples from the patient is an alternative to obtain the APO concentration.

Figure 12.5 shows a set of SERS spectra of APO in aqueous solution, at three different concentrations (10^{-4}, 10^{-5} e 10^{-6} M). SERS signal is recorded onto gold nanostructured surfaces fabricated with $N_{LP} = 3 \times 10^4$ in Ar atmosphere, at a partial pressure of 100 Pa. The experiment, performed in order to maximize the analyte signal, with a *dip time* of 60 min, shows the high sensibility of the sensors used towards the drug molecule, even at low concentrations. SERS signals are in all three cases, and, in particular, it is possible to detect the concentration of interest (10^{-6} M).

In a second experimental phase, it has been possible to verify the ability of the method to discriminate among different drug concentrations, on the base of the quantitative evaluation of SERS spectra intensities. SERS active gold sensors, fabricated with $N_{LP} = 1 \times 10^4$ in Ar atmosphere (100 Pa), have been employed, with promising results, for the detection of APO in aqueous solution in a range of concentrations between 100 and 0.1 μg/mL (3.3×10^{-4}–3.3×10^{-7} M). Such sensors were dipped in a selected volume of analyte solution, for a time drastically reduced compared to the previous experiments. In particular, a *dip time* of 5 min has been chosen as the optimum value.

Figure 12.6 shows SERS spectra of APO solutions at four different concentrations (100, 10, 1, 0.1 μg/mL), obtained with a dip time of 5 min and 5 s acquisition time. The peaks at 1515 and 1353 cm^{-1}, which are among the most intense characteristic SERS peaks of APO, were selected as markers to evaluate the SERS signal dependence of the analyte concentration.

SERS spectrum for the lowest concentration of 0.1 μg/mL shows a weak signature at 1515 cm^{-1}, and longer acquisition times do not result in an improvement in signal

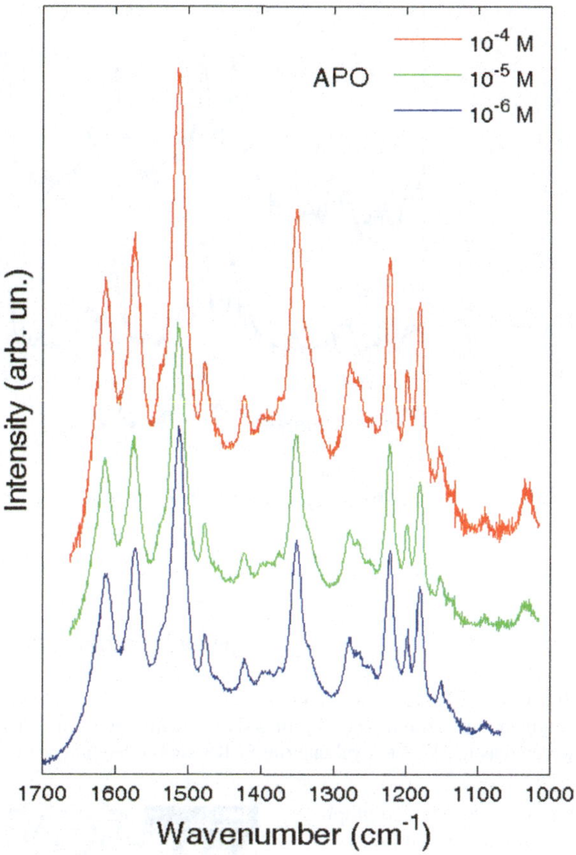

Fig. 12.5 SERS spectra of APO at three different concentrations (10^{-4}; 10^{-5}; 10^{-6} M) adsorbed onto the surface of an Au sensor grown in Ar at 100 Pa with 30,000 laser pulses. Exciting radiation: 785 nm [46]

quality. From the concentration of 1 μg/mL, the intensity of APO peaks increases with its concentration, without any change in the position of all characteristic peaks.

The optimization of the number of laser pulses during the fabrication of nanostructured gold sensors by PLD allowed to obtain sensors with an improved sensitivity to detect with a good S/N APO in aqueous solution at a concentration of 3.3×10^{-7} M.

Figure 12.7 shows a representative micrograph of the surface morphology of this class of SERS active gold sensors for APO detection. The gold thin films consist of almost spherical, mutually aggregated, Au NPs that form islands with irregular shape, separated from each other by channels with irregular shape and length whose features were analyzed and associated to the hot spots required by the SERS mechanism [48]. The inset of Fig. 12.7 shows a typical average size of such inter-island channels.

Figure 12.8 compares the SERS spectra of APO at the concentration of 6.6 10^{-5} M in blood plasma and 3.3 10^{-6} M in blood serum. Spectra are obtained with two different sensors produced with 1×10^4 laser pulses and 2×10^4 respectively.

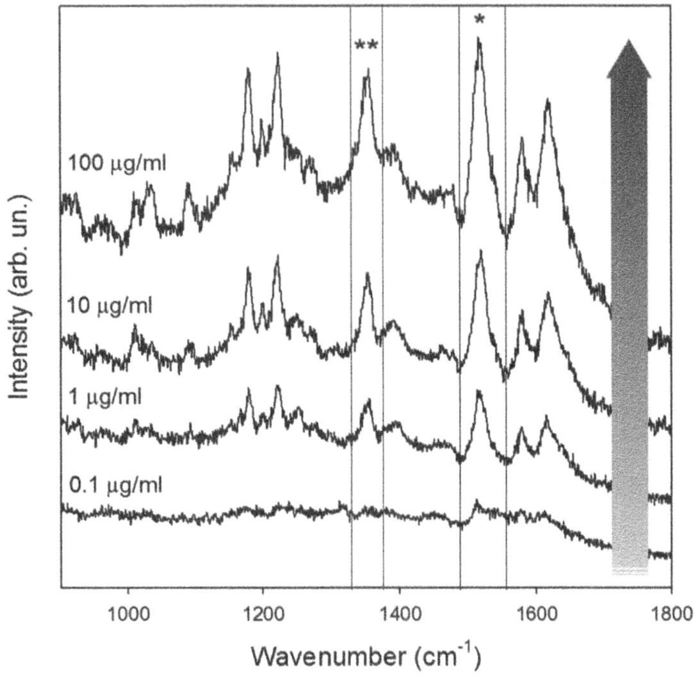

Fig. 12.6 SERS spectra of APO solutions at four different concentrations (100, 10, 1, 0.1 μg/mL) obtained with a dip time of 5 min and (5 s acquisition time). The peaks at 1515 and 1353 cm^{-1} are selected as markers to evaluate the SERS signal dependence on the APO concentration [47]

Fig. 12.7 SEM micrograph representing the surface morphology of a gold sensor deposited at 100 Pa in Ar with 10^4 laser pulses. The inset presents a typical average size of the inter-island channels

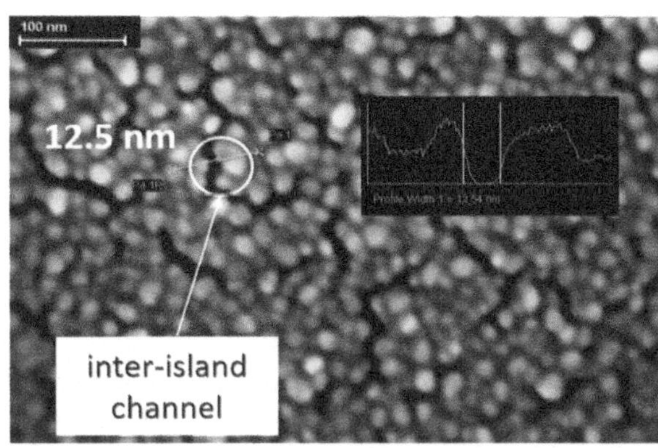

In the sample of blood serum all features of APO SERS spectrum are recognizable. Similarly, to what happens in aqueous solutions, the sensor deposited with the higher number of laser pulses shows much better defined and more intense Raman features of APO (spectrum 4), with respect to those of spectrum 3, even though the latter is

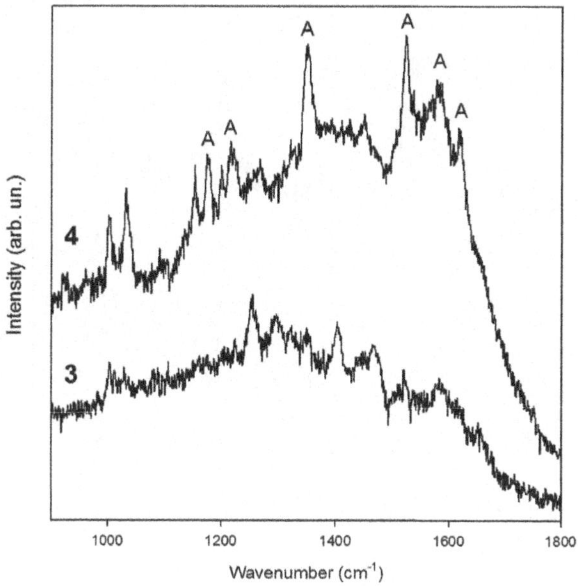

Fig. 12.8 Comparison between the SERS spectrum of APO in blood plasma (6.6×10^{-5} M; curve 3) collected onto Au nanostructured sensor with a number of laser pulses of 1×10^4 (Ar, 100 Pa), and the SERS spectrum of APO in blood serum (3.3×10^{-6} M; curve 4) collected onto Au nanostructured sensor with a number of laser pulses of 2×10^4 (Ar, 100 Pa). Exciting line: 785 nm [49]

recorded with a concentration of APO larger by more than one order of magnitude than the former.

The reduced sensitivity of the method towards APO detection in biological samples, with respect to the results obtained in aqueous solution, can be ascribed to the presence, in blood serum, of proteins (e.g., albumin), which bind APO to a relevant fraction. The complexity of the biological matrix allows understanding the decrease by a factor of 10 of the detectable concentration of APO.

12.3.2 Carbamazepine (CBZ)

The second example deals with the SERS response of CBZ adsorbed on gold nanostructured sensors, and the building of the pertinent response curve (calibration curve), as a function of the CBZ concentration in methanol. CBZ, benzo[b] [1] benzazepine-11-carboxamide (molecular weight 236.27 g/mol) is an anti-epileptic drug widespread throughout the world, in particular in developing Countries. Its chemical structure is shown in the following figure.

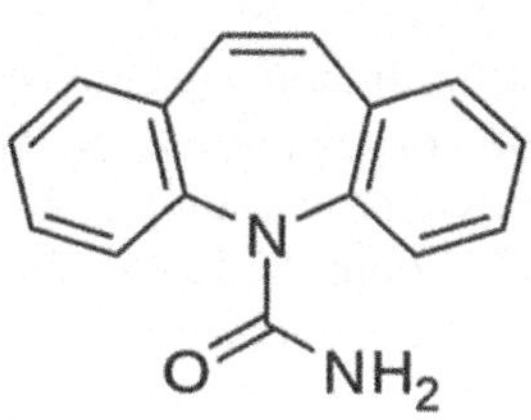

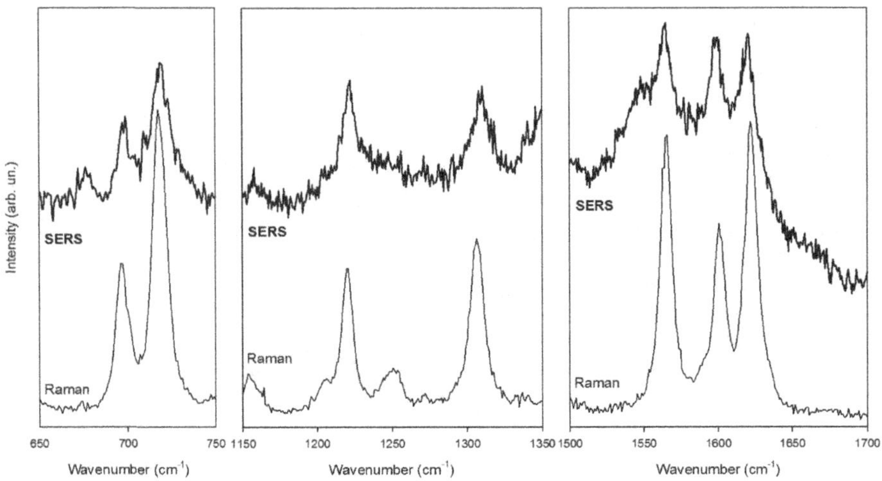

Fig. 12.9 Comparison between Raman spectrum of CBZ in methanol (20 mg/ml) and SERS spectrum of CBZ (4.2×10^{-4} M) taken on gold sensor. Contact time 60 min, solution volume 1 ml [35]

The optimization of the deposition conditions in the production of gold nanostructured sensors on glass supports through the PLD technique (see Case 1) allowed to obtain a good SERS signal of the anti-epileptic drug and to assign the markers on the SERS spectrum. In particular, Fig. 12.9 shows the comparison between CBZ SERS spectrum and its Raman spectrum in methanol (MeOH). CBZ markers are located in the 600–1700 cm^{-1} spectral region. The main SERS-active modes, i.e. peaks located at 1619, 1599, 1564, 1309, 1222, 718, 697 cm^{-1}, correspond to the signals of the Raman spectrum of CBZ in methanol and from Density Functional Theory calculation [35] they can be related to characteristic vibrations of the drug molecule. Slight shifts of a few cm^{-1} in the SERS spectrum are related to the different chemical environment CBZ molecule feels when it is close to gold rather than when it is in solution. These slight differences confirm the weak interaction between CBZ and the SERS-active nanostructured surface.

The weak interaction between the drug molecule and the sensor surface allowed the development of a method based on the use of a washable single sensor to obtain a calibration curve of CBZ in MeOH in the therapeutic range (6–12 mg/l). It was proved that, after proper washing in MeOH, the sensor can be used again and it was verified the linear relation between the CBZ SERS response and its concentration in MeOH in the range 6–50 mg/l.

Figure 12.10 shows two CBZ markers. In both spectral regions, the intensity of the SERS response depends on the CBZ concentration in MeOH. Collecting enough spectra for each concentration it is possible to estimate an average value for the intensity of the two signals (1621 and 717 cm^{-1}). Figure 12.11 shows the trend of such average values as a function of the CBZ concentration.

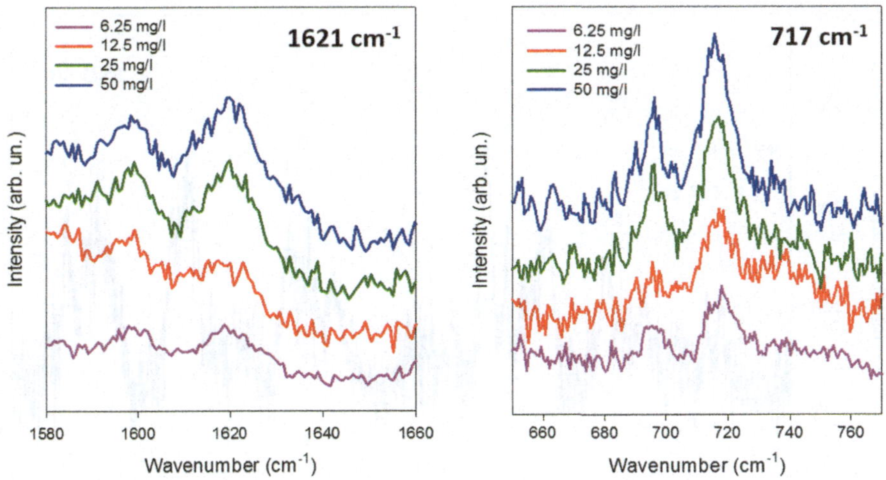

Fig. 12.10 CBZ SERS signals in two different spectral regions, as a function of CBZ concentration in methanol (6.25–50 mg/l) [50]

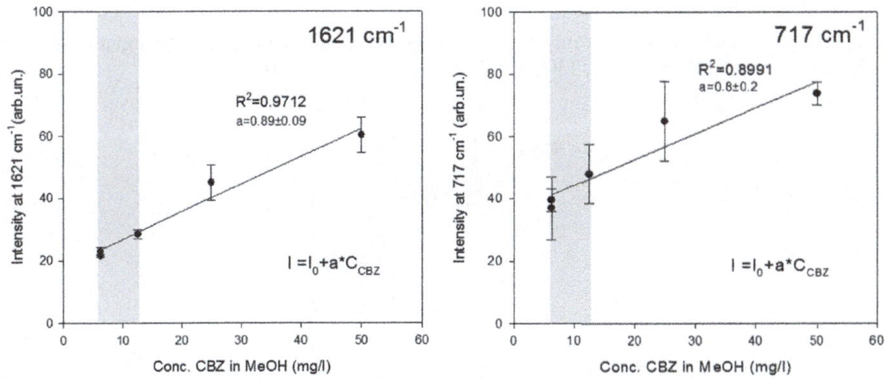

Fig. 12.11 Intensity average values of CBZ signals located at 1621 and 717 cm^{-1}, as a function of the concentration, obtained on single washable sensor [50]. The therapeutic range of concentrations is shaded

Figure 12.12 shows the SERS spectrum of a sample of serum taken from a patient treated with CBZ; arrows show the signals assigned to the presence of CBZ.

We then turned our aim to SERS measurements on clinical samples. Figure 12.13 shows the SERS spectra taken from blood plasma added with CBZ at 2.4 mg/l, after protein precipitation with MeOH. The extraction allows to recover almost all the CBZ from the blood plasma, as evidenced by independent HPLC/MS measurements on the MeOH extract.

As expected we detect the expected SERS marker of CBZ at low wavenumber, where the signal to noise ratio of the Raman spectra retrieved with a portable Raman

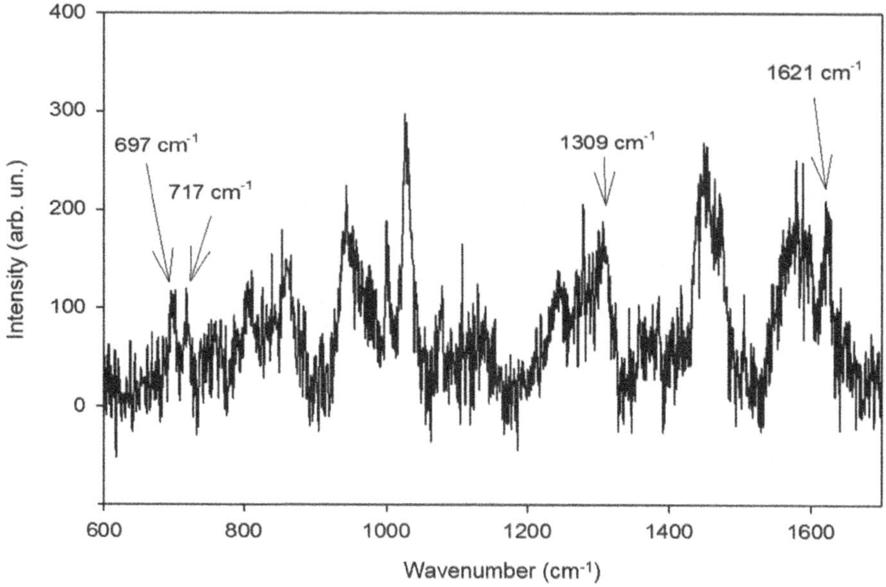

Fig. 12.12 SERS spectrum of serum taken from a patient treated with CBZ. The signals assigned to the presence of CBZ (697, 717, 1309 e 1621 cm^{-1}) are shown [50]

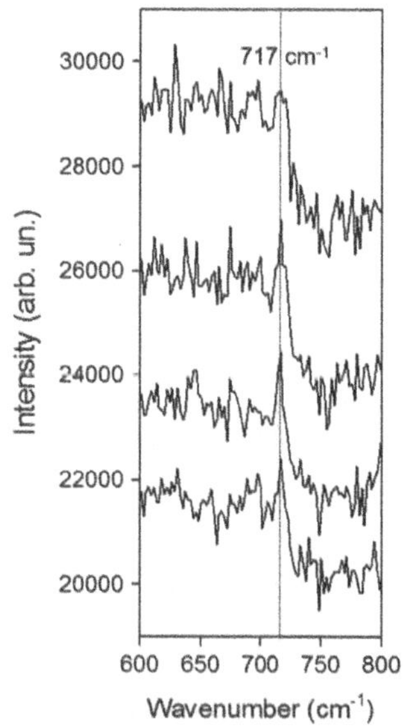

Fig. 12.13 SERS spectra of extract in methanol of blood plasma added with CBZ at 2.4 mg/l. Each spectrum was recorded on a different spot of the sensor, which displays a remarkable uniformity of the Raman signal over a background [51]

equipment (B&WTek i-Raman BWS415; excitation wavelength, 785 nm) is better. Work is in progress to disentangle the other markers of CBZ from the measured spectra and assess the possible role of molecules, such as cholesterol, which could be co-extracted with CBZ.

12.4 Conclusion and Perspectives

Developing quantitative SERS protocols for TDM requires validations against established analytical techniques (HPLC-MS, immunoassay). Present research activity aims at overcoming difficulties associated to weak signals of drugs and to the presence of significant background signals arising from biomolecules co-dissolved in the fluid to be analyzed. A related research direction is devoted to the development of efficient SERS sensors with high sensitivity and spatial uniformity, within a highly repeatable and cost-effective fabrication process.

Acknowledgements We acknowledge the contributes of A. Bombelli, C. Pedrini, and M. Pistaffa to this research.

References

1. E. Boisselier, D. Astruc, Chem. Soc. Rev. **38**, 1759 (2009)
2. E.C. Dreaden, A.M. Alkilany, X. Huang, C.J. Murphy, M.A. El-Sayed, Chem. Soc. Rev. **41**, 2740 (2012)
3. E. Boulais, R. Lachaine, A. Hatef, M. Meunier, J. Photochem. Photobiol. C: Photochem. Rev. **17**, 26 (2013)
4. F. Berorelle, M. Ceccarello, M. Pinto, G. Fracasso, D. Badocco, V. Amendola, P. Pastore, M. Colombatti, M. Meneghetti, J. Phys. Chem. C **118**, 14534 (2014)
5. T. Graham, Philos. Trans. R. Soc. **151**, 183 (1861)
6. C.J.S. Thompson, Alchemy and Alchemists, Mineola (NY, USA), Dover Publ. Inc, 2002
7. V. Biringuccio, The Pirotechnia, Mineola (NY, USA, 1990), Dover Publ. Inc
8. Paracelsus, *On the Diseases that Deprive Man of His Reason*, in ed. by Paracelsus: Four Treatises, H.E. Sigerist (Johns Hopkins University Press, Baltimore, MD, USA, 1941)
9. G. Andrea, Libro de i Secreti con Ricetti. Folio 109r, 1562, Ms. in the rare book collection of the K. Spencer Res. Library, Univ. of Kansas, Lawrence (KS, USA)
10. F. Antonii Philosophi et Medici Londinensis, Panacea Aurea sive tractatus duo de ipsius Auro potabili, 1618, ex Bibliopolio Frobeniano, Hamburg (Germany)
11. J. Knuckels, Neuetliche Observationes oder Anmerkungen von Auro und Argento Potabili, 1676, Schutzens, Hamburg (Germany)
12. Dictionnaire de Chymie, 1769, Lacombe, Paris (France)
13. W. Ostwald, Kolloid Z. **4**, 5 (1909)
14. M. Faraday, Philos. Trans. R. Soc. **147**, 145 (1857)
15. J.V. Shoemaker, *Materia Medica and Therapeutics*, 4th edn. (F.A. Davis Co, Philadelphia, PA, USA, 1898)
16. T. Sollman, *A Manual of Pharmacology* (W. B. Saunders Co, Philadelphia, PA, USA, 1917)
17. J.F. Schamberg, C.S. Wright, Arch. Dermatol. Syphilol. **15**, 119 (1927)

18. T. Sollman, *A Manual of Pharmacology*, 7th edn. (W. B. Saunders Co, Philadelphia, PA, USA, 1949)
19. A.G. Berger, S.M. Restaino, I.M. White, Anal. Chim. Acta **949**, 59 (2017)
20. S. Fornasaro, S.D. Marta, M. Rabusin, A. Bonifacio, V. Sergo, Faraday Discuss. **187**, 485 (2016)
21. B.W. Abou-Khalil, Epilepsy Currents/Am. Epilepsy Soc. **5**, 115 (2004)
22. S. Kasera, L.O. Herrmann, del J. Barrio, J.J. Baumberg, O.A. Scherman, Sci. Rep. **4**, Article number: 6785 (2014) https://doi.org/10.1038/srep06785
23. A. Jaworska, S. Fornasaro, V. Sergo, A. Bonifacio, Biosensors **6**, 47 (2016)
24. D.J. Thurman, E. Beghi, C.E. Begley, A.T. Berg, J.R. Buchhalter, D. Ding, D.C. Hesdorffer, W.A. Hauser, L. Kazis, R. Kobau, B. Kroner, D. Labiner, K. Liow, G. Logroscino, M.T. Medina, C.R. Newton, K. Parko, A. Paschal, P.M. Preux, J.W. Sander, A. Selassie, W. Theodore, T. Tomson, S. Wiebe, Epilepsia **52**, 2 (2011)
25. R. Quintas, A. Raggi, A.M. Giovannetti, M. Pagani, C. Sabariego, A. Cieza, M. Leonardi, Epilepsy Behav. **25**, 60 (2012)
26. A.C. Meyer, T. Dua, J. Ma, S. Saxena, G. Birbeck, Bull. World Health Organ. **88**, 260 (2010)
27. E. Beghi, Lancet Neurol. **3**, 618 (2004)
28. L. Mercolini, R. Mandrioli, M. Amore, M.A. Raggi, J. Pharm. Biomed. Anal. **53**, 62 (2010)
29. S.I. Johannessen, D. Battino, D. Berry, M. Bialer, G. Krämer, T. Tomson, P.N. Patsalos, Ther. Drug Monit. **25**, 347 (2003)
30. J. Parkinson, An Essay on the Shaking Palsy, 1817, London (UK), Sherwood, Neely, and Jones. Reprint available at: J. Neuropsychiatry Clin. Neurosci. 2002, **14**, 223
31. L.V. Kalia, A.E. Lang, Lancet **386**, 896 (2015)
32. A.E. Elia, C. Dollenz, P. Soliveri, A. Albanese, Eur. J. Neurol. **19**, 76 (2012)
33. C. McLaughlin, D. MacMillan, C. McCardle, W.E. Smith, Anal. Chem. **74**, 3160 (2002)
34. I.J. Hidi, A. Mühlig, M. Jahn, F. Liebold, D. Cialla, K. Weberac, J. Popp, Anal. Methods **6**, 3943 (2014)
35. C. Zanchi, A. Lucotti, M. Tommasini, S. Trusso, U. de Grazia, E. Ciusani, P.M. Ossi, Appl. Surf. Sci. **396**, 1866–1874 (2017)
36. J.E. Moore, S.M. Morton, L. Jensen, J. Phys. Chem. Lett. **3**, 2470 (2012)
37. P.M. Ossi, N.R. Agarwal, E. Fazio, F. Neri, S. Trusso, *Laser-mediated Nanoparticle Synthesis and Self-Assembling*, in eds. by M. Castillejo, P.M. Ossi, L. Zhigilei (Eds.). Lasers in Materials Science (Springer, Germany, Heidelberg, 2014)
38. M.C. Spadaro, E. Fazio, F. Neri, S. Trusso, P.M. Ossi, Europhys. Lett. **109**, 25002 (2015)
39. M.C. Spadaro, E. Fazio, F. Neri, P.M. Ossi, S. Trusso, Appl. Phys. A **117**, 137 (2014)
40. P.P. Patil, D.M. Phase, S.A. Kulkarni, S.V. Ghaisas, S.K. Kulkarni, S.M. Kanetkar, S.B. Ogale, V.G. Bhide, Phys. Rev. Lett. **28**, 3 (1987)
41. V. Amendola, M. Meneghetti, Phys. Chem. Chem. Phys. **11**, 3805 (2009)
42. Z. Yan, D.B. Chrisey, J. Photochem. Photobiol. C: Photochem. Rev. **13**, 204 (2012)
43. D. Zhang, B. Gökce, S. Barcikowski, Chem. Rev. **117**, 3990 (2017)
44. S.K.-M. Lai, H.-W. Tang, K.-C. Lau, K.-M. Ng, J. Phys. Chem. C **120**, 20368 (2016)
45. A. Schwenke, P. Wagener, S. Nolte, S. Barcikowski, Appl. Phys. A **104**, 77 (2011)
46. N.R. Agarwal, E. Fazio, F. Neri, S. Trusso, C. Castiglioni, A. Lucotti, N. Santo, P.M. Ossi, Cryst. Res. Technol. **46**, 836 (2011)
47. C. Zanchi, A. Lucotti, M. Tommasini, S. Trusso, U. de Grazia, E. Ciusani, P.M. Ossi, Beilstein J. Nanotechnol. **6**, 2224 (2015)
48. N.R. Agarwal, F. Neri, S. Trusso, P.M. Ossi, Plasmonics **8**, 1707 (2013)
49. S. Trusso, C. Zanchi, A. Bombelli, A. Lucotti, M. Tommasini, de U. Grazia, E. Ciusani, L.M. Romito, P.M. Ossi, J. Instrum. (2016) https://doi.org/10.1088/1748-0221/11/05/c05006 (online only)
50. A. Bombelli, Dosaggio SERS di farmaci con film nanostrutturati di oro, Tesi di laurea magistrale, 2016, Politecnico di Milano
51. M. Pistaffa, SERS assessment of carbamazepine in methanol, chlorophorm and patient samples using Au sensors with optimized surface nanostructure, Master of Science Thesis, Politecnico di Milano, 2017

Chapter 13
Nonlinear Optics in Laser Ablation Plasmas

Mohamed Oujja, Mikel Sanz, Rebeca de Nalda and Marta Castillejo

Abstract This chapter describes current research on the nonlinear optical response, specifically low-order harmonic generation, of nanosecond laser ablation plumes of solid targets. The spatiotemporal control of the driving laser with respect to the position and time of the ablation event allows for the differentiation of distinct populations of species and enables the in situ diagnosis of complex ablation plume environments.

13.1 Introduction

Harmonic generation (HG) is a nonlinear process that results from the interaction of a source of intense optical frequency radiation with a material medium [1]. Lasers provide the necessary intensity for the generation of new frequencies, that are multiple values of the source frequency, in an elastic process that does not involve exchange of energy between the laser and the medium. In media with inversion symmetry, all even-order nonlinear effects are forbidden [2, 3]. This is the reason why a non-centrosymmetric medium, such as a crystal, is necessary for second harmonic generation. For all optical materials, regardless of their symmetry properties, odd-order effects are possible, provided the intensity of the **driving laser** is sufficiently high. Thus, in gases, third harmonic generation is the lowest order nonlinear **frequency up-conversion** process.

Odd harmonic generation of a fundamental driving laser beam in interaction with centro-symmetric gaseous media is used for obtaining coherent light in the **vacuum ultraviolet (VUV)** and **extreme ultraviolet (XUV)** regions of the spectrum. During decades, low-order harmonic generation processes, using laser pulses of nanosecond duration, have served to provide coherent light sources in the UV and VUV spectral regions [4–6]. As only laboratory, table-top set-ups are required for the purpose, these short wavelength sources have found numerous applications in spectroscopy and molecular photodissociation studies [7, 8].

M. Oujja · M. Sanz · R. de Nalda · M. Castillejo (✉)
Instituto de Química Física Rocasolano, CSIC, Madrid, Spain
e-mail: marta.castillejo@iqf.csic.es

P. M. Ossi (ed.), *Advances in the Application of Lasers in Materials Science*,
Springer Series in Materials Science 274,
https://doi.org/10.1007/978-3-319-96845-2_13

The highest achievable nonlinear order depends on the laser intensity in the interaction region. **High-order harmonic generation (HHG)** [9] becomes possible when using near-infrared (NIR), ultrashort laser pulses of femtosecond duration and intensities of 10^{14}–10^{15} W/cm^2 and nonlinear media that are currently atomic and molecular gases and vapours in static cells or gas jets [10–14]. Through these processes it is possible to reach the water window spectral region down to 2.3–4.6 nm.

An efficient **nonlinear optical response** is necessary for using nonlinear harmonic sources in fundamental studies and for applications in areas of biology, medicine and processing and analysis of materials [13]. In the quest for highly efficient nonlinear media, able to sustain a harmonic spectrum towards the short wavelength range, **laser ablation** plasmas have proven to stand as an advantageous alternative to other gaseous media [15–22]. Ganeev and collaborators [17 and references therein] have extensively contributed to develop the technique of HHG in laser ablation plasmas and the corresponding studies have shown that the main advantages of using laser ablation plasmas as nonlinear media include the improved **conversion efficiency** and the extension of the harmonics spectrum towards shorter wavelengths. Also, due to the large variety of species in the laser plume, that include atoms, light molecules, radicals, **clusters** and **nanoparticles**, it is possible to rely on atomic or plasmon resonance effects to enhance the nonlinear optical response of the medium [17–22]. Moreover, laser ablation plasmas offer the possibility to tailor their nonlinear response, in terms of composition, density gradients, spatial extent or degree of ionization, through the selection of the ablation laser parameters, including wavelength, intensity, pulse duration, polarization and ablating beam spatial and temporal profile.

Laser ablation plumes of solid targets are complex media from the point of view of both composition and dynamics. As mentioned, components of ablation plumes range from neutral atoms, ions and **free electrons** to aggregates, such as small molecules, radicals, clusters or nanoparticles [23–28]. The relative abundance of these species and their particular velocities and angular distributions depend on the material and laser pulse properties and also on the ambient conditions in which the plasma plume is expanding. Analysis of composition and dynamics of laser-induced plasmas is carried out through a variety of techniques, where **optical emission spectroscopy (OES)** and **time-of-flight mass spectrometry (TOF-MS)** stand as some of the most used. The information retrieved by OES from the plasma spontaneous emission allows the assessment of important parameters, such as the electron temperature and density, while TOF-MS methods are based on the collection and analysis of the ions present in the laser plasma. Other diagnostic methods are also used like **pulsed laser deposition (PLD)** experiments, where analysis of the morphology, composition and crystallinity of the ejected material, collected on appropriate substrates placed in proximity to the ablation event, allows inference on the nature and dynamics of the plume species [29].

The analysis of the nonlinear response of laser ablation plasmas has emerged as a promising alternative, to be used as stand alone or in combination with OES, to generate information on the different populations of species in the plasma plume. The initial idea that nonlinear optical processes can be used as a diagnostic of complex media like laser ablation plasmas was proposed in the pioneering works of Zheltikov et al. [30]. Recent investigations carried out in laser ablation plumes of different types of target materials have revealed that low-order HG and frequency mixing in laser plumes enables the in situ diagnosis of the complex ablation environment [16, 31–38].

After this Introduction, this Chapter is organized as follows. Section 13.2 presents essential concepts of nonlinear optics, specifically related with HG and frequency mixing in gases and laser ablation plumes, in both the low field intensity, perturbative regime and the high field, non-perturbative approach. In Sect. 13.3, the experimental systems used for studying frequency up-conversion in laser ablation plasmas in the perturbative regime (low-order harmonics) are succinctly described. Section 13.4 overviews the current research on the nonlinear optical response of nanosecond laser ablation plasmas of solid targets. Through the example of various types of target materials, this Section shows how the dependence of harmonic signals with laser ablation and driving parameters (ablation wavelength, pulse energies, position of driving beam focus across the plasma plume, **ablation-driving delay**, polarization, etc.), displays distinct characteristics and how the spatiotemporal control of the driving laser with respect to the position and time of the ablation event allows for the differentiation of distinct populations of species. Finally, Sect. 13.5 states the main conclusions and gives an outlook of the work presented in this Chapter.

13.2 Fundamentals of Harmonic Generation

In HG [1], the nonlinear optical interaction of a fundamental radiation of frequency ω with a medium leads to the generation of coherent light at frequencies $n\omega$, harmonics of the fundamental radiation, The integer number n indicates the nonlinear order of the process. Only laser sources provide sufficient intensity for efficient HG, where the new frequencies collinearly propagate with the fundamental beam retaining its coherence and directionality. The process of HG is illustrated schematically in Fig. 13.1a.

If the fundamental optical field is not too large ($E << 5 \times 10^{11}$ V/m), the polarization $\tilde{P}(t)$ of the medium, defined as the dipole moment per unit volume, can be described by **time dependent perturbation theory** as a power series expansion:

$$\tilde{P}(t) = \epsilon_0 \left[\chi^{(1)} \tilde{E}(t) + \chi^{(2)} \tilde{E}^2(t) + \chi^{(3)} \tilde{E}^3(t) + \ldots \right] \tag{13.1}$$

In this expression, the polarization $\tilde{P}$ and the electric field $\tilde{E}$ are both vectorial quantities, ϵ_0 is the vacuum permittivity and $\chi^{(n)}$ is the nth-order susceptibility. It is assumed that $\chi^{(n)}$ is much smaller than $\chi^{(n-1)}$. The linear term, $\tilde{P}(t) = \epsilon_0 \chi^{(1)} \tilde{E}(t)$, accounts for linear optical effects, such as refraction or absorption, while the subsequent higher order terms are responsible for nonlinear optical effects, including harmonic generation.

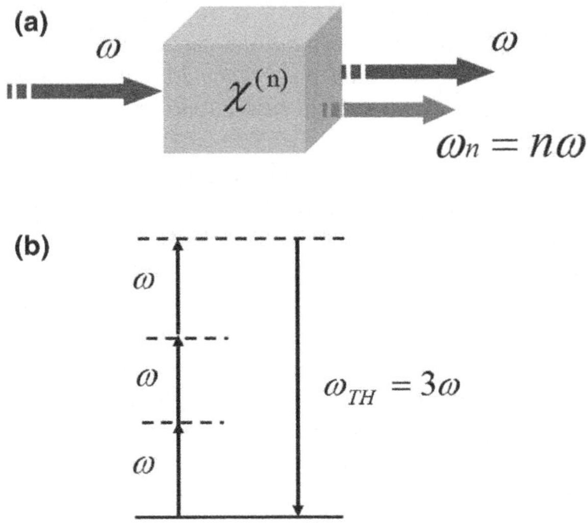

Fig. 13.1 Schematic representation of: **a** a generic harmonic generation process where ω is the frequency of the fundamental radiation, $\chi^{(n)}$ is the nonlinear **susceptibility** of the medium and n the harmonic order; **b** the third harmonic generation process

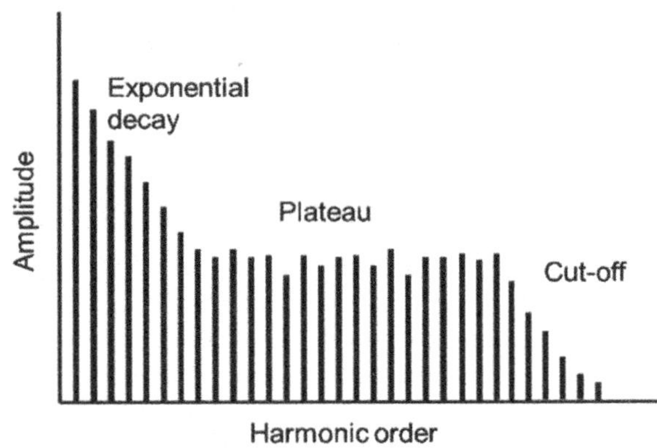

Fig. 13.2 Scheme of a typical HHG spectrum, characterized by an initial region with exponentially decaying amplitude, a *plateau* and a cut-off where the intensity of the harmonic signal vanishes

As stated previously, symmetry considerations forbid even-order nonlinear effects in centro-symmetric media. Therefore in a gas or in an ablation plume, third-order effects based on $\chi^{(3)}$, such as third harmonic generation (Fig. 13.1b), are the lowest-order occurring nonlinear phenomena.

When high intensity laser fields are applied ($E >> 5 \times 10^{11}$ V/m), as those involved in ultrashort femtosecond lasers, generation of high harmonics of the fundamental radiation is possible. In this case the framework of a perturbative regime is no longer adequate to describe the process of HG. Non-perturbative schemes, as the three-step model [39], adequately describe the main characteristic features of the HHG process [9–14]. Figure 13.2 schematizes a typical HHG spectrum using a gas as nonlinear medium. At low orders, the harmonic intensity decays exponentially. This is followed by a *plateau*, for a wide range of harmonic orders, and finally by a cut-off region where the intensity decays abruptly to zero.

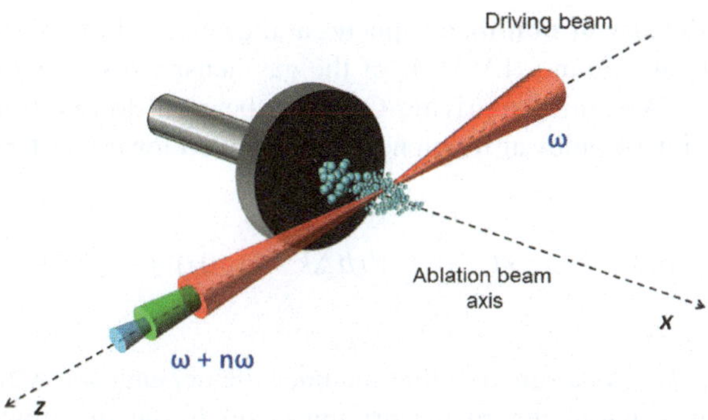

Fig. 13.3 Scheme of the HG process in a laser ablation plasma. The ablation laser beam propagates at normal incidence towards the target surface along the x axis. The driving beam generating harmonic radiation propagates parallel to the surface along the z axis and is focused within the plasma volume. Adapted from [37]

While previous and numerous reports have disseminated the results concerning HHG in laser ablation plasmas [14–22], this Chapter focusses on the process of low-order harmonic generation in this type of nonlinear media. In this case, harmonic radiation results from the interaction of a driving laser, delivering nanosecond pulses, with the expanding laser ablation plasma. Figure 13.3 is a schematic representation of the HG process. Typically, ablation is carried out with laser pulses arriving at normal incidence with respect to the target surface while the driving pulse propagates, at a given delay with respect to the ablation event, parallel to the target surface. The double control of the position of the driving beam focus with respect to the plasma volume and of the ablation-driving delay allows for the spatiotemporal scanning of the nonlinear response of the ablation plume.

As indicated, low-order harmonic generation, using a **fundamental driving beam** peak intensity at focus of ca. 4×10^{11} W/cm^2, can be treated with a perturbative approach. For a centro-symmetric medium, as is the case of an expanding laser ablation plume, the nonlinear interaction of the driving laser, of frequency ω, leads to the generation of odd harmonics of frequencies $n\omega$ as depicted in Fig. 13.1. The formalism described in [4, 37] serves to treat this process, as it includes the influence on HG efficiency of **phase matching** effects in media with inhomogeneous densities, and also serves to discuss the effect of the displacement of the driving laser focus across the centre of the density distribution.

In the case of a laser ablation plume, the normalized inhomogeneous density distribution of species along the direction of propagation of the driving laser, z, at a distance x from the target, $S(x, z)$, can be well represented by a Lorentzian function centred in $z = 0$:

$$S(x, z) = \frac{N(x, z)}{N_0(x)} = \frac{1}{1 + \left[\frac{2z}{L(x)}\right]^2} \tag{13.2}$$

$N_0(x)$ is the **density of nonlinear species** at the centre of the plume and $L(x)$ the full width at half maximum (FWHM) of the gas density distribution at a distance x from the target. Assuming a driving Gaussian beam of lowest order of intensity I_ω, the generated harmonics at frequencies $n\omega$ are also lowest-order Gaussian with intensities:

$$I_{n\omega} = \alpha(n, \omega, b, k_\omega, k_{n\omega})\left|\chi^{(n)}\right|^2 (I_\omega)^n (b\Delta k^{(n)}(x, 0))^2 F^{(n)}\left[b\Delta k^{(n)}(x, z), x, z\right] \tag{13.3}$$

$\alpha(n, \omega, b, k_\omega, k_{n\omega})$ is a function that includes the dependence with the harmonic order, the confocal parameter of the driving beam b, the fundamental frequency and the wave vectors of the fundamental and harmonic light beams, k_ω and $k_{n\omega}$ respectively. $\chi^{(n)}$ is the nth order susceptibility of the nonlinear medium, z the distance of the driving beam focus to the centre of the plume and $F^{(n)}\left[b\Delta k^{(n)}(x, z), x, z\right]$ is the phase-matching function which depends on x, z and $b\Delta k^{(n)}(x, z)$. The term $\Delta k^{(n)}(x, z)$ denotes the **wave-vector mismatch** between the generated radiation and the driving polarization and has the same spatial dependence as the density of nonlinear species:

$$\Delta k^{(n)}(x, z) = k_{n\omega}(x, z) - nk_\omega(x, z) = S(x, z)\Delta k^{(n)}(x, 0), \tag{13.4}$$

where, in turn, $\Delta k^{(n)}(x, 0)$ is proportional to the maximum density $N_0(x)$ of nonlinear scatterers:

$$\Delta k^{(n)}(x, 0) = N_0(x)C^{(n)} \tag{13.5}$$

and $C^{(n)}$ gives the wavelength dependence of the wave-vector mismatch per atom for the nth order HG process.

The phase-matching function is represented by the integral:

$$F^{(n)}\left[b\Delta k^{(n)}(x, z), x, z\right] = \left|\frac{2}{b}\int_{-\infty}^{\infty} \frac{S(x, l)}{(1 + i\varepsilon)^{n-1}} \exp\left[-i\int_{-\infty}^{l} \Delta k^{(n)}(x, l')dl'\right] dl\right|^2 \tag{13.6}$$

with $\varepsilon = 2(l - z)/b$.

The integral in (13.6) can evaluated numerically for the successive nth orders, therefore the described formalism, assuming that both the driving and harmonic beams propagate through the plume with negligible absorption, can be used to analyse the dependence of the intensity of the harmonics generated in a given ablation plume with the ablation pulse energy, the distance to the target x and the position of the focus along the propagation direction of the driving beam, z (see Fig. 13.3). This is represented by the corresponding dependences of the functions $G^{(n)}$ defined as:

$$G^{(n)}\left[b\Delta k^{(n)}(x,z),x,z\right]=(b\Delta k^{(n)}(x,0))^2F^{(n)}\left[b\Delta k^{(n)}(x,z),x,z\right]. \tag{13.7}$$

13.3 Experimental Systems for Frequency up-Conversion in Laser Ablation Plasmas

In a compositionally simple laser ablation plasma (i.e., from a metal target), the nonlinear optical response of a given species can be determined once it has been established that the desired species is the main harmonic emitter [16]. In this case, an external diagnostic for the presence of the species is necessary. This can be achieved by OES, **laser-induced fluorescence (LIF)**, mass spectrometry-based techniques or by analysis of the ablation debris deposited on a nearby substrate [35]. However, in most cases, the ablation plasma contains a variety of components and the study of the nonlinear behaviour of a given species in a multicomponent plume can only be carried out by discrimination through the differing spatiotemporal behaviour of the plasma components.

The typical experimental set-up for measuring HG in nanosecond laser ablation plasmas includes a vacuum ablation chamber and an optical layout for propagation and separation of the harmonic radiation generated by the interaction of a fundamental driving beam with the laser plasma [31–38]. The solid target is mounted on a rotating holder inside the ablation chamber to avoid target cratering induced by repetitive irradiation. Ablation is usually performed at normal incidence with a Q-switched Nd:YAG laser delivering Gaussian pulses of nanosecond duration and operating at its fundamental wavelength of 1064 or their harmonics in the visible-UV region (i.e. 532, 355, 266 nm). The driving laser responsible for HG is a second linearly polarized Q-switched Nd:YAG system operating at 1064 nm that propagates across the plume perpendicularly to the ablating beam and at a given distance (in the range of mm) from the target surface.

Figure 13.4 presents a scheme of a prototypical system used for measuring the frequency up-conversion in laser ablation plasmas, especially adapted to allow for the detection of up to the 9th harmonic of the fundamental 1064 nm radiation [37]. The corresponding wavelength is in the VUV region of the spectrum (9th harmonic at 118.2 nm) and therefore the ablation chamber is connected to a second vacuum chamber that allows for the propagation of the newly generated coherent radiation. In this second chamber, the generated harmonics co-propagate with the driving laser through a lithium fluoride window and are spectrally separated with a 60° lithium fluoride prism which, for each harmonic, is adjusted by rotation to the appropriate incident angle of the incoming beam. Additional spectral filtering of harmonics is achieved in a vacuum monochromator and the harmonic signals are measured with a sodium salicylate coated window-photomultiplier combination attached to the exit slit. If detection is limited to the UV region, (i.e. 3rd and 5th harmonics of the Nd:YAG fundamental radiation), the experimental scheme becomes simpler. As harmonic radiation can propagate in air, thus the separation chamber and the vacuum

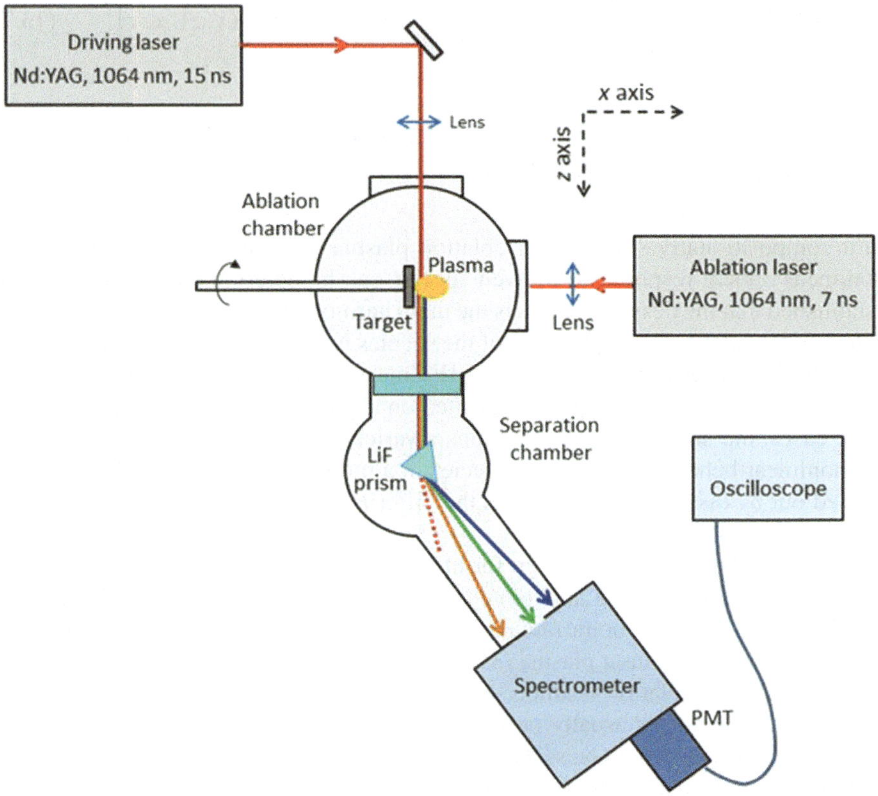

Fig. 13.4 Scheme of the experimental set-up used for measuring low-order (3rd–9th) HG of a fundamental NIR driving laser in a plasma induced by laser ablation at 1064 nm. The detail of the interaction region is shown in Fig. 13.3. Adapted from [37]

monochromator are no longer needed. In this case, the set-up includes a system of lenses and dichroic mirrors to guide the light and a detection system consisting of a monochromator and an intensified charge coupled device (ICCD).

The systems used for measuring HG in ablation plasmas [15–22, 31–38] include the electronic control of the delay between the ablation event and the arrival of the driving laser pulses to the plume. This delay is typically in the 0–100 μs range and control is performed with nanosecond resolution. The driving beam is focused at the centre of the plume with the aid of a lens. Together with the temporal mapping of the ablation plume by HG, it is possible to explore the spatial variation of the nonlinear optical response of the laser plasma by displacing the focus of the driving beam along the propagation direction and along the normal to the target (z and x coordinates respectively, see Figs. 13.3 and 13.4).

Under some conditions, the detection system used to measure the generated harmonics serves for the simultaneous detection of the optical emissions from electronically excited species generated in the plasma. However, in some cases, a dedicated separate OES system is mounted to detect the spontaneous plasma emissions in presence or absence of the driving laser [37, 38].

13.4 Harmonic Generation in Nanosecond Laser Ablation Plasmas of Solid Targets

Current research on the low-order HG on laser ablation plasmas of solids, using a nanosecond laser as driving fundamental radiation, has dealt with a large variety of materials, including, dielectrics, metals and **semiconductors**. These studies have aimed at determining the dependence of various low-order harmonic signals with ablation wavelength, ablation and driving pulse energies and polarizations, ablation-driving delay, position of the driving laser across the plasma plume, etc. This Section illustrates, through selected examples of various materials and laser ablation conditions, the capabilities of the process of HG to assess the **plume composition and dynamics** with spatial resolution and to characterize the nonlinear optical properties of complex plume species, such as aggregates and nanoparticles.

13.4.1 Third and Fifth Harmonic Generation in Nanosecond Laser Ablation Plasmas of Dielectrics

Generation of low-order harmonics of the fundamental radiation of a Q-switched Nd:YAG laser (1064 nm, pulse 15 ns) was investigated in laser ablation plumes of the highly **ionic insulators** $\mathbf{CaF_2}$ [31], $\mathbf{MgF_2}$ and **NaCl** [33]. The ablation process was triggered by a separate Q-switched Nd:YAG laser operating at the wavelength of 532 or 266 nm, selected for each material to keep to a minimum the cratering of the target by repetitive irradiation and to ensure the ejection of a sufficiently intense luminous plume.

The third and fifth harmonics (in what follows TH and FH) of the driving laser are detected as sharp spectral lines at the corresponding wavelengths of 355 and 213 nm respectively. Figure 13.5a displays, for the case of CaF_2 ablated at 266 nm, the spectrum around the TH region, where atomic emission lines from neutral and ionic calcium (Ca I and Ca II, respectively) are also observed. Both harmonics show the expected dependence with the driving laser pulse intensity, $(I_\omega)^n$ as in expression (13.3) above, according with the perturbative regime of low-order HG. This is observed in the log-log plots of Fig. 13.5b. At energies over 100 mJ/pulse, equivalent to an intensity of 2×10^{11} W/cm^2, a saturation regime is reached due to absorption of the newly generated light by the plasma and to plasma modification by the intense driving laser radiation.

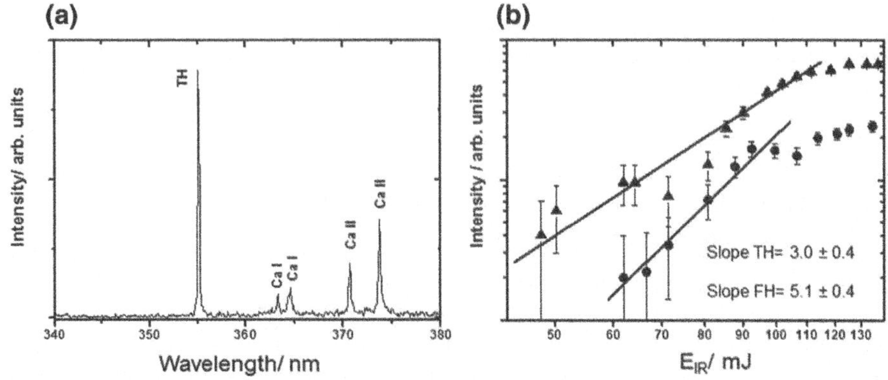

Fig. 13.5 **a** TH of 1064 nm driving pulse generated in CaF_2 ablation plasma at a delay of 200 ns from the arrival of the ablating pulse of 266 nm (fluence of 5 J/cm^2). Assigned spontaneous calcium emissions (neutrals, Ca I, and ions, Ca II) are also marked; **b** behaviour of the TH and FH generated in the CaF_2 ablation plume as a function of the fundamental laser intensity. Both signals show the expected power laws for low intensities, followed by a saturation region above 100 mJ (corresponding to an intensity of 2×10^{11} W/cm^2). Adapted from [31]

In this work, an estimate of the efficiency of the HG process in CaF_2 was obtained yielding values of $\approx 4 \times 10^{-5}$ and 10^{-5}, for the TH and FH respectively, for driving laser pulse energies of 100 mJ. These values correspond to energies per pulse of ca. 1 and 4 μJ, at 355 and 213 nm. Higher values were obtained for other ionic insulators as is the case of the 266 nm laser ablation plasma of NaCl, where the efficiency of TH generation was estimated in $\approx 4 \times 10^{-4}$ [33].

The newly generated coherent radiation comes from the overall contribution of different species in the ablation plume, each having their own nonlinear response to the propagating driving laser. In certain cases, and with the aid of external diagnostics, it is possible to ascertain the nature of the main nonlinear scatterers. For the mentioned dielectric materials, the products of laser ablation include single atoms and diatomic molecular species [31, 40]. Figure 13.6a correlates, for Ca_2F, the temporal behaviour of the TH and FH signals with that of the induced fluorescence emission of the CaF (B $^2\Sigma^+$–X $^2\Sigma^+$) system observed at around 530 nm. This emission is induced by **two-photon absorption** of the driving laser, according with the scheme of Fig. 13.6b. As observed, the emission signal coming from the driving laser excitation of electronically ground state CaF molecules follows a dependence with time delay, between the ablation event and the arrival of the fundamental NIR laser to the surface of the sample, that is almost identical to that of the harmonic signals. This finding allows establishing that harmonics are selectively generated by ground state CaF species through a two-photon resonantly enhanced sum mixing process, exploiting the $(B\,^2\Sigma^+–X\,^2\Sigma^+)$ transition of the molecule.

As summarized above for the CaF_2 target material, the fact that strongly **polar molecules** are the dominating species for HG in an ablation plume has important implications for the generation of high-order harmonics driven from resonant

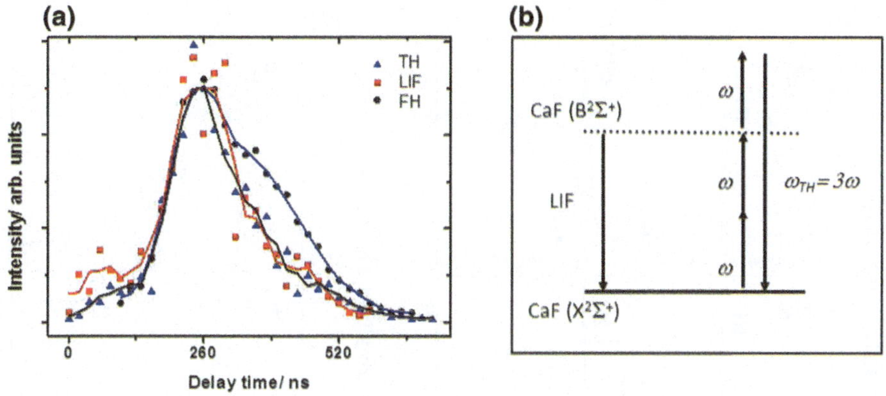

Fig. 13.6 **a** Temporal evolution of the TH, FH and LIF signal of the CaF ($B\,^2\Sigma^+$–$X\,^2\Sigma^+$) system as a function of time delay between the ablation event in CaF_2, induced by 5 J/cm^2 at 266 nm, and the arrival of the fundamental 1064 nm laser beam propagating at 1 mm from the surface of the sample. The lines are visual guides); **b** energy level scheme showing how two-photon resonant absorption of driving frequency ω induces CaF ($B\,^2\Sigma^+$) fluorescence and contributes to resonantly enhance the TH generation process. Adapted from [31]

femtosecond pulses. Alignment of polar molecules, or of molecules with anisotropic polarizability, in the gas phase plasma environment with an additional laser pulse facilitates to study in detail how the HHG yield can be increased in an aligned molecular medium [41].

13.4.2 Low-Order Harmonic Generation in Laser Ablation Plasmas of Metals

In comparison with multicomponent targets, as the dielectric materials mentioned above, metal targets offer, in principle, the possibility to study the HG process in a more compositionally simple environment and, therefore, the characteristics of the harmonic signals can provide information about the spatiotemporal distribution of plasma species under different laser ablation conditions.

The generation of low-order harmonics of a driving 1064 nm laser was investigated in the ablation plumes of various metals using an ablation nanosecond laser of the same wavelength [33, 34]. Figure 13.7a displays, as an example, the emission spectra collected in the presence of the 1064 nm driving beam for the case of aluminium and iron targets. Detected emissions correspond to the TH at 355 nm and the FH observed in the second order of the grating of the monochromator at 426 nm (twice the wavelength of the FH). These spectral features appear as sharp peaks and their dependence with driving pulse energy confirm the order of the nonlinear

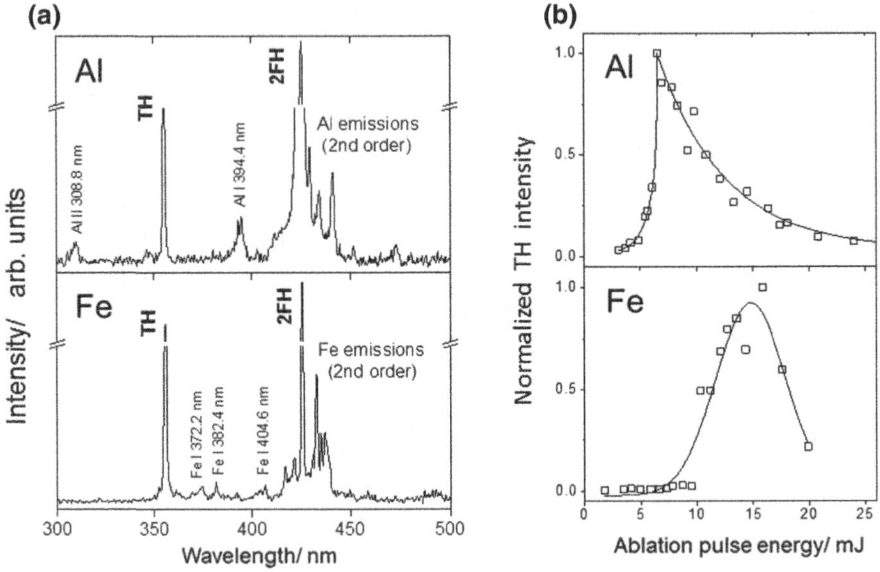

Fig. 13.7 **a** Spectra of emissions detected by 1064 nm ablation of aluminium and iron targets (pulses of 7 and 12 mJ respectively). The driving beam (pulse energy of 560 mJ) propagates at 0.6 mm from the target with a delay of 200 ns. Atomic emissions, together with the lines corresponding to third harmonic (TH) and the second order of the fifth harmonic (2FH) are indicated; **b** dependence of the normalized TH intensity with the ablation pulse energy. The continuous lines are visual guides. Adapted from [33]

process involved (see Fig. 13.5b). Together with the harmonics, **atomic spectral lines** indicate the presence of neutrals and ions of the target metal in the plume.

Low-order HG processes in aluminium and iron plasmas share a range of characteristics with those taking place in laser plasmas of several other metal targets, such as manganese, copper or silver [33, 34]. In all these materials, the yields of TH and FH generation reach a maximum value for a given ablation pulse energy in the range of 5–15 mJ. Particularly, for Al and Fe where the maximum TH yield is achieved at 7 and 15 mJ respectively (Fig. 13.7b).

According to the perturbative treatment of the HG process, harmonic signals are expected to grow with the local density of nonlinear emitters (through the expected quadratic dependence, see expressions (13.5) and (13.6) above) and this explains the increase of signal intensity with ablation pulse energy. The subsequent drop of harmonic yield is partly ascribed to the enhanced density of free electrons in the plume, as these negatively contribute to phase matching. Other factors accounting for the signal decrease at higher ablation pulse energy include the modification of mass distribution and degree of ionization of the plume. Overall, the differences of behaviour with ablation pulse energy for the different metals are related to the different composition and dynamics of the plumes and to dissimilar density and

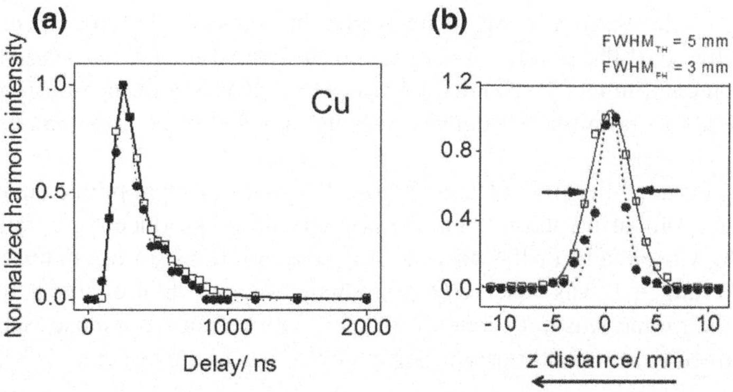

Fig. 13.8 **a** Normalized harmonic signals of the TH (open squares) and FH (solid circles) orders of a nanosecond 1064 nm driving laser in a copper ablation plasma, induced by a 6 ns, 1064 nm ablation laser with pulses of 10 mJ, as a function of the ablation-driving delay. The driving laser propagates at of 0.6 mm from the surface of the target with pulse energy of 560 mJ. The lines are visual guides; **b** z-scan plot of harmonic signals obtained at the optimum delay of 250 ns, where $z = 0$ corresponds to the centre of the ablation plasma in the direction of propagation of the driving laser (indicated by a black arrow). The widths of the Gaussian fits (FWHM) are indicated. Adapted from [34]

phase mismatch conditions of the nonlinear species contributing to the generation of each harmonic.

The dependence of harmonic signal with ablation pulse energy maps the density of species in the plume in the region of interaction with the driving laser. Measuring the intensity of emitted harmonics as a function of the ablation-driving delay and of the position of the focal spot of the driving laser (x and z coordinates of Fig. 13.3) serves to sample the density of nonlinear species across the spatiotemporal region of the plume. An illustrative example of this type of measurements is given in Fig. 13.8. As shown in Fig. 13.8a, the harmonic intensities maximize for a given ablation-driving delay, of 250 ns for the case of copper, when the driving laser propagates parallel to the surface at a distance of 0.6 mm. The temporal behaviour of the harmonics reflects the distribution of flight velocities of plume species responsible for HG. For the particular case exemplified here, the estimated flight velocity is $\approx 2.4 \times 10^5$ cm/s, a value that is typical of plume species expanding under similar ablation conditions [42]. The fact that the relative strength of the TH and FH is independent of the temporal delay, indicates that this ratio is not governed by the instantaneous density of nonlinear plume species.

In relation to the spatial behaviour of the harmonic signals, Fig. 13.8b shows the z-scan measurements performed at the optimum delay of 250 ns. The maximum signal is obtained at $z=0$ which corresponds to the centre of the ablation plasma in the direction of propagation of the driving laser. The TH distribution across z is broader to that of the FH (full with at half maximum of 5 and 3 mm respectively). This effect is expected due to the sharper laser intensity dependence of the latter according with the $(I_\omega)^n$ law in the perturbative approximation of HG [see expression (13.3)].

The width of the harmonic signal measures the density of plume species across the z coordinate of the plume. The variation of this parameter with target material and ablation conditions [31–37] demonstrates the capability of the nonlinear optical measurements to monitor the density and flight velocity of laser ablation plume species.

In [34] the capability of the HG technique to assess different plume environments was tested by comparing plumes of the same metal targets produced by ablation pulses of the same wavelength (1064 nm) but different durations, in the nanosecond and picosecond ranges. It was observed that upon picosecond ablation, the active spatial region of the plume was very much restricted with respect to nanosecond ablation being confined to few hundreds of micrometres close to the target (x-coordinate).

13.4.3 Harmonic Generation by Atomic and Nanoparticle Precursors in Nanosecond Ablation Plasma of Semiconductors

As shown above, generation of low-order harmonics of a nanosecond laser propagating across a laser ablation plume is a universal phenomenon in what regards the nature of the target as any material is susceptible to generate a plasma with a non-negligible degree of optical non-linearity. Harmonic signals are characterized through their dependence with the energy of the nanosecond ablation pulse, the ablation-driving delay and the position of the driving laser focus in the volume of the plume. Harmonic signals provide a spatiotemporal mapping of the local density of plume species and therefore low-order HG can be envisaged as a novel diagnostic tool of the laser ablation process. This can be of interest in the control of plume composition and dynamics in applications related with synthesis of new materials by PLD or elemental analysis of solids by **laser-induced breakdown spectroscopy (LIBS)**.

In some specific cases, the nature of the scatterers contributing to the nonlinear optical response of the plasma can be established. This is the case of materials where either the fundamental driving or the newly generated frequencies are in resonance with single or multiphoton electronic transitions of the plume species (as in 4.1 above). For single-component target materials, such as metals, the situation is in principle much simpler, although, even in this case, the presence of neutrals, ions or aggregates adds complexity to the plume environment. Two-component semiconductors offer a much more complex scenario, as regards the plume composition, where the capabilities of HG probing of the laser plasma can be tested. In ablation plasmas of II-VI semiconductors, materials with a broad range of applications, different techniques, mainly OES, LIF and TOF-MS and analysis of deposits by PLD, have established the presence of neutral and ionized atoms, small molecules and a whole range of middle-sized aggregates [43–45]. The different roles of atomic species and aggregates in the nonlinear response of laser ablation plasmas of semiconductors has been addressed in studies of HHG [17, 46] and low-order HG [32, 37, 38].

TH and FH generation was investigated in laser ablation plasmas of the II–VI semiconductors **ZnS** and **CdS** induced by nanosecond visible and ultraviolet pulses (532 and 266 nm respectively) using a 1064 nm driving laser [32, 36]. Collection of harmonic signals generated across the plasma with temporal and spatial resolution allowed the identification of several distinct populations of harmonic emitters in such plasmas. Particularly, the dependence with the delay between the ablating event and the arrival of the probe laser to the plume revealed interesting information on the plume composition.

Figure 13.9 illustrates two different target and ablation conditions [32]. The intensity of the TH generated in a CdS at 266 nm ablation plume appears in Fig. 13.9a as a function of the delay between the ablating event and the arrival of the driving laser to the centre of plume along the propagation direction ($z = 0$, Fig. 13.3). Data, collected for propagation of the driving laser at two different distances from the target of $x = 0.6$ and 1.0 mm, exhibit two temporal components. The most intense is observed at short delays with maxima at 150 ns for $x = 0.6$ mm and at 250 ns for $x = 1.0$ mm. The weaker (more than one order of magnitude) component temporally extends up to around 2 μs (for 0.6 mm) or 3 μs (for 1.0 mm). This type of behaviour is indicative of the presence of two components of the plasma responsible for HG, either two different species or one species with a bimodal kinetic energy distribution.

The spatiotemporal characteristics of the HG appear to be very sensitive to the target composition and ablation conditions. For 532 nm ablation of ZnS, the delay dependence for three different ablation pulse energies, Fig. 13.9b, shows some particular characteristics, in variance with results obtained upon ablation of CdS at 266 nm. In this case three temporal components are identified (dotted vertical lines in Fig. 13.9b). At 1.0 mm from the target, these components appear centred at around 200 ns, 1 and 10 μs. For moderate laser ablation energies the first component is relatively weak, while the slower species dominate TH generation. However, this tendency reverts for higher ablation energies where the intensity ratio of the fast versus the slow components is higher.

In [32, 37, 38] detailed studies on the spatial dependence of the different temporal components of harmonics generated in nanosecond laser plumes of CdS and ZnS were undertaken. Particularly, the study of HG up to the 9th order was carried out in [37, 38]. Figure 13.10 shows the dependence with the spatial coordinates x and z (defined in Fig. 13.3) for TH and FH harmonics in a 1064 nm ablation plume of ZnS.

The dependence with the distance of the driving beam to the target (Fig. 13.10a) is characterized by a fast decay resulting in a negligible nonlinear response for distances to the target above 3 mm. This decay refers to the early component of the harmonics and is somehow faster for the FH. As already mentioned in 4.2 for the dependence of harmonic signal with ablation pulse energy, the evolution of the signal with the distance to the target reflects that of the local density of nonlinear emitters, as in expressions (13.5) and (13.6) above. Thus, the HG signals can provide a measurement

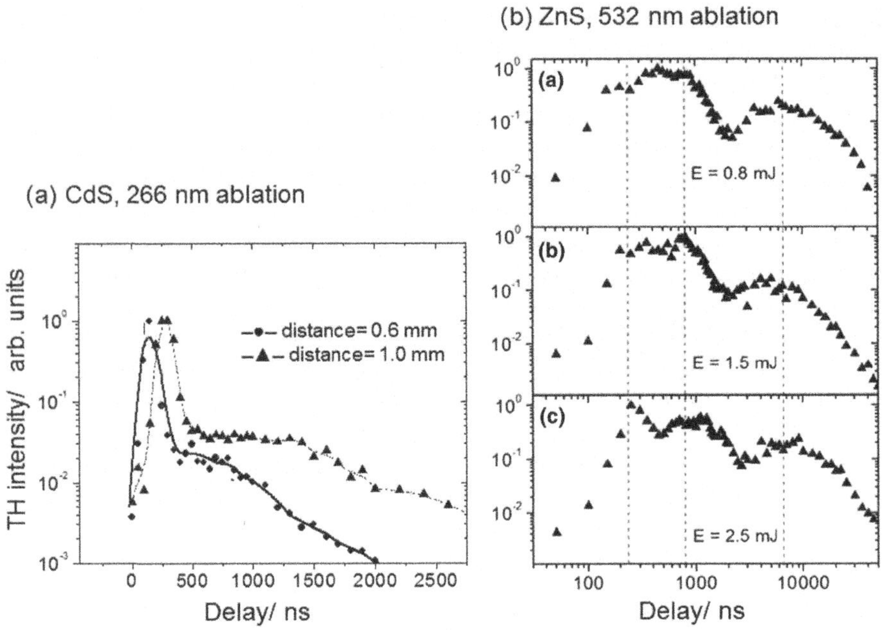

Fig. 13.9 TH of a 1064 nm driving laser as a function of the ablation-driving delay induced in: **a** a 266 nm CdS ablation plasma (ablation pulse energy 3.6 mJ) for two different distances from the surface; **b** a 532 nm ZnS ablation plasma at 1.0 mm from the surface. The three panels correspond to three different ablating laser pulse energies as indicated. Vertical dashed lines mark the central temporal positions of the three populations found. Intensities have been normalized and are plotted in log-log scale for better inspection. Adapted from [32]

of the density of species interacting with the driving beam as it temporally evolves from the arrival to the ablating pulse to the target.

Figure 13.10b illustrates z-scan measurements for the FH generated in the 1064 nm ZnS plasma. These were carried out by monitoring the signal as the position of the driving beam focus moves through the ablation plasma in the laser propagation direction (z axis) at $x = 0.6$ mm. The z-scan was obtained at ablation-driving delays of 250 ns and 10 μs to detect the maximum frequency up-conversion yield from fast and slow plume species respectively (temporal components, see Fig. 13.9). For the early component, the amplitude of the harmonic signal decreases symmetrically when focussing the driving beam away from the centre of the plasma. This is similar to the result presented in Fig. 13.8b for a copper ablation plasma. Contrarily, for the component centred at 10 μs, the behaviour is non-symmetrical with respect to both sides of the centre of the plasma, with a minimum value around $z=0$.

In Fig. 13.10 the continuous lines drawn together with the experimental data correspond to the calculations obtained through the application of the perturbative model of frequency up-conversion as described in Sect. 13.2 [4, 37] and assuming that both the driving and harmonic beams propagate through the plume with negligible

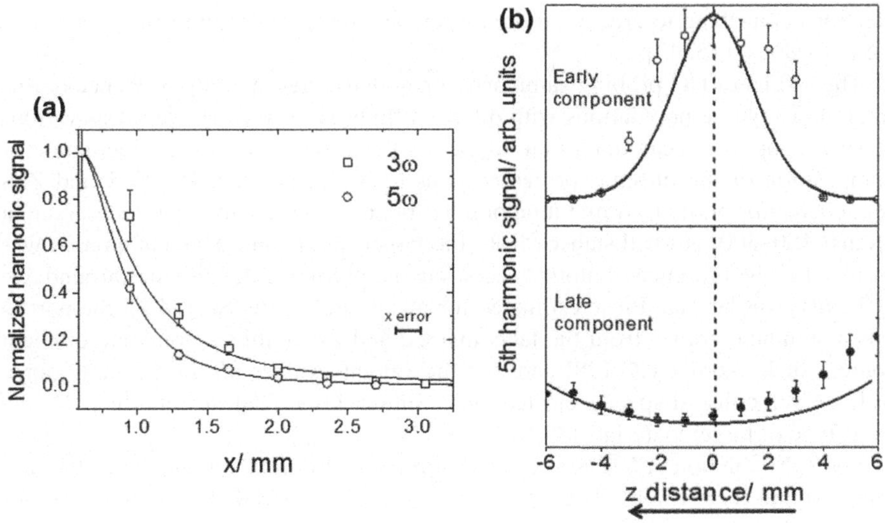

Fig. 13.10 Spatial behaviour of harmonics generated in ZnS nanosecond, 1064 nm ablation plume: **a** early component of the TH and FH signals (normalized to the value at $x=0.6$ mm) as a function of the distance of the driving beam to the target for $z=0$. For each measurement, the ablation-driving laser delay was set at the value for optimum harmonic signal. Squares and circles refer to the early temporal component of the respective harmonics; **b** z-scans of the FH as a function of the position of the driving beam focus with respect to the centre of the ablation plasma ($z=0$) in the propagation direction (shown by a black arrow) for $x=0.6$ mm. The ablation-driving delays are set at 250 ns (and 10 μs), for measuring the early (top) and late (bottom) temporal components of the harmonic signal. In **a** and **b** the continuous lines are the result of the calculations in the perturbative model of HG (see text). Adapted from [37]

absorption. Details of such calculations are given in [37]. A satisfactory agreement was obtained for the dependence of the ZnS harmonics with x, the distance to the target, and to the z coordinate, providing evidence that phase mismatch effects mostly govern the spatial dependence of the harmonic signals. With a driving beam confocal parameter b of 4 mm, the best values of $L(x)$, the FWHM of the plasma density distribution, and of $b\Delta k^{(5)}(x, 0)$, proportional to the maximum density of nonlinear species according to expression (13.5), are 4 and -10 respectively for the fast temporal component. For the slower temporal component, maximizing at 10 μs, good agreement with experimental data was achieved with values of $L(x) = 14$ mm and $b\Delta k^{(5)}(x, 0) = -20$. The different values of the width of the density distribution and of the phase mismatch of early and late components, that provide best fits to the data within the approximations of the model, signpost different spatial extensions along the driving laser propagation direction and dissimilar composition of species at the explored ablation-driving time delays. The distribution of late nonlinear precursors along the driving laser propagation direction is considerably wider than that of the early emitters and at these longer delays higher populations of bound states and free electrons contribute to create a more dispersive nonlinear medium (more negative $b\Delta k^{(n)}$ values) [15, 47].

The results of HG probing summarized here have presented the evidence of different laser plume populations with different flight velocities. However, as various plasma components can contribute to the nonlinear harmonic signal, the nature or composition of the plasma species remains to be ascertained. For CdS and ZnS semiconductor plasmas, a first temporal harmonic signal component is observed at around 100–300 ns for distances from the target of ~1 mm. This component was assigned to the presence of atomic species in the plasma with velocities around 5×10^5 cm/s (similar than those estimated for metal targets, see Sect. 4.2). Support to this assignment comes from the facts that Cd and Zn atomic vapours are efficient sources of low-order HG [48] and that the velocity values are in the same range of flight velocities of species ejected under similar laser ablation conditions for the same type of target materials [49, 50].

For ZnS, ablation at 532 nm resulted in three temporal components for the harmonic signal (Fig. 13.9b). The two later components centred at ~1 and ~10 μs for the same distance, with flight velocities around 10^5 and 10^4 cm/s respectively, are due to the contribution of clusters and nanoparticles formed in the nanosecond laser ablation plume by recombination of atomic and molecular species near the target surface [23, 24, 26–28, 43]. Assuming a totally statistic ablation process, an estimation of the range of clustered species was performed in [32] and was found to be in good agreement with results obtained in TOF-MS experiments [43].

13.4.4 Low-Order HG in Nanosecond Laser Ablation Plasmas of Carbon Containing Materials

Previous sections have described how HG can be used as a probe of different ablation plume populations. Particularly for semiconductors, the dependence of low-order harmonics with ablation-driving delay reveals up to three populations, of atoms, clusters and nanoparticles, contributing to the nonlinear response of the plasma to the incoming fundamental driving beam. Although HG can map out the presence of the plume components with spatiotemporal resolution, the nature of these species has to be identified with the aid of complementary techniques, such as OES or others.

Studies on low-order HG in laser ablation plasmas of carbon containing materials have served to further clarify the role of nanoparticles in the nonlinear optical response of the medium in the perturbative and HHG regimes [35, 36, 51, 52]. In particular, in studies of low-order HG [35, 36], the degree of aggregation of nanosecond laser ablation plumes of **boron carbide (B_4C)** and **graphite** was controlled through the addition of different pressures of buffer gas in the ablation chamber and its effect on the temporal behaviour of the harmonic signal was characterized.

Figure 13.11a describes the dependence with ablation-driving delay of the TH of a fundamental driving beam at 1064 nm generated by laser ablation under different

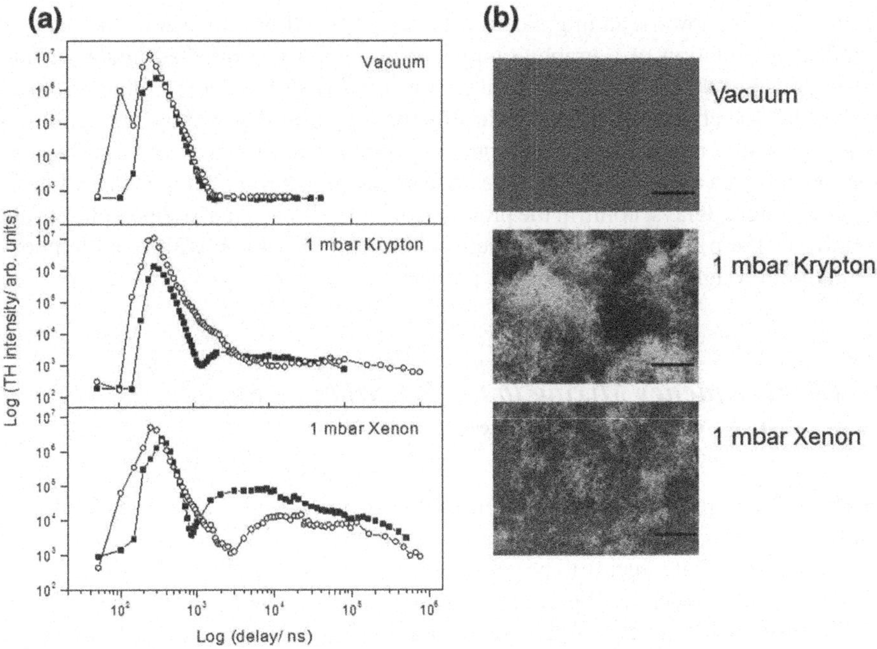

Fig. 13.11 **a** Log-log plot of the TH signal in nanosecond 1064 nm laser ablation plasmas of graphite (solid squares) and boron carbide (open circles) as a function of the ablation-driving delay, measured at 0.6 mm from the target surface under different background atmospheres as indicated; **b** scanning electron microscopy images of deposits fabricated by PLD of graphite (size of bar is 10 μm in the three images). Adapted from [35]

background inert gases for these two materials. If laser ablation is carried out in an inert gas atmosphere, multiple collisions of ablation products ejected from the target with buffer gas atoms lead to an enhancement of the degree of aggregation of the plasma [53, 54]. Under these conditions, the contribution of heavier species, such as clusters and nanoparticles, to HG should be enhanced and this favours the investigation of such contribution. As seen in Fig. 13.11a, the TH signal under vacuum presents a single maximum at a delay of 250 and 350 ns for boron carbide and graphite respectively. This first component is attributed to the contribution to TH of light plume species. Addition of 1 mbar of Krypton and Xenon to the ablation chamber results, for both materials, in the appearance of a temporally wide delayed component. This second component is due to heavier clusters and nanoparticles. The larger size of Xenon atoms increases the probability of collisions and the subsequent aggregation of species, which in turns leads to a TH signal enhancement with respect to that observed under the same pressure of Krypton.

For the above cases, confirmation of the presence of nanoparticles in the ablation plasma was obtained by analysis of deposits collected on monocrystalline silicon (100) wafers simultaneously with HG measurements in a PLD experiment.

Figure 13.11b shows scanning electron microscopy (SEM) images of the deposits obtained by ablation of a graphite target. While deposits collected under vacuum are highly uniform without particulates, those fabricated under 1 mbar of buffer gas display the characteristic dendritic structure of carbon, grown by aggregation of clusters. Similar nanostructured morphology is obtained for boron carbide. The correlation between these results, and the appearance of a second component of the TH signal at longer delays, confirm the presence of aggregates in the graphite plasma and establishes the possibility to study their presence in the plume with spatiotemporal resolution through the HG diagnostic probe.

13.4.5 Frequency Mixing in the Perturbative Regime in Laser Ablation Plasmas

The hypothesis, initially proposed in the pioneering works of Zheltikov et al. [30] that nonlinear optical processes in laser ablation plumes can be used as a diagnostic of these complex media, seems to be confirmed by results, summarized in this Chapter, of studies undertaken on different target materials and ablation conditions.

The investigations discussed here are based on the use of a tight focusing geometry where the driving field probes with temporal resolution the local density in the ablation plasma as a short of 3D microscope. As outlined in several studies [22, 31, 37], resonances between electronic transitions in the medium and the driving and newly generated harmonic fields have been used to enhance the sensitivity for probing particular plasma species.

Symmetry considerations, outlined in Sect. 13.2, prevent the generation of even harmonics in centro-symmetric systems if a single-color driving beam is employed. However, if bichromatic driving fields are applied, this restriction no longer holds and a larger range of new frequencies can be generated, with the additional advantage of permitting the exploitation of a broader range of resonances in the nonlinear species.

Several studies have reported harmonic generation in laser plasmas using bichromatic driving fields i.e. [36, 54–57]. Particularly in [36], the generation of a frequency-mixed beam at the frequency of the 4th harmonic of a 1064 nm fundamental beam was demonstrated. The 4th harmonic resulted from a parametric process in the nanosecond NIR laser-produced plasma of boron carbide. Figure 13.12a is a scheme of the set-up used in such study. While the ablation laser interacts with the target at normal incidence, the driving laser is first frequency-doubled and later split in two arms in order to generate a bichromatic driving beam (1064 nm + 532 nm). This beam interacts with the plasma after an electronically controlled delay, and generates a range of harmonics and frequency-mixed beams that co-propagate with the driving beam. A set of dichroic mirrors discards the fundamental beams and drives the beam of interest toward the detection system. In these conditions, a beam at 266 nm is generated in co-propagation resulting from the frequency mixing of the 1064 nm and the 532 nm beams.

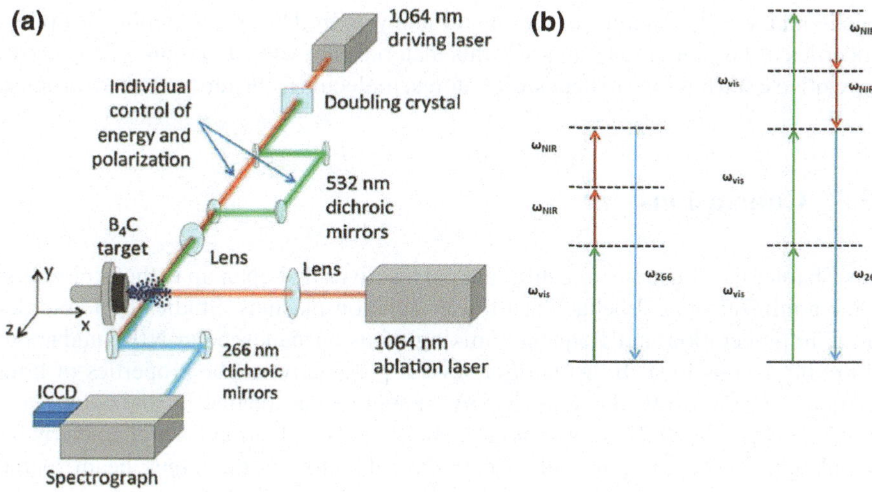

Fig. 13.12 **a** Scheme of the experimental set-up for frequency mixing experiments in a laser ablation plasma of CB_4, see text; **b** sum-frequency mixing (left) and difference frequency mixing (right) schemes for generation of 266 nm coherent light. Adapted from [36]

The newly generated radiation at 266 nm was measured in [36] as a function of the intensity of each of the driving beams to unravel the order in the perturbative limit of the involved nonlinear parametric process. The lowest-order frequency mixing process that can generate 266 nm with the bichromatic driving beam is a sum-frequency process involving four waves according with the scheme $\omega_{vis} + 2\omega_{NIR} \rightarrow \omega_{266}$ (Fig. 13.12b, left panel). This process would imply a linear behaviour as a function of the green (532 nm) beam intensity, and quadratic as a function of the NIR (1064 nm) beam intensity. Surprisingly, the experimental power laws revealed by measurements were found compatible with a quadratic dependence with the energy of the NIR pulses and with a third order dependence of the energy of the green pulses. Since the study indicates that this result is does not correspond to the mentioned four-wave sum frequency mixing process, it is proposed that the lowest order process that would be allowed by symmetry laws would be a six-wave difference frequency mixing process of the form $3\omega_{vis} - 2\omega_{NIR} \rightarrow \omega_{266}$ (Fig. 13.12b, right panel). It is interesting to note here that the parametric process responsible for the newly generated radiation at 266 nm was found highly dependent of the relative polarizations of the NIR and green driving beams, with parallel polarizations yielding the highest conversion efficiency.

Studies outlined in Sects. 13.4.4 and 13.4.5 of this Chapter refer to ablation plumes of boron carbide, one of the hardest materials known, used in numerous industrial applications. Further investigations have addressed HHG schemes in ablation plumes of B_4C targets [58]. Fabrication of thin films of boron carbide by PLD allows exploring a broad range of conditions that, through the control of the laser and plasma parameters, have yielded deposits with different morphologies, crystallinity and stoichiometry [59]. Therefore, the in situ analysis by HG of the nonlinear optical properties of laser ablation plumes of this material is of value to monitor and control the relative contributions to deposits of atoms, molecules, clusters and nanoparticles.

13.5 Conclusions

This Chapter has illustrated the different perspectives that open up in the exploration of the nonlinear optical behaviour of laser ablation plasmas. Studies on low-order harmonic generation and frequency mixing of using nanosecond NIR fundamental driving beams in such media offer several perspectives. The properties of laser ablation plasmas, particularly its density, spatial extent, degree of ionization, etc., can be tuned, through the choice of target material and ablation laser characteristics (wavelength, pulse duration, pulse energy, focalization on the target, beam spatial profile, etc.), to find conditions that optimize a given property with respect to the highly efficient generation of new coherent light of short wavelength. These studies, performed in the framework of the perturbative regime of nonlinear optics, are very important to understand the impeding and restricting process of high-order harmonic generation, that take place at much higher driving laser intensities using femtosecond lasers as fundamental radiation. These investigations also serve to identify the type of solid ablation materials that lead to highly efficient frequency up-conversion in the extreme ultraviolet or X-ray spectral regions and that provide enough number of harmonic photons to be used in fundamental studies and in practical applications.

A second perspective is presented by investigations of the generation of up to the 9th harmonic order of a nanosecond NIR laser beam, performed in pump and probe configurations, where the spatiotemporal control of the driving laser with respect to the position and time of the ablation event allows for the separation of distinct populations in the ablation plasma. Low-order harmonic generation is now accepted as a powerful technique for the detection of a broad range of precursor ablation species in complex laser ablation plasmas, including neutral atoms and ions, small molecules, clusters and nanoparticles. This has been illustrated in this Chapter for materials with different characteristics, such as ionic insulators, metals, semiconductors and carbon containing materials, where it appears that the harmonics generated in their ablation plumes are particularly sensitive to the presence of nanoparticles. Due to the fact that, in principle, all species can emit harmonic radiation when exposed to an intense laser pulse, harmonic generation can be conceived as a highly versatile tool for in situ diagnostic method of laser plasmas, free from the limitations shown by other diagnostic techniques regarding mass or spectral range, and of relevant application in pulsed laser deposition processes or laser plume analysis in conjunction with optical emission spectroscopy based techniques.

The effect of the fundamental driving beam on the laser ablation plasma has not been discussed in this Chapter, although several studies have aimed at its characterization and understanding (see for example [37, 38] and references therein). The delayed nanoparticle populations can experience extensive fragmentation and ionization. Thus, the resulting excited atoms and ions may act as delayed nonlinear species yielding particularly high conversion efficiency in the high harmonic limit.

Finally, application of the harmonic generation technique to laser plasmas with a high population of aggregates, such as clusters and nanoparticles, gives the possibility to study the nonlinear optical properties of such species in environment-free conditions. In that respect, the femtosecond laser plasmas of metals seem to be an adequate environment where these capacities could be tested.

Acknowledgements The authors thank funding from the Spanish Ministerio de Economía, Industria y Competitividad (Projects CTQ2013-43086-P and CTQ2016-75880-P). M.O. thanks CSIC for a contract. Fruitful collaborations with M. Martín, M. López-Arias, A. Benítez-Cañete, I. Lopez-Quintas, J.P. Marangos, and R.A. Ganeev are acknowledged.

References

1. J.F. Reintjes, *Optical Parametric Processes in Liquids and Gases* (Academic Press, New York, 1984)
2. C.J. Joachain, N.J. Kylstra, R.M. Potvliege, *Atoms in Intense Laser Fields*. Cambridge University Press (2011)
3. G.H.C. New, Contemporary Phys. **52**, 281 (2011)
4. A. Lago, G. Hilber, R. Wallenstein, Phys. Rev. A **36**, 3827 (1987)
5. R. Hilbig, R. Wallenstein, IEEE J. Quantum Elec. **17**, 1566 (1981)
6. M. Castillejo, J.Y. Zhou, M.H.R. Hutchinson, Appl. Phys. B **45**, 293 (1988)
7. M. Castillejo, M. Martin, R. de Nalda, M. Oujja, Chem. Phys. Lett. **237**, 367 (1995)
8. A.H. Zanganeh, J.H. Fillion, J. Ruiz, M. Castillejo, J.L. Lemaire, N. Shafizadeh, F. Rostas, J. Chem. Phys. **112**, 5660 (2000)
9. X.F. Li, A. L'Huillier, M. Ferray, L.A. Lompre, G. Mainfray, Phys. Rev. A **39**, 5751 (1989)
10. C. Lyngå, A. L'Huillier, C.-G. Wahlström, J. Phys. B: At. Mol. Opt. Phys. **29**, 3293 (1996)
11. N. Hay, R. de Nalda, T. Halfmann, K.J. Mendham, M.B. Mason, M. Castillejo, J.P. Marangos, Eur. Phys. J. D **14**, 231 (2001)
12. M. Lein, R. De Nalda, E. Heesel, N. Hay, E. Springate, R. Velotta, M. Castillejo, P.L. Knight, J.P. Marangos, J. Mod. Optics **52**, 465 (2005)
13. P. Jaeglé, *Coherent Sources of XUV Radiation: Soft X-Ray Lasers and High-Order Harmonic Generation* (Springer, New York, 2006)
14. S.-Y. Zhao, P.-L. Zhang, G.-Z. Zhang, W.-Z. Zhao, Appl. Opt. **28**, 4521 (1989)
15. P. Lee, D.V. Giovanielli, R.P. Godwin, G.H. McCall, App. Phys. Lett. **24**, 406 (1974)
16. W. Theobald, C. Wülker, F.P. Schäfer, B.N. Chichkov, Opt. Commun. **120**, 177 (1995)
17. R.A. Ganeev, *Plasma Harmonics* (CRC Press, Boca Raton, 2014)
18. R.A. Ganeev, C. Hutchison, I. Lopez-Quintas, F. McGrath, D.Y. Lei, M. Castillejo, J.P. Marangos, Phys. Rev. A **88**, 033803 (2013)
19. R.A. Ganeev, T. Witting, C. Hutchison, V.V. Strelkov, F. Frank, M. Castillejo, I. Lopez-Quintas, Z. Abdelrahman, J.W.G. Tisch, J.P. Marangos, Phys. Rev. A **88**, 033838 (2013)
20. T. Ozaki, L.B. Elouga Bom, J. Abdul-Hadi, R.A. Ganeev, Laser Part. Beams **28**, 69 (2010)
21. H. Singhal, P.A. Naik, M. Kumar, J.A. Chakera, P.D. Gupta, J. App. Phys. **115**, 033104 (2014)
22. R.A. Ganeev, M. Suzuki, S. Yoneya, H. Kuroda, J. Phys. B **48**, 165401 (2015)
23. M. Jadraque, A.B. Evtushenko, D. Ávila-Brande, M. López-Arias, V. Loriot, Y.G. Shukhov, L.S. Kibis, A.V. Bulgakov, M. Martín, J. Phys. Chem. C **117**, 5416 (2013)
24. D. Yadav, V. Gupta, R.K. Thareja, J. App. Phys. **106**, 064903 (2009)

25. A.A. Puretzky, D.B. Geohegan, G.B. Hurst, M.V. Buchanan, B.S. Luk'yanchuk, Phys. Rev. Lett. **83**, 444 (1999)
26. O. Albert, S. Roger, Y. Glinec, J.C. Loulergue, J. Etchepare, C. Boulmer-Leborgne, J. Perrière, E. Millon, App. Phys. A **76**, 319 (2003)
27. F. Claeyssens, M.N.R. Ashfold, E. Sofoulakis, C.G. Ristoscu, D. Anglos, C. Fotakis, J. App. Phys. **91**, 6162 (2002)
28. A. Mohanta, R.K. Thareja, J. App. Phys. **106**, 124909 (2009)
29. R. Eason, *Pulsed Laser Deposition of Thin Films: Applications-Led Growth of Functional Materials* (Wiley, 2007)
30. A.N. Naumov, D.A. Sidorov-Biryukov, A.B. Fedotov, A.M. Zheltikov, Opt. Spectrosc. **90**, 778 (2001)
31. M. Oujja, R. de Nalda, M. López-Arias, R. Torres, J.P. Marangos, M. Castillejo, Phys. Rev. A **81**, 043841 (2010)
32. R. de Nalda, M. Lopez-Arias, M. Sanz, M. Oujja, M. Castillejo, Phys. Chem. Chem. Phys. **13**, 10755 (2011)
33. M. López-Arias, M. Oujja, M. Sanz, R. de Nalda, R.A. Ganeev, M. Castillejo, Mol. Phys. **110**, 1651 (2012)
34. M. López-Arias, M. Oujja, M. Sanz, R.A. Ganeev, G.S. Boltaev, N.K. Satlikov, R.I. Tugushev, T. Usmanov, M. Castillejo, J. App. Phys. **111**, 043111 (2012)
35. I. Lopez-Quintas, M. Oujja, M. Sanz, M. Martín, R.A. Ganeev, M. Castillejo, App. Surf. Sci. **278**, 33 (2013)
36. M. Oujja, A. Benítez-Cañete, M. Sanz, I. Lopez-Quintas, M. Martín, R. de Nalda, M. Castillejo, App. Surf. Sci. **336**, 53 (2015)
37. M. Oujja, I. Lopez-Quintas, A. Benítez-Cañete, R. de Nalda, M. Castillejo, Appl. Surf. Sci. **392**, 572 (2017)
38. I. Lopez-Quintas, Ph.D. Thesis, Universidad Complutense de Madrid (2016)
39. P.B. Corkum, Phys. Rev. Lett. **1993**, 71 (1994)
40. B.S. Zhao, M. Castillejo, D.S. Chung, B. Friedrich, D. Herschbach, Rev. Scien. Instr. **75**, 146 (2004)
41. N. Hay, R. Velotta, M. Lein, R. de Nalda, E. Heesel, M. Castillejo, J.P. Marangos, Phys. Rev. A **65**, 053805 (2002)
42. O. Albert, S. Roger, Y. Glinec, J.C. Loulergue, J. Etchepare, C. Boulmer-Leborgne, J. Perrière, E. Millon, Appl. Phys. A **76**, 319 (2003)
43. J. Álvarez-Ruiz, M. López-Arias, R. Nalda, M. Martín, A. Arregui, L. Bañares, App. Phys. A **95**, 681 (2008)
44. M. Sanz, M. López-Arias, J.F. Marco, R. de Nalda, S. Amoruso, G. Ausanio, S. Lettieri, R. Bruzzese, X. Wang, M. Castillejo, J. Phys. Chem. C **115**, 3203 (2011)
45. I. Lopez-Quintas, V. Loriot, D. Avila, J.G. Izquierdo, E. Rebollar, L. Banares, M. Castillejo, R. de Nalda, M. Martin, Phys. Chem. Chem. Phys. **18**, 3522 (2016)
46. R.A. Ganeev, M. Suzuki, S. Yoneya, H. Kur, J. Appl. Phys. **117**, 023114 (2015)
47. A.B. Fedotov, N.I. Koroteev, A.M. Zheltikov, Laser Phys. **5**, 812 (1995)
48. *Physics of New Laser Sources, N.B. Abraham*, in ed by F.T. Arecchi, A. Mooradian, A. Sona, NATO ASI Series (1985)
49. H. Ezumi, S. Keitoku, Jpn. J. Appl. Phys. **32**, 1783 (1993)
50. A. Klini, A. Manousaki, D. Anglos, C. Fotakis, J. Appl. Phys. **98**, 123301 (2005)
51. R.A. Ganeev, P.A. Naik, H. Singhal, J.A. Chakera, M. Kumar, M.P. Joshi, A.K. Srivastava, P.D. Gupta, Phys. Rev. A **83**, 013820 (2011)
52. R.A. Ganeev, C. Hutchinson, T. Witting, F. Frank, W.A. Okell, A. Zäir, S. Weber, P.V. Redkin, D.Y. Lei, T. Roschuk, S.A. Maier, I. López-Quintás, M. Martín, M. Castillejo, J.W.G. Tisch, J.P. Marangos, J. Phys. B **45**, 165402 (2012)
53. P.M. Ossi, C.E. Bottani, A. Miotello, Thin Solid Films **482**, 2 (2005)
54. P.M. Ossi, A. Miotello, J. Non-Crystalline Solids **2007**, 353 (1860)
55. R.A. Ganeev, H. Singhal, P.A. Naik, I.A. Kulagin, P.V. Redkin, J.A. Chakera, M. Tayyab, R.A. Khan, P.D. Gupta, Phys. Rev. A: At., Mol., Opt. Phys. **80**, 033845 (2009)

56. R.A. Ganeev, H. Singhal, P.A. Naik, J.A. Chakera, H.S. Vora, R.A. Khan, P.D. Gupta, Phys. Rev. A: At., Mol., Opt. Phys. **82**, 053831 (2010)
57. R.A. Ganeev, G.S. Boltaev, T. Usmanov, Opt. Commun. **324**, 114 (2014)
58. R.A. Ganeev, M. Suzuki, H. Kuroda, Opt. Commun. **370**, 6 (2016)
59. I. Lopez-Quintas, M. Oujja, M. Sanz, A. Benitez-Cañete, R.J. Chater, M.V. Cañamares, J.F. Marco, M. Castillejo, Appl. Surf. Sci. **328**, 170 (2014)

Subject Index

Page number in bold=highly relevant in our opinion; Page number in italics=relevant

P. M. Ossi (ed.), *Advances in the Application of Lasers in Materials Science*,
Springer Series in Materials Science274,
https://doi.org/10.1007/978-3-319-96845-2

The manufacturer's authorised representative in the EU is Springer Nature Customer Service Centre GmbH, Europaplatz 3, 69115 Heidelberg, Germany. If you have any concerns regarding our products, please contact ProductSafety@springernature.com

Printed and bound by CPI Group (UK) Ltd, Croydon, CR0 4YY
30/09/2026
02225235-0010